THE FRONTIERS COLLECTION

THE FRONTIERS COLLECTION

The books in this collection are devoted to challenging and open problems at th forefront of modern science, including related philosophical debates. In contrast to typical research monographs, however, they strive to present their topics in a manner accessible also to scientifically literate non-specialists wishing to gain insight into the deeper implications and fascinating questions involved. Taken as a whole, the series reflects the need for a fundamental and interdisciplinary approach to modern science. Furthermore, it is intended to encourage active scientists in all areas to ponder over important and perhaps controversial issues beyond their own speciality. Extending from quantum physics and relativity to entropy, consciousness and complex systems – the Frontiers Collection will inspire readers to push back the frontiers of their own knowledge.

Other Recent Titles

Weak Links
The Universal Key to the Stability of Networks and Complex Systems
By P. Csermely

Entanglement, Information, and the Interpretation of Quantum Mechanics
By G. Jaeger

Homo Novus - A Human Without Illusions
U.J. Frey, C. Störmer, K.P. Willführ

The Physical Basis of the Direction of Time
By H.D. Zeh

Mindful Universe
Quantum Mechanics and the Participating Observer
By H. Stapp

Decoherence and the Quantum-To-Classical Transition
By M.A. Schlosshauer

The Nonlinear Universe
Chaos, Emergence, Life
By A. Scott

Symmetry Rules
How Science and Nature Are Founded on Symmetry
By J. Rosen

Quantum Superposition
Counterintuitive Consequences of Coherence, Entanglement, and Interference
By M.P. Silverman

For all volumes see back matter of the book

Bernhard Graimann · Brendan Allison · Gert Pfurtscheller
Editors

BRAIN–COMPUTER INTERFACES

Revolutionizing Human–Computer Interaction

Editors
Dr. Bernhard Graimann
Otto Bock HealthCare GmbH
Max-Näder-Str. 15
37115 Duderstadt
Germany
graimann@ottobock.de

Dr. Brendan Allison
Institute for Knowledge Discovery
Laboratory of Brain-Computer Interfaces
Graz University of Technology
Krenngasse 37
8010 Graz
Austria
allison@tugraz.at

Prof. Dr. Gert Pfurtscheller
Institute for Knowledge Discovery
Laboratory of Brain-Computer Interfaces
Graz University of Technology
Krenngasse 37
8010 Graz
Austria
pfurtscheller@tugraz.at

Series Editors:

Avshalom C. Elitzur
Bar-Ilan University, Unit of Interdisciplinary Studies, 52900 Ramat-Gan, Israel
email: avshalom.elitzur@weizmann.ac.il

Laura Mersini-Houghton
Dept. Physics, University of North Carolina, Chapel Hill, NC 27599-3255, USA
email: mersini@physics.unc.edu

Maximilian A. Schlosshauer
Niels Bohr Institute, Blegdamsvej 17, 2100 Copenhagen, Denmark
email: schlosshauer@nbi.dk

Mark P. Silverman
Trinity College, Dept. Physics, Hartford CT 06106, USA
email: mark.silverman@trincoll.edu

Jack A. Tuszynski
University of Alberta, Dept. Physics, Edmonton AB T6G 1Z2, Canada
email: jtus@phys.ualberta.ca

Rüdiger Vaas
University of Giessen, Center for Philosophy and Foundations of Science, 35394 Giessen, Germany
email: ruediger.vaas@t-online.de

H. Dieter Zeh
Gaiberger Straße 38, 69151 Waldhilsbach, Germany
email: zeh@uni-heidelberg.de

ISSN 1612-3018
ISBN 978-3-642-02090-2 ISBN 978-3-642-02091-9 (eBook)
DOI 10.1007/978-3-642-02091-9
Springer Heidelberg Dordrecht London New York

Library of Congress Control Number: 2010934515

Cover design: KuenkelLopka GmbH, Heidelberg

Printed on acid-free paper

Springer is part of Springer Science+Business Media (www.springer.com)

Preface

It's an exciting time to work in Brain–Computer Interface (BCI) research. A few years ago, BCIs were just laboratory gadgets that only worked with a few test subjects in highly controlled laboratory settings. Since then, many different types of BCIs have succeeded in providing real-world communication solutions for several severely disabled users. Contributions have emerged from a myriad of research disciplines across academic, medical, industrial, and nonprofit sectors. New systems, components, ideas, papers, research groups, and success stories are becoming more common. Many scientific conferences now include BCI related special sessions, symposia, talks, posters, demonstrations, discussions, and workshops. The popular media and general public have also paid more attention to BCI research.

However, the field remains in its infancy, with many fundamental challenges remaining. BCI success stories are still expensive, time consuming, and excruciatingly infrequent. We still cannot measure nor understand the substantial majority of brain activity, which limits any BCI's speed, usability, and reliability. Communication and collaboration across disciplines and sectors must improve. Despite increased efforts from many groups, you still can't really do very much with a BCI. The increased publicity has also brought some stories that are biased, misleading, confusing, or inaccurate.

All of the above reasons inspired a book about BCIs intended for non-expert readers. There is a growing need for a straightforward overview of the field for educated readers who do not have a background in BCI research nor some of its disciplines. This book was written by authors from different backgrounds working on a variety of BCIs. Authors include experts in psychology, neuroscience, electrical engineering, signal processing, software development, and medicine. The chapters describe different systems as well as common principles and issues. Many chapters present emerging ideas, research, or analysis spanning different disciplines and BCI approaches. The style and content provide a readable and informative overview aimed toward non-specialists.

The first chapter gives a particularly easy introduction to BCIs. The next three chapters cover the foundations of BCIs in more detail. Chapters 4 through 8 describe the four most cited non-invasive BCI systems, and chapters 9 and 10 cover neurorehabilitation. Chapter 11 focuses on BCIs for locked-in patients and presents a unique interview with a locked-in patient. Invasive approaches are addressed in chapters 12 to 14. Chapters 15 and 16 present a freely available BCI framework (BCI 2000) and one of the first commercial BCI systems. Chapters 17 and 18 deal with signal processing. The last chapter gives a look into the future of BCIs.

Graz, Austria
April 2010

Bernhard Graimann
Brendan Allison
Gert Pfurtscheller

Contents

Brain–Computer Interfaces: A Gentle Introduction 1
Bernhard Graimann, Brendan Allison, and Gert Pfurtscheller

Brain Signals for Brain–Computer Interfaces 29
Jonathan R. Wolpaw and Chadwick B. Boulay

Dynamics of Sensorimotor Oscillations in a Motor Task 47
Gert Pfurtscheller and Christa Neuper

Neurofeedback Training for BCI Control 65
Christa Neuper and Gert Pfurtscheller

The Graz Brain-Computer Interface 79
Gert Pfurtscheller, Clemens Brunner, Robert Leeb,
Reinhold Scherer, Gernot R. Müller-Putz and Christa Neuper

BCIs in the Laboratory and at Home: The Wadsworth Research Program 97
Eric W. Sellers, Dennis J. McFarland, Theresa M. Vaughan, and
Jonathan R.Wolpaw

Detecting Mental States by Machine Learning Techniques: The Berlin Brain–Computer Interface 113
Benjamin Blankertz, Michael Tangermann, Carmen Vidaurre,
Thorsten Dickhaus, Claudia Sannelli, Florin Popescu, Siamac Fazli,
Márton Danóczy, Gabriel Curio, and Klaus-Robert Müller

Practical Designs of Brain–Computer Interfaces Based on the Modulation of EEG Rhythms 137
Yijun Wang, Xiaorong Gao, Bo Hong, and Shangkai Gao

Brain–Computer Interface in Neurorehabilitation 155
Niels Birbaumer and Paul Sauseng

Non Invasive BCIs for Neuroprostheses Control of the Paralysed Hand 171
Gernot R. Müller-Putz, Reinhold Scherer, Gert Pfurtscheller,
and Rüdiger Rupp

Brain–Computer Interfaces for Communication and Control in Locked-in Patients 185
Femke Nijboer and Ursula Broermann

Intracortical BCIs: A Brief History of Neural Timing 203
Dawn M. Taylor and Michael E. Stetner

BCIs Based on Signals from Between the Brain and Skull 221
Jane E. Huggins

A Simple, Spectral-Change Based, Electrocorticographic Brain–Computer Interface 241
Kai J. Miller and Jeffrey G. Ojemann

Using BCI2000 in BCI Research 259
Jürgen Mellinger and Gerwin Schalk

The First Commercial Brain–Computer Interface Environment 281
Christoph Guger and Günter Edlinger

Digital Signal Processing and Machine Learning 305
Yuanqing Li, Kai Keng Ang, and Cuntai Guan

Adaptive Methods in BCI Research - An Introductory Tutorial 331
Alois Schlögl, Carmen Vidaurre, and Klaus-Robert Müller

Toward Ubiquitous BCIs 357
Brendan Z. Allison

Index 389

Contributors

Brendan Allison Institute for Knowledge Discovery, Laboratory of Brain-Computer Interfaces, Graz University of Technology, Krenngasse 37, 8010 Graz, Austria, allison@tugraz.at

Kai Keng Ang Institute for Infocomm Research, A*STAR, Singapore, kkang@i2r.a-star.edu.sg

Niels Birbaumer Institute of Medical Psychology and Behavioral Neurobiology, University of Tübingen, Tübingen, Germany, niels.birbaumer@uni-tuebingen.de

Benjamin Blankertz Berlin Institute of Technology, Machine Learning Laboratory, Berlin, Germany; Fraunhofer FIRST (IDA), Berlin, Germany, blanker@cs.tu-berlin.de

Chadwick B. Boulay Wadsworth Center, New York State Department of Health and School of Public Health, State University of New York at Albany, New York, NY 12201, USA, cboulay@wadsworth.org

Ursula Broermann Institute of Medical Psychology and Behavioral Neurobiology, Eberhard Karls University of Tübingen, Tübingen, Germany

Clemens Brunner Institute for Knowledge Discovery, Laboratory of Brain-Computer Interfaces, Graz University of Technology, Krenngasse 37, 8010 Graz, Austria, clemens.brunner@tugraz.at

Gabriel Curio Campus Benjamin Franklin, Charité University Medicine Berlin, Berlin, Germany, gabriel.curio@charite.de

Márton Danóczy Berlin Institute of Technology, Machine Learning Laboratory, Berlin, Germany, marton@cs.tu-berlin.de

Thorsten Dickhaus Berlin Institute of Technology, Machine Learning Laboratory, Berlin, Germany, dickhaus@cs.tu-berlin.de

Günter Edlinger Guger Technologies OG / g.tec medical engineering GmbH, Herbersteinstrasse 60, 8020 Graz, Austria, edlinger@gtec.at

Siamac Fazli Berlin Institute of Technology, Machine Learning Laboratory, Berlin, Germany, fazli@cs.tu-berlin.de

Shangkai Gao Department of Biomedical Engineering, School of Medicine, Tsinghua University, Beijing, China, gsk-dea@tsinghua.edu.cn

Xiaorong Gao Department of Biomedical Engineering, School of Medicine, Tsinghua University, Beijing, China

Bernhard Graimann Strategic Technology Management, Otto Bock HealthCare GmbH, Max-Näder Straße 15, 37115 Duderstadt, Germany, graimann@ottobock.de

Cuntai Guan Institute for Infocomm Research, A*STAR, Singapore, ctguan@i2r.a-star.edu.sg

Christoph Guger Guger Technologies OG / g.tec medical engineering GmbH, Herbersteinstrasse 60, 8020 Graz, Austria, guger@gtec.at

Jane E. Huggins University of Michigan, Ann Arbor, MI, USA, janeh@umich.edu

Bo Hong Department of Biomedical Engineering, School of Medicine, Tsinghua University, Beijing, China

Robert Leeb Institute for Knowledge Discovery, Laboratory of Brain-Computer Interfaces, Graz University of Technology, Krenngasse 37, 8010 Graz, Austria, robert.leeb@tugraz.at

Yuanqing Li School of Automation Science and Engineering, South China University of Technology, Guangzhou 510640, China, auyqli@scut.edu.cn

Dennis J. McFarland Laboratory of Neural Injury and Repair, Wadsworth Center New York State Department of Health, Albany, NY12201-0509, USA, mcfarlan@wadsworth.org

Jürgen Mellinger Institute of Medical Psychology and Behavioral Neurobiology, University of Tübingen, Tübingen, Germany, juergen.mellinger@uni-tuebingen.de

Kai J. Miller Physics, Neurobiology and Behavior, University of Washington, Seattle, WA 98195, USA, kjmiller@u.washington.edu

Klaus-Robert Müller Berlin Institute of Technology, Machine Learning Laboratory, Berlin, Germany, krm@cs.tu-berlin.de

Gernot R. Müller-Putz Institute for Knowledge Discovery, Laboratory of Brain-Computer Interfaces, Graz University of Technology, Krenngasse 37, 8010 Graz, Austria, gernot.mueller@tugraz.at

Christa Neuper Institute for Knowledge Discovery, Graz University of Technology, Graz, Austria; Department of Psychology, University of Graz, Graz, Austria, christa.neuper@uni-graz.at

Femke Nijboer Institute of Medical Psychology and Behavioral Neurobiology, Eberhard Karls University of Tübingen, Tübingen, Germany; Human-Media Interaction, University of Twente, Enschede, the Netherlands, femke.nijboer@utwente.nl

Jeffrey G. Ojemann Neurological Surgery, University of Washington, Seattle, WA 98195, USA, jojemann@u.washington.edu

Gert Pfurtscheller Institute for Knowledge Discovery, Laboratory of Brain-Computer Interfaces, Graz University of Technology, Krenngasse 37, 8010, Graz, Austria, pfurtscheller@tugraz.at

Florin Popescu Fraunhofer FIRST (IDA), Berlin, Germany, florin.popescu@first.fraunhofer.de

Rüdiger Rupp Orthopedic University Hospital of Heidelberg University, Schlierbacher Landstrasse 200a, Heidelberg, Germany, Ruediger.Rupp@ok.uni-heidelberg.de

Claudia Sannelli Berlin Institute of Technology, Machine Learning Laboratory, Berlin, Germany, sannelli@cs.tu-berlin.de

Paul Sauseng Department of Psychology, University Salzburg, Salzburg, Austria, paul.sauseng@sbg.ac.at

Gerwin Schalk Laboratory of Neural Injury and Repair, Wadsworth Center New York State Department of Health, Albany, NY 12201-0509, USA, schalk@wadsworth.org

Reinhold Scherer Institute for Knowledge Discovery, Laboratory of Brain-Computer Interfaces, Graz University of Technology, Krenngasse 37, 8010 Graz, Austria, reinhold.scherer@tugraz.at

Alois Schlögl Institute of Science and Technology Austria (IST Austria), Am Campus 1, A–3400 Klosterneuburg, Austria, alois.schloegl@gmail.com

Eric W. Sellers Department of Psychology, East Tennessee State University, Johnson City, TN 37641, USA, sellers@etsu.edu

Michael E. Stetner Department of Brain and Cognitive Sciences, Massachusetts Institute of Technology, Cambridge, MA 02139, USA, stetner@mit.edu

Michael Tangermann Berlin Institute of Technology, Machine Learning Laboratory, Berlin, Germany, schroedm@cs.tu-berlin.de

Dawn M. Taylor Dept of Neurosciences, The Cleveland Clinic, Cleveland, OH 44195, USA; Department of Veterans Affairs, Cleveland Functional Electrical Stimulation Center of Excellence, Cleveland, OH 44106, USA, dxt42@case.edu

Theresa M. Vaughan Laboratory of Neural Injury and Repair, Wadsworth Center New York State Department of Health, Albany, NY 12201-0509, USA, vaughan@wadsworth.org

Carmen Vidaurre Berlin Institute of Technology, Machine Learning Laboratory, Berlin, Germany, vidcar@cs.tu-berlin.de

Yijun Wang Department of Biomedical Engineering, School of Medicine, Tsinghua University, Beijing, China

Jonathan R. Wolpaw Wadsworth Center, New York State Department of Health and School of Public Health, State University of New York at Albany, New York, NY 12201, USA, wolpaw@wadsworth.org

List of Abbreviations

ADHD	Attention deficit hyperactivity disorder
AEP	Auditory evoked potential
ALS	Amyotrophic lateral sclerosis
AP	Action potential
AR	Autoregressive model
BCI	Brain–Computer Interface
BMI	Brain–Machine Interface
BOLD	Blood oxygenation level dependent
BSS	Blind source separation
CLIS	Completely locked-in state
CNS	Central nervous system
CSP	Common spatial patterns
ECG	Electrocardiogram, electrocardiography
ECoG	Electrocorticogram, electrocorticography
EEG	Electroencephalogram, electroencephalography
EMG	Electromyogram, electromyography
EOG	Electrooculogram
EP	Evoked potential
EPSP	Excitatory postsynaptic potential
ERD	Event-related desynchronization
ERP	Event-related potential
ERS	Event-related synchronization
FES	Functional electrical stimulation
fMRI	Functional magnetic resonance imaging
fNIR	Functional near infrared
HCI	Human–computer interface
ICA	Independent component analysis
IPSP	Inhibitory postsynaptic potential
ITR	Information transfer rate
LDA	Linear discriminant analysis
LFP	Local field potential
LIS	Locked-in state

MEG	Magnetoencephalogram, magentoencephalography
MEP	Movement-evoked potential
MI	Motor imagery
MND	Motor neuron disease
MRI	Magnetic resonance imaging
NIRS	Near-infrared spectroscopy
PCA	Principal component analysis
PET	Positron emission tomography
SCP	Slow cortical potential
SMA	Supplementary motor area
SMR	Sensorimotor rhythm
SNR	Signal-to-noise ratio
SSVEP	Steady-state visual-evoked potential
VEP	Visual evoked potential
VR	Virtual reality

Brain–Computer Interfaces: A Gentle Introduction

Bernhard Graimann, Brendan Allison, and Gert Pfurtscheller

Stardate 3012.4: The U.S.S. Enterprise has been diverted from its original course to meet its former captain Christopher Pike on Starbase 11. When Captain Jim Kirk and his crew arrive, they find out that Captain Pike has been severely crippled by a radiation accident. As a consequence of this accident Captain Pike is completely paralyzed and confined to a wheelchair controlled by his brain waves. He can only communicate through a light integrated into his wheelchair to signal the answers "yes" or "no". Commodore Mendez, the commander of Starbase 11, describes the condition of Captain Pike as follows: "He is totally unable to move, Jim. His wheelchair is constructed to respond to his brain waves. He can turn it, move it forwards, backwards slightly. Through a flashing light he can say 'yes' or 'no'. But that's it, Jim. That is as much as the poor ever can do. His mind is as active as yours and mine, but it's trapped in a useless vegetating body. He's kept alive mechanically. A battery driven heart. . . ."

This episode from the well-known TV series Star Trek was first shown in 1966. It describes a man who suffers from locked-in syndrome. In this condition, the person is cognitively intact but the body is paralyzed. In this case, paralyzed means that any voluntary control of muscles is lost. People cannot move their arms, legs, or faces, and depend on an artificial respirator. The active and fully functional mind is trapped in the body – as accurately described in the excerpt of the Star Trek episode above. The only effective way to communicate with the environment is with a device that can read brain signals and convert them into control and communication signals.

Such a device is called a brain–computer interface (BCI). Back in the 60s, controlling devices with brain waves was considered pure science fiction, as wild and fantastic as warp drive and transporters. Although recording brain signals from the human scalp gained some attention in 1929, when the German scientist Hans Berger recorded the electrical brain activity from the human scalp, the required technologies for measuring and processing brain signals as well as our understanding of brain function were still too limited. Nowadays, the situation has changed. Neuroscience

B. Graimann (✉)
Strategic Technology Management, Otto Bock HealthCare GmbH, Max-Näder Straße 15, 37115 Duderstadt, Germany
e-mail: graimann@ottobock.de

B. Graimann et al. (eds.), *Brain–Computer Interfaces*, The Frontiers Collection,
DOI 10.1007/978-3-642-02091-9_1, © Springer-Verlag Berlin Heidelberg 2010

research over the last decades has led to a much better understanding of the brain. Signal processing algorithms and computing power have advanced so rapidly that complex real-time processing of brain signals does not require expensive or bulky equipment anymore.

The first BCI was described by Dr. Grey Walter in 1964. Ironically, this was shortly before the first Star Trek episode aired. Dr. Walter connected electrodes directly to the motor areas of a patient's brain. (The patient was undergoing surgery for other reasons.) The patient was asked to press a button to advance a slide projector while Dr. Walter recorded the relevant brain activity. Then, Dr. Walter connected the system to the slide projector so that the slide projector advanced whenever the patient's brain activity indicated that he wanted to press the button. Interestingly, Dr. Walter found that he had to introduce a delay from the detection of the brain activity until the slide projector advanced because the slide projector would otherwise advance before the patient pressed the button! Control before the actual movement happens, that is, control without movement – the first BCI!

Unfortunately, Dr. Walter did not publish this major breakthrough. He only presented a talk about it to a group called the Ostler Society in London [1]. There was little progress in BCI research for most of the time since then. BCI research advanced slowly for many more years. By the turn of the century, there were only one or two dozen labs doing serious BCI research. However, BCI research developed quickly after that, particularly during the last few years. Every year, there are more BCI-related papers, conference talks, products, and media articles. There are at least 100 BCI research groups active today, and this number is growing.

More importantly, BCI research has succeeded in its initial goal: proving that BCIs can work with patients who need a BCI to communicate. Indeed, BCI researchers have used many different kinds of BCIs with several different patients. Furthermore, BCIs are moving beyond communication tools for people who cannot otherwise communicate. BCIs are gaining attention for healthy users and new goals such as rehabilitation or hands-free gaming. BCIs are not science fiction anymore. On the other hand, BCIs are far from mainstream tools. Most people today still do not know that BCIs are even possible. There are still many practical challenges before a typical person can use a BCI without expert help. There is a long way to go from providing communication for some specific patients, with considerable expert help, to providing a range of functions for any user without help.

The goal of this chapter is to provide a gentle and clear introduction of BCIs. It is meant for newcomers of this exciting field of research, and it is also meant as a preparation for the remaining chapters of this book. Readers will find answers to the following questions: What are BCIs? How do they work? What are their limitations? What are typical applications, and who can benefit from this new technology?

1 What is a BCI?

Any natural form of communication or control requires peripheral nerves and muscles. The process begins with the user's intent. This intent triggers a complex process in which certain brain areas are activated, and hence signals are sent via the peripheral nervous system (specifically, the motor pathways) to the corresponding muscles, which in turn perform the movement necessary for the communication or control task. The activity resulting from this process is often called motor output or efferent output. Efferent means conveying impulses from the central to the peripheral nervous system and further to an effector (muscle). Afferent, in contrast, describes communication in the other direction, from the sensory receptors to the central nervous system. For motion control, the motor (efferent) pathway is essential. The sensory (afferent) pathway is particularly important for learning motor skills and dexterous tasks, such as typing or playing a musical instrument.

A BCI offers an alternative to natural communication and control. A BCI is an artificial system that bypasses the body's normal efferent pathways, which are the neuromuscular output channels [2]. Figure 1 illustrates this functionality.

Instead of depending on peripheral nerves and muscles, a BCI directly measures brain activity associated with the user's intent and translates the recorded brain activity into corresponding control signals for BCI applications. This translation involves signal processing and pattern recognition, which is typically done by a computer. Since the measured activity originates directly from the brain and not from the peripheral systems or muscles, the system is called a Brain–Computer Interface.

A BCI must have four components. It must record activity directly from the brain (invasively or non-invasively). It must provide feedback to the user, and must do so in realtime. Finally, the system must rely on intentional control. That is, the user must choose to perform a mental task whenever s/he wants to accomplish a goal with the BCI. Devices that only passively detect changes in brain activity that occur without any intent, such as EEG activity associated with workload, arousal, or sleep, are not BCIs.

Although most researchers accept the term "BCI" and its definition, other terms has been used to describe this special form of human–machine interface. Here are some definitions of BCIs found in BCI literature:

> Wolpaw et al.: "A direct brain-computer interface is a device that provides the brain with a new, non-muscular communication and control channel". [2].

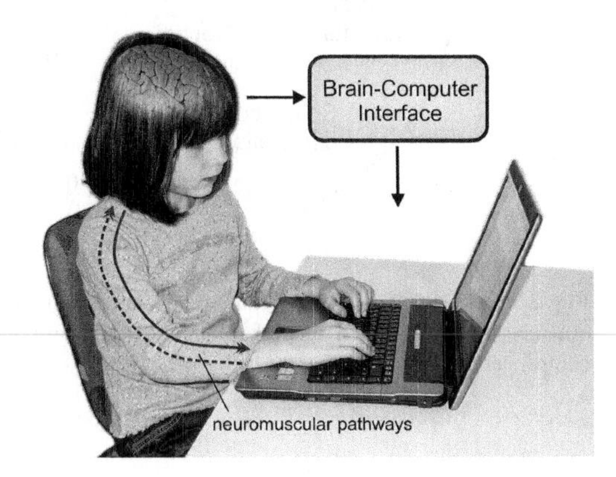

Fig. 1 A BCI bypasses the normal neuromuscular output channels

Donoghue et al.: "A major goal of a BMI (brain-machine interface) is to provide a command signal from the cortex. This command serves as a new functional output to control disabled body parts or physical devices, such as computers or robotic limbs" [3]

Levine et al.: "A direct brain interface (DBI) accepts voluntary commands directly from the human brain without requiring physical movement and can be used to operate a computer or other technologies." [4]

Schwartz et al.: "Microelectrodes embedded chronically in the cerebral cortex hold promise for using neural activity to control devices with enough speed and agility to replace natural, animate movements in paralyzed individuals. Known as cortical neural prostheses (CNPs), devices based on this technology are a subset of neural prosthetics, a larger category that includes stimulating, as well as recording, electrodes." [5]

Brain–computer interfaces, brain–machine interfaces (BMIs), direct brain interfaces (DBIs), neuroprostheses – what is the difference? In fact, there is no difference between the first three terms. BCI, BMI, and DBI all describe the same system, and they are used as synonyms.

"Neuroprosthesis," however, is a more general term. Neuroprostheses (also called neural prostheses) are devices that cannot only receive output from the nervous system, but can also provide input. Moreover, they can interact with the peripheral and the central nervous systems. Figure 2 presents examples of neuroprostheses, such as cochlear implants (auditory neural prostheses) and retinal implants (visual neural prostheses). BCIs are a special category of neuroprostheses.

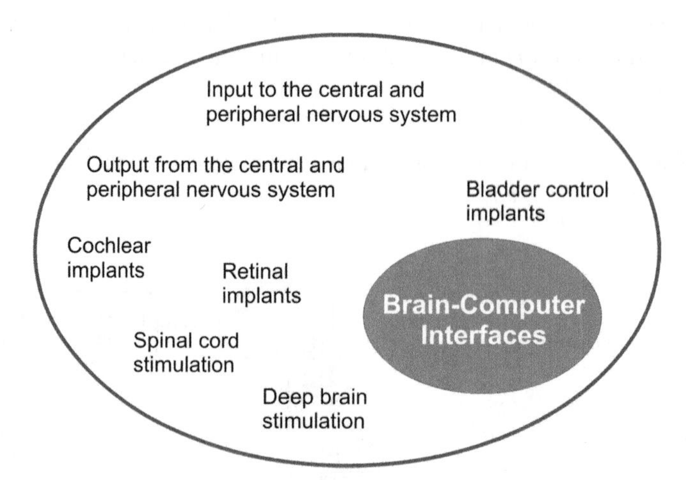

Fig. 2 Neuroprostheses can stimulate and/or measure activity from the central or peripheral nervous system. BCIs are a special subcategory that provides an artificial output channel from the central nervous system

They are, as already described in the definitions above, direct artificial output channels from the brain. Unlike other human–computer interfaces, which require muscle activity, BCIs provide "non-muscular" communication. One of the most important reasons that this is significant is that current BCI systems aim to provide assistive devices for people with severe disabilities that can render people unable to perform physical movements. Radiation accidents like the one in the Star Trek episode described above are unlikely today, but some diseases can actually lead to the locked-in syndrome.

Amyotrophic lateral sclerosis (ALS) is an example of such a disease. The exact cause of ALS is unknown, and there is no cure. ALS starts with muscle weakness and atrophy. Usually, all voluntary movement, such as walking, speaking, swallowing, and breathing, deteriorates over several years, and eventually is lost completely. The disease, however, does not affect cognitive functions or sensations. People can still see, hear, and understand what is happening around them, but cannot control their muscles. This is because ALS only affects special neurons, the large alpha motor neurons, which are an integral part of the motor pathways. Death is usually caused by failure of the respiratory muscles.

Life-sustaining measures such as artificial respiration and artificial nutrition can considerably prolong the life expectancy. However, this leads to life in the locked-in state. Once the motor pathway is lost, any natural way of communication with the environment is lost as well. BCIs offer the only option for communication in such cases. More details about ALS and BCIs can be found in the chapters "Brain–Computer Interface in Neurorehabilitation" and "Brain–Computer Interfaces for Communication and Control in Locked-in Patients" of this book.

So, a BCI is an artificial output channel, a direct interface from the brain to a computer or machine, which can accept voluntary commands directly from the brain without requiring physical movements. A technology that can listen to brain activity that can recognize and interpret the intent of the user? Doesn't this sound like a mind reading machine? This misconception is quite common among BCI newcomers, and is presumably also stirred up by science fiction and poorly researched articles in popular media. In the following section, we explain the basic principles of BCI operation. It should become apparent that BCIs are not able to read the mind.

2 How Do BCIs Work?

BCIs measure brain activity, process it, and produce control signals that reflect the user's intent. To understand BCI operation better, one has to understand how brain activity can be measured and which brain signals can be utilized. In this chapter, we focus on the most important recording methods and brain signals. Chapter "Brain Signals for Brain–Computer Interfaces" of this book gives a much more detailed representation of these two topics.

2.1 Measuring Brain Activity (Without Surgery)

Brain activity produces electrical and magnetic activity. Therefore, sensors can detect different types of changes in electrical or magnetic activity, at different times over different areas of the brain, to study brain activity.

Most BCIs rely on electrical measures of brain activity, and rely on sensors placed over the head to measure this activity. Electroencephalography (EEG) refers to recording electrical activity from the scalp with electrodes. It is a very well established method, which has been used in clinical and research settings for decades. Figure 3 shows an EEG based BCI. EEG equipment is inexpensive, lightweight, and comparatively easy to apply. Temporal resolution, meaning the ability to detect changes within a certain time interval, is very good. However, the EEG is not without disadvantages: The spatial (topographic) resolution and the frequency range are limited. The EEG is susceptible to so-called artifacts, which are contaminations in the EEG caused by other electrical activities. Examples are bioelectrical activities caused by eye movements or eye blinks (electrooculographic activity, EOG) and from muscles (electromyographic activity, EMG) close to the recording sites. External electromagnetic sources such as the power line can also contaminate the EEG.

Furthermore, although the EEG is not very technically demanding, the setup procedure can be cumbersome. To achieve adequate signal quality, the skin areas that are contacted by the electrodes have to be carefully prepared with special abrasive

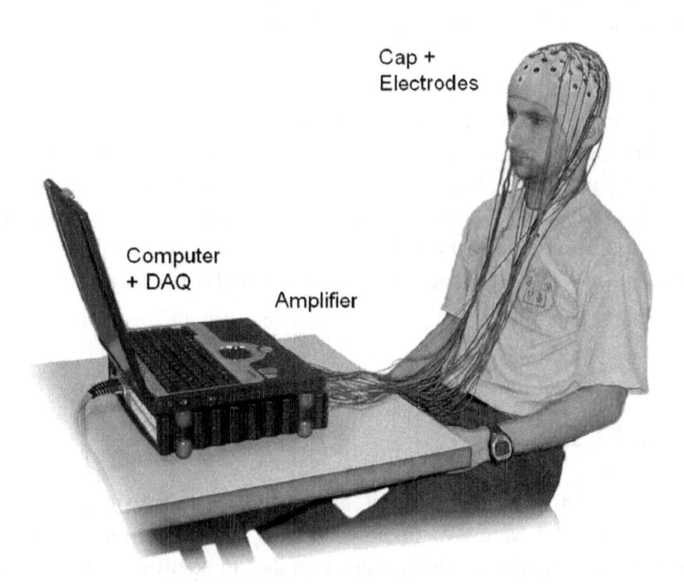

Fig. 3 A typical EEG based BCI consists of an electrode cap with electrodes, cables that transmit the signals from the electrodes to the biosignal amplifier, a device that converts the brain signals from analog to digital format, and a computer that processes the data as well as controls and often even runs the BCI application

electrode gel. Because gel is required, these electrodes are also called wet electrodes. The number of electrodes required by current BCI systems range from only a few to more than 100 electrodes. Most groups try to minimize the number of electrodes to reduce setup time and hassle. Since electrode gel can dry out and wearing the EEG cap with electrodes is not convenient or fashionable, the setting up procedure usually has to be repeated before each session of BCI use. From a practical viewpoint, this is one of largest drawbacks of EEG-based BCIs. A possible solution is a technology called dry electrodes. Dry electrodes do not require skin preparation nor electrode gel. This technology is currently being researched, but a practical solution that can provide signal quality comparable to wet electrodes is not in sight at the moment.

A BCI analyzes ongoing brain activity for brain patterns that originate from specific brain areas. To get consistent recordings from specific regions of the head, scientists rely on a standard system for accurately placing electrodes, which is called the International 10–20 System [6]. It is widely used in clinical EEG recording and EEG research as well as BCI research. The name 10–20 indicates that the most commonly used electrodes are positioned 10, 20, 20, 20, 20, and 10% of the total nasion-inion distance. The other electrodes are placed at similar fractional distances. The inter-electrode distances are equal along any transverse (from left to right) and antero-posterior (from front to back) line and the placement is symmetrical. The labels of the electrode positions are usually also the labels of the recorded channels. That is, if an electrode is placed at site C3, the recorded signal from this electrode is typically also denoted as C3. The first letters of the labels give a hint of the brain region over which the electrode is located: Fp – pre-frontal, F – frontal, C – central, P – parietal, O – occipital, T – temporal. Figure 4 depicts the electrode placement according to the 10–20 system.

While most BCIs rely on sensors placed outside of the head to detect electrical activity, other types of sensors have been used as well [7]. Magnetoencephalography (MEG) records the magnetic fields associated with brain activity. Functional magnetic resonance imaging (fMRI) measures small changes in the blood oxygenation level-dependent (BOLD) signals associated with cortical activation. Like fMRI also near infrared spectroscopy (NIRS) is a hemodynamic based technique for assessment of functional activity in human cortex. Different oxygen levels of the blood result in different optical properties which can be measured by NIRS. All these methods have been used for brain–computer communication, but they all have drawbacks which make them impractical for most BCI applications: MEG and fMRI are very large devices and prohibitively expensive. NIRS and fMRI have poor temporal resolution, and NIRS is still in an early stage of development [7–9].

2.2 Measuring Brain Activity (With Surgery)

The techniques discussed in the last section are all non-invasive recording techniques. That is, there is no need to perform surgery or even break the skin. In contrast, invasive recording methods require surgery to implant the necessary

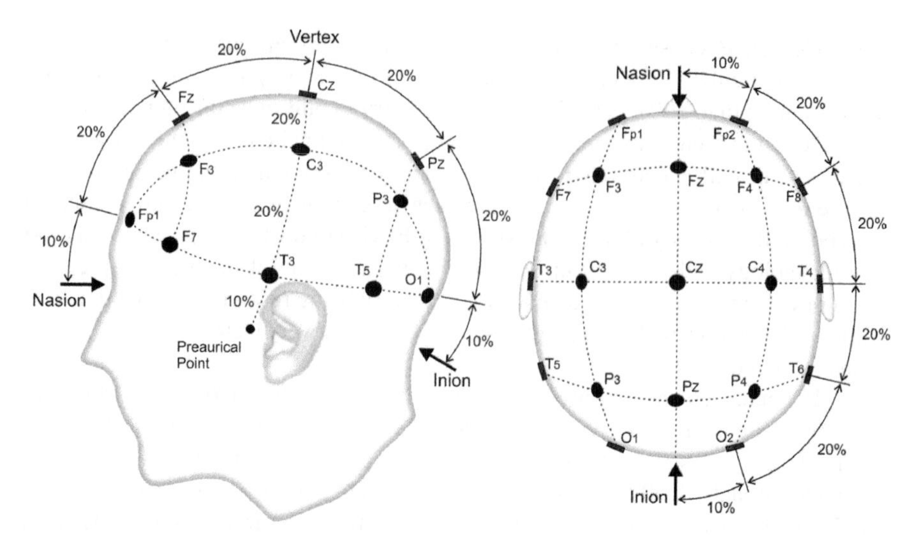

Fig. 4 The international 10–20 system: the left image shows the left side of the head, and the right side presents the view from above the head. The nasion is the intersection of the frontal and nasal bones at the bridge of the nose. The inion is a small bulge on the back of the skull just above the neck

sensors. This surgery includes opening the skull through a surgical procedure called a craniotomy and cutting the membranes that cover the brain. When the electrodes are placed on the surface of the cortex, the signal recorded from these electrodes is called the electrocorticogram (ECoG). ECoG does not damage any neurons because no electrodes penetrate the brain. The signal recorded from electrodes that penetrate brain tissue is called intracortical recording.

Invasive recording techniques combine excellent signal quality, very good spatial resolution, and a higher frequency range. Artifacts are less problematic with invasive recordings. Further, the cumbersome application and re-application of electrodes as described above is unnecessary for invasive approaches. Intracortical electrodes can record the neural activity of a single brain cell or small assemblies of brain cells. The ECoG records the integrated activity of a much larger number of neurons that are in the proximity of the ECoG electrodes. However, any invasive technique has better spatial resolution than the EEG.

Clearly, invasive methods have some advantages over non-invasive methods. However, these advantages come with the serious drawback of requiring surgery. Ethical, financial, and other considerations make neurosurgery impractical except for some users who need a BCI to communicate. Even then, some of these users may find that a noninvasive BCI meets their needs. It is also unclear whether both ECoG and intracortical recordings can provide safe and stable recording over years. Long-term stability may be especially problematic in the case of intracortical recordings. Electrodes implanted into the cortical tissue can cause tissue reactions that lead to deteriorating signal quality or even complete electrode failure. Research on invasive

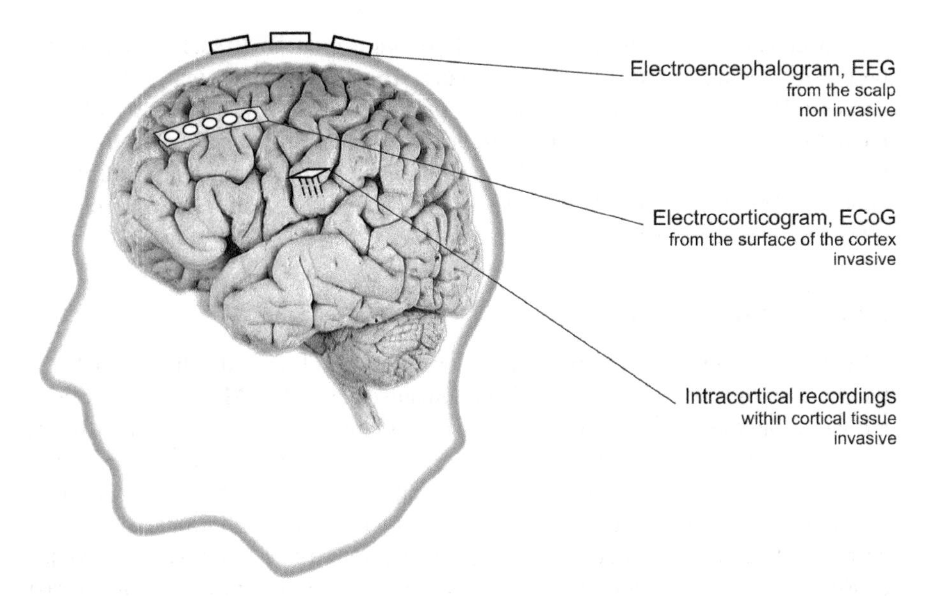

Fig. 5 Three different ways to detect the brain's electrical activity: EEG, ECoG, and intracortical recordings

BCIs is difficult because of the cost and risk of neurosurgery. For ethical reasons, some invasive research efforts rely on patients who undergo neurosurgery for other reasons, such as treatment of epilepsy. Studies with these patients can be very informative, but it is impossible to study the effects of training and long term use because these patients typically have an ECoG system for only a few days before it is removed.

Chapters "Intracortical BCIs: A Brief History of Neural Timing" through "A simple, spectral-change based, electrocorticographic Brain–Computer Interface" in this book describe these difficulties and give a more comprehensive overview of this special branch of BCI research. Figure 5 summarizes the different methods for recording bioelectrical brain activity.

2.3 Mental Strategies and Brain Patterns

Measuring brain activity effectively is a critical first step for brain–computer communication. However, measuring activity is not enough, because a BCI cannot read the mind or decipher thoughts in general. A BCI can only detect and classify specific patterns of activity in the ongoing brain signals that are associated with specific tasks or events. What the BCI user has to do to produce these patterns is determined by the mental strategy (sometimes called experimental strategy or approach) the BCI system employs. The mental strategy is the foundation of any brain–computer communication. The mental strategy determines what the user has to do to volitionally produce brain patterns that the BCI can interpret. The mental strategy also sets certain constraints on the hardware and software of a BCI, such as the signal processing techniques to be employed later. The amount of training required to successfully use a BCI also depends on the mental strategy. The most common mental strategies are selective (focused) attention and motor imagery [2, 10–12]. In the following, we briefly explain these different BCIs. More detailed information about different BCI approaches and associated brain signals can be found in chapter "Brain Signals for Brain–Computer Interfaces".

2.3.1 Selective Attention

BCIs based on selective attention require external stimuli provided by the BCI system. The stimuli can be auditory [13] or somatosensory [14]. Most BCIs, however, are based on visual stimuli. That is, the stimuli could be different tones, different tactile stimulations, or flashing lights with different frequencies. In a typical BCI setting, each stimulus is associated with a command that controls the BCI application. In order to select a command, the users have to focus their attention to the corresponding stimulus. Let's consider an example of a navigation/selection application, in which we want to move a cursor to items on a computer screen and then we want to select them. A BCI based on selective attention could rely on five stimuli. Four stimuli are associated with the commands for cursor movement: left, right, up, and down. The fifth stimulus is for the select command. This system would allow two dimensional navigation and selection on a computer screen. Users operate this BCI by focusing their attention on the stimulus that is associated with the intended command. Assume the user wants to select an item on the computer screen which is one position above and left of the current cursor position. The user would first need to focus on the stimulus that is associated with the up command, then on the one for the left command, then select the item by focusing on the stimulus associated with the select command. The items could represent a wide variety of desired messages or commands, such as letters, words, instructions to move a wheelchair or robot arm, or signals to control a smart home.

A 5-choice BCI like this could be based on visual stimuli. In fact, visual attention can be implemented with two different BCI approaches, which rely on somewhat different stimuli, mental strategies, and signal processing. These approaches are named after the brain patterns they produce, which are called P300 potentials and steady-state visual evoked potentials (SSVEP). The BCIs employing these brain patterns are called P300 BCI and SSVEP BCI, respectively.

A P300 based BCI relies on stimuli that flash in succession. These stimuli are usually letters, but can be symbols that represent other goals, such as controlling a cursor, robot arm, or mobile robot [15, 16]. Selective attention to a specific flashing symbol/letter elicits a brain pattern called P300, which develops in centro-parietal brain areas (close to location Pz, as shown in Fig. 3) about 300 ms after the presentation of the stimulus. The BCI can detect this P300. The BCI can then determine the symbol/letter that the user intends to select.

Like a P300 BCI, an SSVEP based BCI requires a number of visual stimuli. Each stimulus is associated with a specific command, which is associated with an output the BCI can produce. In contrast to the P300 approach, however, these stimuli do not flash successively, but flicker continuously with different frequencies in the range of about 6–30 Hz. Paying attention to one of the flickering stimuli elicits an SSVEP in the visual cortex (see Fig. 5) that has the same frequency as the target flicker. That is, if the targeted stimulus flickers at 16 Hz, the resulting SSVEP will also flicker at 16 Hz. Therefore, an SSVEP BCI can determine which stimulus occupies the user's attention by looking for SSVEP activity in the visual cortex at a specific frequency. The BCI knows the flickering frequencies of all light sources, and when an SSVEP is detected, it can determine the corresponding light source and its associated command.

BCI approaches using selective attention are quite reliable across different users and usage sessions, and can allow fairly rapid communication. Moreover, these approaches do not require significant training. Users can produce P300s and SSVEPs without any training at all. Almost all subjects can learn the simple task of focusing on a target letter or symbol within a few minutes. Many types of P300 and SSVEP BCIs have been developed. For example, the task described above, in which users move a cursor to a target and then select it, has been validated with both P300 and SSVEP BCIs [15, 17]. One drawback of both P300 and SSVEP BCIs is that they may require the user to shift gaze. This is relevant because completely locked-in patients are not able to shift gaze anymore. Although P300 and SSVEP BCIs without gaze shifting are possible as well [10, 18], BCIs that rely on visual attention seem to work best when users can shift gaze. Another concern is that some people may dislike the external stimuli that are necessary to elicit P300 or SSVEP activity.

2.3.2 Motor Imagery

Moving a limb or even contracting a single muscle changes brain activity in the cortex. In fact, already the preparation of movement or the imagination of movement also change the so-called sensorymotor rhythms. Sensorimotor rhythms (SMR) refer to oscillations in brain activity recorded from somatosensory and motor areas (see Fig. 6). Brain oscillations are typically categorized according to specific frequency bands which are named after Greek letters (delta: < 4 Hz, theta: 4–7 Hz, alpha: 8–12 Hz, beta: 12–30 Hz, gamma: > 30 Hz). Alpha activity recorded from sensorimotor areas is also called mu activity. The decrease of oscillatory activity in a specific frequency band is called event-related desynchronization (ERD). Correspondingly, the increase of oscillatory activity in a specific frequency band is called event-related synchronization (ERS). ERD/ERS patterns can be volitionally produced by motor imagery, which is the imagination of movement without actually performing the movement. The frequency bands that are most important for motor imagery are mu and beta in EEG signals. Invasive BCIs often also use gamma activity, which is hard to detect with electrodes mounted outside the head.

Topographically, ERD/ERS patterns follow a homuncular organization. Activity invoked by right hand movement imagery is most prominent over electrode location C3 (see Fig. 4). Left hand movement imagery produces activity most prominent

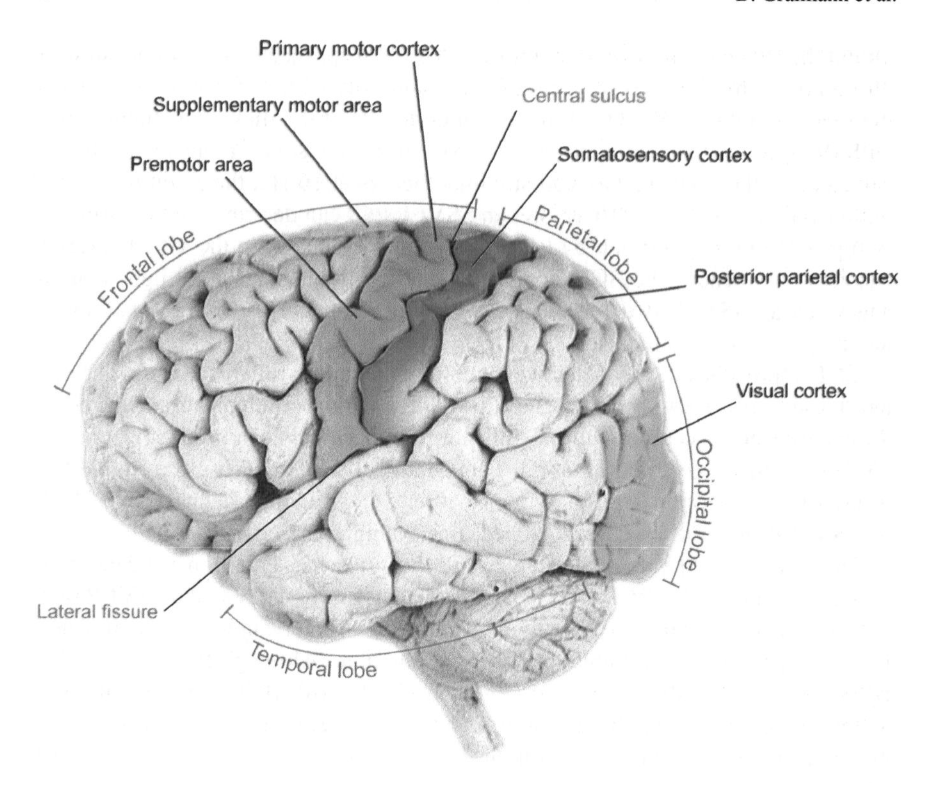

Fig. 6 The cerebrum is subdivided into four lobes: frontal, parietal, occipital, and temporal lobe. The central sulcus divides the frontal lobe from the parietal lobe. It also separates the precentral gyrus (*indicated in red*) and the postcentral gyrus (*indicated in blue*). The temporal lobe is separated from the frontal lobe by the lateral fissure. The occipital lobe lies at the very back of the cerebrum. The following cortical areas are particularly important for BCIs are: motor areas, somatosensory cortex, posterior parietal cortex, and visual cortex

over C4. That is, activity invoked by hand movement imagery is located on the contralateral (opposide) side. Foot movement imagery invokes activity over Cz. A distinction between left and right foot movement is not possible in EEG because the corresponding cortical areas are too close. Similarly, ERD/ERS patterns of individual fingers cannot be discriminated in EEG. To produce patterns that can be detected, the cortical areas involved have to be large enough so that the resulting activity is sufficiently prominent compared to the remaining EEG (background EEG). Hand areas, foot areas, and the tongue area are comparatively large and topographically different. Therefore, BCIs have been controlled by imagining moving the left hand, right hand, feet, and tongue [19].

ERD/ERS patterns produced by motor imagery are similar in their topography and spectral behavior to the patterns elicited by actual movements. And since these patterns originate from motor and somatosensory areas, which are directly connected to the normal neuromuscular output pathways, motor imagery is a particularly suitable mental strategy for BCIs. The way how motor imagery must be performed to best use a BCI can be different. For example, some BCIs can tell if the users are thinking of moving your left hand, right hand, or feet. This can lead to a BCI that allows 3 signals, which might be mapped on to commands to move left, right, and select. Another type of motor imagery BCI relies on more abstract, subject-specific types of movements. Over the course of several training sessions with a BCI, people can learn and develop their own motor imagery strategy. In a cursor movement task, for instance, people learn which types of imagined movements are best for BCI control, and reliably move a cursor up or down. Some subjects can learn to move a cursor in two [20] or even three [21] dimensions with further training.

In contrast to BCIs based on selective attention, BCIs based on motor imagery do not depend on external stimuli. However, motor imagery is a skill that has to be learned. BCIs based on motor imagery usually do not work very well during the first session. Instead, unlike BCIs on selective attention, some training is necessary. While performance and training time vary across subjects, most subjects can attain good control in a 2-choice task with 1–4 h of training (see chapters "The Graz Brain–Computer Interface", "BCIs in the Laboratory and at Home: The Wadsworth Research Program", and "Detecting Mental States by Machine Learning Techniques: The Berlin Brain–Computer Interface" in this book). However, longer training is often necessary to achieve sufficient control. Therefore, training is an important component of many BCIs. Users learn through a process called operant conditioning, which is a fundamental term in psychology. In operant conditioning, people learn to associate a certain action with a response or effect. For example, people learn that touching a hot stove is painful, and never do it again. In a BCI, a user who wants to move the cursor up may learn that mentally visualizing a certain motor task such as a clenching one's fist is less effective than thinking about the kinaesthetic experience of such a movement [22]. BCI learning is a special case of operant conditioning because the user is not performing an action in the classical sense, since s/he does not move. Nonetheless, if imagined actions produce effects, then conditioning can still occur. During BCI use, operant conditioning involves training with feedback that is usually presented on a computer screen. Positive feedback indicates that the brain signals are modulated in a desired way. Negative or no feedback is given when the user was not able to perform the desired task. BCI learning is a type of feedback called neurofeedback. The feedback indicates whether the user performed the mental task well or failed to achieve the desired goal through the BCI. Users can utilize this feedback to optimize their mental tasks and improve BCI performance. The feedback can be tactile or auditory, but most often it is visual. Chapter "Neurofeedback Training for BCI Control" in this book presents more details about neuro-feedback and its importance in BCI research.

2.4 Signal Processing

A BCI measures brain signals and processes them in real time to detect certain patterns that reflect the user's intent. This signal processing can have three stages: preprocessing, feature extraction, and detection and classification.

Preprocessing aims at simplifying subsequent processing operations without losing relevant information. An important goal of preprocessing is to improve signal quality by improving the so-called signal-to-noise ratio (SNR). A bad or small SNR means that the brain patterns are buried in the rest of the signal (e.g. background EEG), which makes relevant patterns hard to detect. A good or large SNR, on the other hand, simplifies the BCI's detection and classification task. Transformations combined with filtering techniques are often employed during preprocessing in a BCI. Scientists use these techniques to transform the signals so unwanted signal components can be eliminated or at least reduced. These techniques can improve the SNR.

The brain patterns used in BCIs are characterized by certain features or properties. For instance, amplitudes and frequencies are essential features of sensorimotor rhythms and SSVEPs. The firing rate of individual neurons is an important feature of invasive BCIs using intracortical recordings. The feature extraction algorithms of a BCI calculate (extract) these features. Feature extraction can be seen as another step in preparing the signals to facilitate the subsequent and last signal processing stage, detection and classification.

Detection and classification of brain patterns is the core signal processing task in BCIs. The user elicits certain brain patterns by performing mental tasks according to mental strategies, and the BCI detects and classifies these patterns and translates them into appropriate commands for BCI applications.

This detection and classification process can be simplified when the user communicates with the BCI only in well defined time frames. Such a time frame is indicated by the BCI by visual or acoustic cues. For example, a beep informs the user that s/he could send a command during the upcoming time frame, which might last 2–6 s. During this time, the user is supposed to perform a specific mental task. The BCI tries to classify the brain signals recorded in this time frame. This type of BCI does not consider the possibility that the user does not wish to communicate anything during one of these time frames, or that s/he wants to communicate outside of a specified time frame.

This mode of operation is called synchronous or cue-paced. Correspondingly, a BCI employing this mode of operation is called a synchronous BCI or a cue-paced BCI. Although these BCIs are relatively easy to develop and use, they are impractical in many real-world settings. A cue-pased BCI is somewhat like a keyboard that can only be used at certain times.

In an asynchronous or self-paced BCI, users can interact with a BCI at their leisure, without worrying about well defined time frames [23]. Users may send a signal, or choose not to use a BCI, whenever they want. Therefore, asynchronous BCIs or self-paced BCIs have to analyse the brain signals continuously. This mode of operation is technically more demanding, but it offers a more natural and convenient form of interaction with a BCI. More details about signal processing and the most frequently used algorithms in BCIs can be found in chapters "Digital Signal Processing and Machine Learning" and "Adaptive Methods in BCI Reaearch – An Introductory Tutorial" of this volume.

3 BCI Performance

The performance of a BCI can be measured in various ways [24]. A simple measure is classification performance (also termed classification accuracy or classification rate). It is the ratio of the number of correctly classified trials (successful attempts to perform the required mental tasks) and the total number of trials. The error rate is also easy to calculate, since it is just the ratio of incorrectly classified trials and the total number of trials.

Although classification or error rates are easy to calculate, application dependent measures are often more meaningful. For instance, in a mental typewriting application the user is supposed to write a particular sentence by performing a sequence of mental tasks. Again, classification performance could be calculated, but the number of letters per minute the users can convey is a more appropriate measure. Letters per minute is an application dependent measure that assesses (indirectly) not only the classification performance but also the time that was necessary to perform the required tasks.

A more general performance measure is the so-called information transfer rate (ITR) [25]. It depends on the number of different brain patterns (classes) used, the time the BCI needs to classify these brain patterns, and the classification accuracy. ITR is measured in bits per minute. Since ITR depends on the number of brain patterns that can be reliably and quickly detected and classified by a BCI, the information transfer rate depends on the mental strategy employed.

Typically, BCIs with selective attention strategies have higher ITRs than those using, for instance, motor imagery. A major reason is that BCIs based on selective attention usually provide a larger number of classes (e.g. number of light sources). Motor imagery, for instance, is typically restricted to four or less motor imagery tasks. More imagery tasks are possible but often only to the expense of decreased classification accuracy, which in turn would decrease in the information transfer rate as well.

There are a few papers that report BCIs with a high ITR, ranging from 30 bits/min [26, 27] to slightly above 60 bits/min [28] and, most recently, over 90 bits per minute [29]. Such performance, however, is not typical for most users in real world settings. In fact, these record values are often obtained under laboratory conditions by individual healthy subjects who are the top performing subjects in a lab. In addition, high ITRs are usually reported when people only use a BCI for short periods. Of course, it is interesting to push the limits and learn the best possible performance of current BCI technology, but it is no less important to estimate realistic performance in more practical settings. Unfortunately, there is currently no study available that investigates the average information transfer rate for various BCI systems over a larger user population and over a longer time period so that a general estimate of average BCI performance can be derived. The closest such study is the excellent work by Kübler and Birbaumer [30].

Furthermore, a minority of subjects exhibit little or no control [11, 26, 31,]. The reason is not clear, but even long sustained training cannot improve performance for those subjects. In any case, a BCI provides an alternative communication channel, but this channel is slow. It certainly does not provide high-speed interaction. It cannot compete with natural communication (such as speaking or writing) or traditional man-machine interfaces in terms of ITR. However, it has important applications for the most severely disabled. There are also new emerging applications for less severely disabled or even healthy people, as detailed in the next section.

4 Applications

BCIs can provide discrete or proportional output. A simple discrete output could be "yes" or "no", or it could be a particular value out of N possible values. Proportional output could be a continuous value within the range of a certain minimum and maximum. Depending on the mental strategy and on the brain patterns used, some BCIs are more suitable for providing discrete output values, while others are more suitable for allowing proportional control [32]. A P300 BCI, for instance, is particularly appropriate for selection applications. SMR based BCIs have been used for discrete control, but are best suited to proportional control applications such as 2-dimensional cursor control.

In fact, the range of possible BCI applications is very broad – from very simple to complex. BCIs have been validated with many applications, including spelling devices, simple computer games, environmental control, navigation in virtual reality, and generic cursor control applications [26, 33, 34].

Most of these applications run on conventional computers that host the BCI system and the application as well. Typically, the application is specifically tailored for a particular type of BCI, and often the application is an integral part of the BCI system. BCIs that can connect and effectively control a range of already existing assistive devices, software, and appliances are rare. An increasing number of systems allow control of more sophisticated devices, including orthoses, prostheses, robotic arms, and mobile robots [35–40]. Figure 7 shows some examples of BCI applications, most of which are described in detail in this book (corresponding references are given in the figure caption).

The concluding chapter discusses the importance of an easy to use "universal" interface that can allow users to easily control any application with any BCI. There is little argument that such an interface would be a boon to BCI research. BCIs can control any application that other interfaces can control, provided these applications can function effectively with the low information throughput of BCIs. On the other hand, BCIs are normally not well suited to controlling more demanding and complex applications, because they lack the necessary information transfer rate. Complex tasks like rapid online chatting, grasping a bottle with a robotic arm, or playing some computer games require more information per second than a BCI can provide. However, this problem can sometimes be avoided by offering short cuts.

For instance, consider an ALS patient using a speller application for communication with her caregiver. The patient is thirsty and wants to convey that she wants to drink some water now. She might perform this task by selecting each individual

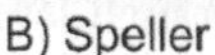

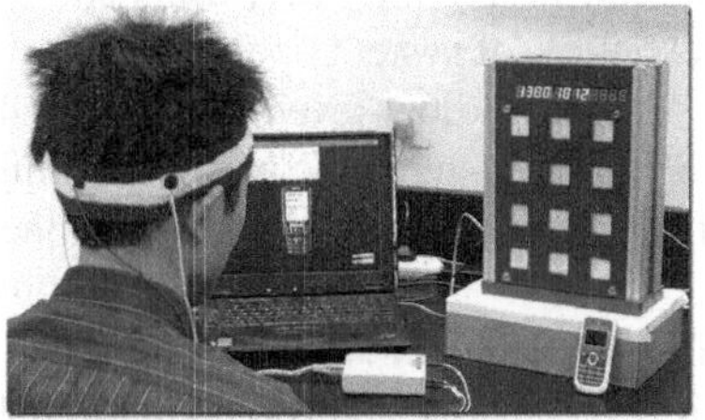

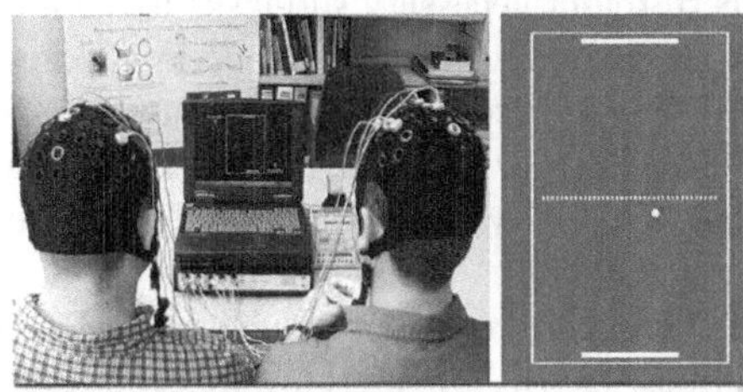

Fig. 7 Examples of BCI applications. (**a**) Environmental control with a P300 BCI (see chapter "The First Commercial Brain–Computer Interface Environment"), (**b**) P300 Speller (see chapter "BCIs in the Laboratory and at Home: The Wadsworth Research Program"), (**c**) Phone number dialling with an SSVEP BCI (see chapter "Practical Designs of Brain–Computer Interfaces Based on the Modulation of EEG Rhythms"), (**d**) Computer game Pong for two players, E) Navigation in a virtual reality environment (see chapter "The Graz Brain–Computer Interface"), (**f**) Restoration of grasp function of paraplegic patients by BCI controlled functional electrical stimulation (see chapter "Non invasive BCIs for neuroprostheses control of the paralysed hand")

letter and writing the message "water, please" or just "water". Since this is a wish the patient may have quite often, it would be useful to have a special symbol or command for this message. In this way, the patient can convey this particular message much faster, ideally with just one mental task. Many more short cuts might allow other tasks, but these short cuts lack the flexibility of writing individual messages. Therefore, an ideal BCI would allow a combination of simple commands to convey information flexibly and short cuts that allow specific, common, complex commands.

In other words, the BCI should allow a combination of process-oriented (or low-level) control and goal-oriented (or high level) control [41, 42]. Low-level control means the user has to manage all the intricate interactions involved in achieving a task or goal, such as spelling the individual letters for a message. In contrast, goal-oriented or high-level control means the users simply communicate their goal to the application. Such applications need to be sufficiently intelligent to autonomously perform all necessary sub-tasks to achieve the goal. In any interface, users should not be required to control unnecessary low-level details of system operation.

This is especially important with BCIs. Allowing low-level control of a wheelchair or robot arm, for example, would not only be slow and frustrating but potentially dangerous. Figure 8 presents two such examples of very complex applications.

The semi-autonomous wheelchair Rolland III can deal with different input modalities, such as low-level joystick control or high-level discrete control. Autonomous and semi-autonomous navigation is supported. The rehabilitation robot FRIEND II (Functional Robot Arm with User Friendly Interface for disabled People) is a semi-autonomous system designed to assist disabled people in activities of daily living. It is system based on a conventional wheelchair equipped with a stereo camera system, a robot arm with 7 degrees-of-freedom, a gripper with force/torque sensor, a smart tray with tactile surface and weight sensors, and a computing unit consisting of three independent industrial PCs. FRIEND II can perform certain operations

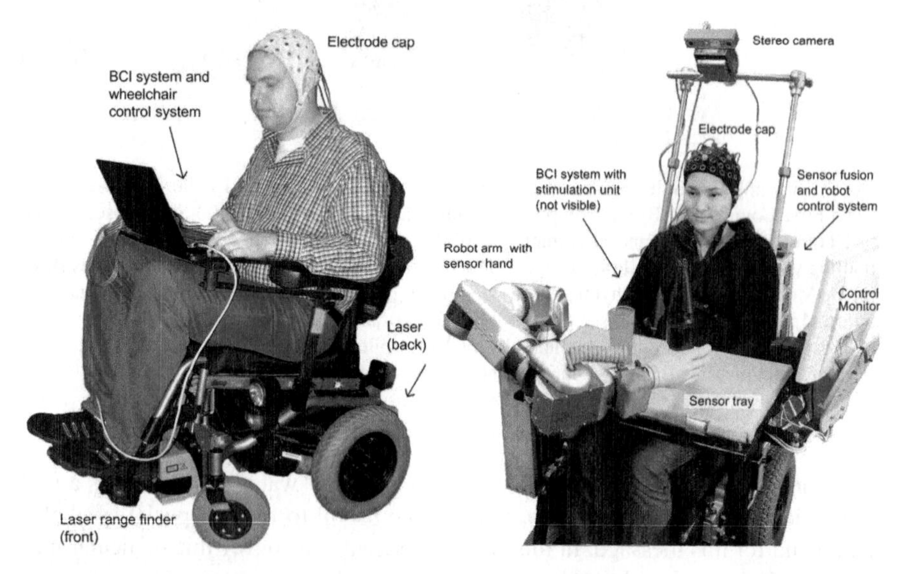

Fig. 8 Semi-autonomous assistive devices developed at the University of Bremen that include high level control: Intelligent wheelchair Rolland III, and rehabilitation robot FRIEND II (modified from [35])

completely autonomously. An example of such an operation is a "pour in beverage" scenario. In this scenario, the system detects the bottle and the glass (both located at arbitrary positions on the tray), grabs the bottle, moves the bottle to the glass while automatically avoiding any obstacles on the tray, fills the glass with liquid from the bottle while avoiding pouring too much, and finally puts the bottle back in its original position – again avoiding any possible collisions.

These assistive devices offload much of the work from the user onto the system. The wheelchair provides safety and high-level control by continuous path planning and obstacle avoidance. The rehabilitation robot offers a collection of tasks which are performed autonomously and can be initiated by single commands. Without this device intelligence, the user would need to directly control many aspects of device operation. Consequently, controlling a wheelchair, a robot arm, or any complex device with a BCI would be almost impossible, or at least very difficult, time consuming, frustrating, and in many cases even dangerous. Such complex BCI applications are not broadly available, but are still topics of research and are being evaluated in research laboratories. The success of these applications, or actually of any BCI application, will depend on their reliability and on their acceptance by users.

Another factor is whether these applications provide a clear advantage over conventional assistive technologies. In the case of completely locked-in patients, alternative control and communication methodologies do not exist. BCI control and communication is usually the only possible practical option. However, the situation is different with less severely disabled or healthy users, since they may be able to communicate through natural means like speech and gesture, and alternative control and communication technologies based on movement are available to them such as keyboards or eye tracking systems. Until recently, it was assumed that users would only use a BCI if other means of communication were unavailable. More recent work showed a user who preferred a BCI over an eye tracking system [43]. Although BCIs are gaining acceptance with broader user groups, there are many scenarios where BCIs remain too slow and unreliable for effective control. For example, most prostheses cannot be effectively controlled with a BCI.

Typically, prostheses for the upper extremities are controlled by electromyographic (myoelectric) signals recorded from residual muscles of the amputation stump. In the case of transradial amputation (forearm amputation), the muscle activity recorded by electrodes over the residual flexor and extensor muscles is used to open, close, and rotate a hand prosthesis. Controlling such a device with a BCI is not practical. For higher amputations, however, the number of degrees-of-freedom of a prostheses (i.e. the number of joints to be controlled) increases, but the number of available residual muscles is reduced. In the extreme case of an amputation of the entire arm (shoulder disarticulation), conventional myoelectric control of the prosthetic arm and hand becomes very difficult. Controlling such a device by a BCI may seem to be an option. In fact, several approaches have been investigated to control prostheses with invasive and non-invasive BCIs [39, 40, 44]. Ideally, the control of prostheses should provide highly reliable, intuitive, simultaneous, and proportional control of many degrees-of-freedom. In order to provide sufficient flexibility, low-level control is required. Proportional control in this case means the user can modulate speed and force of the actuators in the prosthesis. "Simultaneous" means that several degrees-of-freedom (joints) can be controlled at the same time. That is, for instance, the prosthetic hand can be closed while the wrist of the hand is rotated at the same time. "Intuitive" means that learning to control the prosthesis should be easy. None of the BCI approaches that have been currently suggested for controlling prostheses meets these criteria. Non-invasive approaches suffer from limited bandwidth, and will not be able to provide complex, high-bandwidth control in the near future. Invasive approaches show considerable more promise for such control in the near future. However, then these approaches will need to demonstrate that they have clear advantages over other methodologies such as myoelectric control combined with targeted muscle reinnervation (TMR).

TMR is a surgical technique that transfers residual arm nerves to alternative muscle sites. After reinnervation, these target muscles produce myoelectric signals (electromyographic signals) on the surface of the skin that can be measured and used to control prosthetic devices [45]. For example, in persons who have had their arm removed at the shoulder (called "shoulder disarticulation amputees"), residual peripheral nerves of arm and hand are transferred to separate regions of the pectoralis muscles.

Figure 9 shows a prototype of a prosthesis with 7 degrees-of-freedom (7 joints) controlled by such a system. Today, there is no BCI that can allow independent control of 7 different degrees of freedom, which is necessary to duplicate all the movements that a natural arm could make. On the other hand, sufficiently independent control signals can be derived from the myoelectric signals recorded from the

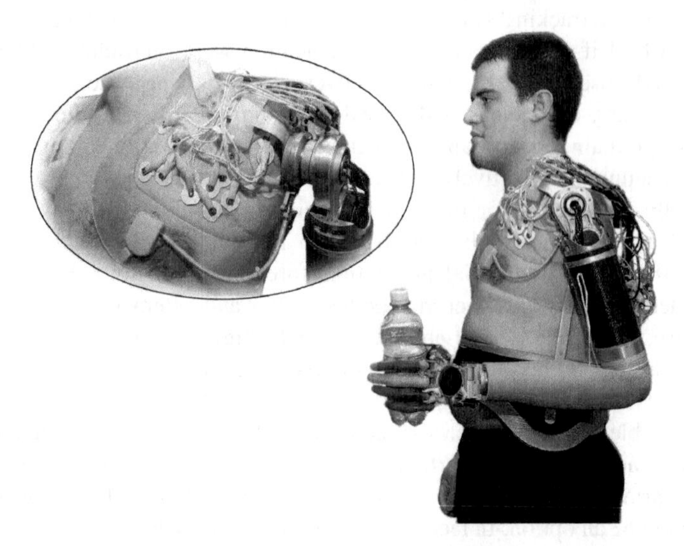

Fig. 9 Prototype of a prosthesis (Otto Bock HealthCare Products, Austria) with 7 degrees-of-freedom fitted to a shoulder disarticulation amputee with targeted muscle reinnervation (TMR). Control signals are recorded from electrodes mounted on the left pectoralis muscle

pectoralis. Moreover, control is largely intuitive, since users invoke muscle activity in the pectoralis in a similar way as they did to invoke movement of their healthy hand and arm. For instance, the users' intent to open the hand of their "phantom limb" results in particular myoelectric activity patterns that can be recorded from the pectoralis, and can be translated into control commands that open the prosthetic hand correspondingly. Because of this intuitive control feature, TMR based prosthetic devices can also be seen as thought-controlled neuroprostheses. Clearly, TMR holds the promise to improve the operation of complex prosthetic systems. BCI approaches (non-invasive and invasive) will need to demonstrate clinical and commercial advantages over TMR approaches in order to be viable.

The example with prostheses underscores a problem and an opportunity for BCI research. The problem is that BCIs cannot provide effective control because they cannot provide sufficient reliability and bandwidth (information per second). Similarly, the bandwidth and reliability of modern BCIs is far too low for many other goals that are fairly easy with conventional interfaces. Rapid communication, most computer games, detailed wheelchair navigation, and cell phone dialing are only a few examples of goals that require a regular interface.

Does this mean that BCIs will remain limited to severely disabled users? We think not, for several reasons. First, as noted above, there are many ways to increase the "effective bandwidth" of a BCI through intelligent interfaces and high level selection. Second, BCIs are advancing very rapidly. We don't think a BCI that is as fast as a keyboard is imminent, but substantial improvements in bandwidth are feasible. Third, some people may use BCIs even though they are slow because they are attracted to the novel and futuristic aspect of BCIs. Many research labs have demonstrated that computer games such as Pong, Tetris, or Pacman can be controlled by BCIs [46] and that rather complex computer applications like Google Earth can be navigated by BCI [47]. Users could control these systems more effectively with a keyboard, but may consider a BCI more fun or engaging. Motivated by the advances in BCI research over the last years, companies have started to consider BCIs as possibility to augment human–computer interaction. This interest is underlined by a number of patents and new products, which are further discussed in the concluding chapter of this book.

We are especially optimistic about BCIs for new user groups for two reasons. First, BCIs are becoming more reliable and easier to apply. New users will need a BCI that is robust, practical, and flexible. All applications should function outside the lab, using only a minimal number of EEG channels (ideally only one channel), a simple and easy to setup BCI system, and a stable EEG pattern suitable for online detection. The Graz BCI lab developed an example of such a system. It uses a specific motor imagery-based BCI designed to detect the short-lasting ERS pattern in the beta band after imagination of brisk foot movement in a single EEG channel [48]. Second, we are optimistic about a new technology called a "hybrid" system, which is composed of 2 BCIs or at least one BCI and another system [48–50]. One example of such a hybrid BCI relies on simple, one-channel ERD BCI to activate the flickering lights of a 4-step SSVEP-based hand orthosis only when the SSVEP system was needed for control [48].

Most present-day BCI applications focus on communication or control. New user groups might adopt BCIs that instead focus on neurorehabilitation. This refers to the goal of using a BCI to treat disorders such as stroke, ADHD, autism, or emotional disorders [51–53].

A BCI for neurorehabilitation is a new concept that uses neurofeedback and operant conditioning in a different way than a conventional BCI. For communication and control applications, neurofeedback is necessary to learn to use a BCI. The ultimate goal for these applications is to achieve the best possible control or communication performance. Neurofeedback is only a means to that end. In neurofeedback and neuro-rehabilitation applications, the situation is different. In these cases, the training itself is the actual application. BCIs are the most advanced neurofeedback systems available. It might be the case that modern BCI technology used in neurofeedback applications to treat neurological or neuropsychological disorders such as epilepsy, autism or ADHD is more effective than conventional neurofeedback. Neuro-rehabilitation of stroke is another possible BCI neurorehabilitation application. Here, the goal is to apply neuro-physiological regultion to foster cortical reorganization and compensatory cerebral activation of brain regions not affected by stroke [54]. Chapter Brain–Computer Interface in Neurorehabilitation of this book discusses this new direction in more detail.

5 Summary

A BCI is new direct artificial output channel. A conventional BCI monitors brain activity and detects certain brain patterns that are interpreted and translated to commands for communication or control tasks. BCIs may rely on different technologies to measure brain activity. A BCI can be invasive or non-invasive, and can be based on electrophysiological signals (EEG, ECoG, intracortical recordings) or other signals such as NIRS or fMRI. BCIs also vary in other ways, including the mental strategy used for control, interface parameters such as the mode of operation (synchronous or asynchronous), feedback type, signal processing method, and application. Figure 10 gives a comprehensive overview of BCI components and how they relate to each other.

BCI research over the last 20 years has focused on developing communication and control technologies for people suffering from severe neuromuscular disorders that can lead to complete paralysis or the locked-in state. The objective is to provide these users with basic assistive devices. Although the bandwidth of present-days BCIs is very limited, BCIs are of utmost importance for people suffering from complete locked-in syndrome, because BCIs are their only effective means of communication and control.

Advances in BCI technology will make BCIs more appealing to new user groups. BCI systems may provide communication and control to users with less severe disabilities, and even healthy users in some situations. BCIs may also provide new means of treating stroke, autism, and other disorders. These new BCI applications and groups will require new intelligent BCI components to address different

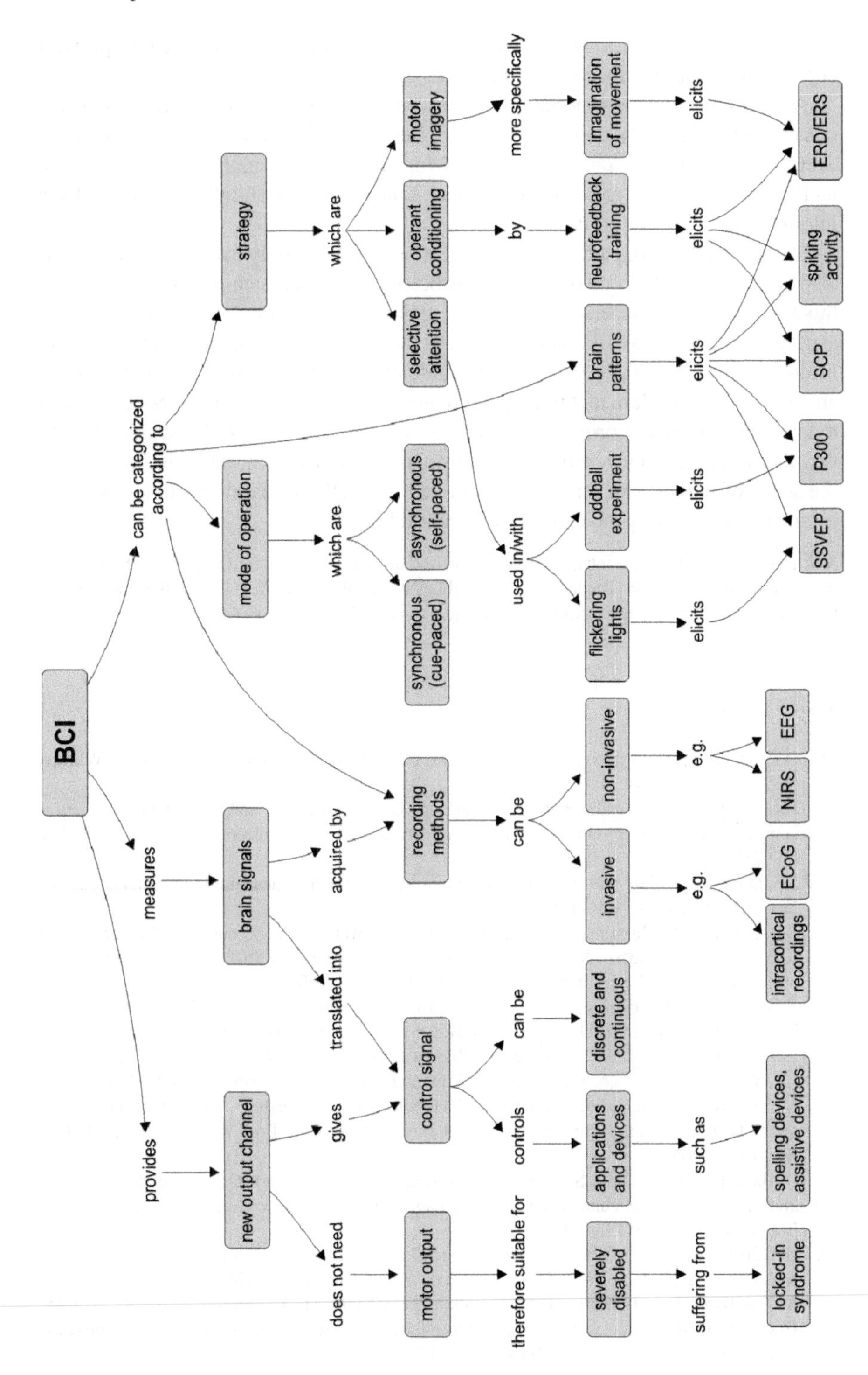

Fig. 10 Brain–computer interface concept-map

challenges, such as making sure that users receive the appropriate visual, proprioceptive, and other feedback to best recover motor function.

As BCIs become more popular with different user groups, increasing commercial possibilities will likely encourage new applied research efforts that will make BCIs even more practical. Consumer demand for reduced cost, increased performance, and greater flexibility and robustness may contribute substantially to making BCIs into more mainstream tools.

Our goal in this chapter was to provide a readable, friendly overview to BCIs. We also wanted to include resources with more information, such as other chapters in this book and other papers. Most of this book provides more details about different aspects of BCIs that we discussed here, and the concluding chapter goes "back to the future" by revisiting future directions. While most BCIs portrayed in science fiction are way beyond modern technology, there are many significant advances being made today, and reasonable progress is likely in the near future. We hope this chapter, and this book, convey not only some important information about BCIs, but also the sense of enthusiasm that we authors and most BCI researchers share about our promising and rapidly developing research field.

Acknowledgement The contribution of the second author was supported in part by the Information and Communication Technologies Collaborative Project "BrainAble" within the Seventh Framework of the European Commission, Project number ICT-2010-247447.

References

1. D.C. Dennett, *Consciousness explained*, Back Bay Books, Lippincott Williams & Wilkins, (1992).
2. J.R. Wolpaw, N. Birbaumer, D.J. McFarland, G. Pfurtscheller, and T.M. Vaughan, Brain-computer interfaces for communication and control. Clin Neurophysiol, 113, Jun., 767–791, (2002).
3. J.P. Donoghue, Connecting cortex to machines: recent advances in brain interfaces. Nat Neurosci. 5 (Suppl), Nov., 1085–1088, (2002).
4. S.P. Levine, J.E. Huggins, S.L. BeMent, R.K. Kushwaha, L.A. Schuh, E.A. Passaro, M.M. Rohde, and D.A. Ross, Identification of electrocorticogram patterns as the basis for a direct brain interface, J Clin Neurophysiol. 16, Sep., 439–447, (1999).
5. A.B. Schwartz, Cortical neural prosthetics. Annu Rev Neurosci, 27, 487–507, (2004).
6. E. Niedermeyer and F.L.D. Silva, *Electroencephalography: Basic principles, clinical applications, and related fields*, Lippincott Williams & Wilkins, (2004).
7. J.R. Wolpaw, G.E. Loeb, B.Z. Allison, E. Donchin, O.F. do Nascimento, W.J. Heetderks, F. Nijboer, W.G. Shain, and J.N. Turner, BCI Meeting 2005 – workshop on signals and recording methods, IEEE Trans Neural Syst Rehabil Eng: A Pub IEEE Eng Med Biol Soc. 14, Jun., 138–141, (2006).
8. G. Bauernfeind, R. Leeb, S.C. Wriessnegger, and G. Pfurtscheller, Development, set-up and first results for a one-channel near-infrared spectroscopy system. Biomedizinische Technik. Biomed Eng. 53, 36–43, (2008).
9. G. Dornhege, J.D.R. Millan, T. Hinterberger, D.J. McFarland, K. Müller, and T.J. Sejnowski, *Toward Brain-Computer Interfacing*, The MIT Press, Cambridge, MA, (2007).
10. B.Z. Allison, D.J. McFarland, G. Schalk, S.D. Zheng, M.M. Jackson, and J.R. Wolpaw, Towards an independent brain-computer interface using steady state visual evoked potentials. Clin Neurophysiol, 119, Feb., 399–408, (2008).

11. C. Guger, S. Daban, E. Sellers, C. Holzner, G. Krausz, R. Carabalona, F. Gramatica, and G. Edlinger, How many people are able to control a P300-based brain-computer interface (BCI)? Neurosci Lett, 462, Oct., 94–98, (2009).
12. G. Pfurtscheller, G. Müller-Putz, B. Graimann, R. Scherer, R. Leeb, C. Brunner, C. Keinrath, G. Townsend, M. Naeem, F. Lee, D. Zimmermann, and E. Höfler, Graz-Brain-Computer Interface: State of Research. In R. Dornhege (Eds.), *Toward brain-computer interfacing*, MIT Press, Cambridge, MA, pp. 65–102, (2007).
13. D.S. Klobassa, T.M. Vaughan, P. Brunner, N.E. Schwartz, J.R. Wolpaw, C. Neuper, and E.W. Sellers, Toward a high-throughput auditory P300-based brain-computer interface. Clin Neurophysiol, 120, Jul., 1252–1261, (2009).
14. G.R. Müller-Putz, R. Scherer, C. Neuper, and G. Pfurtscheller, Steady-state somatosensory evoked potentials: suitable brain signals for brain-computer interfaces? IEEE Trans Neural Syst Rehabil Eng, 14, Mar., 30–37, (2006).
15. L. Citi, R. Poli, C. Cinel, and F. Sepulveda, P300-based BCI mouse with genetically-optimized analogue control. IEEE Trans Neural Syst Rehabil Eng, 16, Feb., 51–61, (2008).
16. C.J. Bell, P. Shenoy, R. Chalodhorn, and R.P.N. Rao, Control of a humanoid robot by a noninvasive brain-computer interface in humans. J Neural Eng, 5, Jun., 214–220, (2008).
17. B. Allison, T. Luth, D. Valbuena, A. Teymourian, I. Volosyak, and A. Graeser, BCI Demographics: How Many (and What Kinds of) People Can Use an SSVEP BCI? IEEE Trans Neural Syst Rehabil Eng: A Pub IEEE Eng Med Biol Soc, 18(2), Jan., 107–116, (2010).
18. S.P. Kelly, E.C. Lalor, R.B. Reilly, and J.J. Foxe, Visual spatial attention tracking using high-density SSVEP data for independent brain-computer communication. IEEE Trans Neural Syst Rehabil Eng, 13, Jun., 172–178, (2005).
19. A. Schlögl, F. Lee, H. Bischof, and G. Pfurtscheller, Characterization of four-class motor imagery EEG data for the BCI-competition 2005. J Neural Eng, 2, L14–L22, (2005).
20. G.E. Fabiani, D.J. McFarland, J.R. Wolpaw, and G. Pfurtscheller, Conversion of EEG activity into cursor movement by a brain-computer interface (BCI). IEEE Trans Neural Syst Rehabil Eng, 12, Sep., 331–338, (2004).
21. D.J. McFarland, D.J. Krusienski, W.A. Sarnacki, and J.R. Wolpaw, Emulation of computer mouse control with a noninvasive brain-computer interface. J Neural Eng, 5, Jun., 101–110, (2008).
22. C. Neuper, R. Scherer, M. Reiner, and G. Pfurtscheller, Imagery of motor actions: differential effects of kinesthetic and visual-motor mode of imagery in single-trial EEG. Brain Res. Cogn Brain Res, 25, Dec., 668–677, (2005).
23. S.G. Mason and G.E. Birch, A brain-controlled switch for asynchronous control applications. IEEE Trans Bio-Med Eng, 47, Oct., 1297–1307, (2000).
24. A. Schlögl, J. Kronegg, J. Huggins, and S. Mason, Evaluation criteria for BCI research, In: *Toward brain-computer interfacing*, MIT Press, Cambridge, MA, pp. 342, 327, (2007).
25. D.J. McFarland, W.A. Sarnacki, and J.R. Wolpaw, Brain-computer interface (BCI) operation: optimizing information transfer rates. Biol Psychol, 63, Jul., 237–251, (2003).
26. B. Blankertz, G. Dornhege, M. Krauledat, K. Müller, and G. Curio, The non-invasive Berlin Brain-Computer Interface: fast acquisition of effective performance in untrained subjects. NeuroImage, 37, Aug., 539–550, (2007).
27. O. Friman, I. Volosyak, and A. Gräser, Multiple channel detection of steady-state visual evoked potentials for brain-computer interfaces. IEEE Trans Bio-Med Eng, 54, Apr., 742–750, (2007).
28. X. Gao, D. Xu, M. Cheng, and S. Gao, A BCI-based environmental controller for the motion-disabled. IEEE Trans Neural Syst Rehabil Eng, 11, Jun., 137–140, (2003).
29. G. Bin, X. Gao, Z. Yan, B. Hong, and S. Gao, An online multi-channel SSVEP-based brain-computer interface using a canonical correlation analysis method. J Neural Eng, 6, Aug., 046002, (2009).
30. A. Kübler and N. Birbaumer, Brain-computer interfaces and communication in paralysis: extinction of goal directed thinking in completely paralysed patients? Clin Neurophysiol, 119, Nov., 2658–2666, (2008).

31. C. Guger, G. Edlinger, W. Harkam, I. Niedermayer, and G. Pfurtscheller, How many people are able to operate an EEG-based brain-computer interface (BCI)? IEEE Trans Neural Syst and Rehabil Eng, 11, Jun., 145–147, (2003).
32. S.G. Mason, A. Bashashati, M. Fatourechi, K.F. Navarro, and G.E. Birch, A comprehensive survey of brain interface technology designs. Ann Biomed Eng, 35, Feb., 137–169, (2007).
33. G. Pfurtscheller, G.R. Müller-Putz, A. Schlögl, B. Graimann, R. Scherer, R. Leeb, C. Brunner, C. Keinrath, F. Lee, G. Townsend, C. Vidaurre, and C. Neuper, 15 years of BCI research at Graz University of Technology: current projects. IEEE Trans Neural Syst Rehabil Eng, 14, Jun., 205–210, (2006).
34. E.W. Sellers and E. Donchin, A P300-based brain-computer interface: initial tests by ALS patients. Clin Neurophysiol: Off J Int Feder Clin Neurophysiol, 117, Mar., 538–548, (2006).
35. B. Graimann, B. Allison, C. Mandel, T. Lüth, D. Valbuena, and A. Gräser, Non-invasive brain-computer interfaces for semi-autonomous assistive devices. Robust Intell Syst, 113–138, (2009).
36. R. Leeb, D. Friedman, G.R. Müller-Putz, R. Scherer, M. Slater, and G. Pfurtscheller, Self-Paced (Asynchronous) BCI control of a wheelchair in virtual environments: A case study with a Tetraplegic. Comput Intell Neurosci, 79642, (2007).
37. G. Pfurtscheller, C. Neuper, G.R. Müller, B. Obermaier, G. Krausz, A. Schlögl, R. Scherer, B. Graimann, C. Keinrath, D. Skliris, M. Wörtz, G. Supp, and C. Schrank, Graz-BCI: state of the art and clinical applications. IEEE Trans Neural Syst Rehabil Eng, 11, Jun., 177–180, (2003).
38. J.D.R. Millán, F. Renkens, J. Mouriño, and W. Gerstner, Noninvasive brain-actuated control of a mobile robot by human EEG, IEEE Trans Biomed Eng, 51, Jun., 1026–1033, (2004).
39. M. Velliste, S. Perel, M.C. Spalding, A.S. Whitford, and A.B. Schwartz, Cortical control of a prosthetic arm for self-feeding. Nature, 453, 1098–1101, (2008).
40. G.R. Müller-Putz and G. Pfurtscheller, Control of an Electrical Prosthesis With an SSVEP-Based BCI, IEEE Trans Biomed Eng, 55, 361–364, (2008).
41. B.Z. Allison, E.W. Wolpaw, and J.R. Wolpaw, Brain-computer interface systems: progress and prospects. Expert Rev Med Devices, 4, Jul., 463–474, (2007).
42. J.R. Wolpaw, Brain-computer interfaces as new brain output pathways, J Physiol, 579, Mar., 613–619, (2007).
43. T. Vaughan, D. McFarland, G. Schalk, W. Sarnacki, D. Krusienski, E. Sellers, and J. Wolpaw, The wadsworth BCI research and development program: at home with BCI. IEEE Trans Neural Syst Rehabil Eng, 14, 229–233, (2006).
44. L.R. Hochberg, M.D. Serruya, G.M. Friehs, J.A. Mukand, M. Saleh, A.H. Caplan, A. Branner, D. Chen, R.D. Penn, and J.P. Donoghue, Neuronal ensemble control of prosthetic devices by a human with tetraplegia. Nature, 442, Jul., 164–171, (2006).
45. T.A. Kuiken, G.A. Dumanian, R.D. Lipschutz, L.A. Miller, and K.A. Stubblefield, The use of targeted muscle reinnervation for improved myoelectric prosthesis control in a bilateral shoulder disarticulation amputee. Prosthet Orthot Int, 28, Dec., 245–253, (2004).
46. R. Krepki, B. Blankertz, G. Curio, and K. Müller, The Berlin Brain-Computer Interface (BBCI) – towards a new communication channel for online control in gaming applications. Multimedia Tools Appl, 33, 73–90, (2007).
47. R. Scherer, A. Schloegl, F. Lee, H. Bischof, J. Jansa, and G. Pfurtscheller, The self-paced Graz Brain-computer interface: Methods and applications. Comput Intell Neurosci, (2007).
48. G. Pfurtscheller, T. Solis-Escalante, R. Ortner, and P. Linortner, Self-Paced operation of an SSVEP-based orthosis with and without an imagery-based brain switch: A feasibility study towards a Hybrid BCI. IEEE Trans Neural Syst Rehabil Eng, 18(4), Feb., 409–414, (2010).
49. B.Z. Allison, C. Brunner, V. Kaiser, G.R. Müller-Putz, C. Neuper, and G. Pfurtscheller, Toward a hybrid brain–computer interface based on imagined movement and visual attention. J Neural Eng, 7, 026007, (2010).
50. C. Brunner, B.Z. Allison, D.J. Krusienski, V. Kaiser, G.R. Müller-Putz, G. Pfurtscheller, and C. Neuper, Improved signal processing approaches in an offline simulation of a hybrid brain-computer interface. J Neurosci Methods, 188(1), 30 Apr., 165–173, (2010).

51. N. Birbaumer and L.G. Cohen, Brain-computer interfaces: communication and restoration of movement in paralysis. J Physiol, 579, Mar., 621–636, (2007).
52. E. Buch, C. Weber, L.G. Cohen, C. Braun, M.A. Dimyan, T. Ard, J. Mellinger, A. Caria, S. Soekadar, A. Fourkas, and N. Birbaumer, Think to move: a neuromagnetic brain-computer interface (BCI) system for chronic stroke. Stroke, 39, Mar., 910–917, (2008).
53. J. Pineda, D. Brang, E. Hecht, L. Edwards, S. Carey, M. Bacon, C. Futagaki, D. Suk, J. Tom, C. Birnbaum, and A. Rork, Positive behavioral and electrophysiological changes following neurofeedback training in children with autism. Res Autism Spect Disord, 2, Jul., 557–581.
54. N. Birbaumer, C. Weber, C. Neuper, E. Buch, K. Haapen, and L. Cohen, Physiological regulation of thinking: brain-computer interface (BCI) research. Prog Brain Res, 159, 369–391, (2006).

Brain Signals for Brain–Computer Interfaces

Jonathan R. Wolpaw and Chadwick B. Boulay

1 Introduction

This chapter describes brain signals relevant for brain–computer interfaces (BCIs). Section 1 addresses the impetus for BCI research, reviews key BCI principles, and outlines a set of brain signals appropriate for BCI use. Section 2 describes specific brain signals used in BCIs, their neurophysiological origins, and their current applications. Finally, Sect. 3 discusses issues critical for maximizing the effectiveness of BCIs.

1.1 The Need for BCIs

People affected by amyotrophic lateral sclerosis (ALS), brainstem stroke, brain or spinal cord injury, cerebral palsy, muscular dystrophies, multiple sclerosis, and numerous other diseases often lose normal muscular control. The most severely affected may lose most or all voluntary muscle control and become totally "locked-in" to their bodies, unable to communicate in any way. These individuals can nevertheless lead lives that are enjoyable and productive if they can be provided with basic communication and control capability [1–4]. Unlike conventional assistive communication technologies, all of which require some measure of muscle control, a BCI provides the brain with a new, non-muscular output channel for conveying messages and commands to the external world.

1.2 Key Principles

A brain–computer interface, or BCI, is a communication and control system that creates a non-muscular output channel for the brain. The user's intent is conveyed

---

J.R. Wolpaw (✉)
Wadsworth Center, New York State Department of Health and School of Public Health, State University of New York at Albany, Albany, NY 12201, USA
e-mail: wolpaw@wadsworth.org

B. Graimann et al. (eds.), *Brain–Computer Interfaces*, The Frontiers Collection,
DOI 10.1007/978-3-642-02091-9_2,

by brain signals (such as electroencephalographic activity (EEG)), instead of being executed through peripheral nerves and muscles. Furthermore, these brain signals do not depend on neuromuscular activity for their generation. Thus, a true, or "independent," BCI is a communication and control system that does not rely in any way on muscular control.

Like other communication and control systems, a BCI establishes a real-time interaction between the user and the outside world. The user encodes his or her intent in brain signals that the BCI detects, analyzes, and translates into a command to be executed. The result of the BCI's operation is immediately available to the user, so that it can influence subsequent intent and the brain signals that encode that intent. For example, if a person uses a BCI to control the movements of a robotic arm, the arm's position after each movement affects the person's intent for the succeeding movement and the brain signals that convey that intent.

BCIs are not "mind-reading" or "wire-tapping" devices that listen in on the brain, detect its intent, and then accomplish that intent directly rather than through neuromuscular channels. This misconception ignores a central feature of the brain's interactions with the external world: that the skills that accomplish a person's intent, whether it be to walk in a specific direction, speak specific words, or play a particular piece on the piano, are mastered and maintained by initial and continuing adaptive changes in central nervous system (CNS) function. From early development throughout later life, CNS neurons and synapses change continually to acquire new behaviors and to maintain those already mastered [5, 6]. Such CNS plasticity is responsible for basic skills such as walking and talking, and for more specialized skills such as ballet and piano, and it is continually guided by the results of performance.

This dependence on initial and continuing adaptation exists whether the person's intent is carried out normally, that is, by peripheral nerves and muscles, or through an artificial interface, a BCI, that uses brain signals rather than nerves and muscles. BCI operation is based on the interaction of two adaptive controllers: the user, who must generate brain signals that encode intent; and the BCI system, which must translate these signals into commands that accomplish the user's intent. Thus, BCI operation is a skill that the user and system acquire and maintain. This dependence on the adaptation of user to system and system to user is a fundamental principle of BCI operation.

1.3 The Origin of Brain Signals Used in BCIs

In theory, a BCI might use brain signals recorded by a variety of methodologies. These include: recording of electric or magnetic fields; functional magnetic resonance imaging (fMRI); positron emission tomography (PET); and functional near-infrared (fNIR) imaging [7]. In reality, however, most of these methods are at present not practical for clinical use due to their intricate technical demands, prohibitive expense, limited real-time capabilities, and/or early stage of development. Only electric field recording (and possibly fNIR [8]) is likely to be of significant practical value for clinical applications in the near future.

Electric field recordings measure changes in membrane potentials of CNS synapses, neurons, and axons as they receive and transmit information. Neurons receive and integrate synaptic inputs and then transmit the results down their axons in the form of action potentials. Synaptic potentials and action potentials reflect changes in the flow of ions across the neuronal membranes. The electric fields produced by this activity can be recorded as voltage changes on the scalp (EEG), on the cortical surface (electrocorticographic activity (ECoG)), or within the brain (local field potentials (LFPs) or neuronal action potentials (spikes)). These three alternative recording methods are shown in Fig. 1. Any voltage change that can be detected with these recording methods might constitute a brain signal feature useful for a BCI. Intracortical electrodes can detect modulations in the spiking frequencies of single neurons; LFP, ECoG, or EEG recording can detect event-related voltage potentials (ERPs) or rhythmic voltage oscillations (such as mu or beta rhythms [9]). ERPs are best observed in the averaged signal time-locked to the evoking event or stimulus, and cortical oscillations are best observed by examining the frequency components of the signal. While the number of brain signal features that might be useful for a BCI is large, only a few have actually been tested for this purpose.

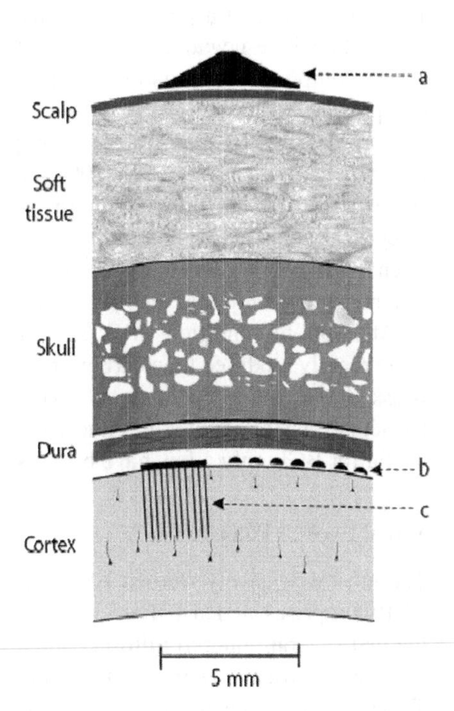

Fig. 1 Recording sites for electrophysiological signals used by BCI systems. (**a**):Electroencephalographic activity (EEG) is recorded by electrodes on the scalp. (**b**): Electrocorticographic activity (ECoG) is recorded by electrodes on the cortical surface. (**c**): Neuronal action potentials (spikes) or local field potentials (LFPs) are recorded by electrode arrays inserted into the cortex (modified from Wolpaw and Birbaumer[120])

2 Brain Signals for BCIs and Their Neurophysiological Origins

This section discusses the principal brain signal features that have been used in BCIs to date. The presentation is organized first by the invasiveness of the recording method and then by whether the feature is measured in the time domain or the frequency domain. It summarizes the characteristics, origin, and BCI uses of each feature.

2.1 Brain Signal Features Measured Noninvasively

Scalp-recorded EEG provides the most practical noninvasive access to brain activity. EEG is traditionally analyzed in the time domain or the frequency domain. In the time domain, the EEG is measured as voltage levels at particular times relative to an event or stimulus. When changes in voltage are time-locked to a particular event or stimulus, they are called "event-related potentials" (ERPs). In the frequency domain, the EEG is measured as voltage oscillations at particular frequencies (e.g., 8–12 Hz mu or 18–26 Hz beta rhythms over sensorimotor cortex). EEG features measured in both the time and the frequency domains have proved useful for BCIs.

2.1.1 Event-related Potentials (ERPs)

Brain processing of a sensory stimulus or another event can produce a time-locked series of positive-negative deflections in the EEG [10]. These event-related potential (ERP) components are distinguished by their scalp locations and latencies. Earlier components with latencies <100 ms originate largely in primary sensory cortices and are determined mainly by the properties of the evoking stimulus. Later ERP components with latencies of 100–>500 msec reflect to a greater extent ongoing brain processes and thus are more variable in form and latency. They originate in cortical areas associated with later and more complex processing. The longest-latency ERPs, or slow cortical potentials (SCPs), have latencies up to several seconds or even minutes and often reflect response-oriented brain activity.

Attempts to condition ERPs in humans have shown that later components are more readily modified than earlier ones. For example, they may change if the subject is asked to attend to the stimulus, to remember the stimulus, or to respond to it in a particular fashion. While it is likely that many different ERP components could be useful for BCI, only a few have been used successfully so far.

Visual Evoked Potentials

The most extensively studied ERP is the visual evoked potential (VEP) [11]. The VEP comprises at least three successive features, or components, that occur in the first several hundred milliseconds after a visual stimulus. The polarities, latencies, and cortical origins of the components vary with the stimulus presented. Typically, an initial negative component at 75 msec originates in primary visual cortex (area V1). It is followed by a positive component at ~100 ms (P1 or P100) and a negative complex at ~145 msec (N1 or N145). While the neural generators for P1 and N1 are still debated, they are probably striate and extrastriate visual areas. The steady-state visual evoked potential (SSVEP) is elicited by repetitive pattern-reversal stimulation [12]. It is thought to arise from the same areas that produce the VEP, plus the motion sensitive area MT/V5.

The VEP and SSVEP depend mainly on the properties of the visual stimulus. They have not been shown to vary with intent on a trial-by-trial basis. Nevertheless, the first BCI systems [13–15] and some modern systems [16, 17] are VEP- or SSVEP-based. For example, the BCI user can modulate the SSVEP by looking at one of several visual stimuli, each with different stimulus properties (i.e., different flash rates). SSVEP features vary according to the stimulus properties and are measured to determine which target is being looked at. The computer then executes the command associated with the target that the user is looking at. This communication system is equivalent to systems that determine gaze direction from the eyes themselves. Since it depends on the user's ability to control gaze direction, it is categorized as a dependent BCI system, that is, it requires muscle control. Recent evidence suggests that it is possible to modulate SSVEP features by shifting attention (as opposed to gaze direction), and thus that the SSVEP might support operation of an independent BCI (i.e., a BCI that does not depend on neuromuscular function) [16, 18, 19].

P300

When an auditory, visual, or somatosensory (touch) stimulus that is infrequent, desirable, or in other ways significant is interspersed with frequent or routine stimuli, it typically evokes a positive peak at about 300 ms after stimulus onset (i.e., a P300 ERP) in the EEG over centroparietal cortex [20–22]. A stimulation protocol of this kind is known as an 'oddball' paradigm [23, 24]. Evocation of the P300 in the oddball paradigm requires that the subject attend to the target stimulus. While the underlying neural generators of the P300 are debated, it is thought that the signal reflects rapid neural inhibition of ongoing activity and that this inhibition facilitates transmission of stimulus/task information from frontal to temporal-parietal cortical areas [25, 26].

In a P300-based BCI system, the user is presented with an array of auditory, visual, or somatosensory stimuli, each of which represents a particular output (e.g., spelling a particular letter), and pays attention to the stimulus that represents the action s/he desires. That attended stimulus elicits a P300 and the other stimuli do not (Fig. 2a). The BCI recognizes the P300 and then executes the output specified by the eliciting stimulus. Since it requires only that a user modulate attention, rather than any muscular output, a P300-based BCI is an independent BCI.

The first P300-based BCI system was developed by Donchin and his colleagues [27, 28]. It presents the user with a 6×6 matrix of letters, numbers, and/or other symbols. The individual rows and columns of the array flash in succession as the user attends to the desired item and counts how many times it flashes. The

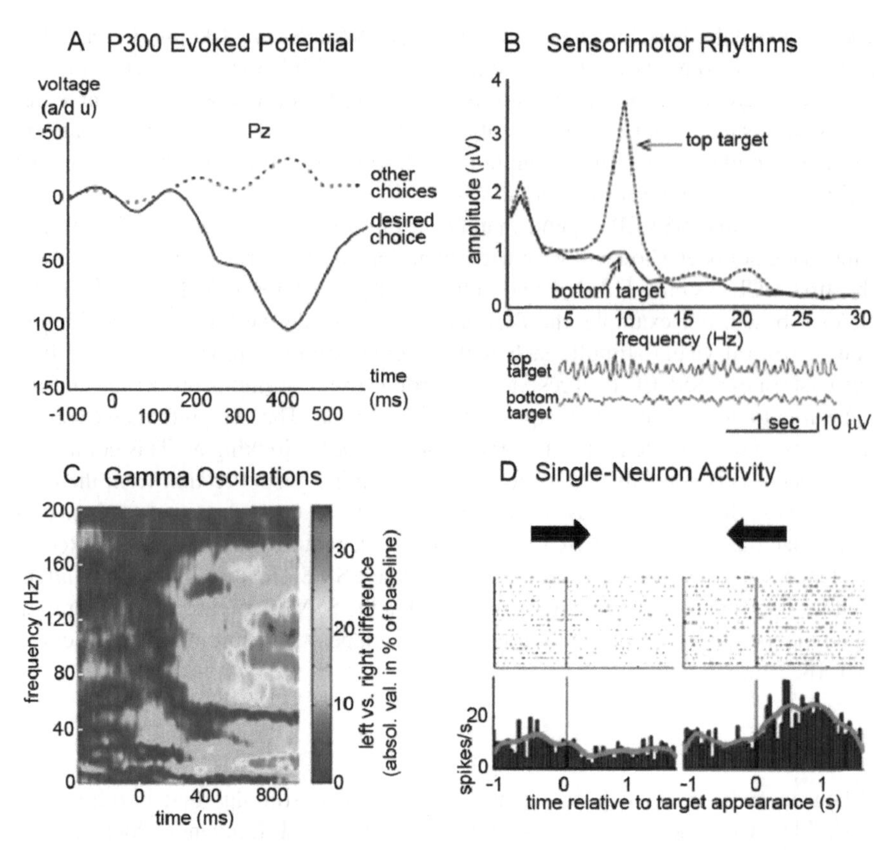

Fig. 2 Examples of brain signals that can be used in BCI systems. (**a**): An attended stimulus interspersed among non-attended stimuli evokes a positive peak at about 300 ms after stimulus onset (i.e., a P300 ERP) in the EEG over centroparietal cortex. The BCI detects the P300 and executes the command associated with the attended stimulus (modified from Donchin [27]). (**b**): Users control the amplitude of a 8–12 Hz mu rhythm (or a 18–26 Hz beta rhythm) in the EEG recorded over sensorimotor cortex to move the cursor to a target at the top of the screen or to a target at the bottom (or to additional targets at intermediate locations) [62, 121–123]. Frequency spectra (*top*) and sample EEG traces (*bottom*) for top and bottom targets show that this user's control is focused in the mu-rhythm frequency band. Users can also learn to control movement in two or three dimensions. (**c**): High-frequency gamma oscillations (>30 Hz) in the ECoG signal increase during leftward joystick movements compared to rightward movements (modified from Leuthardt [85]). Gamma oscillations recorded by ECoG could be useful for cursor control in future BCI systems. (**d**): A tetraplegic BCI-user decreases or increases the firing of a single neuron with a leftward preferred direction to move a computer cursor right or left, respectively (modified from Truccolo [117]). Raster plots show single-neuron activity from individual trials, and the histograms are the sums of 20 trials. The actual cursor trajectory is calculated from the firing rates of many neurons with different preferred directions

intersection of the row and column that produce the largest P300 identifies the target item, and the BCI then produces the output associated with that item. In recent work, improved signal processing methods and presentation parameters have extended this basic protocol to a 9×8 matrix of items and combined it with specific applications such as word-processing with a predictive speller and e-mail [29].

P300-based BCI systems have several important advantages: the P300 is detectable in most potential users; its relatively short latency (as opposed to that of SCPs (see below)) supports faster communication; with an appropriate protocol it does not attenuate significantly [30]; initial system calibration to each user usually requires less than 1 h; and very little user training is needed. In people with visual impairments, auditory or tactile stimuli could potentially be used instead [31, 32].

Slow Cortical Potentials

The slowest features of the scalp-recorded EEG yet used in BCI systems are slow voltage changes generated in cortex. These potential shifts occur over 0.5–10.0 s and are called slow cortical potentials (SCPs). In normal brain function, negative SCPs accompany mental preparation, while positive SCPs probably accompany mental inhibition [33, 34]. Negative and positive SCPs probably reflect an increase and decrease, respectively, in excitation of cortical neurons [35]. This change in activation may be further modulated by dopaminergic and cholinergic systems. For example, the contingent negative variation is a slowly developing negative SCP that occurs between a warning stimulus and a stimulus requiring a response [22]. It reflects increased activation and decreased firing thresholds in cortical networks associated with the preparatory process preceding the response.

In studies spanning more than 30 years, Birbaumer and his colleagues have demonstrated that people can learn to control SCP amplitude through an operant conditioning protocol, and can use this control to operate a BCI system [36–39]. In the standard format, SCP amplitude is displayed as the vertical position of a cursor, and after a 2-s baseline period, the user increases or decreases negative SCP amplitude to select the top or bottom target, respectively. The system can also operate in a mode that translates SCP amplitude into an auditory or tactile output. This SCP-based BCI can support basic word-processing and other simple control tasks. People who are severely disabled by ALS, and have little or no ability to use conventional assistive communication technology, may be able to master SCP control and use it for basic communication [e.g., 40].

2.1.2 Cortical Oscillations

Brain activity is reflected in a variety of oscillations, or rhythms, in scalp-recorded EEG. Each rhythm is distinguished by its frequency range, scalp location, and correlations with ongoing brain activity and behavioral state [41]. While these rhythms are traditionally believed to simply reflect brain activity, it is possible that they play functional roles (e.g., in synchronizing cortical regions) [42–44]. Rhythms that can be modulated independently of motor outputs are potentially useful for BCI applications. Up to the present, mainly sensorimotor rhythms have been applied to this purpose.

Sensorimotor Rhythms

In relaxed awake people, the EEG recorded over primary sensorimotor cortical areas often displays 8–12 Hz (mu-rhythm) and 18–26 Hz (beta-rhythm) activity [9, 41, 45, 46]. These sensorimotor rhythms (SMRs) are thought to be produced by thalamocortical circuits [41, 47] (see also chapter "Dynamics of sensorimotor oscillations in a motor task" in this book). SMR activity comprises a variety of different rhythms that are distinguished from each other by location, frequency, and/or relationship to concurrent sensory input or motor output. Some beta rhythms are harmonics of mu rhythms, while others are separable from mu rhythms by topography and/or timing, and thus are independent EEG features [48–50].

For several reasons, SMR rhythms are likely to be good signal features for EEG-based BCI uses. They are associated with the cortical areas most directly connected to the brain's normal neuromuscular outputs. Movement or preparation for movement is usually accompanied by SMR decrease, especially contralateral to the movement. This decrease has been labeled "event-related desynchronization" or ERD [49, 51]. Its opposite, rhythm increase, or "event-related synchronization" (ERS) occurs after movement and with relaxation [49]. Furthermore, and most pertinent to BCI use, ERD and ERS do not require actual movement, they occur also with motor imagery (i.e., imagined movement) [48, 52]. Thus, they might support independent BCIs. Since the mid-1980s, several mu/beta rhythm-based BCIs have been developed [e.g., 53, 54–59]. Several laboratories have shown that people can learn to control mu or beta rhythm amplitudes in the absence of movement or sensation.

With the BCI system of Wolpaw, McFarland, and their colleagues [59–63], people with or without motor disabilities learn to control mu- and/or beta-rhythm amplitudes. For each dimension of cursor control, a linear equation translates mu or beta amplitude from one or several scalp locations into cursor trajectory 20 times/s (e.g. Fig. 2b). Most users (i.e., about 80%) acquire significant one-dimensional control over 3–5 24-min sessions. Multi-dimensional control requires more extensive training. Users initially employ motor imagery to control the cursor, but, as training proceeds, imagery usually becomes less important, and users learn to move the cursor automatically, as they would perform a conventional motor act. The Wadsworth SMR-based BCI has been used to answer simple yes/no questions with accuracies >95%, to perform basic word processing [64–66], and to move a cursor in two or three dimensions [62, 67] (see also chapter "BCIs in the Laboratory and at Home: The Wadsworth Research Program" in this book).

In the Graz BCI system, a feature-selection algorithm uses data obtained during training sessions to select among a feature set that includes the frequency bands from 5 to 30 Hz and to train a classifier. In subsequent sessions, the classifier translates the user's EEG into a continuous output (e.g. cursor movement) or discrete output (e.g. letter selection). The classifier is adjusted between daily sessions. Features selected by the algorithm are concentrated in the mu- and beta-rhythm bands in EEG over sensorimotor cortex [52]. This BCI system has been use to open and close a prosthetic hand, to select letters, and to move a computer cursor [68–71].

2.2 Brain Signal Features Measured from the Cortical Surface

Short-term studies in hospitalized patients temporarily implanted with electrode arrays on the cortical surface prior to epilepsy surgery have revealed sharply focused ECoG activity associated with movement or sensation, or with motor imagery [e.g., 72]. Compared to scalp-recorded EEG, this ECoG activity has higher amplitude, greater topographical specificity, wider frequency range, and much less susceptibility to artifacts such as EMG activity. Thus, ECoG might be able to provide BCI-based communication and control superior to that practical or perhaps even possible with EEG.

ECoG features in both the time domain and frequency domain are closely related to movement type and direction [73–76]. The local motor potential is an ECoG signal feature identified in the time domain that predicts joystick movement [76, 77]. In the frequency domain, a particularly promising feature of ECoG activity is gamma activity, which comprises the 35–200 Hz frequency range. Only low-frequency gamma is evident in EEG, and only at small amplitudes. Unlike lower-frequency (mu and beta) SMRs, gamma activity increases in amplitude (i.e., displays event-related synchronization (ERS)) with muscle activity. Lower-frequency gamma (30–70 Hz) increases throughout muscle contraction while higher-frequency gamma (>75 Hz) increases with contraction onset and offset only [78]. Gamma activity, particularly at higher frequencies, is somatotopically organized and is more spatially focused than mu/beta changes [78–82]. While most studies have related gamma to actual movement or sensory input, others have shown that gamma is modulated by attention or motor imagery (including speech imagery) [83, 84].

While a gamma-based BCI system has yet to be implemented as a communication device, recent experiments suggest that gamma activity will be a useful brain signal for cursor control and possibly even synthesized speech. High-gamma frequencies up to 180 Hz held substantial information about movement direction in a center-out task [Fig. 2c; 85], and these signals were used to decode two-dimensional joystick kinematics [76] or to control a computer cursor with motor imagery [85]. Subjects learned to control cursor movement more rapidly with ECoG features than with EEG features [76, 86]. More details about ECoG based BCIs are given in chapters "BCIs Based on Signals from Between the Brain and Skull" and "A simple, Spectral-Change Based, Electrocorticographic Brain–Computer Interface" in this book.

2.3 Brain Signal Features Measured Within the Cortex

Intracortical recording (or recording within other brain structures) with microelectrode arrays provides the highest resolution signals, both temporally and spatially. Low-pass filtering (<1 kHz) of the intracortical signal reveals the local field potentials (LFPs) (essentially micro-EEG) which reflect nearby synaptic and neuronal activity [87, 88], while high-pass filtering (>1 kHz) reveals individual action potentials (i.e., spikes) from nearby individual cortical neurons. Both synaptic potentials and action potentials are possible input signals for BCI systems. At the same time, the necessary insertion of electrode arrays within brain tissue faces as yet unresolved problems in minimizing tissue damage and scarring and ensuring long-term recording stability [e.g., 89].

2.3.1 Local Field Potentials (LFPs) in the Time Domain

LFPs change when synapses and neurons within the listening sphere of the electrode tip are active. In primary and supplementary motor areas of cortex, the LFP prior to movement onset is a complex waveform, called the movement-evoked potential (MEP) [90]. The MEP has multiple components, including two positive (P1, P2) and two negative (N1, N2) peaks. Changes in movement direction modulate the amplitudes of these components [91]. It is as yet uncertain whether this signal is present in people with motor disabilities of various kinds, and thus whether it is likely to provide useful signal features for BCI use.

Kennedy and his colleagues [92] examined time-domain LFPs using cone electrodes implanted in the motor cortex of two individuals with locked-in syndrome. Subjects were able to move a cursor in one dimension by modulating LFP amplitude to cross a predetermined voltage threshold. One subject used LFP amplitude to control flexion of a virtual digit.

2.3.2 Local Field Potentials in the Frequency Domain

It is often easier to detect LFP changes in the frequency domain. Engagement of a neural assembly in a task may increase or decrease the synchronization of the synaptic potentials of its neurons. Changes in synchronization of synaptic potentials within the listening sphere of an intracortical electrode appear as changes in amplitude or phase of the frequency components of the LFP. Thus, LFP frequency components are potentially useful features for BCI use.

LFP frequency-domain features have not been used in human BCIs as yet, though some evidence suggests that they may be useful [93]. Research in non-human primates is more extensive. Changes in amplitude in specific frequency bands predict movement direction [91], velocity [94], preparation and execution [95], grip [96], and exploratory behavior [97].

Andersen and colleagues have recorded LFPs from monkey parietal cortex to decode high-level cognitive signals that might be used for BCI control [98–101]. LFP oscillations (25–90 Hz) from the lateral intraparietal area can predict when the monkey is planning or executing a saccade, and the direction of the saccade. Changes in rhythms in the parietal reach region predict intended movement direction or the state of the animal (e.g., planning vs. executing, reach vs. saccade). Further research is needed to determine whether human LFP evoked responses or rhythms are useful input signals for BCI systems.

2.3.3 Single-Neuron Activity

For more than 40 years, researchers have used metal microelectrodes to record action potentials of single neurons in the cerebral cortices of awake animals [e.g. 102]. Early studies showed that the firing rates of individual cortical neurons could be operantly conditioned [103–105]. Monkeys were able to increase or decrease single-neuron firing rates when rewarded for doing so. While such control was typically initially associated with specific muscle activity, the muscle activity tended to drop out as conditioning continued. However, it remains possible that undetected muscle activity and positional feedback from the activated muscles contribute to this control of single-neuron activity in the motor cortex. It also remains unclear to what extent the patterns of neuronal activity would differ if the animal were not capable of movement (due, for example, to a spinal cord injury).

Firing rates of motor cortex neurons correlate in various ways with muscle activity and with movement parameters [106, 107]. Neuronal firing rates may be related to movement velocity, acceleration, torque, and/or direction. In general, onset of a specific movement coincides with or follows an increase in the firing rates of neurons with preference for that movement and/or a decrease in the firing rates of neurons with preference for an opposite movement. Most recent research into the BCI use of such neuronal activity has employed the strategy of first defining the neuronal activity associated with standardized limb movements, then using this activity to control simultaneous comparable cursor movements, and finally showing that the neuronal activity alone can continue to control cursor movements accurately in the absence of continued close correlations with limb movements [108–111].

In addition to real or imagined movements, other cognitive activities may modulate cortical cells. Motor planning modulates firing rates of single neurons in posterior parietal areas [e.g., 99, 112]. Both a specific visual stimulus and imaginative recall of that stimulus activate neurons in sensory association areas [113]. Andersen and colleagues have used firing rates of single neurons in parietal areas of the monkey cortex to decode goal-based target selection [99] or to predict direction and amplitude of intended saccades [114]. The degree to which these signals will be useful in BCI applications is as yet unclear.

Two groups have used single-neuron activity for BCI operation in humans with severe disabilities. Kennedy and colleagues implanted neurotrophic cone electrodes [115] and found that humans could modulate neuronal firing rates to control switch closure or one-dimensional cursor movement [116]. Donoghue, Hochberg and their colleagues have implanted multielectrode arrays in the hand area of primary motor cortex in several severely disabled individuals. A participant could successfully control the position of a computer cursor through a linear filter that related spiking in motor cortex to cursor position [93]. Subsequent analyses of data from two people revealed that neuronal firing was related to cursor velocity and position [Fig. 2d; 117].

3 Requirements for Continued Progress

At present, it is clear that all three classes of electrophysiological signals – EEG, ECoG, and intracortical (i.e., single neurons/LFPs) – have promise for BCI uses. At the same time, it is also clear that each method is still in a relatively early stage of development and that substantial further work is essential.

EEG-based BCIs that use the P300 are already being used successfully at home by a few people severely disabled by ALS [118, 119]. If these systems are to be widely disseminated, further development is needed to make them more robust and to reduce the need for technical support. EEG-based BCIs that use SMRs are still confined to the laboratory. They require reduction in training requirements and increased day-to-day reliability before they can be moved into users' homes.

ECoG-based BCIs can at present be evaluated only for relatively short periods in people implanted prior to epilepsy surgery. These short-term human studies need to establish that ECoG is substantially superior to EEG in the rapidity of user training and/or in the complexity of the control it provides. At the same time, array development and subsequent animal studies are needed to produce ECoG arrays suitable for long-term human use and to demonstrate their safety and long-term effectiveness. When these steps are completed, long-term human implantations may be justified, and full development of the potential of ECoG-based BCIs can proceed.

Intracortical BCIs require essentially the same pre-clinical studies needed for ECoG-based BCIs. Extensive animal studies are needed to determine whether intracortical methods are in fact substantially superior to EEG and ECoG methods in the control they can provide, and to verify their long-term safety and effectiveness. In addition, for both ECoG and intracortical BCIs, wholly implantable, telemetry-based systems will be essential. Systems that entail transcutaneous connections are not suitable for long-term use.

Different BCI methods are likely to prove best for different applications; P300-based BCIs are good for selecting among many options while SMR-based BCIs are good for continuous multi-dimensional control. Different BCI methods may prove better for different individuals; some individuals may be able to use a P300-based BCI but not an SMR-based BCI, or vice-versa. For all methods, issues of convenience, stability of long-term use, and cosmesis will be important. Minimization of the need for ongoing technical support will be a key requirement, since a continuing need for substantial technical support will limit widespread dissemination.

Acknowledgments The authors' brain–computer interface (BCI) research has been supported by the National Institutes of Health, the James S. McDonnell Foundation, the ALS Hope Foundation, the NEC Foundation, the Altran Foundation, and the Brain Communication Foundation.

References

1. F. Maillot, L. Laueriere, E. Hazouard, B. Giraudeau, and P. Corcia, Quality of life in ALS is maintained as physical function declines. Neurology, 57, 1939, (2001).
2. R.A. Robbins, Z. Simmons, B.A. Bremer, S.M. Walsh, and S. Fischer, Quality of life in ALS is maintained as physical function declines. Neurology, 56, 442–444, (2001).
3. Z. Simmons, B.A. Bremer, R.A. Robbins, S.M Walsh, and S. Fischer, Quality of life in ALS depends on factors other than strength and physical function, Neurology, 55, 388–392, (2000).
4. Z. Simmons, S.H. Felgoise, B.A. Bremer, et al., The ALSSQOL: balancing physical and nonphysical factors in assessing quality of life in ALS, Neurology, 67, 1659–1664, (2006).
5. C. Ghez and J. Krakauer, Voluntary movement. In E.R. Kandel, J.H. Schwartz,, T.M. Jessell, (Eds.), *Principles of neural science,* McGraw-Hill, New York, pp. 653–674, (2000).

6. A.W. Salmoni, R.A. Schmidt, and C.B. Walter, Knowledge of results and motor learning: a review and critical reappraisal, Psychol Bull, 95, 355–386, (1984).
7. J.R. Wolpaw, G.E. Loeb, B.Z. Allison, et al., BCI Meeting 2005 – workshop on signals and recording methods, IEEE Trans Neural Syst Rehabil Eng, 14, 138–141, (2006).
8. G. Bauernfeind, R. Leeb, S.C. Wriessnegger, and G. Pfurtscheller, Development, set-up and first results for a one-channel near-infrared spectroscopy system, Biomedizinische Technik, 53, 36–43, (2008).
9. J.W. Kozelka and T.A. Pedley, Beta and mu rhythms, J Clin Neurophysiol, 7, 191–207, (1990).
10. F.H. L da Silva, Event-related potentials: Methodology and quantification. In E. Niedermeyer and F.H.L da Silva, (Eds.), *Electroencephalography: Basic principles, clinical applications, and related fields*, Williams and Wilkins, Baltimore, MD, pp. 991–1002, (2004).
11. G.G. Celesia and N.S. Peachey, Visual Evoked Potentials and Electroretinograms. In E. Niedermeyer and F.H.L da Silva (Eds.), *Electroencephalography: Basic principles, clinical applications, and related fields*, Williams and Wilkins, Baltimore, MA, pp. 1017–1043, (2004).
12. D. Regan, Steady-state evoked potentials. J Opt Soc Am, 67, 1475–1489, (1977).
13. E.E Sutter, The brain response interface: communication through visually guided electrical brain responses. J Microcomput Appl, 15, 31–45, (1992).
14. J.J. Vidal, Toward direct brain-computer communication. Annu Rev Biophys Bioeng, 2, 157–180, (1973).
15. J.J. Vidal, Real-time detection of brain events in EEG. IEEE Proc: Special Issue on Biol Signal Processing and Analysis, 65, 633–664, (1977).
16. B.Z. Allison, D.J. McFarland G. Schalk, S.D. Zheng, M.M. Jackson, and J.R. Wolpaw, Towards an independent brain-computer interface using steady state visual evoked potentials. Clin Neurophysiol, 119, 399–408, (2008).
17. G.R. Muller-Putz and G. Pfurtscheller, Control of an electrical prosthesis with an SSVEP-based BCI. IEEE Trans Biomed Eng, 55, 361–364, (2008).
18. P. Malinowski, S. Fuchs, and M.M. Muller Sustained division of spatial attention to multiple locations within one hemifield. Neurosci Lett, 414, 65–70, (2007).
19. A. Nijholt and D. Tan, Brain-Computer Interfacing for Intelligent Systems. Intell Syst IEEE, 23, 72–79, (2008).
20. J. Polich, Updating P300: an integrative theory of P3a and P3b. Clin Neurophysiol, 118, 2128–2148, (2007).
21. S. Sutton, M. Braren, J. Zubin, and E.R John, Evoked correlates of stimulus uncertainty. Science, 150, 1187–1188, (1965).
22. W.G. Walter, R. Cooper, V.J. Aldridge, W.C. McCallum, and A.L. Winter, Contingent negative variation: An electric sign of sensorimotor association and expectancy in the human brain. Nature, 203, 380–384, (1964).
23. E. Donchin, W. Ritter, and C. McCallum, Cognitive psychophysiology: the endogenous components of the ERP. In P. Callaway, P. Tueting, and S. Koslow (Eds.), *Brain-event related potentials in man*, Academic, New York, pp. 349–411, (1978).
24. W.S. Pritchard, Psychophysiology of P300, Psychol Bull, 89, 506–540, (1981).
25. E. Donchin, Presidential address, 1980. Surprise! ... Surprise? Psychophysiology, 18, 493–513, (1981).
26. J. Polich and J.R. Criado, Neuropsychology and neuropharmacology of P3a and P3b, Int J Psychophysiol 60, 172–185, (2006).
27. E. Donchin, K.M. Spencer, and R. Wijesinghe The mental prosthesis: assessing the speed of a P300-based brain-computer interface, IEEE Trans Rehabil Eng, 8, 174–179, (2000).
28. L.A. Farwell and E. Donchin, Talking off the top of your head: toward a mental prosthesis utilizing event-related brain potentials, Electroencephalogr Clin Neurophysiol, 70, 510–523, (1988).

29. T.M. Vaughan, E.W. Sellers, D.J. McFarland, C.S. Carmack, P. Brunner, P.A. Fudrea, E.M. Braun, S.S. Lee, A. Kübler, S.A. Mackler, D.J. Krusienski, R.N. Miller, and J.R. Wolpaw, *Daily use of an EEG-based brain-computer interface by people with ALS: technical requirements and caretaker training. Program No. 414.6. 2007 Abstract Viewer/Itinerary Planner.* Society for Neuroscience, Washington, DC, (2007). Online.
30. E.W. Sellers, D.J. Krusienski, D.J. McFarland, T.M. Vaughan, and J.R. Wolpaw, A P300 event-related potential brain-computer interface (BCI): The effects of matrix size and inter stimulus interval on performance. Biol Psychol, 73, 242–252, (2006).
31. A.A. Glover, Onofrj, M.C. M.F. Ghilardi, and I. Bodis-Wollner, P300-like potentials in the normal monkey using classical conditioning and the "oddball" paradigm. Electroencephalogr Clin Neurophysiol, 65, 231–235, (1986).
32. B. Roder, F. Rosler, E. Hennighausen, and F. Nacker, Event-related potentials during auditory and somatosensory discrimination in sighted and blind human subjects. Brain Res, 4, 77–93, (1996).
33. N. Birbaumer, Slow cortical potentials: their origin, meaning, and clinical use. In: G.J.M. van Boxtel and K.B.E. Bocker (Eds.), *Brain and behavior past, present, and future,* Tilburg Univ Press, Tilburg, pp 25–39, (1997).
34. B. Rockstroh, T. Elbert, A. Canavan, W. Lutzenberger, and N. Birbaumer, *Slow cortical potentials and behavior*, Urban and Schwarzenberg, Baltimore, MD, (1989).
35. N. Birbaumer, T. Elbert, A.G.M. Canavan, and B. Rockstroh, Slow potentials of the cerebral cortex and behavior. Physiol Rev, 70, 1–41, (1990).
36. N. Birbaumer, T. Hinterberger, A. Kubler, and N. Neumann, The thought-translation device (TTD): neurobehavioral mechanisms and clinical outcome. IEEE Trans Neural Syst Rehabil Eng, 11, 120–123, (2003).
37. T. Elbert, B. Rockstroh, W. Lutzenberger, and N. Birbaumer Biofeedback of slow cortical potentials. I. Electroencephalogr Clin Neurophysiol, 48, 293–301, (1980).
38. W. Lutzenberger, T. Elbert, B. Rockstroh, and N. Birbaumer, Biofeedback of slow cortical potentials. II. Analysis of single event-related slow potentials by time series analysis. Electroencephalogr Clin Neurophysiol 48, 302–311, (1980).
39. M. Pham, T. Hinterberger, N. Neumann, et al., An auditory brain-computer interface based on the self-regulation of slow cortical potentials. Neurorehabil Neural Repair, 19, 206–218, (2005).
40. N. Birbaumer, N. Ghanayim, T. Hinterberger, et al., A spelling device for the paralysed. Nature, 398, 297–298, (1999).
41. E. Niedermeyer, The Normal EEG of the Waking Adult. In: E. Niedermeyer E and F.H.L. da Silva (Eds.), *Electroencephalography: Basic principles, clinical applications, and related fields*, Williams and Wilkins, Baltimore, pp. 167–192, (2004).
42. P. Fries, D. Nikolic, and W. Singer, The gamma cycle. Trends Neurosci, 30, 309–316, (2007).
43. SM. Montgomery and G. Buzsaki, Gamma oscillations dynamically couple hippocampal CA3 and CA1 regions during memory task performance. Proc Natl Acad Sci USA, 104, 14495–14500, (2007).
44. W. Singer, Neuronal synchrony: a versatile code for the definition of relations? Neuron, 24, 49–65, 111–125, (1999).
45. BJ. Fisch, *Fisch and Spehlmann's third revised and enlarged EEG primer,* Elsevier, Amsterdam, (1999).
46. H. Gastaut, Étude electrocorticographique de la réacivité des rythmes rolandiques. Rev Neurol, 87, 176–182, (1952).
47. F.H.L. da Silva, Neural mechanisms underlying brain waves: from neural membranes to networks. Electroencephalogr Clin Neurophysiol, 79, 81–93, (1991).
48. D.J. McFarland L.A. Miner T.M. Vaughan J.R., and Wolpaw Mu and Beta rhythm topographies during motor imagery and actual movements. Brain Topogr 12, 177–186, (2000).
49. G. Pfurtscheller, EEG event-related desynchronization (ERD) and event-related synchronization (ERS). In: E. Niedermeyer and L.F.H. da Silva (Eds.), *Electroencephalography: Basic principles, clinical aapplications and related fields.* Williams and Wilkins, Baltimore, MD, pp. 958–967, (1999).

50. G. Pfurtscheller and A. Berghold, Patterns of cortical activation during planning of voluntary movement. Clin Neurophysiol, 72, 250–258, (1989).
51. G. Pfurtscheller and F.H.L. da Silva, Event-related EEG/MEG synchronization and desynchronization: basic principles. Clin Neurophysiol, 110, 1842–1857, (1999).
52. G. Pfurtscheller and C. Neuper, Motor imagery activates primary sensorimotor area in humans. Neurosci Lett, 239, 65–68, (1997).
53. O. Bai, P. Lin, S. Vorbach, M.K. Floeter, N. Hattori, and M. Hallett, A high performance sensorimotor beta rhythm-based brain-computer interface associated with human natural motor behavior. J Neural Eng, 5, 24–35, (2008).
54. B. Blankertz, G. Dornhege, M. Krauledat, K.R. Muller, and G. Curio, The non-invasive Berlin Brain-Computer Interface: fast acquisition of effective performance in untrained subjects. NeuroImage, 37, 539–550, (2007).
55. F. Cincotti, D. Mattia, F. Aloise, et al., Non-invasive brain-computer interface system: towards its application as assistive technology. Brain Res Bull, 75, 796–803, (2008).
56. A. Kostov and M. Polak, Parallel man-machine training in development of EEG-based cursor control. IEEE Trans Rehab Engin, 8, 203–205, (2000).
57. G. Pfurtscheller, B. Graimann J.E. Huggins, and S.P. Levine, Brain-computer communication based on the dynamics of brain oscillations. Supplements to Clin Neurophysiol, 57, 583–591, (2004).
58. J.A. Pineda, D.S. Silverman, A. Vankov, and J. Hestenes, Learning to control brain rhythms: making a brain-computer interface possible. IEEE Trans Neural Syst Rehabil Eng, 11, 181–184, (2003).
59. J.R. Wolpaw, D.J. McFarland, G.W. Neat, and C.A. Forneris, An EEG-based brain-computer interface for cursor control. Clin Neurophysiol, 78, 252–259, (1991).
60. D.J. McFarland, D.J. Krusienski, W.A. Sarnacki, and J.R. Wolpaw, Emulation of computer mouse control with a noninvasive brain-computer interface. J Neural Eng, 5, 101–110, (2008).
61. D.J. McFarland, T. Lefkowicz, and J.R. Wolpaw, Design and operation of an EEG-based brain-computer interface (BCI) with digital signal processing technology. Beh Res Meth, 29, 337–345, (1997).
62. J.R. Wolpaw, and D.J. McFarland, Control of a two-dimensional movement signal by a noninvasive brain-computer interface in humans. Proc Natl Acad Sci USA, 101, 17849–17854, (2004).
63. J.R. Wolpaw, D.J. McFarland, and T.M. Vaughan, Brain-computer interface research at the Wadsworth Center. IEEE Trans Rehabil Eng 8, 222–226, (2000).
64. L.A. Miner, D.J. McFarland, J.R. Wolpaw, Answering questions with an electroencephalogram-based brain-computer interface. Arch Phys Med Rehabil, 79, 1029–1033, (1998).
65. T.M. Vaughan, D.J. McFarland, G. Schalk, W.A. Sarnacki, L. Robinson, and J.R. Wolpaw, EEG-based brain-computer-interface: development of a speller. Soc Neurosci Abst 27, 167, (2001).
66. J.R. Wolpaw, H. Ramoser, D.J. McFarland, and G. Pfurtscheller, EEG-based communication: improved accuracy by response verification. IEEE Trans Rehab Eng, 6, 326–333, (1998).
67. D.J. McFarland, W.A. Sarnacki, and J.R. Wolpaw, Electroencephalographic (EEG) control of three-dimensional movement. J. Neural Eng., 11, 7(3):036007, (2010).
68. C. Neuper, A. Schlogl, and G. Pfurtscheller, Enhancement of left-right sensorimotor EEG differences during feedback-regulated motor imagery. J Clin Neurophysiol, 16, 373–382, (1999).
69. G. Pfurtscheller D. Flotzinger, and J. Kalcher, Brain-computer interface – a new communication device for handicapped persons. J Microcomput Appl, 16, 293–299, (1993).
70. G. Pfurtscheller, C. Guger, G. Muller, G. Krausz, and C. Neuper, Brain oscillations control hand orthosis in a tetraplegic. Neurosci Lett, 292, 211–214, (2000).
71. G. Pfurtscheller, C. Neuper, C. Guger, et al., Current trends in Graz Brain-Computer Interface (BCI) research. IEEE Trans Rehabil Eng, 8, 216–219, (2000).

72. NE. Crone, A. Sinai, and A. Korzeniewska, High-frequency gamma oscillations and human brain mapping with electrocorticography. Prog Brain Res, 159, 275–295, (2006).
73. SP. Levine J.E. Huggins, S.L. BeMent, et al., Identification of electrocorticogram patterns as the basis for a direct brain interface. J Clin Neurophysiol, 16, 439–447, (1999).
74. C. Mehring, MP. Nawrot, S.C. de Oliveira, et al., Comparing information about arm movement direction in single channels of local and epicortical field potentials from monkey and human motor cortex. J Physiol, Paris 98, 498–506, (2004).
75. T. Satow, M. Matsuhashi, A. Ikeda, et al., Distinct cortical areas for motor preparation and execution in human identified by Bereitschaftspotential recording and ECoG-EMG coherence analysis. Clin Neurophysiol, 114, 1259–1264, (2003).
76. G. Schalk, J. Kubanek, K.J. Miller, et al., Decoding two-dimensional movement trajectories using electrocorticographic signals in humans. J Neural Eng, 4, 264–275, (2007).
77. G. Schalk, P. Brunner, L.A. Gerhardt, H. Bischof, J.R. Wolpaw, Brain-computer interfaces (BCIs): Detection instead of classification. J Neurosci Methods, (2007).
78. N.E. Crone, D.L. Miglioretti, B. Gordon, and R.P. Lesser, Functional mapping of human sensorimotor cortex with electrocorticographic spectral analysis. II. Event-related synchronization in the gamma band. Brain, 121(Pt 12), 2301–2315, (1998).
79. E.C. Leuthardt, K. Miller, N.R. Anderson, et al., Electrocorticographic frequency alteration mapping: a clinical technique for mapping the motor cortex. Neurosurgery, 60, 260–270; discussion 270–261, (2007).
80. K.J. Miller, E.C. Leuthardt, G, Schalk, et al., Spectral changes in cortical surface potentials during motor movement. J Neurosci, 27, 2424–2432, (2007).
81. S. Ohara, A. Ikeda, T. Kunieda, et al., Movement-related change of electrocorticographic activity in human supplementary motor area proper. Brain, 123(Pt 6), 1203–1215, (2000).
82. G. Pfurtscheller, B. Graimann, J.E. Huggins, S.P. Levine, and L.A. Schuh, Spatiotemporal patterns of beta desynchronization and gamma synchronization in corticographic data during self-paced movement. Clin Neurophysiol, 114, 1226–1236, (2003).
83. J. Kubanek, K.J. Miller, J.G. Ojemann, J.R. Wolpaw, and G. Schalk, Decoding flexion of individual fingers using electrocorticographic signals in humans. J Neural Eng, 6(6), 66001, (2009).
84. G. Schalk, N. Anderson, K. Wisneski, W. Kim, M.D. Smyth, J.R. Wolpaw, D.L. Barbour, and E.C. Leuthardt, *Toward brain-computer interfacing using phonemes decoded from electrocorticography activity (ECoG) in humans. Program No. 414.11. 2007 Abstract Viewer/Itinerary Planner*. Society for Neuroscience, Washington, DC, (2007). Online.
85. E.C. Leuthardt, G. Schalk, J.R. Wolpaw, J.G. Ojemann, and D.W. Moran, A brain-computer interface using electrocorticographic signals in humans. J Neural Eng 1, 63–71, (2004).
86. G. Schalk, K.J. Miller, N.R. Anderson, et al., Two-dimensional movement control using electrocorticographic signals in humans. J Neural Eng, 5, 75–84, (2008).
87. U. Mitzdorf, Current source-density method and application in cat cerebral cortex: investigation of evoked potentials and EEG phenomena. Physiol Rev, 65, 37–100, (1985).
88. U. Mitzdorf, Properties of cortical generators of event-related potentials. Pharmacopsychiatry, 27, 49–51, (1994).
89. K.J. Otto, M.D. Johnson, and D.R. Kipke, Voltage pulses change neural interface properties and improve unit recordings with chronically implanted microelectrodes. IEEE Trans Biomed Eng, 53, 333–340, (2006).
90. O. Donchin, A. Gribova, O. Steinberg, H. Bergman, S. Cardoso de Oliveira, and E. Vaadia, Local field potentials related to bimanual movements in the primary and supplementary motor cortices. Exp Brain Res, 140, 46–55, (2001).
91. J. Rickert, S.C. Oliveira, E. Vaadia, A. Aertsen, S. Rotter, and C. Mehring, Encoding of movement direction in different frequency ranges of motor cortical local field potentials. J Neurosci, 25, 8815–8824, (2005).
92. P.R. Kennedy, M.T. Kirby, M.M. Moore, B. King, and A. Mallory, Computer control using human intracortical local field potentials. IEEE Trans Neural Syst Rehabil Eng, 12, 339–344, (2004).

93. L.R. Hochberg, M.D. Serruya, G.M. Friehs, et al., Neuronal ensemble control of prosthetic devices by a human with tetraplegia. Nature 442, 164–171, (2006).
94. D.A. Heldman, W. Wang, S.S. Chan, and D.W. Moran, Local field potential spectral tuning in motor cortex during reaching. IEEE Trans Neural Syst Rehabil Eng, 14, 180–183, (2006).
95. J.P. Donoghue, J.N. Sanes, N.G. Hatsopoulos, and G. Gaal, Neural discharge and local field potential oscillations in primate motor cortex during voluntary movements. J Neurophysiol, 79, 159–173, (1998).
96. S.N. Baker, J.M. Kilner E.M. Pinches, and R.N. Lemon, The role of synchrony and oscillations in the motor output. Exp Brain Res, 128, 109–117, (1999).
97. V.N. Murthy and E.E. Fetz, Coherent 25- to 35-Hz oscillations in the sensorimotor cortex of awake behaving monkeys. Proc Natl Acad Sci USA, 89, 5670–5674, (1992).
98. R.A. Andersen, S. Musallam, and B. Pesaran, Selecting the signals for a brain-machine interface. Curr Opin Neurobiol, 14, 720–726, (2004).
99. S. Musallam, B.D. Corneil, B. Greger, H. Scherberger, and R.A. Andersen, Cognitive control signals for neural prosthetics. Science, 305, 258–262, (2004).
100. B. Pesaran, J.S. Pezaris, M. Sahani, P.P. Mitra, and R.A. Andersen, Temporal structure in neuronal activity during working memory in macaque parietal cortex. Nat Neurosci, 5, 805–811, (2002).
101. H. Scherberger, M.R. Jarvis, and R.A. Andersen, Cortical local field potential encodes movement intentions in the posterior parietal cortex. Neuron, 46, 347–354, (2005).
102. E.E. Fetz, Operant conditioning of cortical unit activity. Science, 163, 955–958, (1969).
103. E.E., Fetz and D.V. Finocchio, Correlations between activity of motor cortex cells and arm muscles during operantly conditioned response patterns. Exp Brain Res, 23, 217–240, (1975).
104. E.M. Schmidt, Single neuron recording from motor cortex as a possible source of signals for control of external devices. Ann Biomed Eng, 8, 339–349, (1980).
105. A.R. Wyler and K.J. Burchiel, Factors influencing accuracy of operant control of pyramidal tract neurons in monkey. Brain Res, 152, 418–421, (1978).
106. E. Stark, R. Drori, I. Asher, Y. Ben-Shaul, and M. Abeles, Distinct movement parameters are represented by different neurons in the motor cortex. Eur J Neurosci, 26, 1055–1066, (2007).
107. W.T. Thach, Correlation of neural discharge with pattern and force of muscular activity, joint position, and direction of intended next movement in motor cortex and cerebellum. J Neurophysiol, 41, 654–676, (1978).
108. J. Carmena, M. Lebedev, R. Crist, et al., Learning to control a brain-machine interface for reaching and grasping by primates. PLoS Biol, 1, 193–208, (2003).
109. J.K. Chapin, K.A. Moxon, R.S. Markowitz, and M.A. Nicolelis, Real-time control of a robot arm using simultaneously recorded neurons in the motor cortex. Nat Neurosci, 2, 664–670, (1999).
110. M. Serruya, N.G. Hastopoulos, L. Paminski, Fel M.R. lows, and J.P. Donoghue, Instant neural control of a movement signal. Nature, 416, 141–142, (2002).
111. M. Velliste, S. Perel, M.C. Spalding, A.S. Whitford, and A.B. Schwartz, Cortical control of a prosthetic arm for self-feeding. Nature, 453, 1098–1101, (2008).
112. K. Shenoy, D. Meeker, S. Cao, et al., Neural prosthetic control signals from plan activity. Neuroreport, 14, 591–596, (2003).
113. G. Kreiman, C. Koch, and I. Fried, Imagery neurons in the human brain. Nature 408, 357–361, (2000).
114. J.W. Gnadt and R.A. Andersen, Memory related motor planning activity in posterior parietal cortex of macaque. Exp Brain Res 128, 70, 216–220, (1988).
115. P.R. Kennedy, The cone electrode: a long-term electrode that records from neurites grown onto its recording surface. J Neurosci Meth, 29, 181–193, (1989).
116. P.R. Kennedy, R.A. Bakay, M.M. Moore, and J. Goldwaithe, Direct control of a computer from the human central nervous system. IEEE Trans Rehabil Eng, 8, 198–202, (2000).

117. W. Truccolo, G.M. Friehs, J.P. Donoghue, L.R. Hochberg, Primary motor cortex tuning to intended movement kinematics in humans with tetraplegia. J Neurosci, 28, 1163–1178, (2008).
118. E.W. Sellers, T.M. Vaughan, and J.R. Wolpaw, A brain-computer interface for long-term independent home use. Amyotrophic lateral sclerosis, 11(5), 449–455, (2010).
119. E.W Sellers, T.M. Vaughan, D.J. McFarland, D.J. Krusienski, S.A. Mackler, R.A. Cardillo, G. Schalk, S.A. Binder-Macleod, and J.R. Wolpaw, *Daily use of a brain-computer interface by a man with ALS. Program No. 256.1.2006 Abstract Viewer/Itinerary Planner.* Society for Neuroscience, Washington, DC, (2006). Online.
120. J.R. Wolpaw and N. Birbaumer, Brain-computer interfaces for communication and control. In: M.E. Selzer, S. Clarke, L.G. Cohen, P. Duncan, and F.H. Gage (Eds.), *Textbook of neural repair and rehabilitation: Neural repair and plasticity,* Cambridge University Press, Cambridge, pp. 602–614, (2006).
121. J.R. Wolpaw, N. Birbaumer, D.J. McFarland, G Pfurtscheller, and T.M. Vaughan, Brain-computer interfaces for communication and control. Clin Neurophysiol, 113, 767–791, (2002).
122. J.R. Wolpaw and D.J. McFarland, Multichannel EEG-based brain-computer communication. Clin Neurophysiol, 90, 444–449, (1994).
123. J.R. Wolpaw, D.J. McFarland, T.M. Vaughan, and G. Schalk, The Wadsworth Center brain-computer interface (BCI) research and development program. IEEE Trans Neural Syst Rehabil Eng, 11, 204–207, (2003).

Dynamics of Sensorimotor Oscillations in a Motor Task

Gert Pfurtscheller and Christa Neuper

1 Introduction

Many BCI systems rely on imagined movement. The brain activity associated with real or imagined movement produces reliable changes in the EEG. Therefore, many people can use BCI systems by imagining movements to convey information. The EEG has many regular rhythms. The most famous are the occipital alpha rhythm and the central mu and beta rhythms. People can desynchronize the alpha rhythm (that is, produce weaker alpha activity) by being alert, and can increase alpha activity by closing their eyes and relaxing. Sensory processing or motor behavior leads to EEG desynchronization or blocking of central beta and mu rhythms, as originally reported by Berger [1], Jasper and Andrew [2] and Jasper and Penfield [3]. This desynchronization reflects a decrease of oscillatory activity related to an internally or externally-paced event and is known as Event–Related Desynchronization (ERD, [4]). The opposite, namely the increase of rhythmic activity, was termed Event-Related Synchronization (ERS, [5]). ERD and ERS are characterized by fairly localized topography and frequency specificity [6]. Both phenomena can be studied through topographic maps, time courses, and time-frequency representations (ERD maps, [7]).

Sensorimotor areas have their own intrinsic rhythms, such as central beta, mu, and gamma oscillations. The dynamics of these rhythms depend on the activation and deactivation of underlying cortical structures. The existence of at least three different types of oscillations at the same electrode location over the sensorimotor hand area during brisk finger lifting has been described by Pfurtscheller and Lopes da Silva [8, 9–11]. In addition to mu ERD and post-movement beta ERS, induced gamma oscillations (ERS) close to 40 Hz can also be recorded. These 40–Hz oscillations are strongest shortly before a movement begins (called movement onset), whereas the beta ERS is strongest after a movement ends (called movement offset).

G. Pfurtscheller (✉)
Institute for Knowledge Discovery, Graz University of Technology, Graz, Austria
e-mail: pfurtscheller@tugraz.at

B. Graimann et al. (eds.), *Brain–Computer Interfaces*, The Frontiers Collection,
DOI 10.1007/978-3-642-02091-9_3,

Further reports on movement-related gamma oscillations in man can be found in Pfurtscheller et al. [12] and Salenius et al. [13].

Gamma oscillations in the frequency range from 60 to 90 Hz associated with movement execution were observed in subdural recordings (ECoG) [14, 15]. Of interest are the broad frequency band, short duration, and embedding of gamma oscillations in desynchronized alpha band and beta rhythms. There is strong evidence that the desynchronization of alpha band rhythms may be a prerequisite for the development of gamma bursts.

2 Event–Related Potentials Versus ERD/ERS

There are two ways to analyze the changes in the electrical activity of the cortex that accompany brain activities, such as sensory stimulation and motor behavior. One change is time-locked and phase-locked (evoked) and can be extracted from the ongoing activity by simple linear methods such as averaging. The other is time-locked but not phase-locked (induced) and can only be extracted through some non-linear methods such as envelope detection or power spectral analysis. Which mechanisms underlie these types of responses? The time- and phase-locked response can easily be understood in terms of the response of a stationary system to the external stimulus, the result of the existing neuronal networks of the cortex. The induced changes cannot be evaluated in such terms. The latter can be understood as a change in the ongoing activity, resulting from changes in the functional connectivity within the cortex.

A typical example of both phase-locked (evoked) and non-phase-locked (induced) EEG activities is found during preparation for voluntary thumb movement. Both, negative slow cortical potential shifts (known as Bereitschaftspotential) and mu ERD start about 2 s prior to movement onset [4, 16]. Slow cortical potential shifts at central electrode positions have also been reported after visually cued imagined hand movements [17] in parallel to the desynchronization of central beta and mu rhythms [18]. A BCI could even use both slow cortical potential shifts and ERD/ERS for feature extraction and classification.

3 Mu and Beta ERD in a Motor Task

Voluntary movement results in a circumscribed desynchronization in the upper alpha and lower beta band oscillations, localized over sensorimotor areas [4, 19–25]. The desynchronization is most pronounced over the contralateral central region and becomes bilaterally symmetrical with execution of movement.

The time course of the contralateral mu desynchronization is almost identical for brisk and slow finger movements, starting more than 2 s prior to movement onset [22]. Finger movement of the dominant hand is accompanied by a pronounced ERD in the contralateral hemisphere and by a very low ERD in the ipsilateral side, whereas movement of the non-dominant finger is preceded by a less lateralized ERD [22]. Different reactivity patterns have been observed with mu rhythm components in the lower and upper alpha frequency band [26].

Imagining and preparing movement can both produce replicable EEG patterns over primary sensory and motor areas (e.g. [14, 27, 28]). This is in accordance with the concept that motor imagery is realized via the same brain structures involved in programming and preparing of movements [29, 30]. For example, imagery of right and left hand movements results in desynchronization of mu and beta rhythms over the contralateral hand area (see Fig. 1), very similar in topography to planning and execution of real movements [31].

Mu desynchronization is not a unitary phenomenon. If different frequency bands within the range of the extended alpha band are distinguished, two distinct patterns of desynchronization occur. Lower mu desynchronization (in the range of about 8–10 Hz) is obtained during almost any type of motor task. It is topographically

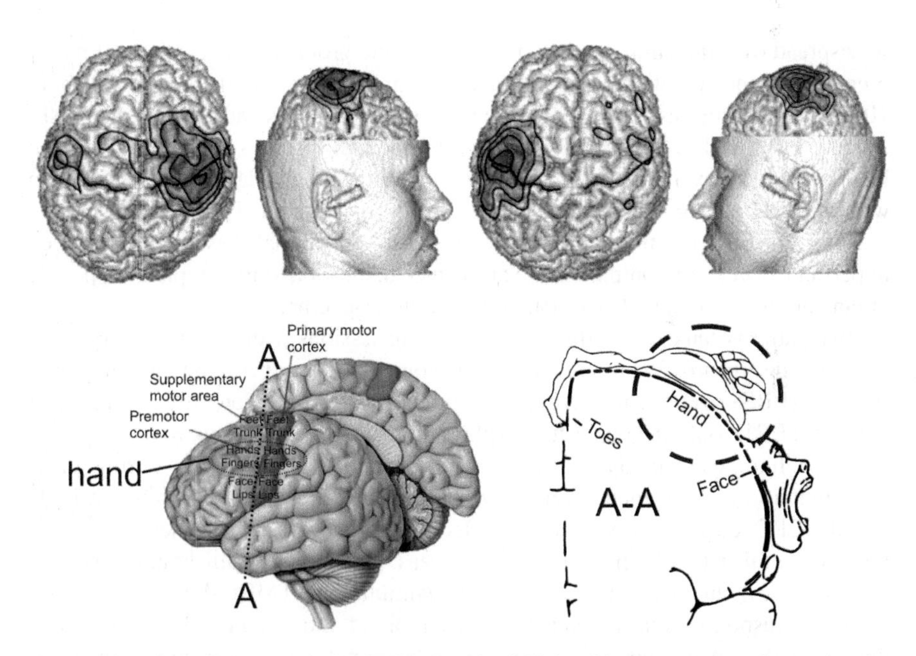

Fig. 1 The top four images show mu ERS in the 10–12 Hz band in the contralateral central region resulting from imagined finger movement. Some ipsilateral (same side) activity is also apparent. The *left* two images show activity in the right central region, which would occur with left hand movement imagery. The *right* two images show activity in the *left* Rolandic region, which would occur with *right* hand movement imagery. The *bottom left* image shows which motor and sensory areas manage different body parts. The *bottom right* image shows the "motor homunculus," which reflects the amount of brain tissue devoted to different body areas. Areas such as the hands and mouth have very large representations, since people can make very fine movements with these areas and are very sensitive there. Other areas, such as the back and legs, receive relatively little attention from the brain. The A-A axis reflects that the bottom right image is a coronal section through the brain, meaning a vertical slice that separates the front of the brain from the back

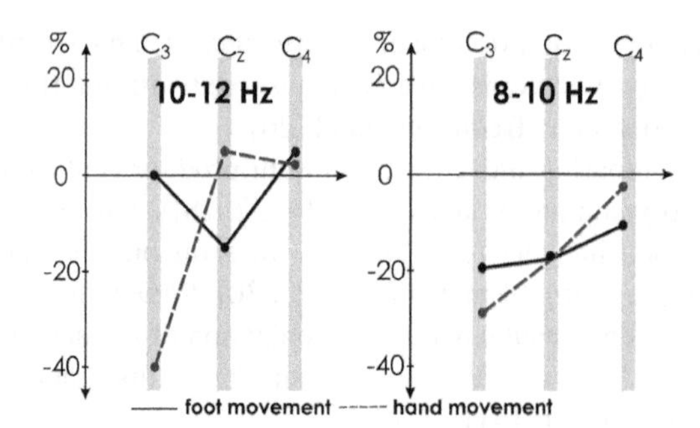

Fig. 2 Relationship between movement type band power changes in the 8–10 Hz (*right side*) and the 10–12 Hz frequency bands (*left side*) over electrode locations C3, Cz and C4. A percentage band power increase marks ERS, and a decrease reflects ERD. The band power was calculated in a 500 ms window prior to movement-onset (modified from [26])

widespread over the entire sensorimotor cortex and probably reflects general motor preparation and attention. Upper mu desynchronization (in the range of about 10–12 Hz), in contrast, is topographically restricted and is rather related to task-specific aspects. The lower mu frequency component displays a similar ERD with hand and foot movement, while the higher components display a different pattern with an ERD during hand and an ERS with foot movement (Fig. 2). This type of reactivity of different mu components suggests a functional dissociation between upper and lower mu components. The former displays a somatotopically specific organization, while the latter is somatotopically unspecific.

Two patterns can develop during BCI training sessions with motor imagery: contralateral desynchronization of upper mu components and a concomitant power increase (ERS) over the ipsilateral side. In contrast to the bilaterally symmetrical lower mu ERD, only the upper mu displays an ipsilateral ERS (Fig. 3).

These findings strongly indicate primary motor cortex activity during mental simulation of movement. Hence, we can assume that the pre-movement ERD and the ERD during motor imagery reflect a similar type of readiness or presetting of neural networks in sensorimotor areas. Functional brain imaging studies (e.g. [32–36]) and transcranial magnetic stimulation (TMS) show an increase of motor responses during mental imagination of movements [37], which further supports involvement of the primary sensorimotor cortex in motor imagery. The observation that movement imagination triggers a significant ipsilateral ERS along with the contralateral ERD supports the concept of antagonistic behavior of neural networks ("focal ERD/surrounding ERS", [38]) described in the next section.

In a recent study, several motor imagery tasks were investigated, such as cue-based left hand, right hand, foot, or tongue motor imagery [39]. Basically, hand motor imagery activates neural networks in the cortical hand representation area, which is manifested in the hand area mu ERD. Such a mu ERD was found in all able-bodied subjects with a clear contralateral dominance. The midcentral ERD, in

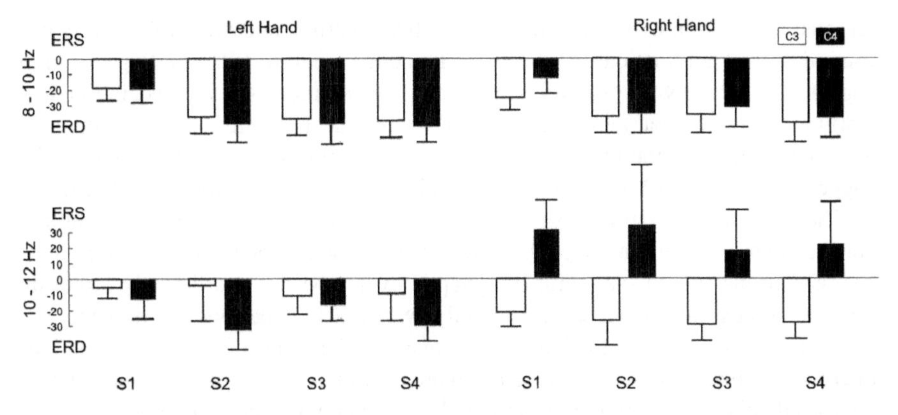

Fig. 3 Mean ERD/ERS values (i.e. mean and standard error) obtained for left hand *(left side)* versus right hand imagery (*right side*) for the 8–10 Hz (*upper row*) and 10–12 Hz frequency bands (*lower row*). Data were obtained from the left (C3, *white bar*) versus right (C4, *black bar*) sensorimotor derivation and are shown separately for 4 BCI training sessions (S1–S4) performed with each of the (N=10) participants. This figure is modified from [9]

the case of foot motor imagery, was weak and not found in every subject. The reason for this may be that the foot representation area is hidden within the mesial wall. The mesial wall refers to the tissue between the left and right hemispheres of the brain. This area is difficult to detect with electrode caps outside the head, since electrode caps are best at measuring signals near the outer surface of the brain just underneath the skull.

Noteworthily, both foot and tongue motor imagery enhanced the hand area mu rhythm. In addition, tongue motor imagery induced mu oscillations in or close to the foot representation area. This antagonistic pattern of a focal ERD accompanied by a lateral (surrounding) ERS will be described in more detail in the following sections.

4 Interpretation of ERD and ERS

Connections between the thalamus and cortex are called thalamo-cortical systems. Increased cellular excitability (that is, an increase in the likelihood that neurons will fire) in thalamo-cortical systems results in a low amplitude desynchronized EEG [40]. Therefore, ERD can be understood as an electrophysiological correlate of activated cortical areas involved in processing sensory or cognitive information or producing motor behavior [5]. An increased and/or more widespread ERD could result from a larger neural network or more cell assemblies in information processing.

Explicit learning of a movement sequence, e.g. key pressing with different fingers, is accompanied by an enhancement of the ERD over the contralateral central regions. Once the movement sequence has been learned and the movement is performed more "automatically," the ERD is reduced. These ERD findings strongly suggest that activity in primary sensorimotor areas increases in association with learning a new motor task and decreases after the task has been learned [41]. The involvement of the primary motor area in learning motor sequences was also suggested by Pascual-Leone [42], who studied motor output maps using TMS.

The opposite of desynchronization is synchronization, when the amplitude enhancement is based on the cooperative or synchronized behavior of a large number of neurons. When the summed synaptic events become sufficiently large, the field potentials can be recorded not only with macro electrodes within the cortex, but also over the scalp. Large mu waves on the scalp need coherent activity of cell assemblies over at least several square centimeters [43, 44]. When patches of neurons display coherent activity in the alpha frequency band, active processing of information is very unlikely and probably reflects an inhibitory effect.

However, inhibition in neural networks is very important, not only to optimize energy demands but also to limit and control excitatory processes. Klimesch [45] suggested that synchronized alpha band rhythms during mental inactivity (idling) are important to introduce powerful inhibitory effects, which could block a memory search from entering irrelevant parts of neural networks. Combined EEG and TMS studies demonstrated that the common pathway from motor cortex to the target hand muscle was significantly inhibited during the movement inhibition conditions, and the upper mu components simultaneously displayed synchronization in the hand area. In contrast, the TMS response was increased during execution of the movement sequence, and a mu ERD was observed in the hand representation area [46].

Adrian and Matthews [47] described a system that is neither receiving nor processing sensory information as an idling system. Localized synchronization of 12–14 Hz components in awake cats was interpreted by Case and Harper [48] as a result of idling cortical areas. Cortical idling can thus denote a cortical area of at least some cm^2 that is not involved in processing sensory input or preparing motor output. In this sense, occipital alpha rhythms can be considered idling rhythms of the visual areas, and mu rhythms as idling rhythms of sensorimotor areas [49]. Also, sleep spindles during early sleep stages can occur when signals through the thalamus are blocked [40].

5 "Focal ERD/Surround ERS"

Localized desynchronization of the mu rhythms related to a specific sensorimotor event does not occur in isolation, but can be accompanied by increased synchronization in neighboring cortical areas that correspond to the same or to another information processing modality. Lopes da Silva [38] introduced the term "focal ERD/surround ERS" to describe this phenomenon. Gerloff et al. [50] reported an antagonistic behavior with desynchronization of central mu rhythm and synchronization of parieto-occipital alpha rhythms during repetitive brief finger movement.

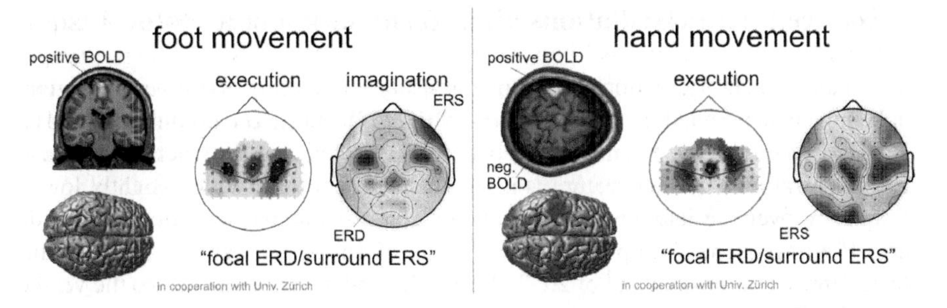

Fig. 4 Examples of activation patterns obtained with functional magnetic resonance imaging (fMRI) and EEG (ERD/ERS) during execution and imagination of hand (*right panel*) and foot movements (*left panel*). Note the correspondence between the focus of the positive BOLD signal and ERD on the one side, and between the negative BOLD signal and ERS on the other side

The opposite phenomenon, the enhancement of central mu rhythm and blocking of occipital alpha rhythm during visual stimulation, was documented by Koshino and Niedermeyer [51] and Kreitmann and Shaw [52]. Figure 4 shows further examples demonstrating the intramodal interaction, in terms of a hand area ERD and foot area ERS during hand movement and hand area ERS and foot area ERD during foot movement, respectively.

The focal mu desynchronization in the upper alpha frequency band may reflect a mechanism responsible for focusing selective attention on a motor subsystem. This effect of focal motor attention may be accentuated when other cortical areas, not directly involved in the specific motor task, are "inhibited". In this process, the interplay between thalamo-cortical modules and the inhibitory reticular thalamic nucleus neurons may play an important role ([44, 53]).

Support for the "focal ERD/surround ERS" phenomenon comes from regional blood flow (rCBF) and functional magnetic resonance imaging (fMRI) studies. Cerebral blood flow decreases in the somatosensory cortical representation area of one body part whenever attention is directed to a distant body part [54]. The BOLD (Blood Oxygen Level Dependent) signal also decreased in the hand motor area when the subject imagined or executed toe movements [36]. In this case, attention was withdrawn from the hand and directed to the foot zones, resulting in a positive BOLD response in the foot representation area. Figure 4 shows this antagonistic behavior in hemodynamic (fMRI) and bioelectrical (EEG) responses during hand and foot movements. Imagined movements are also called covert movements, and real movements are called overt movements. A task-related paradigm (covert movements over 2 s) was used for the fMRI, and an event-related paradigm (overt or covert movements in intervals of approximately 10 s) was used for the EEG. An fMRI study during execution of a learned complex finger movement sequence and inhibition of the same sequence (inhibition condition) is also of interest. Covert movement was accompanied by a positive BOLD signal and a focal mu ERD in the hand area, while movement inhibition resulted in a negative BOLD signal and a mu ERS in the same area [55, 56].

6 Induced Beta Oscillations after Termination of a Motor Task

The induced beta oscillation after somatosensory stimulation and motor behavior, called the beta rebound, is also of interest. MEG [57] and EEG recordings [58] have shown a characteristic feature of the beta rebound, which is its strict somatotopical organization. Another feature is its frequency specificity, with a slightly lower frequency over the lateralized sensorimotor areas as compared to the midcentral area [59]. Frequency components in the range of 16–20 Hz were reported for the hand representation area and of 20–24 Hz for the midcentral area close to the vertex (Fig. 5). The observation that a self-paced finger movement can activate neuronal networks in hand and foot representation areas with different frequency in both areas [26] further shows that the frequency of these oscillations may be characteristic for the underlying neural circuitry.

The beta rebound is found after both active and passive movements [23, 25, 60]. This indicates that proprioceptive afferents play important roles for the desynchronization of the central beta rhythm and the subsequent beta rebound. However, electrical nerve stimulation [59] and mechanical finger stimulation [61] can also induce a beta rebound. Even motor imagery can induce a short-lasting beta ERS or beta rebound [27, 62, 63, 10], which is of special interest for BCI research. Figure 6 shows examples of a midcentral induced beta rebound after foot motor imagery. The occurrence of a beta rebound related to mental motor imagery implies that this activity does not necessarily depend on motor cortex output and muscle activation.

A general explanation for the induced beta bursts in the motor cortex after movement, somatosensory stimulation, and motor imagery could be the transition of the beta generating network from a highly activated to an inhibited state. In the deactivated state, sub-networks in the motor area may reset, motor programs may be cancelled and/or updated by a somatosensory feedback. The function of the beta rebound in the sensorimotor area could be therefore understood as a "resetting function", in contrast to the "binding function" of gamma oscillations [64].

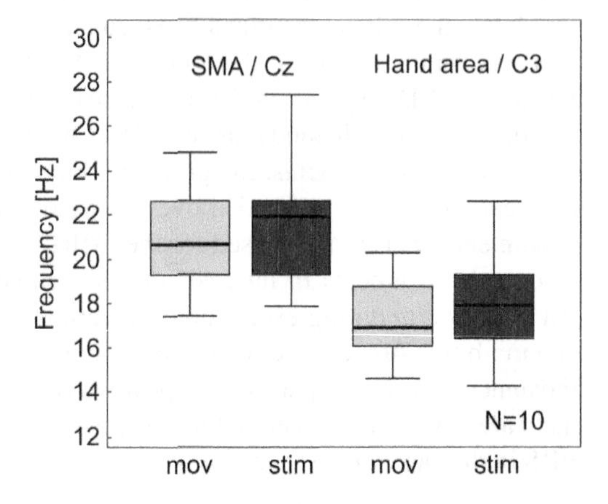

Fig. 5 Frequency of the post-movement beta rebound measured over the foot area/SMA and the hand area in response to movement and (electrical) stimulation of the respective limb. The boxplots represent the distribution of the peak frequencies in the beta band (14–32 Hz) at electrode positions Cz and C3 (modified from [59])

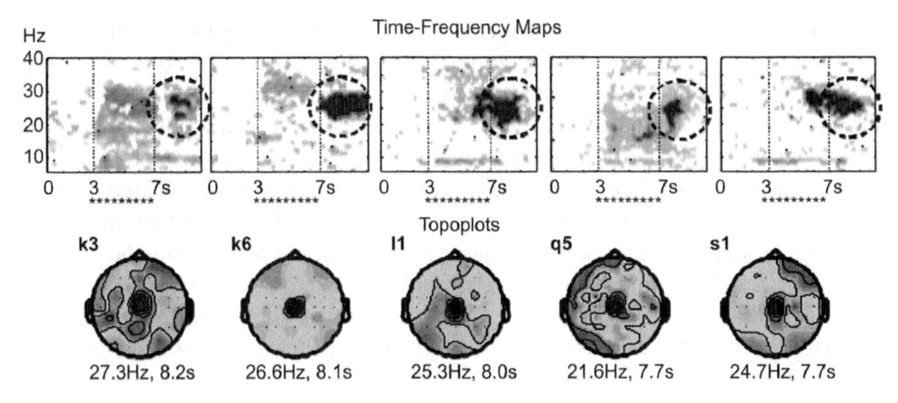

Fig. 6 Time-frequency maps (*upper panel*) and topoplots (*lower panel*) of the beta rebound in a foot motor imagery task. For each subject, the dominant frequency of the beta rebound and the latency until its maximum are indicated. For all subjects, the beta rebound (ERS) is located around the vertex electrode Cz. The subjects started imagery at second 3 and stopped at the end of the trials at second 7. This figure is modified from [10]

Studies that applied TMS during self-paced movement or median nerve stimulation showed that the excitability of motor cortex neurons was significantly reduced in the first second after termination of movement and stimulation, respectively [65]. This suggests that the beta rebound might represent a deactivated or inhibited cortical state. Experiments with periodic electrical median nerve stimulation (e.g. in interstimulus intervals of 1.5 s) and evaluation of the beta rebound further support this hypothesis. Cube manipulation with one hand during nerve stimulation suppressed the beta rebound in MEG [66] and EEG [67]. Sensorimotor cortex activation, such as by cube manipulation, is not only accompanied by an intense outflow from the motor cortex to the hand muscles, but also by an afferent flow from mechanic and proprioreceptors (neurons that detect touch) to the somatosensory cortex. The activation of the sensorimotor cortex could compensate for the short-lasting decrease of motor cortex excitability in response to electrical nerve stimulation and suppress the beta rebound.

7 Short-Lived Brain States

Research in both neuroscience and BCIs may benefit from studying short-lived brain states. A short-lived brain state is a stimulus induced activation pattern detectable through electrical potential (or magnetic field) recordings and last for fractions of 1 s. One example is the "matching process" hypothesis [68] stating that the motor cortex is engaged in the chain of events triggered by the appearance of the target as early as 60–80 ms after stimulus onset, leading to the aimed movement. Other examples are the short-lasting, somatotopically-specific ERD patterns induced by a cue stimulus indicating the type of motor imagery to be performed within the next seconds.

Brain activation with multichannel EEG recordings during two conditions can be studied with a method called common spatial patterns (CSP, [69, 70]). This CSP-method produces spatial filters that are optimal in that they extract signals that are most discriminative between two conditions. The CSP-method helps to distinguish two cue-based motor imagery conditions (e.g. left vs. right hand, left hand vs. foot or right hand vs. foot) with a high time resolution (see chapter "Digital Signal Processing and Machine Learning" in this book for details about CSP). In a group of 10 naive subjects, the recognition rate had a clear initial peak within the first second after stimulation onset, starting about 300 ms after stimulation. Examples for the separability of 2 motor imagery (MI) tasks with a short-lasting classification accuracy of 80–90% are displayed in Fig. 7 (left panels). In some subjects, the

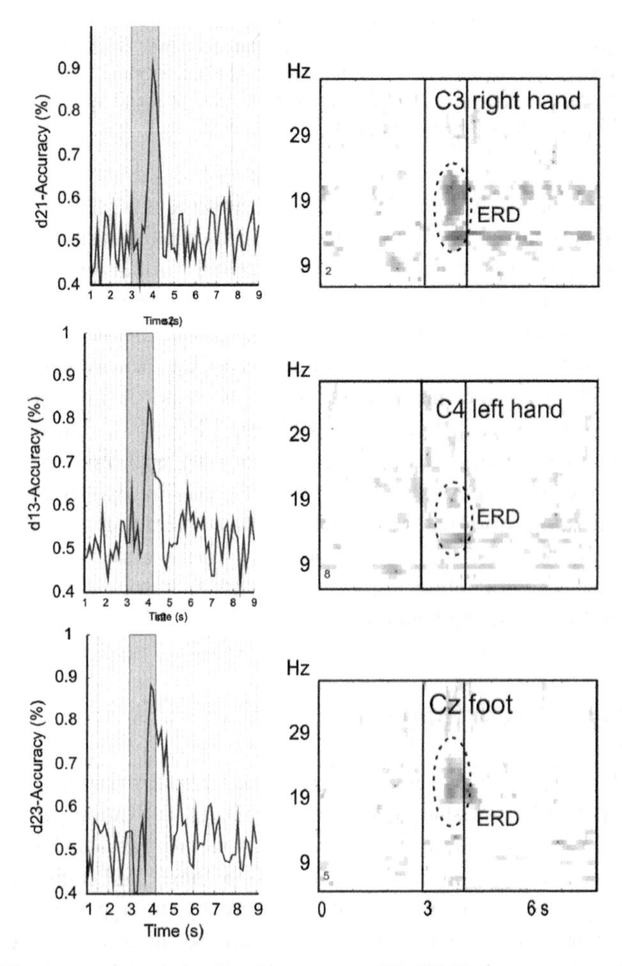

Fig. 7 Left side: time course of classification accuracy (40–100%) for separation of two visually-cued motor imagery tasks (*right vs. left hand, left hand vs. foot and right hand vs. foot MI*) in one naïve subject. Right side: corresponding ERD maps induced by visually-cued motor imagery at somatotopically specific electrode locations in the same subject. This figure is modified from [11]

initial, short-lasting peak was followed by a broad-banded peak within the next seconds. This later peak is very likely the result of the conscious executed motor imagery task. An initial recognition peak after visual cue presentation was already reported in a memorized delay movement experiment with left/right finger and foot movements [71] and after right/left hand movement imagination [72].

The initial, short-lasting separability peak suggests that the EEG signals display different spatio-temporal patterns in the investigated imagery tasks in a small time window of about 500–800 ms length. So, for instance, short-lasting mu and/or beta ERD patterns at somatotopically specific electrode locations are responsible for the high classification accuracy of the visually cued motor imagery tasks. The right side of Fig. 7 presents examples of such ERD maps obtained in one subject, and the left side shows examples of separability curves for the discrimination between left vs. right hand, left hand vs. foot and right hand vs. foot motor imagery from the same subject. The great inter-subject stability and reproducibility of this early separability peak shows that, in nearly every subject, such a somatotopically-specific activation (ERD) pattern can be induced by the cue stimulus. In other words, the cue can induce a short-lived brain state as early as about 300 ms after cue onset. This process is automatic and probably unconscious, may a type of priming effect, and could be relevant to a motor imagery-based BCI.

We hypothesize, therefore, that the short-lived brain states probably reflect central input concerning the motor command for the type of the upcoming MI task. This may be the result of a certain "motor memory", triggered by the visual cue. It has been suggested that such "motor memories" are stored in cortical motor areas and the cerebellum motor systems [73], and play a role when memories of previous experiences are retrieved during the MI process [74].

8 Observation of Movement and Sensorimotor Rhythms

There is increasing evidence that observing movements reduces the mu rhythm and beta oscillations recorded from scalp locations C3 and C4. Gastaut and Bert [75] and later Cochin [76] and Muthukumavaswauny [77] reported an attenuation of the central mu rhythm by observation of experimental hand grasp. Altschuler et al. [78] found that the mu rhythm desynchronized when subjects observed a moving person, but not when they viewed equivalent non-biological motion such as bouncing balls. Also, Cochin [76] reported a larger mu and beta power decrease during observation of a moving person than during observation of flowing water.

Consistent with the reported findings, a previous study in our laboratory [79] showed that the processing of moving visual stimuli depends on the type of the moving object. Viewing a moving virtual hand resulted in a stronger desynchronization of the central beta rhythm than viewing a moving cube (Fig. 8). Moreover, the presence of an object, indicating a goal-directed action, increases the mu rhythm suppression compared to meaningless actions [77].

Modulation of sensorimotor brain rhythms in the mu and beta frequency band during observation of movement has been linked to the activity of the human mirror

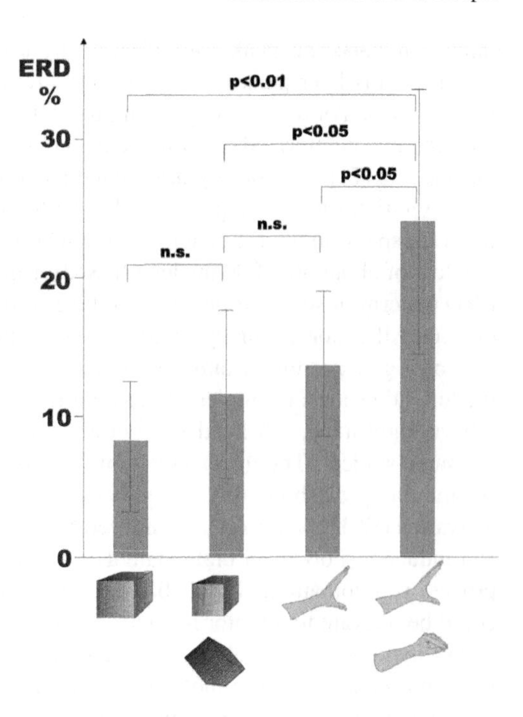

Fig. 8 Mean ERD ± SD over central regions during viewing a static versus moving hand or cube (modified from [80])

neuron system. This system is an action observation/execution matching system capable of performing an internal simulation of the observed action (for a review see [81, 82]). This line of research started on the discovery of so-called mirror neurons in cortical area F5 of macaque monkeys, which are active both in observing and executing a movement [83–85]. Based on functional imaging studies, evidence for a comparable mirror neuron system has been also demonstrated in humans [86], showing a functional correspondence between action observation, internal simulation or motor imagery, and execution of the motor action [87].

Specifically, it has been proposed that the mu rhythm may reflect the downstream modulation of primary sensorimotor neurons by mirror neurons in the inferior frontal gyrus [88, 89]. The underlying idea is that activation of mirror neurons by either executed, imagined or observed motor actions produces asynchronous firing, and is therefore associated with a concomitant suppression or desynchronization of the mu rhythm [90]. This is supported by MEG findings about activation of the viewer's motor cortex [82]. Hari and coworker showed that the 20-Hz rebound after median-nerve stimulation is not only suppressed when the subject moves the fingers or manipulates a small object, but also – although significantly weaker – when the subject just views another person's manipulation movements [91]. Interestingly, the suppression of the rebound is stronger for motor acts presented live than those seen on a video [92].

The fact that similar brain signals, i.e. oscillations in the mu and beta frequency bands, react to both motor imagery and observation of biological movement may play a critical role when using a BCI for neuroprosthesis control [93–97]. In this case, i.e. during BCI-controlled grasping, the feedback is provided by the observation of the own moving hand. Apart from single case studies, little is known about the impact of viewing grasping movements of one's own hand on closed-loop BCI operation. A recent study [9] showed that a realistic grasping movement on a video screen does not disturb the user's capability to control a BCI with motor imagery involving the respective body part.

If BCIs are used to produce grasping movements, such as opening and closing an orthosis, then more research should explore the effect of grasp observation on brain activity. This research should ideally assess the effects of training with BCIs, which are closed-loop systems with feedback, since results might otherwise be less informative.

9 Conclusion

ERD/ERS research can reveal how the brain processes different types of movement. Such research has many practical implications, such as treating movement disorders and building improved BCI systems.

The execution, imagination, or observation of movement strongly affects sensorimotor rhythms such as central beta and mu oscillations. This could mean that similar overlapping neural networks are involved in different motor tasks. Motor execution or motor imagery are generally accompanied by an ERD in attended body-part areas and at the same time in an ERS in non-attended areas. This type of antagonistic pattern is known as "focal ERD/surround ERS", whereby the ERD can be seen as electrophysiological correlate of an activated cortical area and the ERS as correlate of a deactivated (inhibited) area.

The short-lasting beta rebound (beta ERS) is a specific pattern in the beta band seen after a motor task ends. This beta ERS displays a somatotopically-specific organization, and is interpreted as a short-lasting state of deactivation of motor cortex circuitry.

Acknowledgment This research was partly financed by PRESENCCIA, an EU-funded Integrated Project under the IST program (Project No. 27731) and by NeuroCenter Styria, a grant of the state government of Styria (Zukunftsfonds, Project No. PN4055).

References

1. H. Berger, Uber das Elektrenkephalogramm des Menschen II. J Psychol Neurol, 40, 160–179, (1930).
2. H.H. Jasper and H.L. Andrew, Electro encephalography III. Normal differentiation of occipital and precentral regions in man. Arch Neurol Psychiatry, 39, 96–115, (1938).
3. H.H. Jasper and W. Penfield, Electrocorticograms in man: effect of the voluntary movement upon the electrical activity of the precentral gyrus. Arch Psychiat Z. Neurol, 183, 163–174, (1949).

4. G. Pfurtscheller and A. Aranibar, Evaluation of event-related desynchronization (ERD) preceding and following voluntary self-paced movements. Electroencephalogr Clin Neurophysiol, 46, 138–146, (1979).
5. G. Pfurtscheller, Event-related synchronization (ERS): an electrophysiological correlate of cortical areas at rest. Electroencephalogr Clin Neurophysiol, 83, 62–69, (1992).
6. G. Pfurtscheller and F.H. Lopes da Silva, Event-related EEG/MEG synchronization and desynchronization: basic principles. Clin Neurophysiol, 110, 1842–1857, (1999).
7. B. Graimann, J.E. Huggins, S.P. Levine, et al., Visualization of significant ERD/ERS patterns multichannel EEG and ECoG data. Clin Neurophysiol, 113, 43–47, (2002).
8. G. Pfurtscheller and F. H. L. da Silva, *Handbook of electroencephalography and clinical neurophysiology*, vol. 6, 1st edn, 1999 ed, Elsevier, New York, (1999).
9. C. Neuper, R. Scherer, S.C. Wriessnegger, et al., Motor imagery and action observation: modulation of sensorimotor brain rhythms during mental control of a brain computer interface, Clin Neurophysiol, 120, 239–47, (2009).
10. G. Pfurtscheller and T. Solis-Escalante, Could the beta rebound in the EEG be suitable to realize a "brain switch"? Clin Neurophysiol, 120, 24–9, (2009).
11. G. Pfurtscheller, R. Scherer, G.R. Müller-Putz, F.H. Lopes da Silva, Short-lived brain state after cued motor imagery in naive subjects. Eur J Neurosci,28, 1419–26, (2008).
12. G. Pfurtscheller and C. Neuper, Event–related synchronization of mu rhythm in the EEG over the cortical hand area in man. Neurosci Lett, 174, 93–96, (1994).
13. S. Salenius, R. Salmelin, C. Neuper, et al., Human cortical 40 Hz rhythm is closely related to EMG rhythmicity. Neurosci Lett, 213, 75–78, (1996).
14. N.E. Crone, D.L. Miglioretti, B. Gordon, et al., Functional mapping of human sensorimotor cortex with electrocorticographic spectral analysis. II. Event-related synchronization in the gamma band. Brain, 121, 2301–2315, (1998).
15. G. Pfurtscheller, B. Graimann, J.E. Huggins, et al., Spatiotemporal patterns of beta desynchronization and gamma synchronization in corticographic data during self-paced movement. Clin Neurophysiol, 114, 1226–1236, (2003).
16. J.E. Guieu, J.L. Bourriez, P. Derambure, et al., Temporal and spatial aspects of event-related desynchronization nand movement-related cortical potentials, Handbook Electroencephalogr Clin Neurophysiol, 6, 279–290, (1999).
17. R. Beisteiner, P. Höllinger, G. Lindinger, et al., Mental representations of movements. Brain potentials associated with imagination of hand movements. Electroencephalogr Clin Neurophysiol, 96, 83–193, (1995).
18. G. Pfurtscheller and C. Neuper, Motor imagery activates primary sensimotor area in humans. Neurosci Lett, 239, 65–68, (1997).
19. G. Pfurtscheller and A. Berghold, Patterns of cortical activation during planning of voluntary movement. Electroencephalogr Clin Neurophysiol, 72, 250–258, (1989).
20. P. Derambure, L. Defebvre, K. Dujardin, et al., Effect of aging on the spatio temporal pattern of event related desynchronization during a voluntary movement. Electroencephalogr Clin Neurophysiol, 89, 197–203, (1993).
21. C. Toro, G. Deuschl, R. Thatcher, et al., Event–related desynchronization and movement related cortical potentials on the ECoG and EEG. Electroencephalogr Clin Neurophysiol, 93, 380–389, (1994).
22. A. Stancàk Jr and G. Pfurtscheller, Event-related desynchronization of central beta-rhythms during brisk and slow self-paced finger movements of dominant and nondominant hand. Cogn Brain Res, 4, 171–183, (1996).
23. F. Cassim, C. Monaca, W. Szurhaj, et al., Does post-movement beta synchronization reflect an idling motor cortex? Neuroreport, 12, 3859–3863, (2001).
24. C. Neuper and G. Pfurtscheller, Event-related dynamics of cortical rhythms: frequency-specific features and functional correlates. Int J Psychophysiol, 43, 41–58, (2001).
25. M. Alegre, A. Labarga, I.G. Gurtubay, et al., Beta electroencephalograph changes during passive movements: sensory afferences contribute to beta event-related desynchronization in humans. Neurosci Lett, 331, 29–32, (2002).

26. G. Pfurtscheller, C. Neuper, and G. Krausz, Functional dissociation of lower and upper frequency mu rhythms in relation to voluntary limb movement. Clin Neurophysiol, 111, 1873–1879, (2000).
27. G. Pfurtscheller, C. Neuper, D. Flotzinger, et al., EEG-based discrimination between imagination of right and left hand movement. Electroencephalogr Clin Neurophysiol, 103, 642–651, (1997).
28. L. Leocani, G. Magnani, and G. Comi, Event-related desynchronization during execution, imagination and withholding of movement. In: G. Pfurtscheller and F. Lopes da Silva (Eds.), Event-related desynchronization. Handbook of electroenceph and clinical neurophysiology, vol. 6, Elsevier, pp. 291–301, (1999).
29. M. Jeannerod, Mental imagery in the motor context. Neuropsychologia, 33 (11), 1419–1432, (1995).
30. J. Decety, The neurophysiological basis of motor imagery, Behav Brain Res, 77, 45–52, (1996).
31. C. Neuper and G. Pfurtscheller, Motor imagery and ERD. In: G. Pfurtscheller and F. Lopes da Silva (Eds.), Event-related desynchronization. Handbook of electroenceph and clinical neurophysiology, vol. 6, Elsevier, pp. 303–325, (1999).
32. M. Roth, J. Decety, M. Raybaudi, et al., Possible involvement of primary motor cortex in mentally simulated movement: a functional magnetic resonance imaging study. Neuroreport, 7, 1280–1284, (1996).
33. C.A. Porro, M.P. Francescato, V. Cettolo, et al., Primary motor and sensory cortex activation during motor performance and motor imagery: a functional magnetic resonance imaging study. Int J Neurosci Lett, 16, 7688–7698, (1996).
34. M. Lotze, P. Montoya, M. Erb, et al., Activation of cortical and cerebellar motor areas during executed and imagined hand movements: an fMRI study. J Cogn Neurosci, 11, 491–501, (1999).
35. E. Gerardin, A. Sirigu, S. Lehéricy, et al., Partially overlapping neural networks for real and imagined hand movements. Cerebral Cortex, 10, 1093–1104, (2000).
36. H.H. Ehrsson, S. Geyer, and E. Naito, Imagery of voluntary movement of fingers, toes, and tongue activates corresponding body-part-specific motor representations. J Neurophysiol, 90, 3304–3316, (2003).
37. S. Rossi, P. Pasqualetti, F. Tecchio, et al., Corticospinal excitability modulation during mental simulation of wrist movements in human subjects. Neurosci Lett, 243, 147–151, (1998).
38. P. Suffczynski, P.J.M. Pijn, G. Pfurtscheller, et al., Event-related dynamics of alpha band rhythms: A neuronal network model of focal ERD/surround ERS, In: G. Pfurtscheller and F. Lopes da Silva (Eds.), Event-related desynchronization. Handbook of electroenceph and clinical neurophysiology, vol. 6, Elsevier, pp. 67–85, (1999).
39. G. Pfurtscheller, C. Brunner, A. Schlögl, et al., Mu rhythm (de)synchronization and EEG single-trial classification of different motor imagery tasks. NeuroImage, 31, 153–159, (2006).
40. M. Steriade and R. Llinas, The functional states of the thalamus and the associated neuronal interplay. Phys Rev, 68, 649–742, (1988).
41. P. Zhuang, C. Toro, J. Grafman, et al., Event–related desynchronization (ERD) in the alpha frequency during development of implicit and explicit learning. Electroencephalogr Clin Neurophysiol, 102, 374–381, (1997).
42. A.P. Leone, N. Dang, L.G. Cohen, et al., Modulation of muscle responses evoked by transcranial magnetic stimulation during the acquisition of new fine motor skills. J Neurophysiol, 74, 1037–1045, (1995).
43. R. Cooper, A.L. Winter, H.J. Crow, et al., Comparison of subcortical, cortical and scalp activity using chronically indwelling electrodes in man. Electroencephalogr Clin Neurophysiol, 18, 217–228, (1965).
44. F.L. da Silva, Neural mechanisms underlying brain waves: from neural membranes to networks. Electroencephalogr Clin Neurophysiol, 79, 81–93, (1991).

45. W. Klimesch, Memory processes, brain oscillations and EEG synchronization. J Psychophysiol, 24, 61–100, (1996).
46. F. Hummel, F. Andres, E. Altenmuller, et al., Inhibitory control of acquired motor programmes in the human brain. Brain, 125, 404–420, (2002).
47. E.D. Adrian and B.H. Matthews, The Berger rhythm: Potential changes from the occipital lobes in man. Brain, 57, 355–385, (1934).
48. M.H. Case and R.M. Harper, Somatomotor and visceromotor correlates of operantly conditioned 12–14 c/s sensorimotor cortical activity. Electroencephalogr Clin Neurophysiol, 31, 85–92, (1971).
49. W.N. Kuhlman, Functional topography of the human mu rhythm. Electroencephalogr Clin Neurophysiol, 44, 83–93, (1978).
50. C. Gerloff, J. Hadley, J. Richard, et al., Functional coupling and regional activation of human cortical motor areas during simple, internally paced and externally paced finger movements. Brain, 121, 1513–1531, (1998).
51. Y. Koshino and E. Niedermeyer, Enhancement of rolandic mu rhythm by pattern vision. Electroencephalogr Clin Neurophysiol, 38, 535–538, (1975).
52. N. Kreitmann and J.C. Shaw, Experimental enhancement of alpha activity. Electroencephalogr Clin Neurophysiol, 18, 147–155, (1965).
53. C. Neuper and W. Klimesch, Event-related dynamics of brain oscillations: Elsevier, (2006).
54. W.C. Drevets, H. Burton, T.O. Videen, et al., Blood flow changes in human somatosensory cortex during anticipated stimulation. Nature, 373, 249–252, (1995).
55. F. Hummel, R. Saur, S. Lasogga, et al., To act or not to act. Neural correlates of executive control of learned motor behavior. Neuroimage, 23, 1391–1401, (2004).
56. F. Hummel and C. Gerloff, Interregional long-range and short-range synchrony: a basis for complex sensorimotor processing. Progr Brain Res, 159, 223–236, (2006).
57. R. Salmelin, M. Hamalainen, M. Kajola, et al., Functional segregation of movement related rhythmic activity in the human brain. NeuroImage, 2, 237–243, (1995).
58. G. Pfurtscheller and F.H.L. da Silva, Event-related EEG/MEG synchronization and desynchronization: basic principles. Clin Neurophysiol, 110, 1842–1857, (1999).
59. C. Neuper and G. Pfurtscheller, Evidence for distinct beta resonance frequencies in human EEG related to specific sensorimotor cortical areas. Clin Neurophysiol, 112, 2084–2097, (2001).
60. G.R. Müller, C. Neuper, R. Rupp, et al., Event-related beta EEG changes during wrist movements induced by functional electrical stimulation of forearm muscles in man. Neurosci Lett, 340, 143–147, (2003).
61. G. Pfurtscheller, G. Krausz, and C. Neuper, Mechanical stimulation of the fingertip can induce bursts of beta oscillations in sensorimotor areas. J Clin Neurophysiol, 18, 559–564, (2001).
62. G. Pfurtscheller, C. Neuper, C. Brunner, et al., Beta rebound after different types of motor imagery in man. Neurosci Lett, 378, 156–159, (2005).
63. C. Neuper and G. Pfurtscheller, Motor imagery and ERD, In: G. Pfurtscheller and F. H. L. da Silva (Eds.), *Event-related desynchronization. Handbook of electroenceph and clinical neurophysiology*, vol. 6, Elsevier, Amsterdam,, pp. 303–325, (1999).
64. W. Singer, Synchronization of cortical activity and its putative role in information processing and learning. Annu Rev Psychophysiol, 55, 349–374, (1993).
65. R. Chen, B. Corwell, and M. Hallett, Modulation of motor cortex excitability by median nerve and digit stimulation. Expert Rev Brain Res, 129, 77–86, (1999).
66. A. Schnitzler, S. Salenius, R. Salmelin, et al., Involvement of primary motor cortex in motor imagery: a neuromagnetic study. NeuroImage, 6, 201–208, (1997).
67. G. Pfurtscheller, M. Wörtz, G.R. Müller, et al., Contrasting behavior of beta event-related synchronization and somatosensory evoked potential after median nerve stimulation during finger manipulation in man. Neurosci Lett, 323, 113–116, (2002).
68. A.P. Georgopoulos, J.F. Kalaska, R. Caminiti, et al., On the relations between the direction of two-dimensional arm movements and cell discharge in primate motor cortex. J NeuroSci Lett, 2, 1527–1537, (1982).

69. Z.J. Koles, M.S. Lazar, and S.Z. Zhou, Spatial patterns underlying population differences in the background EEG. Brain Topogr, 2, 275–284, (1990).
70. J. Müller-Gerking, G. Pfurtscheller, and H. Flyvbjerg, Designing optimal spatial filters for single-trial EEG classification in a movement task. Clin Neurophysiol, 110, 787–798, (1999).
71. J. Müller-Gerking, G. Pfurtscheller, and H. Flyvbjerg, Classification of movement-related EEG in a memorized delay task experiment. Clin Neurophysiol, 111, 1353–1365, (2000).
72. G. Pfurtscheller, C. Neuper, H. Ramoser, et al., Visually guided motor imagery activates sensorimotor areas in humans. Neurosci Lett, 269, 153–156, (1999).
73. E. Naito, P.E. Roland, and H.H. Ehrsson, I feel my hand moving: a new role of the primary motor cortex in somatic perception of limb movement, Neuron, 36, 979–988, (2002).
74. J. Annett, Motor imagery: perception or action? Neuropsychologia, 33, 1395–1417, (1995).
75. H.J. Gastaut and J. Bert, EEG changes during cinematographic presentation; moving picture activation of the EEG. Electroencephalogr Clin Neurophysiol, 6, 433–444, (1954).
76. S. Cochin, C. Barthelemy, B. Lejeune, et al., Perception of motion and qEEG activity in human adults. Electroencephalogr Clin Neurophysiol, 107, 287–295, (1998).
77. S.D. Muthukumaraswamy, B.W. Johnson, and N.A. McNair, Mu rhythm modulation during observation of an object-directed grasp. Cogn Brain Res, 19, 195–201, (2004).
78. E.L. Altschuler, A. Vankov, E.M. Hubbard, et al., Mu wave blocking by observation of movement and its possible use as a tool to study theory of other minds. Soc Neurosci Abstr, 26, 68, (2000).
79. G. Pfurtscheller, R.H. Grabner, C. Brunner, et al., Phasic heart rate changes during word translation of different difficulties. Psychophysiology, 44, 807–813, (2007).
80. G. Pfurtscheller, R. Scherer, R. Leeb, et al., Viewing moving objects in Virtual Reality can change the dynamics of sensorimotor EEG rhythms. Presence-Teleop Virt Environ, 16, 111–118, (2007).
81. J.A. Pineda, The functional significance of mu rhythms: translating seeing and hearing into doing. Brain Res, 50, 57–68, (2005).
82. R. Hari, Action–perception connection and the cortical mu rhythm. Prog Brain Res, 159, 253–260, (2006).
83. V. Gallese, L. Fadiga, L. Fogassi, et al., Action recognition in the premotor cortex. Brain, 119, 593–609, (1996).
84. G. Rizzolatti, L. Fadiga, V. Gallese, et al., Premotor cortex and the recognition of motor actions. Cogn Brain Res, 3, 131–141, (1996).
85. G. Rizzolatti, L. Fogassi, and V. Gallese, Neurophysiological mechanisms underlying the understanding and imitation of action. Nat Rev Neurosci, 2, 661–670, (2001).
86. G. Buccino, F. Binkofski, G.R. Fink, et al., Action observation activates premotor and parietal areas in a somatotopic manner: an fMRI study. Eur J Neurosci, 13, 400–404, (2001).
87. J. Grézes and J. Decety, Functional anatomy of execution, mental simulation, observation, and verb generation of actions: a meta-analysis. Human Brain Mapp, 12, 1–19, (2001).
88. N. Nishitani and R. Hari, Temporal dynamics of cortical representation for action, Proc Natl Acad Sci, 97, 913–918, (2000).
89. J.M. Kilner and C.D. Frith, A possible role for primary motor cortex during action observation, Proc Natl Acad Sci, 104, 8683–8684, (2007).
90. F.L. da Silva, Event-related neural activities: what about phase? Prog Brain Res, 159, 3–17, (2006).
91. R. Hari, N. Forss, S. Avikainen, et al., Activation of human primary motor cortex during action observation: a neuromagnetic study. Proc Natl Acad Sci, 95, 15061–15065, (1998).
92. J. Järveläinen, M. Schürmann, S. Avikainen, et al., Stronger reactivity of the human primary motor cortex during observation of live rather than video motor acts. Neuroreport, 12, 3493–3495, (2001).
93. G.R. Müller-Putz, R. Scherer, G. Pfurtscheller, et al., Brain-computer interfaces for control of neuroprostheses: from synchronous to asynchronous mode of operation, Biomedizinische Technik, 51, 57–63, (2006).

94. G. Pfurtscheller, C. Guger, G. Müller, et al., Brain oscillations control hand orthosis in a tetraplegic, Neuroscience Letters, 292, 211–214, (2000).
95. G. Pfurtscheller, G.R. Müller, J. Pfurtscheller, et al., "Thought"-control of functional electrical stimulation to restore handgrasp in a patient with tetraplegia, Neurosci Lett, 351, 33–36, (2003).
96. G.R. Müller-Putz, R. Scherer, G. Pfurtscheller, et al., EEG-based neuroprosthesis control: a step towards clinical practice, Neurosci Lett, 382, 169–174, (2005).
97. C. Neuper, R. Scherer, S. Wriessnegger, and G. Pfurtscheller. Motor imagery and action observation: Modulation of sensorimotor brain rhythms during mental control of a brain-computer interface. Clin. Neurophysiol, 121(8), 239–247, (2009).

Neurofeedback Training for BCI Control

Christa Neuper and Gert Pfurtscheller

1 Introduction

Brain–computer interface (BCI) systems detect changes in brain signals that reflect human intention, then translate these signals to control monitors or external devices (for a comprehensive review, see [1]). BCIs typically measure electrical signals resulting from neural firing (i.e. neuronal action potentials, Electroencephalogram (ECoG), or Electroencephalogram (EEG)). Sophisticated pattern recognition and classification algorithms convert neural activity into the required control signals. BCI research has focused heavily on developing powerful signal processing and machine learning techniques to accurately classify neural activity [2–4].

However, even with the best algorithms, successful BCI operation depends significantly on how well users can voluntarily modulate their neural activity. People need to produce brain signals that are easy to detect. This may be particularly important during "real-world" device control, when background mental activity and other electrical noise sources often fluctuate unpredictably. Learning to operate many BCI-controlled devices requires repeated practice with feedback and reward. BCI training hence engages learning mechanisms in the brain.

Therefore, research that explores BCI training is important, and could benefit from existing research involving neurofeedback and learning. This research should consider the specific target application. For example, different training protocols and feedback techniques may be more or less efficient depending on whether the user's task is to control a cursor on a computer screen [5–7], select certain characters or icons for communication [8–11], or control a neuroprosthesis to restore grasping [12, 13].

Even though feedback is an integral part of any BCI training, only a few studies have explored how feedback affects the learning process, such as when people learn to use their brain signals to move a cursor to a target on a computer

C. Neuper (✉)
Department of Psychology, University of Graz, Graz, Austria; Institute for Knowledge Discovery, Graz University of Technology, Graz, Austria
e-mail: christa.neuper@uni-graz.at

B. Graimann et al. (eds.), *Brain–Computer Interfaces*, The Frontiers Collection,
DOI 10.1007/978-3-642-02091-9_4,

screen. BCI operation is a type of neurofeedback application, and understanding the underlying principles of neurofeedback allows the BCI researcher to adapt the training process according to operant learning principles. Although empirical data concerning an "optimal" training setting do not exist, BCI researchers may benefit from experiences with neurofeedback training. In this chapter, we (i) shortly describe the underlying process involved in neurofeedback, (ii) review evidence for feedback effects from studies using neurofeedback and BCI, and (iii) discuss their implications for the design of BCI training procedures.

2 Principles of Neurofeedback

Almost all definitions of neurofeedback include the learning process and operant/instrumental conditioning as basic elements of the theoretical model. It is well established that people can learn to control various parameters of the brain's electrical activity through a training process that involves the real-time display of ongoing changes in the EEG (for a review, see [14, 15]). In such a neurofeedback paradigm, feedback is usually presented visually, by representing the brain signal on a computer monitor, or via the auditory or tactile modalities. This raises the question of how to best represent different brain signals (such as sensorimotor rhythm (SMR), slow cortical potential (SCP), or other EEG activity). Feedback signals are often presented in a computerized game-like format [42, 50, 51]. These environments can help maintain the user's motivation and attention to the task and guide him/her to achieve a specified goal (namely, specific changes in EEG activity) by maintaining a certain "mental state". Changes in brain activity that reflect successful neurofeedback training are rewarded or positively reinforced.

Typically, thresholds are set for maximum training effectiveness, and the reward criteria are based on learning models. This means that the task should be challenging enough that the user feels motivated and rewarded, and hence does not become bored or frustrated. Thus, in therapeutic practice, reward thresholds are set so that the reward is received about 60–70% of the time [16]. Thresholds should be adjusted as the user's performance improves.

Neurofeedback methods have been widely used for clinical benefits associated with the enhancement and/or suppression of particular features of the EEG that have been shown to correlate with a "normal" state of brain functioning. EEG feedback has been most intensively studied in epilepsy and attention deficit disorders. For instance, the control of epileptic seizures through learned enhancement of the 12–15 Hz sensorimotor rhythm (SMR) over the sensorimotor cortex, and through modulation of slow cortical potentials (SCPs), has been established in a number of controlled studies [17–19]. Other studies showed that children with attention deficit hyperactivity disorder (ADHD) improved behavioural and cognitive variables after frequency (e.g., theta/beta) training or SCP training. In a paradigm often applied in ADHD, the goal is to decrease activity in the theta band and to increase activity in the beta band (or to decrease theta/beta ratio) [20, 21].

A complete neurofeedback training process usually comprises 25–50 sessions that each lasts 45–60 min (cf. [22]). However, the number of sessions required appears to vary from individual to individual. At the beginning, a blocked training schedule (e.g., on a daily basis, or at least 3 times per week) may be preferable to facilitate the neuromodulatory learning process. For improving attentional abilities in healthy people, for instance, Gruzelier and coworkers established that the benefits with neurofeedback training can be achieved within only 10 sessions [15, 23]. Further, the learning curve for EEG self-regulation is usually not linear, but often shows a typical time course (see e.g. [24, 42]). That is, users tend to show the greatest improvement early in training, just as 10 h of piano practice could make a big difference to a novice, but probably not to an expert pianist.

2.1 Training of Sensorimotor Rhythms

Many BCI systems utilize sensorimotor EEG activity (i.e. mu and central beta rhythms), since it can be modulated by voluntary imagination of limb movements [25, 26] and is known to be susceptible to operant conditioning [5, 17, 27]. Experiments have confirmed that EEG frequency components recorded from central areas (mu, central beta, SMR) can be enhanced during long-term feedback training [5, 27, 28]. A further reason to use sensorimotor EEG in BCIs is that it is typically modulated by both overt and covert motor activity (that is, both actual and imagined movements), but unaffected by changes in visual stimulation [29]. Therefore, people can use an SMR BCI while watching a movie, focusing on a friend, browsing the web, or performing other visual tasks.

The operant conditioning of this type of EEG activity was discovered in the late 1960s in work with animals and replicated later in humans (for a review, see [17]). Sterman and coworkers observed that, when cats learned to suppress a movement, a particular brain rhythm in the range of 12–15 Hz emerged at the sensorimotor cortex. They successfully trained the cats to produce this "sensory motor rhythm" (SMR) through instrumental learning, by providing rewards only when the cats produced SMR bursts. Since SMR bursts occurred when the cat did not move, the SMR was considered the "idle rhythm" of the sensorimotor cortex. This is similar to the alpha rhythm for the visual system, which is strongest when people are not using their visual systems [30]. An unexpected observation was that cats that had been trained to produce SMR more often also showed higher thresholds to the onset of chemically induced seizures. A number of later studies in humans established that SMR training resulted in decreased seizure activity in epileptic subjects [17].

Other studies suggested that the human mu rhythm is analogous to the SMR found in cats, in terms of cortical topography, relationship to behavior, and reactivity to sensory stimulation [27, 31]. However, it is not clear whether the neurophysiological basis of the two phenomena is really identical (for a recent review of oscillatory potentials in the motor cortex, see [32] as well as chapter "Dynamics of Sensorimotor Oscillations in a Motor Task" in this book).

2.2 How Neurofeedback Works

The effects of neurofeedback techniques can be understood in terms of basic neurophysiological mechanisms like neuromodulation (e.g. ascending brain stem modulation of thalamic and limbic systems) and long-term potentiation (LTP) (for a review see [35]). LTP is one way that the brain permanently changes itself to adapt to new situations. Neurons learn to respond differently to input. For example, a neuron's receptor could become more sensitive to signals from another neuron that provides useful information. LTP is common in the hippocampus and cortex.

Some authors emphasize that neurofeedback augments the brain's capacity to regulate itself, and that this self-regulation (rather than any particular state) forms the basis of its clinical efficacy [16]. This is based on the idea that, during EEG feedback training, the participant learns to exert neuromodulatory control over the neural networks mediating attentional processes. Over time, LTP in these networks consolidates those processes into stable states. This can be compared to learning a motor task like riding a bicycle or typing. As a person practices the skill, sensory and proprioceptive (awareness of body position) input initiates feedback regulation of the motor circuits involved. Over time, the skill becomes more and more automatic. Hence, a person who learns to move a cursor on a computer screen that displays his/her EEG band power changes is learning through the same mechanisms as someone learning to ride a bicycle. In a typical neurofeedback or BCI paradigm, such as the "basket game" described in more detail below, subjects must direct a falling ball into the highlighted target (basket) on a computer screen via certain changes of EEG band power features. Confronted with this task, the participant probably experiences a period of "trial-and-error" during which various internal processes are "tried" until the right mental strategies are found to produce the desired movement of the ball. As the ball moves in the desired direction, the person watches and "feels" him/herself moving it. Rehearsal of these activities during ongoing training sessions can then stabilize the respective brain mechanisms and resulting EEG patterns. Over a number of sessions, the subject probably acquires the skill of controlling the movement of the ball without being consciously aware of how this is achieved.

3 Training Paradigms for BCI Control

A specific type of mental activity and strategy is necessary to modify brain signals to use a BCI effectively. Different approaches to training subjects to control particular EEG signals have been introduced in BCI research. One approach aims to train users to automatically control EEG components through operant conditioning [36]. Feedback training of slow cortical potentials (SCPs) was used, for instance, to realize a communication system for completely paralyzed (locked-in) patients [8]. Other research groups train subjects to control EEG components by the performance of specific cognitive tasks, such as mental motor imagery (e.g. [37]). A third approach views BCI research as mainly a problem of machine learning, and emphasizes detecting some types of brain signals that do not require neurofeedback training (see e.g. [38] and chapter "Brain–Computer Interface in Neurorehabilitation" in this book).

BCIs often rely on motor imagery. Users imagine movements of different body parts, such as the left hand, right hand, or foot [25]; for a review see [37]). This method has been shown to be particularly useful for mental control of neuroprostheses (for a review, see [13]). There is strong evidence that motor imagery activates cortical areas similar to those activated by the execution of the same movement [25]. Consequently, EEG electrodes are placed over primary sensorimotor areas. Characteristic ERD/ERS patterns are associated with different types of motor imagery (see chapter "Dynamics of Sensorimotor Oscillations in a Motor Task" for details) and are also detectable in single trials using an online system.

3.1 Training with the Graz-BCI

Before most "motor imagery" BCIs can be efficiently used, users have to undergo training to obtain some control of their brain signals. Users can then produce brain signals that are easier to detect, and hence a BCI can more accurately classify different brain states. Prior to starting online feedback sessions with an individual, his/her existing brain patterns (e.g. related to different types of motor imagery) must be known. To this end, in the first session of the Graz-BCI standard protocol, users must repeatedly imagine different kinds of movement (e.g., hand, feet or tongue movement) in a cue-based mode while their EEG is recorded. Optimally, this would entail a full-head recording of their EEG, with topographical and time-frequency analyses of ERD/ERS patterns, and classification of the individual's brain activity in different imagery conditions. By applying, for example, the distinction sensitive learning vector quantization (DSLVQ) [39] to the screening data, the frequency components that discriminate between conditions may be identified for each participant. Classification accuracy can also be calculated. This shows which mental states may be distinguished, as well as the best electrode locations. Importantly, specific individualized EEG patterns may be used in subsequent training sessions, where the user receives on-line feedback of motor imagery related changes in the EEG.

In a typical BCI paradigm, feedback about performance is provided by (i) a continuous feedback signal (e.g., cursor movement) and (ii) by the outcome of the trial (i.e. discrete feedback about success or failure). Noteworthily, BCI control in trained subjects is not dependent on the sensory input provided by the feedback signal. For example, McFarland et al. [6] reported that well-trained subjects still displayed EEG control when feedback (cursor movement) was removed for some time. Further, this study showed that visual feedback can not only facilitate, but also impair EEG control, and that this effect varies across individuals. This highlights the need for displays that are immersive, informative, and engaging without distracting or annoying the user.

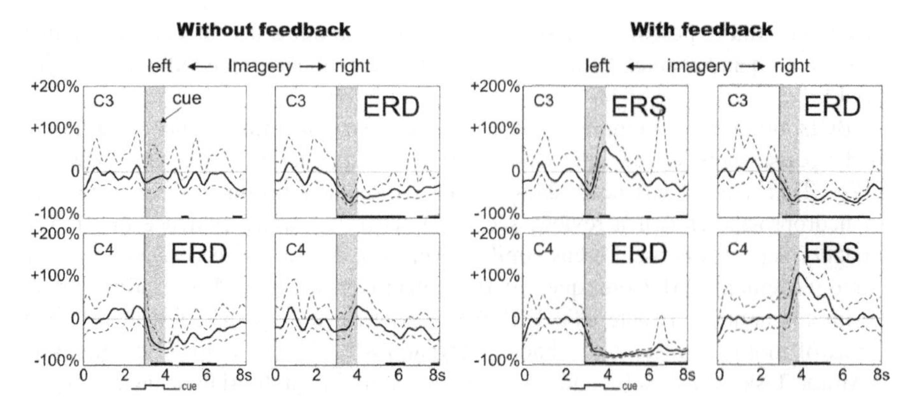

Fig. 1 Band power (11–13 Hz) time courses ±95% confidence interval displaying ERD and ERS from training session without feedback (*left*) and session with feedback (*right*). These data are from one able-bodied subject while he imagined left and right hand movement. Grey areas indicate the time of cue presentation. Sites C3 and C4 are located over the left and right sensorimotor areas of the brain, respectively

When a naïve user starts to practice hand motor imagery, a contralaterally dominant desynchronization pattern is generally found. Changes in relevant EEG patterns usually occur if the user is trained via feedback about the mental task performed. Figure 1 shows an example of band power time courses of 11–13 Hz EEG activity of one subject obtained at two times: during the initial screening without feedback; and during a subsequent training session while a continuous feedback signal (moving bar) was presented. The ERD/ERS curves differ during right versus left motor imagery, with a significant band power decrease (ERD) over the contralateral (opposite) hand area, and a band power increase (ERS) over the ipsilateral (same) side. The feedback enhanced the difference between both patterns and therewith the classification accuracy (see also [40]). The enhancement of oscillatory EEG activity (ERS) during motor imagery is very important in BCI research, since larger ERS leads to more accurate classification of single EEG trials.

BCI training can be run like a computer game to make participants more engaged and motivated. In the "basket-game" paradigm, for example, the user has to mentally move a falling ball into the correct goal ("basket") marked on the screen ([41], see also Fig. 2, left side). If the ball hits the correct basket, it is highlighted and points are earned. The horizontal position of the ball is controlled via the BCI output signal, and the ball's velocity can be adjusted by the investigator.

Four male volunteers with spinal-cord injuries participated in a study using this paradigm. None of them had any prior experience with BCI. Two bipolar EEG signals were recorded from electrode positions close to C3 and C4, respectively. Two different types of motor imagery (either right vs. left hand motor imagery or hand vs. foot motor imagery) were used, and band power within the alpha band and the beta band were classified. Based on each subject's screening data, the best motor imagery tasks were selected, and the classifier output (position of the ball) was weighted to adjust the mean deflection to the middle of the target basket. This way, the BCI

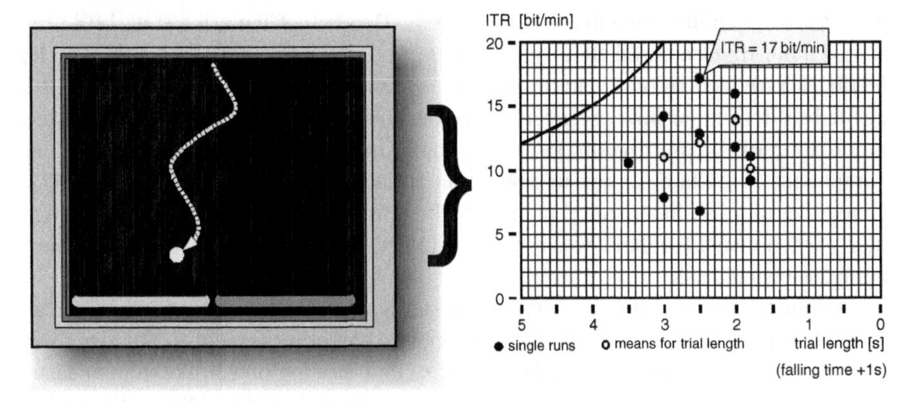

Fig. 2 *Left side*: Graphical display of the "basket-paradigm". The subject has to direct the ball to the indicated goal ("basket"). The trial length varies across the different runs. *Right side*: Information transfer rate (ITR) for one subject in relation to trial length. The *black* line represents the maximum possible ITR for an error-free classification (modified from Krausz et al. [41])

output was adapted to each patient. The participant's task was to hit the highlighted basket (which changed side randomly from trial to trial) as often as possible. The speed was increased run by run until the person considered it too fast. This way, we attempted to find the trial length that maximized the information transfer rate. After each run, users were asked to rate their performance and suggest whether the system operated too slow or too fast. The highest information transfer rate of 17 bits/min was reached with a trial length of 2.5 s ([41], see also Fig. 2 right side).

3.2 Impact of Feedback Stimuli

A well thought-out training protocol and helpful feedback signals are essential to keep the training period as short as possible. The feedback provides the user with information about the efficiency of his/her strategy and enables learning. Two aspects of feedback are crucial. The first aspect is how the brain signal is translated into the feedback signal (for advantages of continuous versus discrete feedback, see [40]). The second aspect is how the feedback is presented. The influence of feedback on the user's attention, concentration and motivation are closely related to the learning process, and should be considered (see also [42]).

As mentioned above, some BCI studies use different feedback modalities. In the auditory modality, Hinterberger et al. [43] and Pham et al. [44] coded slow cortical potential (SCP) amplitude shifts in the ascending and descending pitches on a major tone scale. Rutkowski et al. [45] implemented an auditory representation of people's mental states. Kübler et al. [66] and Furdea et al. [67] showed that P300 BCIs could also be implemented with auditory rather than visual feedback. A BCI using only auditory (rather than visual) stimuli could help severely paralyzed patients with visual impairment. Although the above studies showed that BCI communication using only auditory stimuli is possible, visual feedback turned out to be superior to auditory feedback in most BCIs. Recently, Chatterjee et al. [46] presented a mu-rhythm based BCI using a motor imagery paradigm and haptic (touch) feedback provided by vibrotactile stimuli to the upper limb. Further work will be needed to determine how the neural correlates of vibrotactile feedback affect the modulation of the mu rhythm. However, haptic information may become a critical component of BCIs designed to control an advanced neuroprosthetic device [47]. Cincotti et al. [48] documented interesting benefits of vibrotactile feedback, particularly when visual feedback was not practical because the subjects were performing complex visual tasks.

Despite the success of non-visual BCI systems, most BCIs present feedback visually [1]. Typical visual feedback stimuli comprise cursor movement [6, 7], a moving bar of varying size [40, 46], and the trajectory of a moving object like in the basket game [7, 41]. Other interesting variants include colour signalling [49] and complex virtual reality environments [42, 50, 51].

There is some evidence that a rich visual representation of the feedback signal, such as a 3-dimensional video game or virtual reality environment, may facilitate learning to use a BCI [42, 50–53]. Combining BCI and virtual reality (VR) technologies may lead to highly realistic and immersive BCI feedback scenarios. [51] was as an important step in this direction. This article showed that EEG recording and single trial processing with adequate classification results are possible in a CAVE system (a virtual reality environment that our group used during BCI research), and that subjects could even control events within a virtual environment in real-time (see chapter "The Graz Brain–Computer Interface" for details).

Figure 3 shows that simple bar feedback in a cued hand vs. feet imagery task on a monitor was accompanied by a moderate ERD at electrode position Cz and a heart rate deceleration. In contrast, feedback in form of a virtual street resulted in induced beta oscillations (beta ERS) at Cz and a heart rate acceleration. The mutual interactions between brain and heart activity during complex virtual feedback are especially interesting.

These results suggest that improving the visual display in a BCI could improve a person's control over his/her brain activity. In particular, the studies mentioned above support realistic and engaging feedback scenarios, which are closely related to the specific target application. For example, observing a realistic moving hand should have a greater effect on sensorimotor rhythms than watching abstract feedback in the form of a moving bar [55].

However, the processing of such a realistic feedback stimulus may interfere with the mental motor imagery task, and thus might impair the development of EEG control. Because similar brain signals, namely sensorimotor rhythms in the mu and beta frequency bands, react to both motor imagery [25] and observation of biological movement (e.g. [55–60]; for a review see [61]), a realistic feedback presentation showing (for instance) a moving visual scene may interfere with the motor imagery related brain signals used by the BCI. The mutual interaction between a mentally simulated movement and simultaneous watching of a moving limb should be further investigated.

In a recent study, we explored how different types of visual feedback affect EEG activity (i.e. ERD/ERS patterns) during BCI control [62]. We showed two different

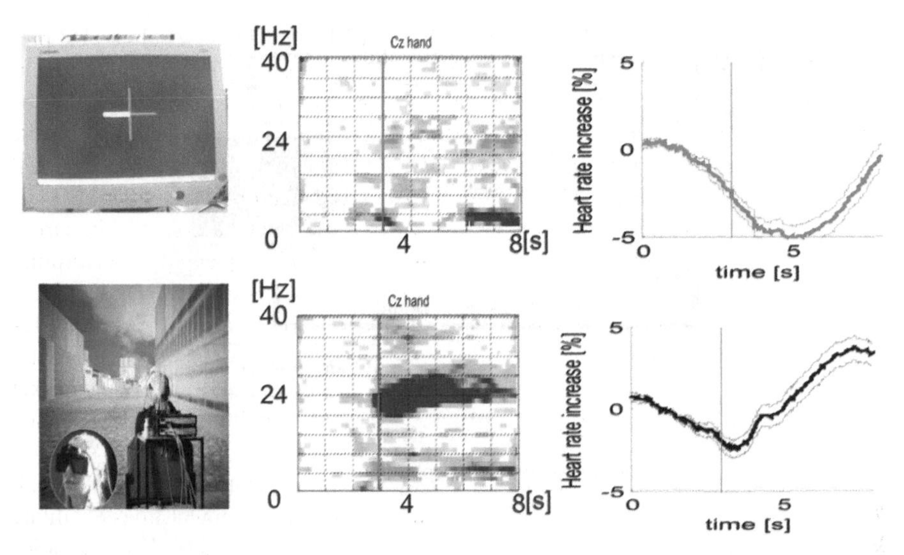

Fig. 3 Effect of different BCI feedback displays on oscillatory activity and heart rate. The *top* images depict a horizontal bar feedback condition, and the *bottom* images present a virtual street feedback condition. The two *left panels* each show the monitor or VR display. The *middle panels* display the corresponding ERD maps computed at electrode position Cz for hand motor imagery during bar and VR feedback. The *right panels* show corresponding changes in the heart rate (modified from Pfurtscheller et al. [54])

presentation types, namely abstract versus realistic feedback, to two experimental groups, while keeping the amount of information provided by the feedback equivalent. The "abstract feedback" group was trained to use left or right hand motor imagery to control a moving bar (varying in size) on a computer monitor. This is the standard protocol of the Graz-BCI, cf. [37, 40]. The "realistic feedback group", in contrast, used their online EEG parameters to control a video presentation showing an object-directed grasp from the actor's perspective. The results suggest that, when feedback provides comparable information on the continuous and final outcomes of mental actions, the type of feedback (abstract vs. realistic) does not influence performance. Considering the task-related EEG changes (i.e. ERD/ERS patterns), there was a significant difference between screening and feedback sessions in both groups, but the "realistic feedback" group showed more significant effects due to the feedback. Another important question is whether VR can reduce the number of training sessions [50].

4 Final Considerations

The BCI paradigm (see Fig. 4) basically corresponds to an operant conditioning neurofeedback paradigm. However, important differences exist between the training (e.g. with the basket paradigm) and the application (e.g. controlling a hand orthosis)

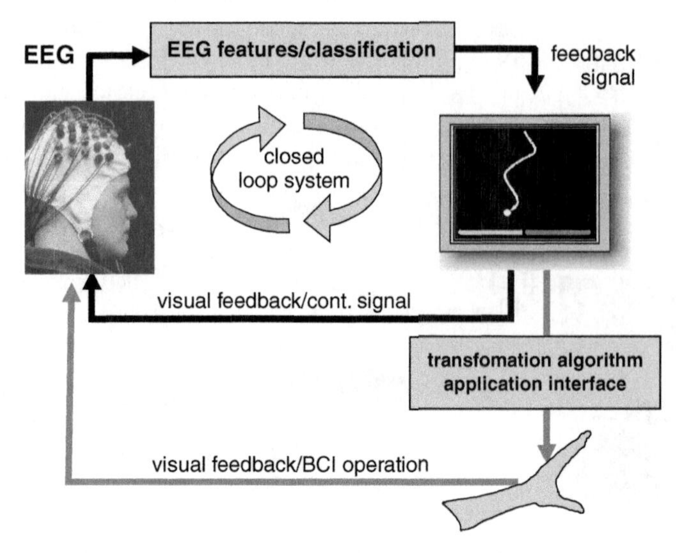

Fig. 4 Basic components of the BCI paradigm during training *(upper loop)* and during controlling a device (application, *lower loop*). The closed loop system indicated by black lines corresponds to an operant conditioning neurofeedback paradigm. BCIs also have other components, including feature selection and classification, an output mechanism such as controlling continuous cursor movement on the screen, and an additional transform algorithm to convert the classifier output to a suitable control signal for device control

in the form of the visual feedback. In the latter case, moving distractions such as a moving hand can interfere with the imagery task.

Unlike the direct visualization of the relevant EEG parameters in classical neurofeedback paradigms, the use of a classifier entails controlling e.g. continuous cursor movement on the screen according to the outcome of a classification procedure (such as a time-varying distance function in the case of a linear discriminant analysis (LDA)). This procedure provides the user with information about the separability of the respective brain patterns, rather than representing a direct analogue of the brain response.

Adaptive classification methods [63, 64] can add another challenge to improving BCI control. Although such adaptive algorithms are intended to automatically optimize control, they create a kind of moving target for self-regulation of brain patterns. The neural activity pattern that worked at one time may become less effective over time, and hence users may need to learn new patterns. Similarly, the translation algorithm (application interface), which converts the classifier output into control parameters to operate the specific device, introduces an additional processing stage. This may further complicate the relationship between neural activity and the final output control. The complex transformation of neural activity to output parameters may make it hard to learn to control neural activity. This explains why some BCI studies report unsuccessful learning [62]. For further discussion of these issues, see [65].

Classifier-based BCI training aims to simultaneously take advantage of the learning capability of both the system and the human user [37]. Therefore, critical aspects of neurofeedback training, as established in clinical therapy, should be considered when designing BCI training procedures. In the neurotherapy approach, neuroregulation exercises aim to modulate brain activity in a desired way, such as by increasing or reducing certain brain parameters (e.g. frequency components). By adjusting criteria for reward presented to the individual, one can "shape" the behaviour of his/her brain as desired.

Considering the user and the BCI system as two interacting dynamic processes [1], the goals of the BCI system are to emphasize those signal features that the user can reliably produce, and control and optimize the translation of these signals into device control. Optimizing the system facilitates further learning by the user.

Summarizing, BCI training and feedback can have a major impact on a user's performance, motivation, engagement, and training time. Many ways to improve training and feedback have not been well explored, and should consider principles in the skill learning and neurofeedback literature. The efficiency of possible training paradigms should be investigated in a well-controlled and systematic manner, possibly tailored for specific user groups and different applications.

Acknowledgments This research was partly financed by PRESENCCIA, an EU-funded Integrated Project under the IST program (Project No. 27731) and by NeuroCenter Styria, a grant of the state government of Styria (Zukunftsfonds, Project No.PN4055).

References

1. J.R. Wolpaw, et al., Brain-computer interfaces for communication and controls. Clin Neurophysiol, 113, 767–791, (2002).
2. K.R. Müller, C.W. Anderson, and G.E. Birch, Linear and nonlinear methods for brain–computer interfaces. IEEE Trans Neural Syst Rehabil Eng, 11, 165–169, (2003).
3. M. Krauledat, et al., The Berlin brain-computer interface for rapid response. Biomedizinische Technik, 49, 61–62, (2004).
4. G. Pfurtscheller, B. Graimann, and C. Neuper, EEG-based brain-computer interface systems and signal processing. In M. Akay (Ed.), *Encyclopedia of biomedical engineering*, Wiley, New Jersey, pp. 1156–1166, (2006).
5. J.R. Wolpaw, D.J. McFarland, and G.W. Neat, An EEG-based brain-computer interface for cursor control. Electroencephalogr Clin Neurophysiol, 78, 252–259, (1991).
6. D.J. McFarland, L.M. McCane, and J.R. Wolpaw, EEG-based communication and control: short-term role of feedback. IEEE Trans Neural Syst Rehabil, 6, 7–11, (1998).
7. B. Blankertz, et al., The non-invasive Berlin brain-computer interface: fast acquisition of effective performance in untrained subjects. NeuroImage, 37, 539–550, (2007).
8. N. Birbaumer, et al., A spelling device for the paralysed. Nature, 398, 297–298, (1999).
9. A. Kübler, et al., Brain-computer communication: Self-regulation of slow cortical potentials for verbal communication. Arch Phys Med Rehabil, 82, 1533–1539, (2001).
10. C. Neuper, et al., Clinical application of an EEG-based brain–computer interface: a case study in a patient with severe motor impairment. Clin Neurophysiol, 114, 399–409, (2003).
11. R. Scherer, et al., An asynchronously controlled EEG-based virtual keyboard: improvement of the spelling rate. IEEE Trans Neural Syst Rehabil Eng, 51, 979–984, (2004).
12. G.R. Müller-Putz, et al., EEG-based neuroprosthesis control: A step into clinical practice. Neurosci Lett, 382, 169–174, (2005).

13. C. Neuper, et al., Motor imagery and EEG-based control of spelling devices and neuroprostheses. In C. Neuper and W. Klimesch (Eds.), *Event-related dynamics of brain oscillations*, Elsevier, Amsterdam, pp. 393–409, (2006).
14. J. Gruzelier and T. Egner, Critical validation studies of neurofeedback. Child Adol Psychiatr Clin N Am, 14, 83–104, (2005).
15. J. Gruzelier, T. Egner, and D. Vernon, Validating the efficacy of neurofeedback for optimising performance. In C. Neuper and W. Klimesch (Eds.), *Event-related dynamics of brain oscillations*, Elsevier, pp. 421–431, (2006).
16. S. Othmer, S.F. Othmer, and D.A. Kaiser, EEG biofeedback: An emerging model for ist global efficacy. In J. Evans and A. Abarbanel (Eds.), *Quantitative EEG and neurofeedback*, Academic, pp. 243–310, (1999).
17. M.B. Sterman, Basic concepts and clinical findings in the treatment of seizure disorders with EEG operant conditioning. Clin Electroencephalogr, 31, 45–55, (2000).
18. U. Strehl, et al., Predictors of seizure reduction after self-regulation of slow cortical potentials as a treatment of drug-resistant epilepsy. Epilepsy Behav, 6, 156–166, (2005).
19. J.E. Walker, and G.P. Kozlowski, Neurofeedback treatment of epilepsy. Child Adol Psychiatr Clin N Am, 14, 163–176, (2005).
20. J.F. Lubar, Neurofeedback for the management of attention-deficit/hyperactivity disorders. In M. Schwartz and F. Andrasik (Eds.), *Biofeedback: A practitioners guide*, Guilford, pp. 409–437, (2003).
21. V.J. Monastra, Overcoming the barriers to effective treatment for attention-deficit/hyperactivity disorder: a neuro-educational approach. Int J Psychophysiol, 58, 71–80, (2005).
22. H. Heinrich, G. Holger, and U. Strehl, Annotation: Neurofeedback train your brain to train behaviour. J Child Psychol Psychiatr, 48, 3–16, (2007).
23. T. Egner. and J.H. Gruzelier, Learned self-regulation of EEG frequency components affects attention and event-related brain potentials in humans. Neuroreport, 21, 4155–4159, (2001).
24. J.V. Hardt, The ups and downs of learning alpha feedback. Biofeedback Res Soc, 6, 118, (1975).
25. G. Pfurtscheller and C. Neuper, Motor imagery activates primary sensimotor area in humans. Neurosci Lett, 239, 65–68, (1997).
26. G. Pfurtscheller, et al., EEG-based discrimination between imagination of right and left hand movement. Electroencephalogr Clin Neurophysiol, 103, 642–651, (1997).
27. W.N. Kuhlman, EEG feedback training: enhancement of somatosensory cortical activity. Electroencephalogr Clin Neurophysiol, 45, 290–294, (1978).
28. S. Waldert, et al., Hand movement direction decoded from MEG and EEG. J Neurosci, 28, 1000–1008, (2008).
29. M.B. Sterman, and L. Friar, Suppression of seizures in an epileptic following sensorimotor EEG feedback training. Electroencephalogr Clin Neurophysiol, 33, 89–95, (1972).
30. M.H. Case and R.M. Harper, Somatomotor and visceromotor correlates of operantly conditioned 12–14 c/s sensorimotor cortical activity. Electroencephalogra Clin Neurophysiol, 31, 85–92, (1971).
31. H. Gastaut, Electrocorticographic study of the reactivity of rolandic rhythm. Revista de Neurologia, 87, 176–182, (1952).
32. McKay, Wheels of motion: oscillatory potentials in the motor cortex. In E. Vaadia and A. Riehle (Eds.), *Motor cortex in voluntary movements: a distributed system for distributed functions*. Series: Methods and New Frontiers in Neuroscience, CRC Press, pp. 181–212, (2005).
33. M. Steriade and R. Llinas, The functional states of the thalamus and the associated neuronal interplay. Phys Rev, 68, 649–742, (1988).
34. F.H. Lopes da Silva, Neural mechanisms underlying brain waves: from neural membranes to networks. Electroencephalogr Clin Neurophysiol, 79, 81–93, (1991).

35. A. Abarbanel, The neural underpinnings of neurofeedback training. In J. Evans and A. Abarbanel (Eds.), *Quantitative EEG and neurofeedback*, Academic, pp. 311–340, (1999).
36. N. Birbaumer, et al., The thought-translation device (TTD): neurobehavioral mechanisms and clinical outcome. IEEE Trans Neural Syst Rehabil Eng, 11, 120–123, (2003).
37. G. Pfurtscheller and C. Neuper, Motor imagery and direct brain–computer communication. Proc IEEE, 89, 1123–1134, (2001).
38. B. Blankertz, et al., Boosting bit rates and error detection for the classification of fast-paced motor commands based on single-trial EEG analysis. IEEE Trans Neural Syst Rehabi Eng, 11, 127–131, (2003).
39. M. Pregenzer, G. Pfurtscheller, and D. Flotzinger, Automated feature selection with a distinction sensitive learning vector quantizer. Neurocomputing, 11, 19–29, (1996).
40. C. Neuper, A. Schlögl, and G. Pfurtscheller, Enhancement of left-right sensorimotor EEG differences during feedback-regulated motor imagery. J Clin Neurophysiol, 16, 373–382, (1999).
41. G. Krausz, et al., Critical decision-speed and information transfer in the "Graz Brain-Computer Interface". Appl Psychophysiol Biofeedback, 28, 233–240, (2003).
42. J.A. Pineda, et al., Learning to control brain rhythms: making a brain-computer interface possible. IEEE Trans Neural Syst Rehabil Eng, 11, 181–184, (2003).
43. T. Hinterberger, et al., Auditory feedback of human EEG for direct brain-computer communication. Proceedings of ICAD 04-Tenth Meeting of the International Conference on Auditory Display, 6–9 July, (2004).
44. M. Pham, et al., An auditory brain-computer interface based on the self-regulation of slow cortical potentials. Neurorehabil Neural Repair, 19, 206–218, (2005).
45. T.M. Rutkowski, et al., Auditory feedback for brain computer interface management An EEG data sonification approach. In B. Gabrys, R.J. Howlett, and L.C. Jain (Eds.), Knowledge-based intelligent information and engineering systems, Lecture Notes in Computer Science, Springer, Berlin Heidelberg (2006).
46. A. Chatterjee, et al., A brain-computer interface with vibrotactile biofeedback for haptic information. J Neuroeng Rehabil, 2007. 4, (2007).
47. M.A. Lebedev. and M.A.L. Nicolelis, Brain–machine interfaces: past, present and future. Trends Neurosci, 29, 536–546, (2006).
48. F. Cincotti, et al., Vibrotactile feedback for brain-computer interface operation. Comput Intell Neurosci, 2007, 48937, (2007).
49. A.Y. Kaplan, et al., Unconscious operant conditioning in the paradigm of brain-computer interface based on color perception. Int J Neurosci Lett, 115, 781–802, (2005).
50. R. Leeb, et al., Walking by thinking: the brainwaves are crucial, not the muscles! Presence: Teleoper Virt Environ, 15, 500–514, (2006).
51. G. Pfurtscheller, et al., Walking from thought. Brain Res, 1071, 145–152, (2006).
52. R. Ron-Angevin, A. Diaz-Estrella, and A. Reyes-Lecuona, Development of a brain-computer interface based on virtual reality to improve training techniques. Cyberpsychol Behav, 8, 353–354, (2005).
53. R. Leeb, et al., Brain-computer communication: motivation, aim and impact of exploring a virtual apartment. IEEE Trans Neural Syst Rehabil Eng, 15, 473–482, (2007).
54. G. Pfurtscheller, R. Leeb, and M. Slater, Cardiac responses induced during thought-based control of a virtual environment. Int J Psychophysiol, 62, 134–140, (2006).
55. G. Pfurtscheller, et al., Viewing moving objects in virtual reality can change the dynamics of sensorimotor EEG rhythms. Presence: Teleop Virt Environ, 16, 111–118, (2007).
56. R. Hari, et al., Activation of human primary motor cortex during action observation: a neuromagnetic study. Proc Natl Acad Sci, 95, 15061–15065, (1998).
57. S. Cochin, et al., Perception of motion and qEEG activity in human adults. Electroencephalogr Clin Neurophysiol, 107, 287–295, (1998).
58. C. Babiloni, et al., Human cortical electroencephalography (EEG) rhythms during the observation of simple aimless movements: a high-resolution EEG study. Neuroimage, 17, 559–572, (2002).

59. S.D. Muthukumaraswamy, B.W. Johnson, and N.A. McNair, Mu rhythm modulation during observation of an object-directed grasp. Cogn Brain Res, 19, 195–201, (2004).
60. L.M. Oberman, et al., EEG evidence for mirror neuron activity during the observation of human and robot actions: Toward an analysis of the human qualitities of interactive robots. Neurocomputing, 70, 2194–2203, (2007).
61. R. Hari, Action-perception connection and the cortical mu rhythm. In C. Neuper and W. Klimesch (Eds.), *Event-related dynamics of brain oscillations*, Elsevier, pp. 253–260, (2006).
62. C. Neuper, et al., Motor imagery and action observation: modulation of sensorimotor brain rhythms during mental control of a brain computer interface. Clin Neurophysiol, 120(2), pp. 239–247, Feb, (2009).
63. P. Shenoy, et al., Towards adaptive classification for BCI. J Neural Eng, 3, 13–23, (2006).
64. C. Vidaurre, et al., A fully on-line adaptive BCI. IEEE Trans Biomed Eng, 53, 1214–1219, (2006).
65. E.E. Fetz, Volitional control of neural activity: implications for brain-computer interfaces. J Psychophysiol, 15, 571–579, (2007).
66. A. Kübler et al., A brain-computer interface controlled auditory event-related potential (p300) spelling system for locked-in patients. Ann N Y Acad Sci, 1157, 90–100, (2009).
67. A. Furdea et al., An auditory oddball (P300) spelling system for brain-computer interfaces. Psychophysiology, 46(3), 617–25, (2009).

The Graz Brain-Computer Interface

Gert Pfurtscheller, Clemens Brunner, Robert Leeb, Reinhold Scherer, Gernot R. Müller-Putz and Christa Neuper

1 Introduction

Brain-computer interface (BCI) research at the Graz University of Technology started with the classification of event-related desynchronization (ERD) [36, 38] of single-trial electroencephalographic (EEG) data during actual (overt) and imagined (covert) hand movement [9, 18, 40]. At the beginning of our BCI research activities we had a cooperation with the Wadsworth Center in Albany, New York State, USA, with the common interest to control one-dimensional cursor movement on a monitor through mental activity [69]. With such a cursor control it is in principle possible to select letters of the alphabet, create words and sentences and realize a thought-based spelling system for patients in a complete or incomplete "locked-in" state [68]. At that time we already analyzed 64-channel EEG data from three patients who had accomplished a number of training sessions with the aim to search for optimal electrode positions and frequency components [38]. Using the distinction sensitive learning vector quantizer (DSLVQ) [54] it was found that for each subject there exist optimal electrode positions and frequency components for on-line EEG-based cursor control. This was confirmed recently by BCI studies in untrained subjects [2, 58].

2 The Graz BCI

The Graz BCI uses the EEG as input signal, motor imagery (MI) as mental strategy and two types of operation and data processing, respectively. In one mode of operation the data processing is restricted to predefined time windows of a few seconds length following the cue stimulus (synchronous or cue-based BCI). In the other

G. Pfurtscheller (✉)
Laboratory of Brain-Computer Interfaces, Institute for Knowledge Discovery, Graz University of Technology, Krenngasse 37, 8010, Graz, Austria
e-mail: pfurtscheller@tugraz.at

B. Graimann et al. (eds.), *Brain–Computer Interfaces*, The Frontiers Collection,
DOI 10.1007/978-3-642-02091-9_5, © Springer-Verlag Berlin Heidelberg 2010

mode, the processing is performed continuously on a sample-by-sample basis (asynchronous, uncued or self-paced BCI). The cue contains either information for the user (e. g., indication for the type of MI to be executed) or is neutral. In the latter case, the user is free to choose one of the predefined mental tasks after the cue. A synchronous BCI system is not available for control outside the cue-based processing window. In the asynchronous mode no cue is necessary; hence, the system is continuously available to the users for control.

The basic scheme of a BCI is illustrated in Fig. 1. The user creates specific brain patterns which are recorded by suitable data acquisition devices. In the subsequent signal processing stage, the data is preprocessed, features are extracted and classified. Then, a control signal is generated, which is connected to an application through a well-defined interface. Finally, the user receives feedback from the application.

Before a BCI can be used successfully, the users have to perform a number of training runs where suitable classifiers specific for the user's brain patterns are set up. In general, the training starts with a small number of predefined tasks repeated periodically in a cue-based mode. The brain signals are recorded and analyzed offline in time windows of a specified length after the cue. That way the machine learning algorithms can adapt to the specific brain patterns and eventually they can learn to recognize mental task-related brain activity. This process is highly subject-specific, which means that this training procedure has to be carried out for every

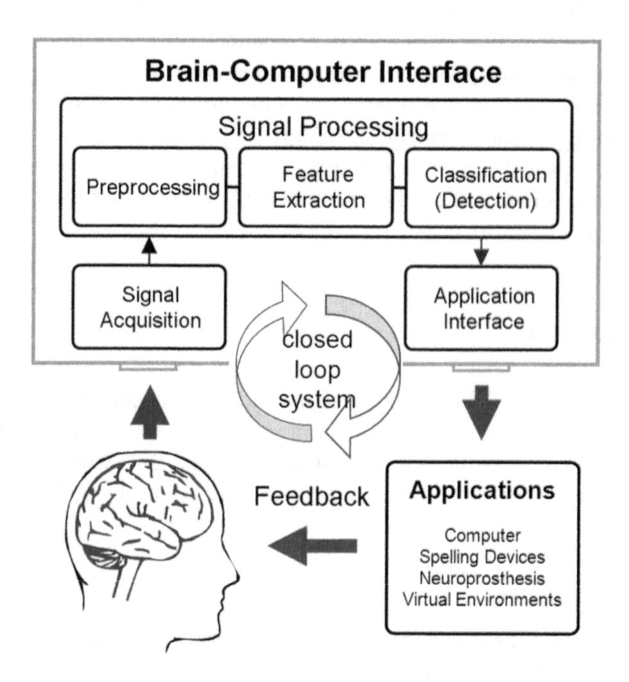

Fig. 1 General scheme of a brain-computer interface (modified from [12])

subject. Once this initial learning phase has resulted in a classifier, the brain patterns can be classified online and suitable feedback can be provided.

The following three steps are characteristic for the Graz BCI:

1. Cue-based multichannel EEG recording (e. g., 30 EEG signals) during three or four MI tasks (right hand, left hand, foot and/or tongue MI) and selection of those MI tasks with the most significant imagery-correlated EEG changes.
2. Selection of a minimal number of EEG channels (electrodes) and frequency bands with the best discrimination accuracy between two MI tasks or one MI task and rest, respectively.
3. Cue-based training without and with feedback to optimize the classification procedure applicable for self-paced, uncued BCI operation.

For detailed references about the Graz BCI see [45, 47, 48, 51].

3 Motor Imagery as Mental Strategy

Motor imagery is a conscious mental process defined as mental simulation of a specific movement [17]. The Graz BCI analyses and classifies the dynamics of brain oscillations in different frequency bands. In detail, two imagery-related EEG phenomena are of importance, either a decrease or increase of power in a given frequency band. The former is called event-related desynchronization (ERD) [36], and the latter event-related synchronization (ERS) [35]. By means of quantification of temporal-spatial ERD and ERS patterns [44] it has been shown that MI can induce different types of EEG changes:

1. Desynchronization (ERD) of sensorimotor rhythms (mu and central beta oscillations) during MI [50],
2. synchronization (ERS) of mu and beta oscillations in non-attended cortical body-part areas during MI [50],
3. synchronization of beta oscillations in attended body-part areas during MI [39], and
4. short-lasting beta oscillations after termination of MI [50, 49].

For the control of an external device based on single-trial classification of brain signals it is essential that MI-related brain activity can be detected in ongoing EEG and classified in real-time. Even though it has been documented that the imagination of simple body-part movements elicits predictable temporally stable changes in the sensorimotor rhythms with a small intra-subject variability [37], there are also participants who do not show the expected imagination-related EEG changes. Moreover, a diversity of time-frequency patterns (i. e., high inter-subject variability), especially with respect to the reactive frequency components, was found when studying the dynamics of oscillatory activity during movement imagination. In general, an amplitude increase in the EEG activity in the form of an ERS can be more easily and accurately detected in single trials as an amplitude decrease (ERD).

Therefore, our interest is focused on finding somatotopically specific ERS patterns in the EEG.

Inter-subject variability refers to differences in the EEG signals across different subjects. No two brains are exactly the same, and so any BCI must be able to adapt accordingly. For example, extensive work from the Graz BCI group and the Wadsworth group has shown that ERD and ERS are best identified at different frequencies. One subject might perform best if a frequency of 11 Hz is considered, while another subject might perform best at 12 Hz. Similarly, some subjects exhibit the strongest ERD/ERS over slightly different areas of the brain. Hence, for most subjects, a classifier that is customized to each user will perform better than a generic one.

3.1 Induced Oscillations in Non-attended Cortical Body Part Areas

Brain activation patterns are very similar during movement execution (ME) and imagination. This is not surprising because we have clear evidence that real and imagined limb movements involve partially overlapping neuronal networks and cortical structures, respectively. This is confirmed by a number of functional magnetic resonance imaging (fMRI) studies (e. g., [6, 11, 26]). For example, covert and overt hand movement results in a hand area ERD and simultaneously in a foot area ERS and foot movement results in an opposite pattern (Fig. 2). This antagonistic behaviour is an example for "focal ERD/surround ERS" (details see Chapter 3 in this book and [44]). When attention is focused to one cortical body part area (e. g., foot), attention is withdrawn from neighbouring cortical areas (e. g., hand). This process can be accompanied in the former case by a "focused ERD" and in the latter case by a "surround ERS".

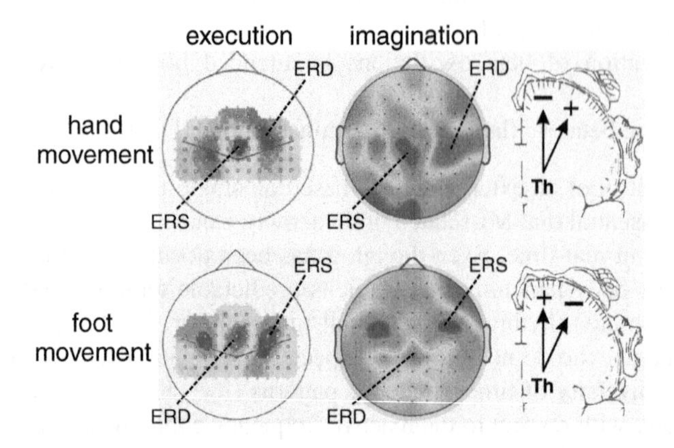

Fig. 2 Examples of topographic maps displaying simultaneous ERD and ERS patterns during execution and imagination of foot and hand movement, respectively. Note the similarity of patterns during motor execution and motor imagery

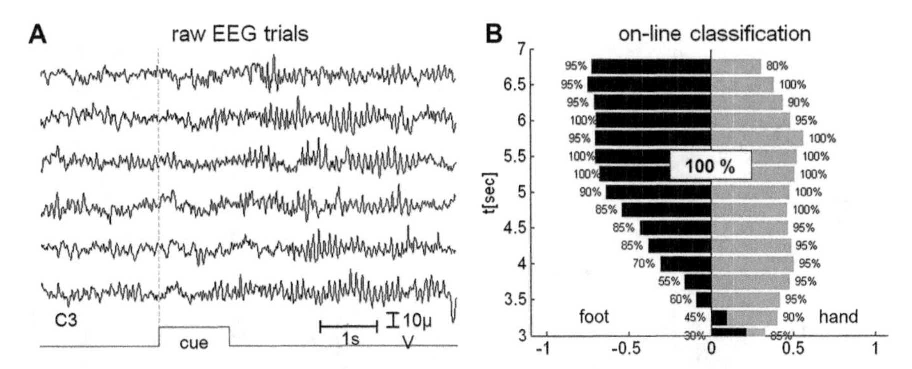

Fig. 3 (**a**) Examples of cue-based raw EEG trials with induced mu and beta oscillations during foot MI recorded at electrode C3. (**b**) On-line classification result for discrimination between foot and right hand MI. Note the 100 % classification accuracy due to the induced mu and beta oscillations

Induced oscillations in non-attended body-part areas are necessary to achieve a classification accuracy close to 100 %. An example for this is given in Fig. 3. In this case hand versus foot MI was used for cue-based control of walking in a virtual street. From 3 bipolar EEG channels placed over hand and foot area representation areas band power was calculated and used for control. In individual runs (40 cues for hand and 40 for foot MI) a classification accuracy of 100 % was achieved (Fig. 3b). The reason for this optimal performance was an increase in mu and beta power (induced beta oscillations) in the non-attended body-part area. In this case the beta and mu power increased in the left hand representation area (electrode C3) during foot MI (Fig. 3a). Details of the experimental paradigm are found in [42].

3.2 Induced Beta Oscillations in Attended Cortical Body Part Areas

Here we report on a tetraplegic patient, born 1975. He has a complete motor and sensory paralysis below C7 and an incomplete lesion below C5 and is only able to move his left arm. During intensive biofeedback training using the Graz BCI, he learned to induce beta bursts with a dominant frequency of 17 Hz by imagining moving his feet [39, 46]. These mentally induced beta oscillations or bursts (beta ERS) are very stable phenomena, limited to a relatively narrow frequency band (15–19 Hz), and focused at electrodes overl the supplementary motor and foot representation areas.

An important question is whether fMRI demonstrates activation in the foot representation and related areas of a tetraplegic patient. Therfore, imaging was carried out on a 3.0 T MRI system. A high resolution T-weighted structural image was acquired to allow functional image registration for precise localisation of activations. For more details, see [7].

The subject was instructed to imagine smooth flexion and extension of the right (left) foot paced by a visual cue at 1 Hz. Imagery of movement of either side alternated with interspersed periods of absolute rest. Each block of imagined foot

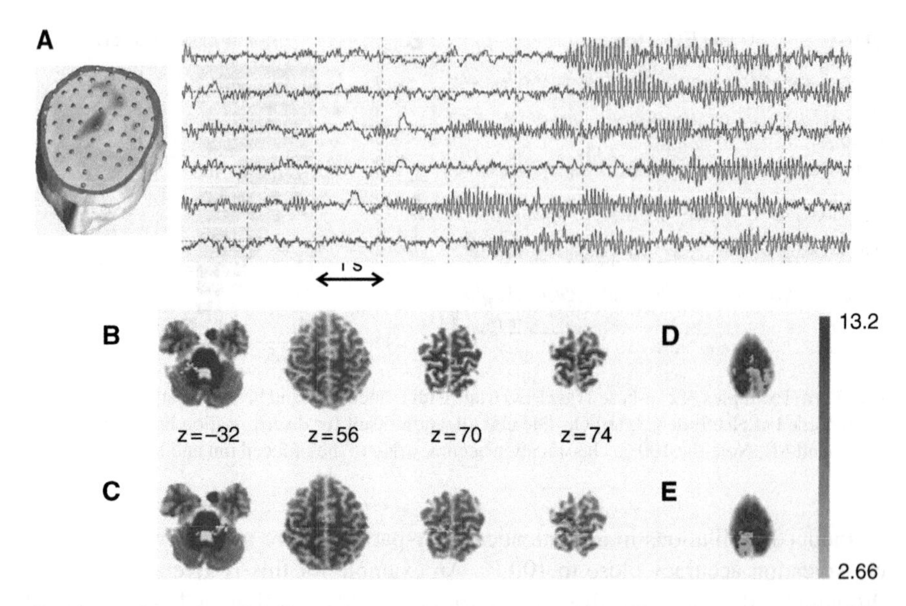

Fig. 4 (**a**) Examples of induced beta oscillations during foot MI at electrode position Cz. Activation patterns through the blood oxygenation level-dependent BOLD signal during right (**b**, **d**) and left foot movement imagination (**c**, **e**) in the tetraplegic patient. Strong activation was found in the contralateral primary motor foot area. Right side on the image corresponds to the left hemisphere. *Z*-coordinates correspond to Talairach space [64]

movement was 30 s long, with five blocks of each type of foot MI. Significant activation with imagery versus rest was detected in the primary foot area contralateral to imagery, premotor and supplementary motor areas, and in the cerebellum ipsilaterally (Fig. 4). This strong activation is at first glance surprising, but may be explained by the long-lasting MI practice and the vividness of MI by the tetraplegic patient.

3.3 The Beta Rebound (ERS) and its Importance for BCI

Of importance for the BCI community is that the beta rebound is not only characteristic for the termination of movement execution but also of MI (for further details see Chapter 3 in this book). Self-paced finger movements induce such a beta rebound not only in the contralateral hand representation area but also with slightly higher frequencies and an earlier onset in midcentral areas overlaying the supplementary motor area (SMA) [52]. This midcentrally induced beta rebound is especially dominant following voluntary foot movement [33] and foot MI [49]. We speculate, therefore, that the termination of motor cortex activation after execution or imagination of a body part movement may involve at least two neural networks, one in the primary motor area and another in the SMA. Foot movement may involve

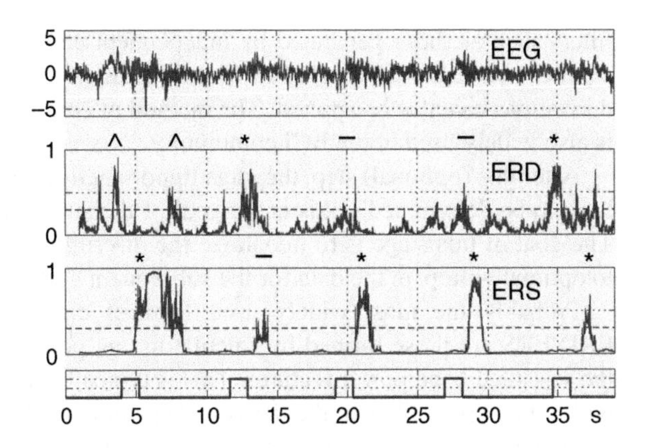

Fig. 5 Spontaneous EEG, classification time courses for ERD and ERS and timing of the cue presentations during brisk foot MI (from *top* to *bottom*). TPs are marked with an asterisk, FPs with a triangle and FNs with a dash. Modified from [53]

both the SMA and the cortical foot representation areas. Considering the close proximity of these cortical areas [16], and the fact that the response of the corresponding networks in both areas may be synchronized, it is likely that a large-amplitude beta rebound (beta ERS) occurs after foot MI. This beta rebound displays a high signal-to-noise ratio and is therefore especially suitable for detection and classification in single EEG trials.

A group of able-bodied subjects performed cue-based brisk foot ME in intervals of approximately 10 s. Detection and classification of the post-movement beta ERS in unseen, single-trial, one-channel EEG signals recorded at Cz (Laplacian derivation; compared to rest) revealed a classification accuracy of 74 % true positives (TPs) and false positives (FPs) of 6 % [62]. This is a surprising result, because the classification of the ERD during movement in the same data revealed a TP rate of only 21 %.

Due to the similarity of neural structures involved in motor execution and in MI, the beta rebound in both motor tasks displays similar features with slightly weaker amplitudes in the imagery task [32]. We speculate, therefore, that a classifier set up and trained with data obtained from an experiment with either cue-based foot movement executions or motor imagery and applied to unseen EEG data obtained from a foot MI task is suitable for an asynchronously operating "brain switch". First results of such an experiment are displayed in Fig. 5. The classification rate of asynchronously performed post-imagery ERS classification (simulation of an asynchronous BCI) was for TP 92.2 and FP 6.3 %.

4 Feature Extraction and Selection

In a BCI, the raw EEG signals recorded from the scalp can be preprocessed in several ways to improve signal quality. Typical preprocessing steps involve filtering in the frequency domain, reducing or eliminating artifacts such as EOG removal or EMG detection, or the generation of new signal mixtures by suitable spatial filters such as those generated by independent component analysis (ICA) or principal component analysis (PCA). Simple spatial filters such as bipolar or Laplacian derivations can also be applied. The method of common spatial patterns (CSP) [57] is also widely used in the BCI community.

After this (optional) step, the most important descriptive properties of the signals have to be determined – this is the goal of the subsequent feature extraction stage. The goal of this stage is to maximize the discriminative information and therefore to optimally prepare the data for the subsequent classification step.

Probably the most widely used features closely related to the concept of ERD/ERS are those formed by calculating the power in specific frequency bands, the so-called band power features. This is typically achieved by filtering the signals with a band pass filter in the desired frequency band, squaring of the samples and averaging over a time window to smooth the result. As a last step, the logarithm is calculated in order to transform the distribution of this feature to a more Gaussian-like shape, because many classifiers such as Fisher's linear discriminant analysis (LDA) in the next stage assume normally distributed features. Band power features have been extensively used in the Graz BCI system. The initial band pass filter is typically implemented with an infinite impulse response filter in order to minimize the time lag between the input and output. The averaging block usually calculates the mean power within the last second, which is most often implemented with a moving average finite impulse response filter.

Other popular features also used by the Graz BCI are parameters derived from an autoregressive (AR) model. This statistical model describes a time series (corresponding to an EEG signal in the case of a BCI) by using past observations in order to predict the current value. These AR parameters were found to provide suitable features to describe EEG signals for subsequent classification in a BCI system and can also be estimated in an adaptive way [61].

Both feature types mentioned above neglect the relationships between single EEG channels. Such relations have proved to be useful in analyzing various neurophysiological problems conducted in numerous studies (see [27, 34, 56, 63, 66] for example) and could provide important information for BCIs. A specific coupling feature, namely the phase synchronization value (or phase locking value), has already been implemented and used in the Graz BCI system. It measures the stability of the phase difference between two signals stemming from two electrodes and could be a suitable measure to assess long-range synchrony in the brain [20]. In contrast to classical coherence, the signal amplitudes do not influence the result, and thus this measure is thought to reflect true brain synchrony [66].

After the feature extraction stage, an important step in the signal processing chain is determining which features contribute most to a good separation of the different classes. An optimal way to find such an optimal feature set is to search in all possible combinations, known as exhaustive search. However, since this method is too time-consuming for even a small number of features, it cannot be used in practical applications and therefore, suboptimal methods have to be applied.

A popular and particularly fast algorithm that yields good results is the so-called sequential floating forward selection (SFFS) [55]. Another extremely fast feature selection algorithm frequently used in feature selection tasks in the Graz BCI is the so-called distinction sensitive learning vector quantization (DSLVQ) [54], an extension of the learning vector quantization method [19]. In contrast to many classical feature selection methods such as SFFS or genetic algorithms, DSLVQ does not evaluate the fitness of many candidate feature subsets.

5 Frequency Band and Electrode Selection

One aim of the Graz group is to create small, robust and inexpensive systems. The crucial issue in this context is the number of electrodes. Generally, wet EEG electrodes are used in BCI research. Wet sensors, however, require scalp preparation and the use of electrolytic gels, which slows down the sensor placement process. There is also the fact that EEG sensors need to be re-applied frequently. For practical issues it is consequently advantageous to minimize the number of EEG sensors.

We studied the correlation between the number of EEG sensors and the achievable classification accuracies. Thirty mastoid-referenced EEG channels were recorded from 10 naive volunteers during cue-based left hand (LMI), right hand (RMI) and foot MI (FMI). A running classifier with a sliding 1-s window was used to calculate the classification accuracies [30]. An LDA classifier was used each time to discriminate between two out of the three MI tasks. For reference, classification accuracies were computed by applying the common spatial pattern (CSP) method [13, 29, 57]. For comparison, LDAs were trained by individual band power estimates extracted from single spatially filtered EEG channels. Bipolar derivations were computed by subtracting the signals of two electrodes, and we derived orthogonal source (Laplacian) derivations by subtracting the averaged signal of the four nearest-neighboring electrodes from the electrode of interest [14]. Band power features were calculated over the 1-s segment. Finally, the DSLVQ method was used to identify the most important individual spectral components. For a more detailed description see [58].

The results are summarized in Table 1. As expected, CSP achieved the best overall performance because the computed values are based on the 30-channel EEG data. Laplacian filters performed second best and bipolar derivations performed worst. Of interest was that LMI versus RMI performed slightly worse than LMI versus FMI and RMI versus FMI. Although the statistical analyses showed no significant

Table 1 Mean (median)±SD (standard deviation) LDA classification accuracies. The values for Laplacian and bipolar derivations are based on single most discriminative band power feature; CSP on the 2 most important spatial patterns (4 features) (modified from [58])

Spatial filter	LMI vs. RMI	LMI vs. FMI	RMI vs. FMI
Bipolar	68.4 (67.8) ± 6.6 %	73.6 (74.6) ± 9.2 %	73.5 (74.9) ± 10.4 %
Laplacian	72.3 (73.5) ± 11.7 %	80.4 (83.2) ± 9.7 %	81.4 (82.8) ± 8.7 %
CSP	82.6 (82.6) ± 10.4 %	87.0 (87.1) ± 7.6 %	88.8 (87.7) ± 5.5 %

differences, only a trend, these findings, consistent over all types of spatial filters, suggest that foot MI in combination with hand MI is a good choice when working with naive BCI users the first time. Frequency components between 10–14 Hz (upper mu components) are induced frequently in the hand representation area (mu ERS) during foot MI and proved to be very important to achieve high classification accuracies [37].

6 Special Applications of the Graz BCI

Three applications are reported. The first two applications involve control of immersive virtual environments, and the third application let users operate Google Earth through thought.

6.1 Self-Paced Exploration of the Austrian National Library

An interesting question is whether it is possible to navigate through a complex Virtual Environment (VE) without using any muscle activity, such as speech and limb movement. Here we report on an experiment in the Graz DAVE (Definitely Affordable Virtual Environment [15]). The goal of this experiment was to walk (self-paced) through a model of the Austrian National Library (see Fig. 6a) presented in the DAVE with three rear-projected active stereo screens and a front-projected screen on the floor.

Subjects started with a cue-based BCI training with 2 MI classes. During this training they learned to establish two different brain patterns by imagining hand or foot movements (for training details see [24, 31]). After offline LDA output analysis, the MI which was not preferred (biased) by the LDA was selected for self-paced training. Each time the LDA output exceeded a selected threshold for a predefined dwell time [65], the BCI replied to the DAVE request with a relative coordinate change. Together with the current position of the subject within the VE and the tracking information of the subject's head (physical movements), the new position within the VE was calculated. The whole procedure resulted in a smart forward movement through the virtual library whenever the BCI detected the specified MI. For further details see [25].

The task of the subject within the VE was to move through MI towards the end of the main hall of the Austrian National Library along a predefined pathway. The subject started at the entrance door and had to stop at five specific points. The experiment was divided in 5 activity times (movement through thought) and 5 pause times (no movement). After a variable pause time of approximately 1 min, the experimenter gave a command ("experimenter-based") and subject started to move as fast as possible towards the next point. From the 5 activity and 5 pause times, the true positives (TP, correct moving) and false negatives (FN, periods of no movement

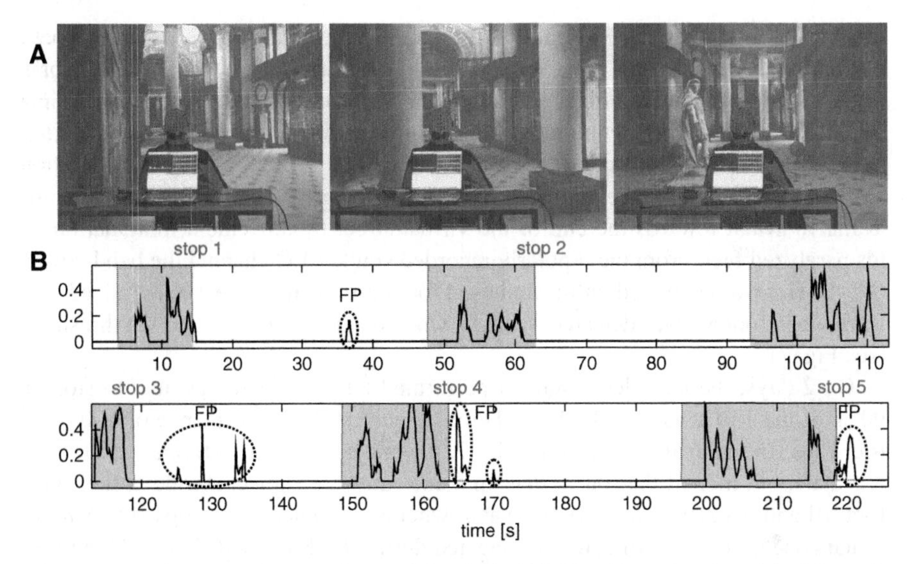

Fig. 6 (**a**) Participant with electrode cap sitting in the DAVE inside a virtual model of the main hall of the Austrian National Library. (**b**) Online performance of the first run of one subject. The rectangles mark the periods when the subject should move to the next point. The thick line is the actual BCI output after the post-processing. Whenever the line is not zero, a movement occurred. Periods of FP are indicated with dotted circles (modified from [21])

during activity time), as well as false positives (FP, movements during the pause time) and true negatives (TN, correct stopping during pause time) were identified. The online performance of the first run of one subject is given in detail in Fig. 6b with a TP ratio of 50.1 % and a FP ratio of 5.8 %.

This experiment demonstrated a successful application of the Graz BCI in a self-paced (asynchronous processing) moving experiment with a small number of EEG channels. Optimizing the threshold and dwell time to distinguish between intentional and non-intentional brain states remains a challenge, and would improve TP and FP ratios.

6.2 Simulation of Self-Paced Wheel Chair Movement in a Virtual Environment

Virtual reality (VR) provides an excellent training and testing environment for rehearsal of scenarios or events that are otherwise too dangerous or costly – or even currently impossible in physical reality. An example is wheelchair control through a BCI [28]. In this case, the output signal of the Graz BCI is used to control not the wheel chair movement itself but the movement of the immersive virtual environment in the form of a street with shops and virtual characters (avatars), and simulates therewith the real wheel chair movement as realistically as possible.

In this experiment, the 33-year-old tetraplegic patient was trained to induce beta oscillation during foot MI (see Section 3.2). The participant sat in his wheelchair in the middle of a multi-projection based stereo and head-tracked VR system commonly known as CAVE (computer animated virtual environment, [4]), in a virtual street with shops on both sides and populated with 15 avatars, which were lined up along the street [10, 23] (see Fig. 7). The task of the participant was to "move" from avatar to avatar towards the end of the virtual street by movement imagination of his paralyzed feet. From the bipolarly recorded single EEG channel the band power (15–19 Hz) was estimated online and used for control. The subject only moved forward when foot MI was detected – that is, when band power exceeded the threshold (see Fig. 7).

On 2 days, the tetraplegic subject performed ten runs, and was able to stop at 90 % of the 150 avatars and talk to them. He achieved a performance of 100 % in four runs. In the example given in Fig. 7, the subject correctly stopped at the first 3 avatars, but missed the fourth one. In some runs, the subject started earlier with foot MI and walked straight to the avatar, whereby in other runs, stops between the avatars occurred. Foot MI could be detected during 18.2 % ± 6.4 % of the run time. The averaged duration of MI periods was 1.6 s ± 1.1 s (see [22]).

This was the first work that showed that a tetraplegic subject, sitting in a wheelchair, could control his movements in a virtual environment using a self-paced

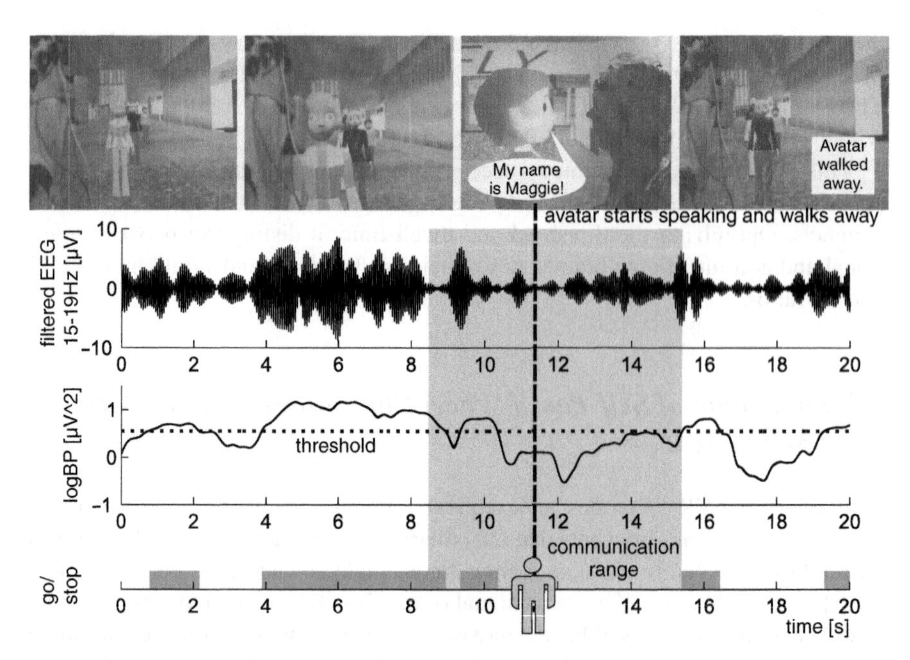

Fig. 7 Picture sequence before, during and after the contact with an avatar (*upper panel*). Example of a band-pass filtered (15–19 Hz) EEG, logarithmic band power time course with threshold (Th) and go/stop signal used for VE control (*lower panels*) (modified from [21])

(asynchronous) BCI based on one single EEG channel and one beta band. The use of a visual-rich VE with talking avatars ensured that the experiment was diversified and engaging but contained ample distractions, as would occur in a real street. Controlling a VE (e. g., the virtual wheelchair) is the closest possible scenario for controlling the real wheelchair in a real street, so virtual reality allows patients to perform movements in a safe environment. Hence, the next step of transferring the BCI from laboratory conditions to real-world control is now possible.

6.3 Control of Google Earth

Google Earth (Google Inc., Mountain View, CA, USA) is a popular virtual globe program that allows easy access to the world's geographic information. In contrast to the previous applications, the range of functions needed to comfortably operate the software is much higher. Common, however, is that a self-paced operational protocol is needed to operate the application. Consequently, two different issues had to be solved. First, a BCI operated according to the self-paced operational paradigm was needed that could detect most different MI patterns, and second a useful translation of the brain patterns into control commands for Google Earth had to be found. We solved the problem by developing a self-paced 3-class BCI and a very intuitive graphical user interface (GUI).

The BCI consisted of two independent classifiers. The first classifier (CFR1), consisting of three pair-wise trained LDAs with majority voting, was trained to discriminate between left hand, right hand and foot/tongue MI in a series of cue-based feedback experiments. After online accuracy exceeded 75 % (chance level around 33 %), a second classifier CFR2, a subject-specific LDA, was trained to detect any of those 3 MI patterns in the ongoing EEG. The output of the BCI was computed such that each time CFR2 detected MI in the ongoing EEG, the class identified by CFR1 was indicated at the output. Otherwise the output was "0". For more details see [60] and the Aksioma homepage at http://www.aksioma.org/brainloop.

The BCI and the GUI were combined by means of the user data protocol (UDP). Here, left hand, right hand and foot MI were used to move the cursor to the left, to the right or downwards, respectively. For selection, the user had to repeatedly imagine the appropriate movement until the desired item was picked. The feedback cursor disappeared during periods of no intentional control.

Figure 8 shows pictures of the experimental setup taken during the Wired Nextfest 2007 festival in Los Angeles, CA, USA. The picture shows the user wearing an electrode cap, the Graz BCI, consisting of a laptop and a biosignal amplifier, the GUI interface and Google Earth. Despite considerable electrical noise from the many other devices at this event, as well as distractions from the audience and other exhibits, the user succeeded in operating Google Earth. Since it is difficult to estimate the performance of a self-paced system without knowing the user's intention, the user was interviewed. He stated that, most of the time, the BCI correctly detected the intended MI patterns as well as the non-control state. The total amount of time

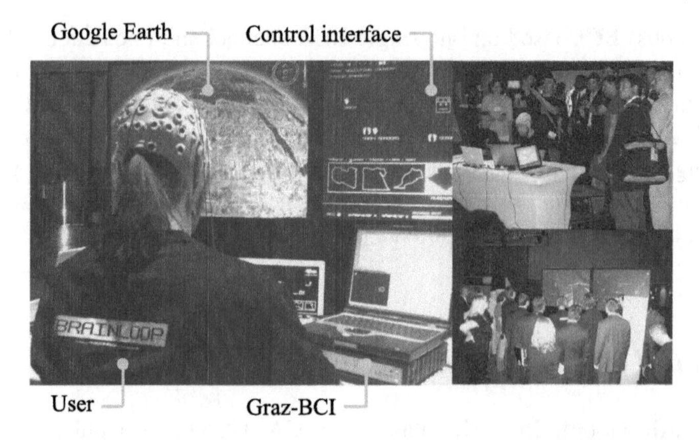

Fig. 8 Photographs of the experimental setup and demonstration at the Wired Nextfest 2007

needed to undergo cue-based training (CFR1) and gain satisfying self-paced control (CFR1 and CFR2) was about 6 h.

7 Future Aspects

There are a number of concepts and/or ideas to improve BCI performance or to use other input signals. Two of these concepts will be discussed shortly. One is to incorporate the heart rate changes in the BCI, the other is to realize an optical BCI.

A hybrid BCI is a system that combines and processes signals from the brain (e. g., EEG, MEG, BOLD, hemoglobin changes) with other signals [70, 71]. The heart rate (HR) is an example of such a signal. It is easy to record and analyze and it is strongly modulated by brain activity. It is well known that the HR is not only modified by brisk respiration [59], but also by preparation of a self-paced movement and by MI [8, 43]. This mentally induced HR change can precede the transient change of EEG activity and can be detectable in single trials as shown recently in [41].

Most BCIs measure brain activity via electrical means. However, metabolic parameters can also be used as input signals for a BCI. Such metabolic parameters are either the BOLD signal obtained with fMRI or the (de)oxyhemoglobin signal obtained with the near-infrared spectroscopy (NIRS). With both signals a BCI has been already realized [67, 3]. The advantage of the NIRS method is that a low-cost BCI can be realized suitable for home application.

First results obtained with one- and multichannel NIRS system indicate that "mental arithmetic" gives a very stable (de)oxyhemoglobin pattern [1, 72] (Fig. 9). At this time, the classification of single patterns still exhibits many false positives. One reason for this is the interference between the hemodynamic response and the spontaneous blood pressure waves of third order known as Mayer-Traube-Hering (MTH) waves [5]. Both phenomena have a duration of approximately 10 s and are therefore in the same frequency range.

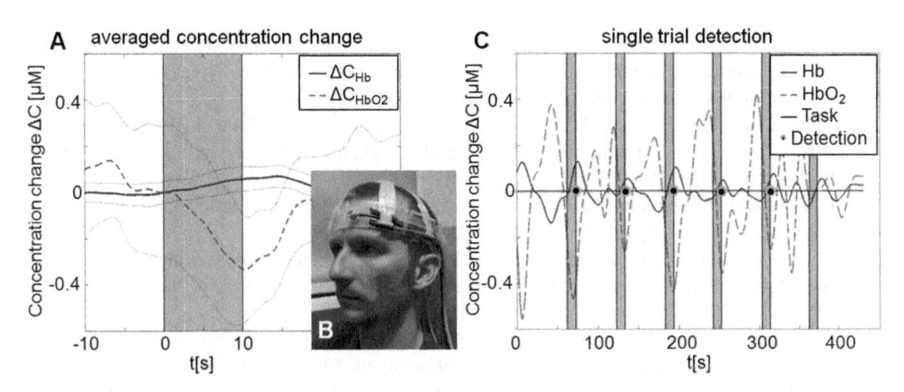

Fig. 9 (**a**) Mean hemodynamic response (mean ± standard deviation) during 42 mental arithmetic tasks. (**b**) Picture of the developed optode setting. (**c**) Single trial detection in one run. The shaded vertical bars indicate the time windows of mental activity

Summarizing, it can be stated that both hybrid and optical BCIs are new approaches in the field of BCIs, but need extensive basic and experimental research.

Acknowledgments The research was supported in part by the EU project PRESENCCIA (IST-2006-27731), EU cost action B27, Wings for Life and the Lorenz-Böhler Foundation. The authors would like express their gratitude to Dr. C. Enzinger for conducting the fMRI recordings and Dr. G. Townsend and Dr. B. Allison for proofreading the manuscript.

References

1. G. Bauernfeind, R. Leeb, S. Wriessnegger, and G. Pfurtscheller, Development, set-up and first results of a one-channel near-infrared spectroscopy system. Biomed Tech, 53, 36–43, (2008).
2. B. Blankertz, G. Dornhege, M. Krauledat, K.-R. Müller, and G. Curio, The non-invasive Berlin brain-computer interface: fast acquisition of effective performance in untrained subjects. NeuroImage, 37, 539–550, (2007).
3. S. Coyle, T. Ward, C. Markham, and G. McDarby, On the suitability of near-infrared (NIR) systems for next-generation brain-computer interfaces. Physiol Mea, 25, 815–822, (2004).
4. C. Cruz-Neira, D.J. Sandin, and T.A. DeFanti, Surround-screen projection-based virtual reality: the design and implementation of the CAVE. Proceedings of the 20th Annual Conference on Computer Graphics and Interactive Techniques, Anaheim, CA, USA, (1993).
5. R.W. de Boer, J.M. Karemaker, and J. Strackee, On the spectral analysis of blood pressure variability. Am J Physiol Heart Circ Physiol, 251, H685–H687, (1986).
6. H.H. Ehrsson, S. Geyer, and E. Naito, Imagery of voluntary movement of fingers, toes, and tongue activates corresponding body-part-specific motor representations. J Neurophysiol, 90, 3304–3316, (2003).
7. C. Enzinger, S. Ropele, F. Fazekas, M. Loitfelder, F. Gorani, T. Seifert, G. Reiter, C. Neuper, G. Pfurtscheller, and G. Müller-Putz, Brain motor system function in a patient with complete spinal cord injury following extensive brain-computer interface training. Exp Brain Res, 190, 215–223, (2008).
8. G. Florian, A. Stancák, and G. Pfurtscheller, Cardiac response induced by voluntary self-paced finger movement. Int J Psychophysiol, 28, 273–283, (1998).
9. D. Flotzinger, G. Pfurtscheller, C. Neuper, J. Berger, and W. Mohl, Classification of non-averaged EEG data by learning vector quantisation and the influence of signal preprocessing. Med Biol Eng Comp, 32, 571–576, (1994).

10. D. Friedman, R. Leeb, A. Antley, M. Garau, C. Guger, C. Keinrath, A. Steed, G. Pfurtscheller, and M. Slater, Navigating virtual reality by thought: first steps. *Proceedings of the 7th Annual International Workshop PRESENCE*, Valencia, Spain, (2004).
11. E. Gerardin, A. Sirigu, S. Lehéricy, J.-B. Poline, B. Gaymard, C. Marsault, Y. Agid, and D. Le Bihan, Partially overlapping neural networks for real and imagined hand movements. Cereb Cortex, 10, 1093–1104, (2000).
12. B. Graimann, Movement-related patterns in ECoG and EEG: visualization and detection. PhD thesis, Graz University of Technology, (2002).
13. C. Guger, H. Ramoser, and G. Pfurtscheller, Real-time EEG analysis with subject-specific spatial patterns for a brain-computer interface (BCI). IEEE Trans Neural Sys Rehabil Eng, 8, 447–450, (2000).
14. B. Hjorth, An on-line transformation of EEG scalp potentials into orthogonal source derivations. Electroencephalogr Clin Neurophysiol, 39, 526–530, (1975).
15. A. Hopp, S. Havemann, and D. W. Fellner, A single chip DLP projector for stereoscopic images of high color quality and resolution. Proceedings of the 13th Eurographics Symposium on Virtual Environments, 10th Immersive Projection Technology Workshop, Weimar, Germany, pp. 21–26, (2007).
16. A. Ikeda, H.O. Lüders, R.C. Burgess, and H. Shibasaki, Movement-related potentials recorded from supplementary motor area and primary motor area – role of supplementary motor area in voluntary movements. *Brain*, 115, 1017–1043, (1992).
17. M. Jeannerod, Neural simulation of action: a unifying mechanism for motor cognition. *NeuroImage*, 14, S103–S109, (2001).
18. J. Kalcher, D. Flotzinger, C. Neuper, S. Gölly, and G. Pfurtscheller, Graz brain-computer interface II: towards communication between humans and computers based on online classification of three different EEG patterns. Med Biol Eng Comput, 34, 382–388 (1996).
19. T. Kohonen, The self-organizing map. Proc IEEE, 78, 1464–1480, (1990).
20. J.-P. Lachaux, E. Rodriguez, J. Martinerie, and F.J. Varela, Measuring phase synchrony in brain signals. Hum Brain Mapp, 8, 194–2008, (1999).
21. R. Leeb, Brain-computer communication: the motivation, aim, and impact of virtual feedback. PhD thesis, Graz University of Technology, (2008).
22. R. Leeb, D. Friedman, G.R. Müller-Putz, R. Scherer, M. Slater, and G. Pfurtscheller, Self-paced (asynchronous) BCI control of a wheelchair in virtual environments: a case study with a tetraplegics. Comput Intell Neurosc, 2007, 79642, (2007).
23. R. Leeb, C. Keinrath, D. Friedman, C. Guger, R. Scherer, C. Neuper, M. Garau, A. Antley, A. Steed, M. Slater, and G. Pfurtscheller, Walking by thinking: the brainwaves are crucial, not the muscles! Presence: Teleoperators Virtual Environ, 15, 500–514, (2006).
24. R. Leeb, F. Lee, C. Keinrath, R. Scherer, H. Bischof, and G. Pfurtscheller, Brain-computer communication: motivation, aim and impact of exploring a virtual apartment. IEEE Trans Neural Sys Rehabil Eng, 15, 473–482, (2007).
25. R. Leeb, V. Settgast, D.W. Fellner, and G. Pfurtscheller, Self-paced exploring of the Austrian National Library through thoughts. Int J Bioelectromagn, 9, 237–244, (2007).
26. M. Lotze, P. Montoya, M. Erb, E. Hülsmann, H. Flor, U. Klose, N. Birbaumer, and W. Grodd, Activation of cortical and cerebellar motor areas during executed and imagined hand movements: an fMRI study. J Cogn Neurosc, 11, 491–501, (1999).
27. L. Melloni, C. Molina, M. Pena, D. Torres, W. Singer, and E. Rodriguez, Synchronization of neural activity across cortical areas correlates with conscious perception. J Neuros, 27, 2858–2865, (2007).
28. J. del R. Millán, F. Renkens, J. Mourino, and W. Gerstner, Noninvasive brain-actuated control of a mobile robot by human EEG. IEEE Trans Biomed Eng, 51, 1026–1033, (2004).
29. J. Müller-Gerking, G. Pfurtscheller, and H. Flyvbjerg, Designing optimal spatial filters for single-trial EEG classification in a movement task. Clin Neurophysiol, 110, 787–798, (1999).
30. J. Müller-Gerking, G. Pfurtscheller, and H. Flyvbjerg, Classification of movement-related EEG in a memorized delay task experiment. Clin Neurophysiol, 111, 1353–1365, (2000).

31. G.R. Müller-Putz, R. Scherer, and G. Pfurtscheller, Control of a two-axis artificial limb by means of a pulse width modulated brain switch. European Conference for the Advancement of Assistive Technology, San Sebastian, Spain, (2007).
32. G.R. Müller-Putz, D. Zimmermann, B. Graimann, K. Nestinger, G. Korisek, and G. Pfurtscheller, Event-related beta EEG-changes during passive and attempted foot movements in paraplegic patients. Brain Res, 1137, 84–91, (2006).
33. C. Neuper and G. Pfurtscheller, Evidence for distinct beta resonance frequencies in human EEG related to specific sensorimotor cortical areas. Clin Neurophysiol, 112, 2084–2097, (2001).
34. G. Nolte, O. Bai, L. Wheaton, Z. Mari, S. Vorbach, and M. Hallett, Identifying true brain interaction from EEG data using the imaginary part of coherency. Clin Neurophysiol, 115, 2292–2307, (2004).
35. G. Pfurtscheller, Event-related synchronization (ERS): an electrophysiological correlate of cortical areas at rest. Electroencephalogr Clin Neurophysiol, 83, 62–69, (1992).
36. G. Pfurtscheller and A. Aranibar, Evaluation of event-related desynchronization (ERD) preceding and following voluntary self-paced movements. Electroencephalogr Clin Neurophysiol, 46, 138–146, (1979).
37. G. Pfurtscheller, C. Brunner, A. Schlögl, and F. H. Lopes da Silva, Mu rhythm (de)synchronization and EEG single-trial classification of different motor imagery tasks. NeuroImage, 31, 153–159, (2006).
38. G. Pfurtscheller, D. Flotzinger, M. Pregenzer, J. R. Wolpaw, and D. J. McFarland, EEG-based brain computer interface (BCI) - search for optimal electrode positions and frequency components. Med Prog Technol, 21, 111–121, (1996).
39. G. Pfurtscheller, C. Guger, G. Müller, G. Krausz, and C. Neuper, Brain oscillations control hand orthosis in a tetraplegic. Neurosci Lett, 292, 211–214, (2000).
40. G. Pfurtscheller, J. Kalcher, C. Neuper, D. Flotzinger, and M. Pregenzer, On-line EEG classification during externally-paced hand movements using a neural network-based classifier. Electroencephalogr Clin Neurophysiol, 99, 416–425, (1996).
41. G. Pfurtscheller, R. Leeb, D. Friedman, and M. Slater, Centrally controlled heart rate changes during mental practice in immersive virtual environment: a case study with a tetraplegic. Int J Psychophysiol, 68, 1–5, (2008).
42. G. Pfurtscheller, R. Leeb, C. Keinrath, D. Friedman, C. Neuper, C. Guger, and M. Slater, Walking from thought. Brain Res, 1071, 145–152, (2006).
43. G. Pfurtscheller, R. Leeb, and M. Slater, Cardiac responses induced during thought-based control of a virtual environment. Int J Psychophysiol, 62, 134–140, (2006).
44. G. Pfurtscheller and F. H. Lopes da Silva, Event-related EEG/MEG synchronization and desynchronization: basic principles. Clin Neurophysiol, 110, 1842–1857, (1999).
45. G. Pfurtscheller and F.H. Lopes da Silva, Event-related desynchronization (ERD) and event-related synchronization (ERS). *Electroencephalography: basic principles, clinical applications and related fields*. Williams & Wilkins, pp. 1003–1016, (2005).
46. G. Pfurtscheller, G.R. Müller, J. Pfurtscheller, H.J. Gerner, and R. Rupp, "Thought"-control of functional electrical stimulation to restore handgrasp in a patient with tetraplegia. Neurosci Lett, 351, 33–36, (2003).
47. G. Pfurtscheller, G. R. Müller-Putz, A. Schlögl, B. Graimann, R. Scherer, R. Leeb, C. Brunner, C. Keinrath, F. Lee, G. Townsend, C. Vidaurre, and C. Neuper, 15 years of BCI research at Graz University of Technology: current projects. IEEE Trans Neural Sys Rehabil Eng, 14, 205–210, (2006).
48. G. Pfurtscheller and C. Neuper, Motor imagery and direct brain-computer communication. Proc IEEE, 89, 1123–1134, (2001).
49. G. Pfurtscheller, C. Neuper, C. Brunner, and F.H. Lopes da Silva, Beta rebound after different types of motor imagery in man. Neurosc Lett, 378, 156–159, (2005).
50. G. Pfurtscheller, C. Neuper, D. Flotzinger, and M. Pregenzer, EEG-based discrimination between imagination of right and left hand movement. Electroencephalogr Clin Neurophysiol, 103, 642–651, (1997).

51. G. Pfurtscheller, C. Neuper, C. Guger, W. Harkam, H. Ramoser, A. Schlögl, B. Obermaier, and M. Pregenzer, Current trends in Graz brain-computer interface (BCI) research. IEEE Transa Rehabil Eng, 8, 216–219, (2000).
52. G. Pfurtscheller, M. Wörtz, G. Supp, and F.H. Lopes da Silva, Early onset of post-movement beta electroencephalogram synchronization in the supplementary motor area during self-paced finger movement in man. Neurosci Lett, 339, 111–114, (2003).
53. G. Pfurtscheller and T. Solis-Escalante, Could the beta rebound in the EEG be suitable to realize a "brain switch"? Clin Neurophysiol, 120, 24–29, (2009).
54. M. Pregenzer, G. Pfurtscheller, and D. Flotzinger, Automated feature selection with a distinction sensitive learning vector quantizer. Neurocomput, 11, 19–29, (1996).
55. P. Pudil, J. Novovičova, and J. Kittler, Floating search methods in feature selection. Pattern Recognit. Lett., 15, 1119–1125, (1994).
56. M. Le Van Quyen, Disentangling the dynamic core: a research program for a neurodynamics at the large-scale. Biol Res, 36, 67–88, (2003).
57. H. Ramoser, J. Müller-Gerking, and G. Pfurtscheller, Optimal spatial filtering of single trial EEG during imagined hand movement. IEEE Trans Rehabil Eng, 8, 441–446, (2000).
58. R. Scherer, Towards practical brain-computer interfaces: self-paced operation and reduction of the number of EEG sensors. PhD thesis, Graz University of Technology, (2008).
59. R. Scherer, G.R. Müller-Putz, and G. Pfurtscheller, Self-initiation of EEG-based brain-computer communication using the heart rate response. J Neural Eng, 4, L23–L29, (2007).
60. R. Scherer, A. Schlögl, F. Lee, H. Bischof, J. Janša, and G. Pfurtscheller, The self-paced Graz brain-computer interface: methods and applications. Comput Intell Neurosci, 2007, 79826, (2007).
61. A. Schlögl, D. Flotzinger, and G. Pfurtscheller, Adaptive autoregressive modeling used for single-trial EEG classification. Biomed Tech, 42, 162–167, (1997).
62. T. Solis-Escalante, G.R. Müller-Putz, and G. Pfurtscheller, Overt foot movement detection in one single Laplacian EEG derivation. J Neurosci Methods, 175(1), 148–153, (2008).
63. G. Supp, A. Schlögl, N. Trujillo-Barreto, M. M. Müller, and T. Gruber, Directed cortical information flow during human object recognition: analyzing induced EEG gamma-band responses in brain's source space. PLoS ONE, 2, e684, (2007).
64. J. Talairach and P. Tournoux, *Co-planar stereotaxic atlas of the human brain*. New York, NY Thieme, 1998.
65. G. Townsend, B. Graimann, and G. Pfurtscheller, Continuous EEG classification during motor imagery – simulation of an asynchronous BCI. IEEE Trans Neural Sys Rehabil Eng, 12, 258–265, (2004).
66. F.J. Varela, J.-P. Lachaux, E. Rodriguez, and J. Martinerie, The brainweb: phase synchronization and large-scale integration. Nat Rev Neurosci, 2, 229–239, (2001).
67. N. Weiskopf, K. Mathiak, S.W. Bock, F. Scharnowski, R. Veit, W. Grodd, R. Goebel, and N. Birbaumer, Principles of a brain-computer interface (BCI) based on real-time functional magnetic resonance imaging (fMRI). IEEE Trans Biomed Eng, 51, 966–970, (2004).
68. J.R. Wolpaw, N. Birbaumer, D.J. McFarland, G. Pfurtscheller, and T.M. Vaughan, Brain-computer interfaces for communication and controls. Clin Neurophysiol, 113, 767–791, (2002).
69. J.R. Wolpaw and D.J. McFarland, Multichannel EEG-based brain-computer communication. Electroencephalogr Clin Neurophysiol, 90, 444–449, (1994).
70. G. Pfurtscheller, B.Z. Allison, C. Brunner, G. Bauernfeind, T. Solis-Escalante, R. Scherer, T.O. Zander, G. Mueller-Putz, C. Neuper, and N. Birbaumer, The hybrid BCI. Front Neurosci, 4, 30, (2010).
71. G. Pfurtscheller, T. Solis Escalante, R. Ortner, P. Linortner, and G. Müller-Putz, Self-paced operation of an SSVEP-based orthosis with and without an imagery-based "brain switch": a feasibility study towards a hybrid BCI. IEEE Trans Neural Syst Rehabil Eng, 18(4), 409–414, (2010).
72. G. Pfurtscheller, G. Bauernfeind, S. Wriessnegger, and C. Neuper, Focal frontal (de)oxyhemoglobin responses during simple arithmetic. Int J Psychophysiol, 76, 186–192, (2010).

BCIs in the Laboratory and at Home: The Wadsworth Research Program

Eric W. Sellers, Dennis J. McFarland, Theresa M. Vaughan, and Jonathan R.Wolpaw

1 Introduction

Many people with severe motor disabilities lack the muscle control that would allow them to rely on conventional methods of augmentative communication and control. Numerous studies over the past two decades have indicated that scalp-recorded electroencephalographic (EEG) activity can be the basis for non-muscular communication and control systems, commonly called brain–computer interfaces (BCIs) [55]. EEG-based BCI systems measure specific features of EEG activity and translate these features into device commands. The most commonly used features are rhythms produced by the sensorimotor cortex [38, 55, 56, 59], slow cortical potentials [4, 5, 23], and the P300 event-related potential [12, 17, 46]. Systems based on sensorimotor rhythms or slow cortical potentials use oscillations or transient signals that are spontaneous in the sense that they are not dependent on specific sensory events. Systems based on the P300 response use transient signals in the EEG that are elicited by specific stimuli.

BCI system operation has been conceptualized in at least three ways. Blankertz et al. (e.g., [7]) view BCI development mainly as a problem of machine learning (see also chapter 7 in this book). In this view, it is assumed that the user produces a signal in a reliable and predictable fashion, and the particular signal is discovered by the machine learning algorithm. Birbaumer et al. (e.g., [6]) view BCI use as an operant conditioning task, in which the experimenter guides the user to produce the desired output by means of reinforcement (see also chapter 9 in this book). We (e.g., [27, 50, 57] see BCI operation as the continuing interaction of two adaptive controllers, the user and the BCI system, which adapt to each other. These three concepts of BCI operation are illustrated in Fig. 1. As indicated later in this review, mutual adaptation is critical to the success of our sensorimotor rhythm (SMR)-based BCI

E.W. Sellers (✉)
Department of Psychology, East Tennessee State University, Box 70649, Johnson City, TN 37641, USA; Laboratory of Neural Injury and Repair, Wadsworth Center, New York State Department of Health, Albany, NY 12201-0509, USA
e-mail: sellers@etsu.edu

B. Graimann et al. (eds.), *Brain–Computer Interfaces*, The Frontiers Collection,
DOI 10.1007/978-3-642-02091-9_6, © Springer-Verlag Berlin Heidelberg 2010

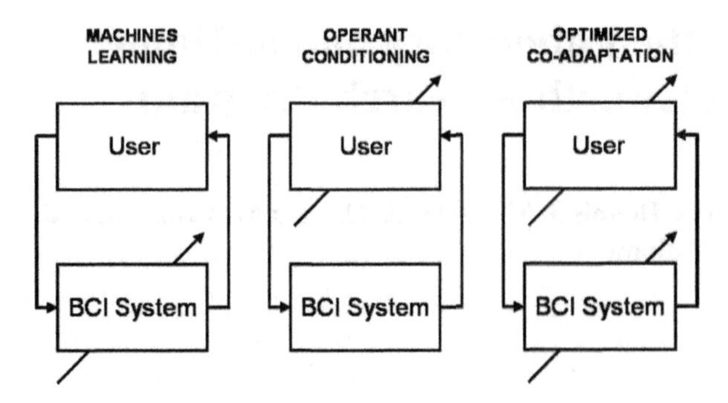

Fig. 1 Three concepts of BCI operation. The *arrows* through the user and/or the BCI system indicate which elements adapt in each concept

system. While the machine learning concept appears most applicable to P300-based BCIs, mutual adaptation is likely to play a role in this system as well, given that periodically updating the classification coefficients tends to improve classification accuracy. Others have applied the machine learning concept to SMR control using an adaptive strategy. In this approach, as the user adapts or changes strategy, the machine learning algorithm adapts accordingly.

At the Wadsworth Center, one of our primary goals is to develop a BCI that is suitable for everyday, independent use by people with severe disabilities at home or elsewhere. Toward that end, over the past 15 years, we have developed a BCI that allows people, including those who are severely disabled, to move a computer cursor in one, two, or three dimensions using mu and/or beta rhythms recorded over sensorimotor cortex. More recently, we have expanded our BCI system to be able to use the P300 response as originally described by Farwell and Donchin [17].

2 Sensorimotor Rhythm-Based Cursor Control

Sensorimotor rhythms (SMRs) are recorded over central regions of the scalp above the sensorimotor cortex. They are distinguished by their changes with movement and sensation. When the user is at rest there are rhythms that occur in the frequency ranges of 8–12 Hz (mu rhythms) and 18–26 Hz (beta rhythms). When the user moves a limb these rhythms are reduced in amplitude (i.e., desynchronized). These SMRs are thus considered to be idling rhythms of sensorimotor cortex that are desynchronized with activation of the motor system [39] and chapter 3 in this book). The changes in these rhythms with imagined movement are similar to the changes with actual movement [28]. Figure 2 shows a spectral analysis of the EEG recorded over the area representing the right hand during rest and during motor imagery, and the corresponding waveforms. It illustrates how the EEG is modulated in a narrow frequency band by movement imagery. Users can employ motor imagery as an initial strategy to control sensorimotor rhythm amplitude. Since different imagined movements produce different spatial patterns of desynchronization

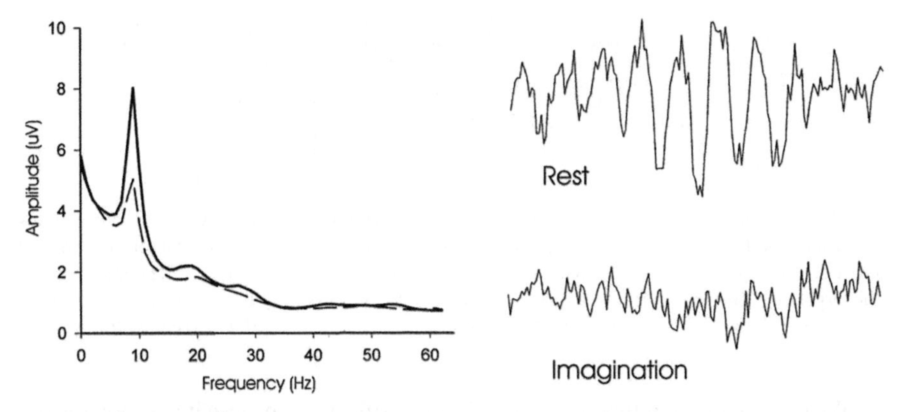

Fig. 2 Effects of motor imagery on sensorimotor rhythms. On the left are frequency spectra of EEG recorded over left sensorimtor cortex from an individual during rest (solid line) and during imagination of right-hand movement. Note that the prominent peak at 10 Hz during rest is attenuated with imagery. On the right are 1-sec. segments of EEG recorded during rest and during right-hand imagery from this same individual. The rest segment shows prominent 10-Hz activity

on the scalp, SMRs from different locations and/or different frequency bands can be combined to provide several independent channels for BCI use. With continued practice, this control tends to become automatic, as is the case with many motor skills [3, 13] and imagery becomes unnecessary.

Users learn over a series of training sessions to control SMR amplitudes in the mu (8–12 Hz) and/or beta (18–26 Hz) frequency bands over left and/or right sensorimotor cortex to move a cursor on a video screen in one, two, or three dimensions [31, 32, 34, 58]. This is not obviously a normal function of these brain signals, but rather the result of training. The SMR-based system uses spectral features extracted from the EEG that are spontaneous in the sense that the stimuli presented to the subject only provide the possible choices. The contingencies (i.e, the causative relationships between rhythm amplitudes and the commands that control cursor movements or other outputs) are arbitrary.

The SMR-based system relies on improvement in user performance through practice [31]. This approach views the user and system as two adaptive controllers that interact (e.g., [50, 57]). By this view, the user's goal is to modulate the EEG to encode commands in signal features that the BCI system can decode, and the BCI system's goal is to vest device control in those signal features that the user can most accurately modulate and to optimize the translation of the signals into device control. This optimization is presumed to facilitate further learning by the user.

Our first reports of SMR-based BCI used a single feature to control cursor movement in one dimension to hit targets on a video monitor [29, 59]. Subsequently, we used two channels of EEG to control cursor movement independently in two dimensions so that users could hit targets located anywhere on the periphery of the screen [56, 58].

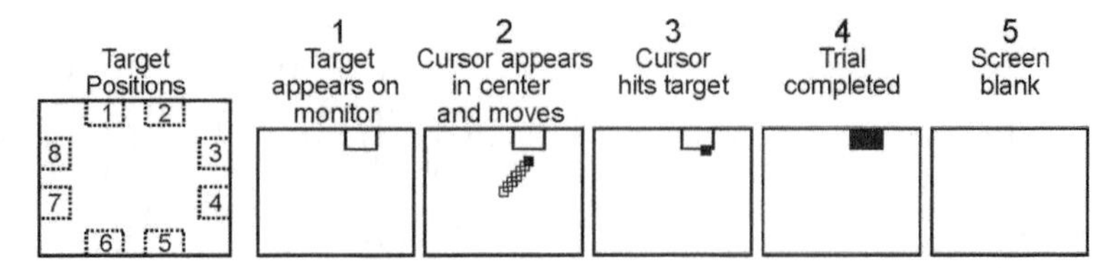

Fig. 3 Two-dimensional SMR control task with 8 possible target positions. Adapted from Wolpaw and McFarland [56]. (1) A target is presented on the screen for 1 s. (2) The cursor appears and moves steadily across the screen with its movement controlled by the user. (3) The cursor reaches the target. (4) The target flashes for 1.5 s when it is hit by the cursor. If the cursor misses the target, the screen is blank for 1.5 s. (5) The screen is blank for a 1-s interval prior to the next trial

We use a regression function rather than classification because it is simpler given multiple targets and generalizes more readily to different target configurations [33]. We use adaptive estimates of the coefficients in the regression functions. The cursor movement problem is modeled as one of minimizing the squared distance between the cursor and the target for a given dimension of control. For one-dimensional movement we use a single regression function. For two- or three-dimensional movement we use separate functions for each dimension of movement.

We found that a regression approach is well suited to SMR cursor movement control since it provides continuous control in one or more dimensions and generalizes well to novel target configurations. The utility of a regression model is illustrated in the recent study of SMR control of cursor movement in two dimensions by Wolpaw and McFarland [56]). A sample trial is shown in Fig. 3. Each trial began when a target appeared at one of eight locations on the periphery of the screen. Target location was block-randomized (i.e., each occurred once every eight trials). One second later, the cursor appeared in the middle of the screen and began to move in two dimensions with its movement controlled by the user's EEG activity. If the cursor reached the target within 10 s, the target flashed as a reward. If it failed to reach the target within 10 s, the cursor and the target simply disappeared. In either case, the screen was blank for one s, and then the next trial began. Users initially learned cursor control in one dimension (i.e., horizontal) based on a regression function. Next they were trained on a second dimension (i.e., vertical) using a different regression function. Finally the two functions were used simultaneously for full two-dimensional control. Topographies of Pearson's r correlation values (a common measure of the linear relationship between two variables) for one user are shown in Fig. 4. It is clear that two distinct patterns of activity controlled cursor movement. Horizontal movement was controlled by a weighted difference of 12-Hz mu rhythm activity between the left and right sensorimotor cortex (see Fig. 4, left topography). Vertical movement was controlled by a weighted sum of activity located over left and right sensorimotor cortex in the 24-Hz beta rhythm band (see Fig. 4, right topography). This study illustrated the generalizability of regression functions to varying target configurations.

This 2004 study also showed that users could move the cursor to novel locations with equal facility. These results showed that ordinary least-squares regression

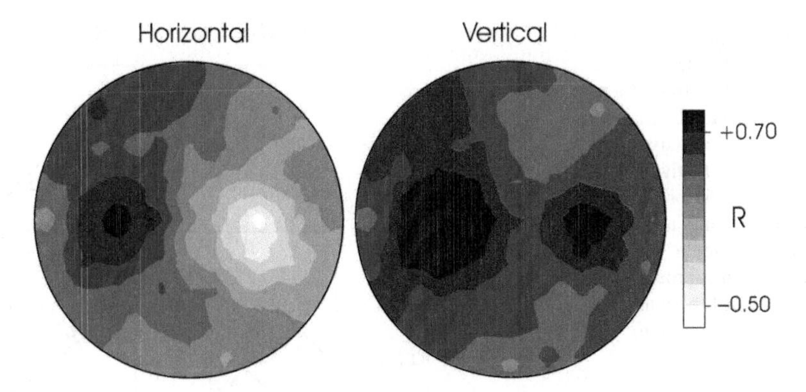

Fig. 4 Scalp topographies (nose at top) of Pearson's r values for horizontal (x) and vertical (y) target positions. In this user, horizontal movement was controlled by a 12-Hz mu rhythm and vertical movement by a 24-Hz beta rhythm. Horizontal correlation is greater on the right side of the head, whereas vertical correlation is greater on the left side. The topographies are for R rather than R^2 to show the opposite (i.e., positive and negative, respectively) correlations of right and left sides with horizontal target level. Adapted from Wolpaw and McFarland [56]

procedures provide efficient models that generalize to novel target configurations. Regression provides an efficient method to parameterize the translation algorithm in an adaptive manner. This method transfers smoothly to different target configurations during the course of multi-step training protocols. This study clearly demonstrated strong simultaneous independent control of horizontal and vertical movement. As documented in the paper [56], this control was comparable in accuracy and speed to that reported in studies using implanted intracortical electrodes in monkeys.

EEG-based BCIs have the advantage of being noninvasive. However, it has been assumed by many that they have a limited capacity for movement control. For example, Hochberg et al (2006) stated without supporting documentation that EEG-based BCIs are limited to 2-D control. In fact, we have recently demonstrated simultaneous EEG-based control of three dimensions of cursor movement [32]. The upper limits of the control possible with noninvasive recording are unknown at present.

We have also evaluated various regression models for controlling cursor movement in a four-choice, one-dimensional cursor movement task [33]. We found that using EEG features from more than one electrode location and more than one frequency band improved performance (e.g., C4 at 12 Hz and C3 at 24 Hz). In addition, we evaluated non-linear models with linear regression by including cross-product (i.e., interaction) terms in the regression function. While the translation algorithm could be based on a classifier or a regression function, we have found that a regression approach is better for the cursor movement task. Figure 5 compares the classification and regression approaches to selecting targets arranged along a single dimension. For the two-target case, both the regression approach and the classification approach require that the parameters of a single function be determined. For the five-target case, the regression approach still requires only a single

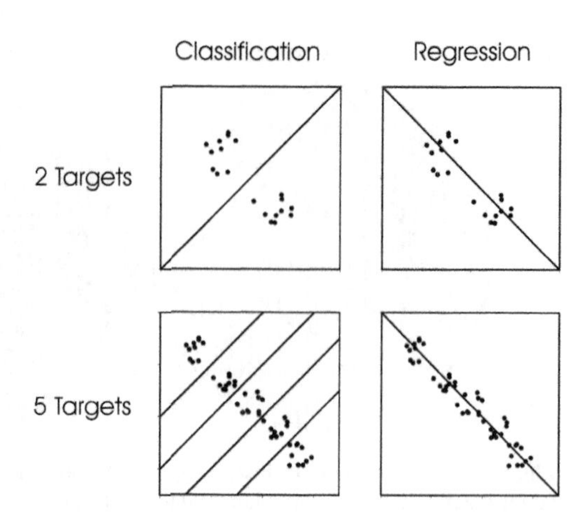

Fig. 5 Comparison of regression and classification for EEG feature translation. For the two-target case, both methods require only one function. For the five-target case, the regression approach still requires only a single function, while the classification approach requires four functions. (See text for full discussion)

function. In contrast, for the five-target case the classification approach requires that four functions be parameterized. With even more targets, and with variable targets, the advantage of the regression approach becomes increasingly apparent. For example, the positioning of icons in a typical mouse-based graphical user interface would require a bewildering array of classifying functions, while with the regression approach, two dimensions of cursor movement and a button selection can access multiple icons, however many there are and wherever they are.

We have conducted preliminary studies that suggest that users are also able to accurately control a robotic arm in two dimensions just as they control cursor movement (34]. In another recent study [32] we trained users on a task that emulated computer mouse control. Multiple targets were presented around the periphery of a computer screen, with one designated as the correct target. The user's task was to use EEG to move a cursor from the center of the screen to the correct target and then use an additional EEG feature to select the target. If the cursor reached an incorrect target, the user was instructed not to select it. Users were able to select or reject the target by performing or withholding hand-grasp imagery [33]. This imagery evokes a transient EEG response that can be detected. It can serve to improve overall accuracy by reducing unintended target selections. The results indicate that users could use brain signals for sequential multidimensional movement and selection. As these results illustrate, SMR-based BCI operation has the potential to be extended to a variety of applications, and the control obtained for one task can transfer directly to another task.

Our current efforts toward improving SMR-based BCI operation are focused on improving accuracy and reliability and on three-dimensional control [32]. This depends on identifying and translating EEG features so that the resulting control signals are as independent, trainable, stable, and predictable as possible. With control signals possessing these traits, the mutual user and system adaptations are effective, the required training time is reduced, and overall performance is improved.

3 P300-Based Item Selection

We have also been developing a BCI system based on the P300 event-related potential. Farwell and Donchin [17] first introduced the BCI paradigm in which the user is presented with a 6 × 6 matrix containing 36 symbols. The user focuses attention on a symbol he/she wishes to select while the rows and columns of the matrix are highlighted in a random sequence of flashes. Each time the desired symbol flashes, a P300 response occurs. To identify the desired symbol, the classifier (typically based on a stepwise linear discriminant analysis (SWLDA)) determines the row and the column to which the user is attending (i.e., the symbol that elicits a P300) by weighting and combining specific spatiotemporal features that are time-locked to the stimulus. The classifier determines the row and the column that produced the largest discriminant values and the intersection of this row and column defines the selected symbol. Figure 6 shows a 6 × 6 P300 matrix display and the average event-related potential responses to the flashing of each symbol. The letter O was the target symbol, and it elicited the largest P300 response. The other characters in the row and the column containing the O elicited a smaller P300 because these symbols flashed each time the O flashed.

The focus of our P300 laboratory studies has been on improving classification accuracy. We have examined variables related to stimulus properties and presentation rate [47], classification methods [21], and classification parameters [22]. Sellers et al. [47] examined the effects of inter-stimulus interval (ISI) and matrix size on classification accuracy using two ISIs (175- and 350-ms) , and two matrices (3 × 3

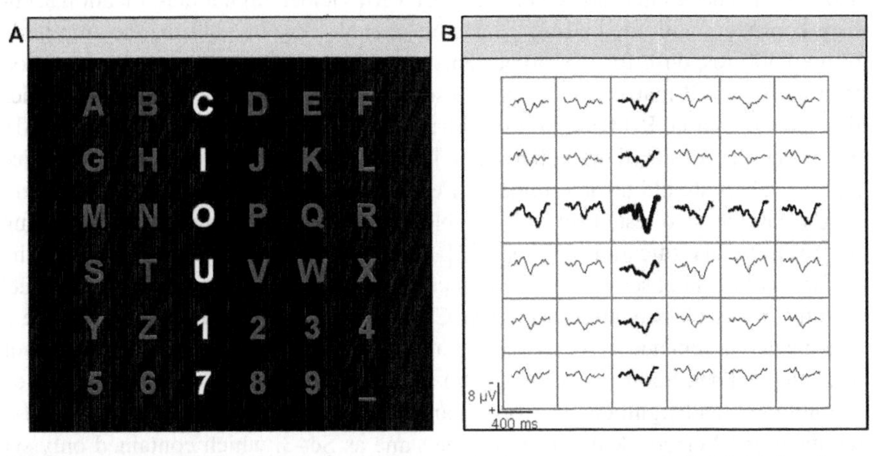

Fig. 6 (A) A 6 × 6 P300 matrix display. The rows and columns are randomly highlighted as shown for column 3. (B) Average waveforms at electrode Pz for each of the 36 symbols in the matrix. The target letter "O" (thick waveform) elicited the largest P300 response, and a smaller P300 response is evident for the other symbols in column 3 and row 3 (medium waveforms) because these stimuli are highlighted simultaneously with the target. All other responses are to non-target symbols (thin waveforms). Each response is the average of 30 stimulus presentations

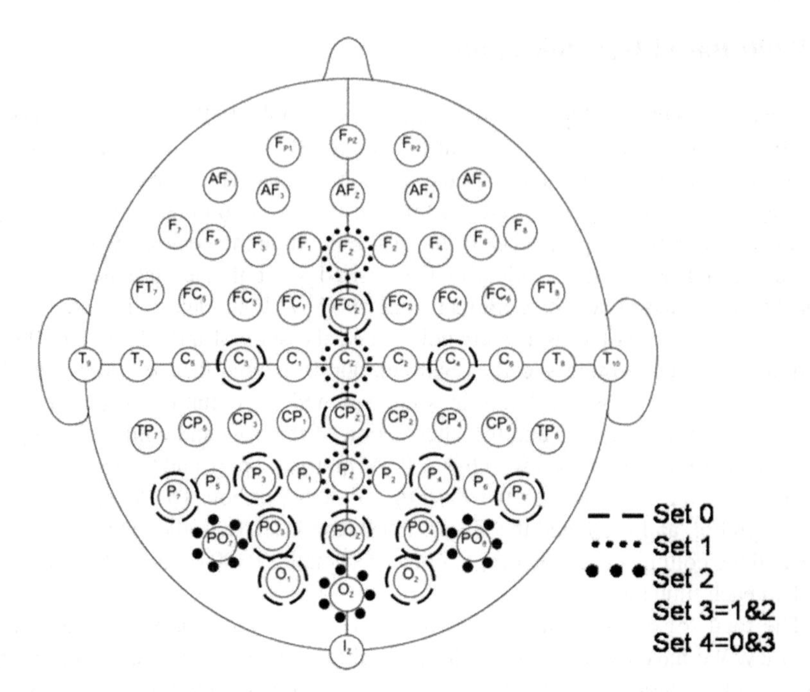

Fig. 7 Montages used to derive SWLDA classification coefficients. Data was collected from all 64 electrodes; only the indicated electrodes were used to derive coefficients (see text). Adapted from Krusienski et al. [22]

and 6 × 6). The results showed that the faster ISI yielded higher classification accuracy, consistent with the findings of Meinicke et al. [35]. In addition, the amplitude of the P300 response for the target items was larger in the 6 × 6 matrix condition than in the 3 × 3 matrix condition. These results are consistent with many studies that show increased P300 amplitude with reduced target probability (e.g., [1, 2, 14]). Moreover, the results of the Sellers et al. [47] study suggest that optimal matrix size and ISI values should be determined for each new user.

Our lab has also tested several variables related to classification accuracy using the SWLDA classification method [22]. Krusienski et al. [22] examined how the variables of channel set, channel reference, decimation factor, and number of model features affect classification accuracy. Channel set was the only factor to have a statistically significant effect on classification accuracy. Figure 7 shows examples of each channel set. Set-1 (Fz, Cz, and Pz) and Set-2 (PO7, PO8, and Oz) performed equally well, and significantly worse than Set-3 (Set-1 and Set-2 combined). Set-4 (containing 19 electrodes) performed the same as Set-3, which contained only six electrodes.

These results of the Krusienski et al. [22] study demonstrate two important points: First, using 19 electrode locations does not provide more useful information than the six electrodes contained in Set-3, in terms of classification accuracy. Second, electrode locations other than those traditionally associated with the P300

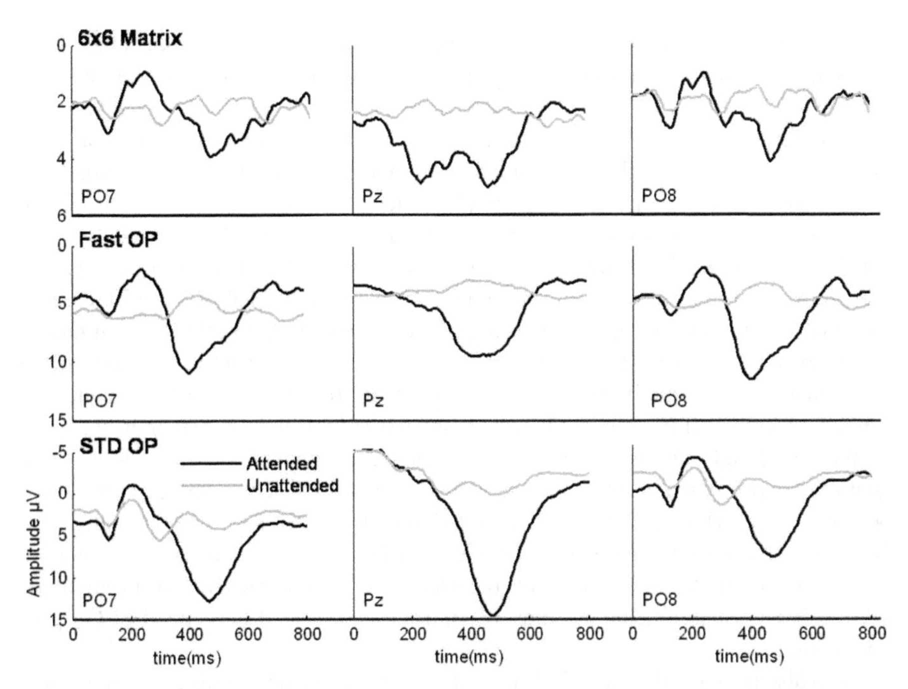

Fig. 8 Sample waveforms for attended (*black*) and unattended (*gray*) stimuli for electrodes PO7, Pz, and PO8. Row 1: Data were collected using a 6 × 6 P300 matrix paradigm with stimulation every 175 ms. Rows 2 and 3: Data were collected using a two-target oddball paradigm with stimulation every 200 ms or every 1.5 s, respectively. The P300 response is evident at Pz (at about 400 ms), and a negative deflection at approximately 200 ms is evident at locations PO7 and PO8

response (i.e., Fz, Cz, and Pz) provide valuable information for classifying matrix data. This is consistent with previous data showing that occipital electrodes improve classification [20, 35]. These electrode locations discriminate attended from non-attended stimuli, as measured by r^2, the squared value of Pearson's r that measures the relationship between two variables [52]. Examination of the waveforms suggests that a negative deflection preceding the actual P300 response provides this additional unique information (see Fig. 8, row 1).

The unique classification information provided by the occipital electrodes is probably not due simply to the user fixating the target symbol. Several previous studies have established that attentional factors increase the amplitude of the occipital P1, N1, and N2 components of the waveform [15, 18, 19, 26]. For example, Mangun et al. [26] had participants maintain fixation on a cross while stimuli were flashed in a random order to the four visual quadrants. The participants' task was to attend to the stimuli in one quadrant and ignore the other three quadrants. The amplitudes of the P1, N1, and N2 responses to stimuli in a given location were significantly larger when the location was attended than when the location was ignored, even though fixation remained on a central cross throughout. In addition, eye movements and blinks were monitored by recording vertical and horizontal EOG, and fixation was verified by vertical and horizontal infrared corneal reflectance [26]. This implies that spatial (also called covert) attention contributes to the observed effects.

Furthermore, it is clear that fixation alone is not sufficient to elicit a P300 response. Evidence for this is provided by numerous studies that present target and non-target items at fixation in a Bernoulli series (e.g.,[16]). In Fig. 8, Rows 2 and 3 show average responses to a standard oddball experiment. The stimuli "X" and "O" were presented at fixation, with a probability of .2 and .8, respectively. The responses shown in Row 2 had a stimulation rate similar to that of P300 BCI experiments (every 175 ms); the responses in Row 3 used a stimulation rate similar to standard oddball experiments (every 1.5 s). The negative peak around 200 ms is present in these oddball conditions, as it is in the 6×6 matrix speller condition shown in Row 1. If fixation alone were responsible for this response, in Rows 2 and 3 the target and non-target items would produce equivalent responses because all stimuli are presented at fixation in the oddball conditions. The responses are clearly not the same. This implies that a visual P300 BCI is not simply classifying gaze direction in a fashion analogous to the Sutter (48) visual evoked potential communication system. A BCI that requires the user to move their eyes may be problematic if they have lost significant eye muscle control, regardless of the type of EEG signal being used.

It is also possible that occipital component represents the negative part of a dipole in the temporal-parietal junction (the area of cortex where the temporal and parietal lobes meet) that is related to the P300 [11, 42]. By this view, neither the positive component at the midline nor the negative occipital component is generated directly below their spatial peaks on the surface. Rather, both would be generated at the temporal-parietal junction, an area known to be closely associated with the regulation of attention.

A P300 BCI must be accurate to be a useful option for communication. Accurate classification depends on effective feature extraction and on the translation algorithm used for classification. Recently, we tested several alternative classification methods including SWLDA, linear support vector machines, Gaussian support vector machines, Pearson's correlation method, and Fisher's linear discriminant [21]. The results indicated that, while all methods attained useful levels of classification performance, the SWLDA and Fisher's linear discriminant methods performed significantly better than the other three methods.

4 A BCI System for Home Use

In addition to working to improve SMR- and P300-based BCI performance, we are also focusing on developing clinically practical BCI systems that can be used by severely disabled people in their homes, on a daily basis, in a largely unsupervised manner. The primary goals of this project are to demonstrate that the BCI system can be used for everyday communication, and that using a BCI has a positive impact on the user's quality of life [54]. In collaboration with colleagues at the University of Tübingen, and the University of South Florida, we have been studying home use (e.g. [24, 37, 44–46, 51]. This initial work has identified critical factors essential for moving out of the lab and into a home setting where people use the BCI in an autonomous fashion. The most pressing needs are to develop a more compact system, to streamline operational characteristics to simplify the process for caregivers, and to provide users with effective and reliable communication applications [51].

Our prototype home system includes a laptop computer, a flat panel display, an eight-channel electrode cap and an amplifier with a built-in A/D board. We have addressed making the system more user friendly by automating several of the BCI2000 software processes, and enabling the caregiver to start the program with a short series of mouse clicks. The caregiver's major tasks are to place the electrode cap and inject gel into the electrodes, a process that takes about 5 min. The software has also been modified to include a menu-driven item selection structure that allows the BCI user to navigate various hierarchical menus to perform specific tasks (e.g., basic communication, basic needs, word processing and environmental controls) more easily and rapidly than earlier versions of the SMR [53] and P300 [44] software. In addition, a speech output option has been added for users who desire this ability. A more complete description of the system is provided in Vaughan et al. [54].

Most recently, we have begun to provide severely disabled users with in-home P300-based BCI systems to use for daily communication and control tasks [45]. The system presents an 8×9 matrix of letters, numbers, and function calls that operate as a keyboard to make the Windows-based programs (e.g., Eudora, Word, Excel, PowerPoint, Acrobat) completely accessible via EEG control. The system normally uses an ISI of 125 ms (a 62.5 ms flash followed by a 62.5 ms blank period), and each series of intensifications lasts for 12.75 s. The first user has now had his system for more than 3 years, and three others have been given systems more recently. Ongoing average accuracies for the 6×6 or 8×9 matrix have ranged from 51 to 83% (while chance performance for the two matrices is 2.7 and 1.4%, respectively). Each user's data are uploaded every week via the Internet and analyzed automatically using our standard SWLDA procedure, so that classification coefficients can be updated as needed [21, 47]. Using a remote access protocol, we can also monitor performance in real-time and update system parameters as needed. In actual practice, the home system usually functions well from week to week and even month to month with little or no active intervention on our part.

These initial results suggest that a P300-BCI can be useful to individuals with severe motor disabilities, and that their caregivers can learn to support its operation without excessive technical oversight [44, 45, 51]. We are concentrating on further increasing the system's functionality and decreasing its need for technical oversight. Furthermore, together with colleagues at Wharton School of Business (University of Pennsylvania), we have established a non-profit foundation (www.braincommunication.org) to enable the dissemination and support of home BCI systems for the severely disabled people who need them.

5 SMR-Based Versus P300-Based BCIs

The goal of the Wadsworth BCI development program is to provide a new mode of communication and control for severely disabled people. As shown above, the SMR-based and P300-based BCI systems employ very different approaches to achieve this goal. The SMR system uses EEG features that are not elicited by specific sensory stimuli. In contrast, the P300 system uses EEG features that are elicited when the user attends to a specific stimulus, among a defined set of stimuli, and the stimuli are presented within the constraints of the oddball paradigm [16, 43]. Another major difference is that the Wadsworth SMR system uses frequency-domain features and the P300 system uses time-domain features. While it is possible to describe P300 in the frequency domain (e.g.[8], this has not (to our knowledge) been done for a P300-based BCI.

Our SMR system uses a regression-based translation algorithm, while our P300 system uses a SWLDA classification-based translation algorithm. A regression approach is well suited to SMR cursor movement applications because it provides continuous control in one or more dimensions and generalizes to novel target configurations [33]. In contrast, the classification approach is well suited to a P300 system because the target item is regarded as one class and all other alternatives are regarded as the other class. In this way it is possible to generalize a single discriminant function to matrices of differing sizes.

Finally, the SMR- and P300-based BCI systems differ in the importance of user training. With the SMR system, the user learns to control SMRs to direct the cursor to targets located on the screen. This ability requires a significant amount of training, ranging from 60 min or so for adequate one-dimensional control to many hours for two- or three-dimensional control. In contrast, the P300 system requires only a few minutes of training. With the SMR system, user performance continues to improve with practice [31], while with the P300 system user performance is relatively stable from the beginning in terms of P300 morphology [10, 16, 41], and in terms of classification accuracy [46, 47]. (At the same time, it is conceivable that, with the development of appropriate P300 training methods, users may be able to increase the differences between their responses to target and non-target stimuli, and thereby improve BCI performance.) Finally, an SMR-based BCI system is more suitable for continuous control tasks such as cursor movement, although a P300-based BCI can provide slow cursor movement control [9, 40].

In sum, both the characteristics of the EEG features produced by a potential BCI user and the functional capabilities the user needs from the system should be considered when configuring a BCI system. Attention to these variables should help to yield the most effective system for each user. These and other user-specific factors are clearly extremely important for successfully translating BCI systems from the laboratory into the home, and for ensuring that they provide their users with effective and reliable communication and control capacities.

Acknowledgements This work was supported in part by grants from the National Institutes of Health (HD30146, EB00856, EB006356), The James S. McDonnell Foundation, The Altran Foundation, The ALS Hope Foundation, The NEC Foundation, and The Brain Communication Foundation.

References

1. B.Z. Allison and J.A. Pineda, ERPs evoked by different matrix sizes: Implications for a brain computer interface (BCI) system, IEEE Trans Neural Syst Rehab Eng, 11, 110–113, (2003).
2. B.Z. Allison and J.A. Pineda, Effects of SOA and flash pattern manipulations on ERPs, performance, and preference: Implications for a BCI system. Int J Psychophysiol, 59, 127–140, (2006).
3. B.Z. Allison, E.W. Wolpaw, and J.R. Wolpaw, Brain-computer interface systems: progress and prospects. Expert Rev Med Devices, 4, 463–474, (2007).
4. N. Birbaumer, et al., A spelling device for the paralyzed. Nature, 398, 297–298, (1999).
5. N. Birbaumer, et al., The thought translation device (TTD) for completely paralyzed patients. IEEE Trans Rehabil Eng, 8, 190–193, (2000).
6. N. Birbaumer, T. Hinterberger A. Kübler, and N. Neumann, The thought translation device (TTD): neurobehavioral mechanisms and clinical outcome. IEEE Trans Neural Syst Rehab Eng, 11, 120–3, (2003).
7. B. Blankertz et al., Boosting bit rates and error detection for the classification of fast-paced motor commands based on single-trial EEG analysis. IEEE Trans Neural Syst Rehab Eng, 11, 127–31, (2003).
8. B. Cacace, D.J. and McFarland, Spectral dynamics of electroencephalographic activity during auditory information processing. Hearing Res, 176, 25–41, (2003).
9. L. Citi, R. Poli, C. Cinel, and F. Sepulveda, P300-based brain computer interface with genetically-optimised analogue control. IEEE Trans Neural Syst Rehab Eng, 16, 51–61, (2008).
10. J. Cohen, and J. Polich, On the number of trials needed for P300. Int J Psychophysiol, 25, 249–55, (1997).
11. J. Dien, K.M Spencer, and E. Donchin, Localization of the event-related potential novelty response as defined by principal components analysis. Cognitive Brain Research, 17, 637–650, (2003).
12. E. Donchin, K.M. Spencer, R. Wijesinghe, The mental prosthesis: Assessing the speed of a P300-based brain-computer interface. IEEE Trans Rehabil Eng, 8, 174–179, (2000).
13. J. Doyon, V. Penhue, and L.G .Ungerleider, Distinct contribution of the cortico-striatal and cortico-cerebellar systems to motor skill learning. Neuropsychologia, 41, 252–262, (2003).
14. C. Duncan – Johnson, E. Donchin, On quantifying surprise: The variation of event-related potentials with subjective probability. Psychophysiology, 14, 456–467, (1977).
15. R.G. Eason, Visual evoked potential correlates of early neural filtering during selective attention, Bull Psychonomic Soc, 18, 203–206, (1981).
16. M. Fabiani, G. Gratton, D. Karis, and E. Donchin, Definition, identification and reliability of measurement of the P300 component of the event-related brain potential. Adv Psychophysiol, 2, 1–78, (1987).
17. L.A. Farwell and E. Donchin, Talking off the top of your head: toward a mental prosthesis utilizing event-related brain potentials, Electroencephalogr Clin Neurophysiol, 70, 510–523, (1988).
18. M.R. Harter, C. Aine, and C. Schroder, Hemispheric differences in the neural processing of stimulus location and type: Effects of selective attention on visual evoked potentials. Neuropsychologia, 20, 421–438, (1982).
19. S.A. Hillyard and T.F. Munte, Selective attention to color and location cues: An analysis with event-related brain potentials. Perception Psychophys, 36, 185–198, (1984).
20. M. Kaper, P. Meinicke, U. Grossekathoefer, T. Lingner, and H. Ritter, BCI Competition 2003-Data set Iib: Support vector machines for the P300 speller paradigm. IEEE Trans Biomed Eng, 51, 1073–1076, (2004).
21. D.J. Krusienski, E.W. Sellers, , F. Cabestaing, S. Bayoudh, D.J. McFarland, T.M. Vaughan, and J.R. Wolpaw, A comparison of classification techniques for the P300 speller. J Neural Eng, 3, 299–305, (2006).

22. D.J Krusienski, E.W Sellers, D.J McFarland, T.M Vaughan, and J.R. Wolpaw, Toward enhanced P300 speller performance. J Neurosci Methods, 167, 15–21, (2008).
23. A. Kübler, et al., Self-regulation of slow cortical potentials in completely paralyzed human patients. Neurosci Lett, 252, 171–174, (1998).
24. A. Kübler, et al., Patients with ALS can use sensorimotor rhythms to operate a brain-computer interface. Neurology, 64, 1775–1777, (2005).
25. L.R. Hochberg, M.D. Serruya, G.M. Friehs, J.A. Mukand, M. Saleh, A.H. Caplan, A. Branner, D. Chen, R.D. Penn, and J.P. Donoghue, Neuronal ensemble control of prosthetic devices by a human with tetraplegia, Nature, 442(7099) 164–171, (2006, Juli 13).
26. G.R. Mangun, S.A. Hillyard, and S.J. Luck, Electrocortical substrates of visual selective attention, In D. Meyer, and S. Kornblum (Eds.), *Attention and performance XIV,* MIT Press, Cambridge, MA, pp. 219–243, (1993).
27. D.J. McFarland, D.J. Krusienski, WA. Sarnacki, and J.R. Wolpaw, Emulation of computer mouse control with a noninvasive brain-computer interface. J Neural Eng, 5, 101–110, (2008).
28. D.J. McFarland, D.J. Krusienski, W.A. Sarnacki, and J.R. Wolpaw, Brain-computer interface signal processing at the Wadsworth Center: mu and sensorimotor beta rhythms. Prog Brain Res, 159, 411–419, (2006)
29. D.J. McFarland, L.A. Miner, T.M. Vaughan, and J.R. Wolpaw, Mu and beta rhythm topographies during motor imagery and actual movements. Brain Topogr, 12, (2000).
30. D.J. McFarland, G. W. Neat, R.F. Read, and J.R. Wolpaw, An EEG-based method for graded cursor control. Psychobiology, 21, 77–81, (1993).
31. D.J. McFarland, W.A Sarnacki, T.M Vaughan, and J.R. Wolpaw, Brain-computer interface (BCI) operation: signal and noise during early training sessions. Clin Neurophysiol, 116, 56–62, (2005).
32. D.J. McFarland, W.A. Sarnacki, and J.R. Wolpaw, Brain-computer interface (BCI) operation: optimizing information transfer rates. Biol Psychol, 63, 237–251, (2003).
33. D.J. McFarland, W.A. Sarnacki, and J.R. Wolpaw, Electroencephalographic (EEG) control of three-dimensional movement. Program No. 778.4. Abstract Viewer/Itinerary Planner. Washington, DC: Society for Neuroscience. Online, (2008).
34. D.J. McFarland and J.R. Wolpaw, Sensorimotor rhythm-based brain-computer interface (BCI): Feature selection by regression improves performance, IEEE Trans Neural Syst Rehabil Eng, 13, 372–379, (2005).
35. D.J. McFarland and J.R. Wolpaw, Brain-computer interface operation of robotic and prosthetic devices. IEEE Comput, 41, 52–56, (2008)
36. P. Meinicke, M. Kaper, F. Hoppe, M. Huemann, and H. Ritter, Improving transfer rates in brain computer interface: A case study. Neural Inf Proc Syst, 1107–1114, (2002).
37. K.R. Muller, C.W. Anderson, and G E. Birch, Linear and nonlinear methods for brain-computer interfaces. IEEE Trans Neural Syst Rehabil Eng, 11, 165–169, (2003).
38. F. Nijboer, E.W. Sellers, J. Mellinger, M.A. Jordan, Matuz, T. A. Furdea, U. Mochty, D.J. Krusienski, T.M. Vaughan, J.R. Wolpaw, N. Birbaumer, and A. Kübler, A brain-computer interface for people with amyotrophic lateral sclerosis. Clin Neurophysiol, 119, 1909–1916. (2008).
39. G. Pfurtscheller, D. Flotzinger, and J. Kalcher, Brain-computer interface- a new communication device for handicapped persons. J Microcomput Appl, 16, 293–299, (1993).
40. G. Pfurtscheller, and F.H. Lopes da Silva, Event-related EEG/MEG synchronization and desynchronization: basic principles, Clin Neurophysiol, 110, 1842–1857, (1999).
41. F. Piccione, et al., P300-based brain computer interface: Reliability and performance in healthy and paralysed participants. Clin Neurophysiol, 117, 531–537, (2006).
42. J. Polich, Habituation of P300 from auditory stimuli, Psychobiology, 17, 19–28, (1989).
43. J. Polich, Updating P300: An integrative theory of P3a and P3b. Clinical Neurophysiology, 118, 2128–2148, (2007).
44. W. Pritchard, The psychophysiology of P300. Psychol Bull, 89, 506–540, (1981).
45. E.W. Sellers, et al., A P300 brain-computer interface (BCI) for in-home everyday use, Poster presented at the Society for Neuroscience annual meeting, Atlanta, GA, (2006).

46. E.W. Sellers et al., Brain-Computer Interface for people with ALS: long-term daily use in the home environment, Program No. 414.5. Abstract Viewer/Itinerary Planner. Washington, DC: Society for Neuroscience. Online, (2007).
47. E.W. Sellers and M. Donchin, A P300-based brain-computer interface: Initial tests by ALS patients, Clinical Neurophysiology. Clin Neurophysiology, 117, 538–548, (2006).
48. E.W Sellers, D.J Krusienski, D.J McFarland, T.M Vaughan, and J.R. Wolpaw, A P300 event-related potential brain-computer interface (BCI): The effects of matrix size and inter stimulus interval on performance. Biol Psychol, 73, 242–252, (2006).
49. E.E. Sutter, The brain response interface: communication through visually guided electrical brain responses. J Microcomput Appl, 15, 31–45, (1992).
50. S. Sutton M. Braren J. Zubin, and E.R. John, Evoked-potential correlates of stimulus uncertainty. Science, 150, 1187–1188, (1965).
51. D.M Taylor, S.I Tillery, A.B. Schwartz, Direct cortical control fo 3D reuroprosthetic devices. Science, 296, 1817–1818, (2002).
52. T.M. Vaughan, et al., aily use of an EEG-based brain-computer interface by people with ALS: technical requirements and caretaker training, Program No. 414.6. Abstract Viewer/Itinerary Planner, Society for Neuroscience, Washington, DC, online, (2007).
53. T.M Vaughan D.J. McFarland G. Schalk E. Sellers, and J.R. Wolpaw, Multichannel data from a brain-computer interface (BCI) speller using a P300 (i.e., oddball) protocol, Society for Neuroscience Abstracts, 28, (2003).
54. T.M Vaughan D.J McFarland G. Schalk W.A Sarnacki L. Robinson, and J.R Wolpaw EEG-based brain-computer interface: development of a speller application. Soc Neuroscie Abs, 26, (2001).
55. T.M.Vaughan et al., The Wadsworth BCI research and development program: At home with BCI. IEEE Trans Rehabil Eng, 1(14), 229–233, (2006).
56. J.R. Wolpaw, N. Birbaumer, D.J. McFarland, G. Pfurtscheller, and T.M. Vaughan, Brain-computer interfaces for communication and control. Clin Neurophysiol, 113, 767–791, (2002).
57. J.R. Wolpaw and D.J. McFarland, Control of a two-dimensional movement signal by a non-invasive brain-computer interface in humans. Proc Natl Acad Sci USA, 101, 17849–17854, (2004).
58. JR. Wolpaw, et al., Brain-computer interface technology: a review of the first international meeting, IEEE Trans Rehabil Eng, 8, 164–173, (2000).
59. J.R. Wolpaw and D.J. McFarland, Multichannel EEG-based brain-computer communication. Electroencephalogr Clin Neurophysiol, 90, 444–449, (1994).
60. J.R. Wolpaw, D.J. McFarland, G.W. Neat, and C.A. Forneris, An EEG-based brain-computer interface for cursor control. Electroencephalogr Clin Neurophysiol, 78, 252–259, (1991).

Detecting Mental States by Machine Learning Techniques: The Berlin Brain–Computer Interface

Benjamin Blankertz, Michael Tangermann, Carmen Vidaurre, Thorsten Dickhaus, Claudia Sannelli, Florin Popescu, Siamac Fazli, Márton Danóczy, Gabriel Curio, and Klaus-Robert Müller

1 Introduction

The Berlin Brain-Computer Interface (BBCI) uses a machine learning approach to extract user-specific patterns from high-dimensional EEG-features optimized for revealing the user's mental state. Classical BCI applications are brain actuated tools for patients such as prostheses (see Section 4.1) or mental text entry systems ([1] and see [2–5] for an overview on BCI). In these applications, the BBCI uses natural motor skills of the users and specifically tailored pattern recognition algorithms for detecting the user's intent. But beyond rehabilitation, there is a wide range of possible applications in which BCI technology is used to monitor other mental states, often even covert ones (see also [6] in the fMRI realm). While this field is still largely unexplored, two examples from our studies are exemplified in Sections 4.3 and 4.4.

1.1 The Machine Learning Approach

The advent of machine learning (ML) in the field of BCI has led to significant advances in real-time EEG analysis. While early EEG-BCI efforts required neurofeedback training on the part of the user that lasted on the order of days, in ML-based systems it suffices to collect examples of EEG signals in a so-called *calibration* during which the user is cued to perform repeatedly any one of a small set of mental tasks. This data is used to adapt the system to the specific brain signals of each user (*machine training*). This step of adaption seems to be instrumental for effective BCI performance due to a large inter-subject variability with respect to the brain signals [7]. After this preparation step, which is very short compared to the subject

B. Blankertz (✉)
Machine Learning Laboratory, Berlin Institute of Technology, Berlin, Germany; Fraunhofer FIRST (IDA), Berlin,Germany
e-mail: blanker@cs.tu-berlin.de

B. Graimann et al. (eds.), *Brain–Computer Interfaces*, The Frontiers Collection,
DOI 10.1007/978-3-642-02091-9_7,

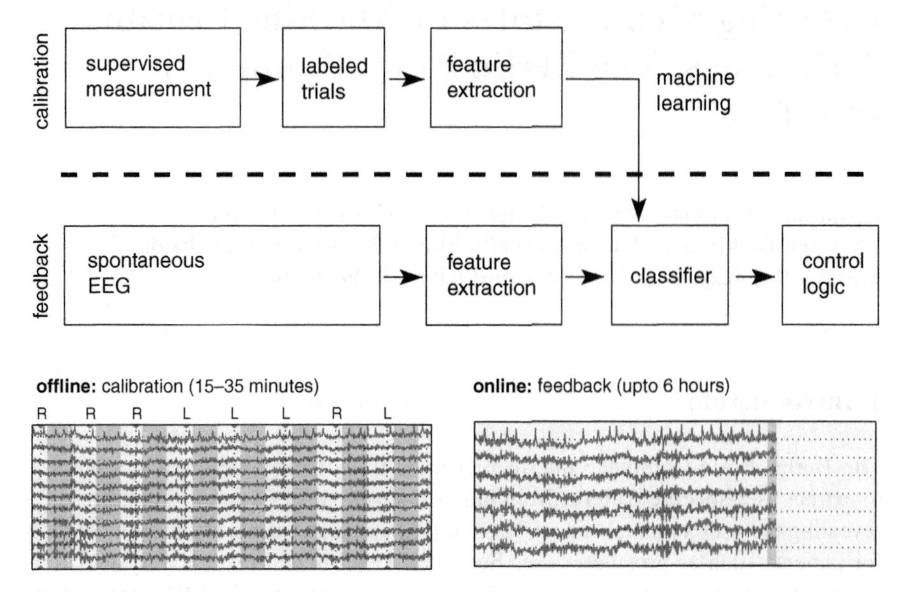

Fig. 1 *Overview of a machine-learning-based BCI system.* The system runs in two phases. In the calibration phase, we instruct the participants to perform certain tasks and collect short segments of labeled EEG (trials). We train the classifier based on these examples. In the feedback phase, we take sliding windows from a continuous stream of EEG; the classifier outputs a real value that quantifies the likeliness of class membership; we run a feedback application that takes the output of the classifier as input. Finally, the user receives the feedback on the screen as, e.g., cursor control

training in the operant conditioning approach [8, 9], the feedback application can start. Here, the users can actually transfer information through their brain activity and control applications. In this phase, the system is composed of the classifier that discriminates between different mental states and the control logic that translates the classifier output into control signals, e.g., cursor position or selection from an alphabet.

An overview of the whole process in an ML-based BCI is sketched in Fig. 1. Note that in alternative applications of BCI technology (see Sections 4.3 and 4.4), the calibration may need novel nonstandard paradigms, as the sought-after mental states (like lack of concentration, specific emotions, workload) might be difficult to induce in a controlled manner.

1.2 Neurophysiological Features

There is a variety of other brain potentials, that are used for brain-computer interfacing, see Chapter 2 in this book for an overview. Here, we only introduce those brain potentials, which are important for this review. New approaches of the Berlin BCI project also exploit the attention-dependent modulation of the P300 component (in visual, auditory [10] and tactile modality), steady state visual evoked potentials (SSVEP) and auditory steady-state responses (ASSR).

1.2.1 Readiness Potential

Event-related potentials (ERPs) are transient brain responses that are time-locked to some event. This event may be an external sensory stimulus or an internal state signal, associated with the execution of a motor, cognitive, or psychophysiologic task. Due to simultaneous activity of many sources in the brain, ERPs are typically not visible in single trials (i.e., the segment of EEG related to *one* event) of raw EEG. For investigating ERPs, EEG is acquired during many repetitions of the event of interest. Then short segments (called epochs or trials) are cut out from the continuous EEG signals around each event and are averaged across epochs to reduce event-unrelated background activity. In BCI applications based on ERPs, the challenge is to detect ERPs in single trials.

The *readiness potential* (RP, or Bereitschaftspotential) is an ERP that reflects the intention to move a limb, and therefore precedes the physical (muscular) initiation of movements. In the EEG, it can be observed as a pronounced cortical negativation with a focus in the corresponding motor area. In hand movements, the RP is focused in the central area contralateral to the performing hand, cf. [11–13] and references therein for an overview. See Fig. 2 for an illustration. Section 4.2 shows an application of BCI technology using the readiness potential. Further details about our BCI-related studies involving RP can be found in [7, 14–16].

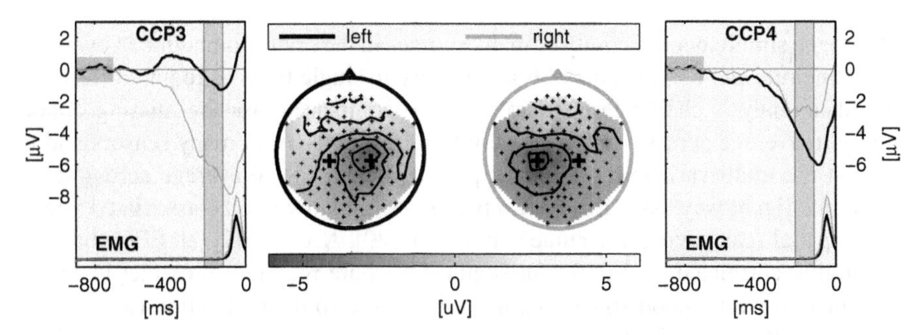

Fig. 2 Response *averaged* event-related potentials (ERPs) of a right-handed volunteer in a left vs. right hand finger tapping experiment ($N = 275$ resp. 283 trials per class). Finger movements were executed in a self-paced manner, i.e., without any external cue, using an approximate inter-trial interval of 2. The two scalp plots show the topographical mapping of scalp potentials averaged within the interval –220 to –120 relative to keypress (time interval vertically shaded in the ERP plots; initial horizontal shading indicates the baseline period). Larger crosses indicate the position of the electrodes CCP3 and CCP4 for which the ERP time course is shown in the subplots at both sides. For comparison time courses of EMG activity for left and right finger movements are added. EMG activity starts after -120 ms and reaches a peak of 70 μV at –50 ms. The readiness potential is clearly visible, a predominantly contralateral negativation starting about 600 before movement and raising continuously until EMG onset

1.2.2 Sensorimotor Rhythms

Apart from transient components, EEG comprises rhythmic activity located over various areas. Most of these rhythms are so-called idle rhythms, which are generated by large populations of neurons in the respective cortex that fire in rhythmical synchrony when they are not engaged in a specific task. Over motor and sensorimotor areas in most adult humans, oscillations with a fundamental frequency between 9 and 13 Hz can be observed, the so called μ-rhythm. Due to its comb-shape, the μ-rhythm is composed of several harmonics, i.e., components of double and sometimes also triple the fundamental frequency [17] with a fixed phase synchronization, cf. [18]. These sensorimotor rhythms (SMRs) are attenuated when engagement with the respective limb takes place. As this effect is due to loss of synchrony in the neural populations, it is termed event-related desynchronization (ERD), see [19]. The increase of oscillatory EEG (i.e., the reestablishment of neuronal synchrony after the event) is called event-related synchronization (ERS). The ERD in the motor and/or sensory cortex can be observed even when an individual is only thinking of a movement or imagining a sensation in the specific limb. The strength of the sensorimotor idle rhythms as measured by scalp EEG is known to vary strongly between subjects. See Chapter. 3 of this book for more details on ERD/ERS.

Sections 3.1 and 3.2 show results of BCI control exploiting the voluntary modulation of sensorimotor rhythm.

2 Processing and Machine Learning Techniques

Due to the simultaneous activity of many sources in the brain, compounded by noise, detecting relevant components of brain activity in single trials as required for BCIs is a data analysis challenge. One approach to compensate for the missing opportunity to average across trials is to record brain activity from many sensors and to exploit the multi-variateness of the acquired signals, i.e., to average across space in an intelligent way. Raw EEG scalp potentials are known to be associated with a large spatial scale owing to volume conduction [20]. Accordingly, all EEG channels are highly correlated, and powerful spatial filters are required to extract localized information with a good signal to noise ratio (see also the motivation for the need of spatial filtering in [21]).

In the case of detecting ERPs, such as RP or error-related potentials, the extraction of features from one source is mostly done by linear processing methods. In this case, the spatial filtering can be accomplished implicitly in the classification step (interchangeability of linear processing steps). For the detection of modulations of SMRs, the processing is non-linear (e.g. calculation of band power). In this case, the prior application of spatial filtering is extremely beneficial. The methods used for BCIs range from simple fixed filters like Laplacians [22], and data driven unsupervised techniques like independent component analysis (ICA) [23] or model based approaches [24], to data driven supervised techniques like common spatial patterns (CSP) analysis [21].

In this Section we summarize the two techniques that we consider most important for classifying multi-variate EEG signals, CSP and regularized linear discriminant analysis. For a more complete and detailed review of signal processing and pattern recognition techniques see [7, 25, 26] and Chapters 17 and 18 of this book.

2.1 Common Spatial Patterns Analysis

The CSP technique (see [27]) allows identification of spatial filters that maximize the variance of signals of one condition and at the same time minimize the variance of signals of another condition. Since variance of band-pass filtered signals is equal to band-power, CSP filters are well suited to detect amplitude modulations of sensorimotor rhythms (see Section 1.2) and consequently to discriminate mental states that are characterized by ERD/ERS effects. As such it has been well used in BCI systems [14, 28] where CSP filters are calculated individually for each user on calibration data.

The CSP technique decomposes multichannel EEG signals in the sensor space. The number of spatial filters equals the number of channels of the original data. Only a few filters have properties that make them favorable for classification. The discriminative value of a CSP filter is quantified by its generalized eigenvalue. This eigenvalue is relative to the sum of the variances in both conditions. An eigenvalue of 0.9 for class 1 means an average ratio of 9:1 of variances during condition 1 and 2. See Fig. 3 for an illustration of CSP filtering.

For details on the technique of CSP analysis and its extensions, please see [21, 29–32] and Chapter 17 of this book.

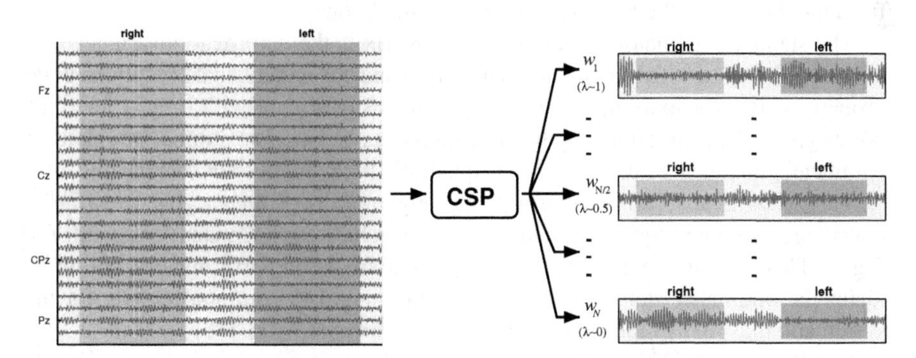

Fig. 3 The inputs of CSP analysis are (band-pass filtered) multi-channel EEG signals which are recorded for two conditions (here "left" and "right" hand motor imagery. The result of CSP analysis is a sequence of spatial filters. The number of filters (here *N*) is equal to the number of EEG channels. When these filters are applied to the continuous EEG signals, the (average) relative variance in the two conditions is given by the eigenvalues. An eigenvalue near 1 results in a large variance of signals of condition 1, and an eigenvalue near 0 results in a small variance for condition 1. Most eigenvalues are near 0.5, such that the corresponding filters do not contribute relevantly to the discrimination

2.2 Regularized Linear Classification

There is some debate about whether to use linear or non-linear methods to classify single EEG trials; see the discussion in [33]. In our experience, linear methods perform well, if an appropriate preprocessing of the data is performed. E.g., band-power features itself are far from being Gaussian distributed due to the involved squaring and the best classification of such features is nonlinear. But applying the logarithm to those features makes their distribution close enough to Gaussian such that linear classification typically works well. Linear methods are easy to use and robust. But there is one caveat that applies also to linear methods. If the number of dimensions of the data is high, simple classification methods like Linear Discriminant Analysis (LDA) will not work properly. The good news is that there is a remedy called *shrinkage* (or regularization) that helps in this case. A more detailed analysis of the problem and the presentation of its solution is quite mathematical. Accordingly, the subsequent subsection is only intended for readers interested in those technical details.

2.2.1 Mathematical Part

For known Gaussian distributions with the same covariance matrix for all classes, it can be shown that Linear Discriminant Analysis (LDA) is the optimal classifier in the sense that it minimizes the risk of misclassification for new samples drawn from the same distributions [34]. Note that LDA is equivalent to Fisher Discriminant Analysis and Least Squares Regression [34]. For EEG classification, the assumption of Gaussianity can be achieved rather well by appropriate preprocessing of the data. But the means and covariance matrices of the distributions have to be estimated from the data, since the true distributions are not known.

The standard estimator for a covariance matrix is the empirical covariance (see equation (1) below). This estimator is unbiased and has under usual conditions good properties. But for extreme cases of high-dimensional data with only a few data points given, the estimation may become inprecise, because the number of unknown parameters that have to be estimated is quadratic in the number of dimensions.

This leads to a systematic error: Large eigenvalues of the original covariance matrix are estimated too large, and small eigenvalues are estimated too small; see Fig. 4. This error in the estimation degrades classification performance (and invalidates the optimality statement for LDA). Shrinkage is a common remedy for the systematic bias [35] of the estimated covariance matrices (e.g. [36]):

Let $\mathbf{x}_1, \ldots, \mathbf{x}_n \in \mathbb{R}^d$ be n feature vectors and let

$$\hat{\mathbf{\Sigma}} = \frac{1}{n-1} \sum_{i=1}^{n} (\mathbf{x}_i - \hat{\mu})(\mathbf{x}_i - \hat{\mu})^\top \quad (1)$$

be the unbiased estimator of the covariance matrix. In order to counterbalance the estimation error, $\hat{\mathbf{\Sigma}}$ is replaced by

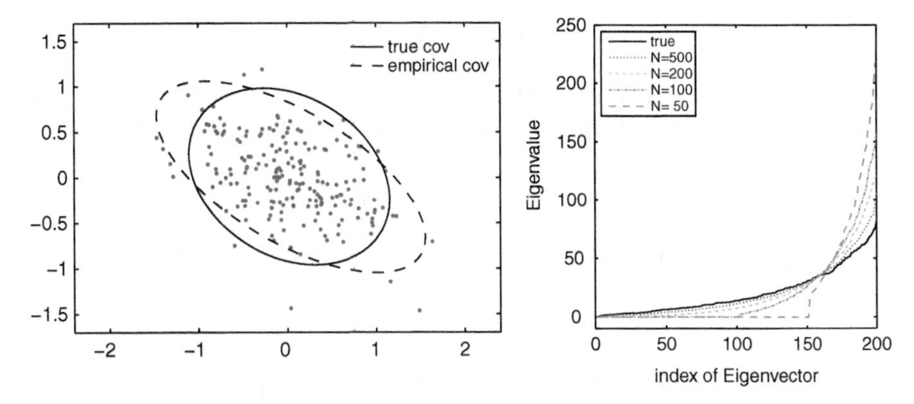

Fig. 4 *Left:* Data points drawn from a Gaussian distribution (*gray dots;* $d = 200$ dimensions, two dimensions selected for visualization) with true covariance matrix indicated by an ellipsoid in *solid line*, and estimated covariance matrix in *dashed line. Right:* Eigenvalue spectrum of a given covariance matrix (*bold line*) and eigenvalue spectra of covariance matrices estimated from a finite number of samples drawn ($N=$ 50, 100, 200, 500) from a corresponding Gaussian distribution

$$\tilde{\boldsymbol{\Sigma}}(\gamma) := (1-\gamma)\hat{\boldsymbol{\Sigma}} + \gamma\nu\mathbf{I} \tag{2}$$

for a tuning parameter $\gamma \in [0, 1]$ and ν defined as average eigenvalue trace$(\hat{\boldsymbol{\Sigma}})/d$ of $\hat{\Sigma}$ with d being the dimensionality of the feature space and $\mathbf{I}$ being the identity matrix. Then the following holds. Since $\hat{\boldsymbol{\Sigma}}$ is positive semi-definite we have an eigenvalue decomposition $\hat{\boldsymbol{\Sigma}} = \mathbf{V}\mathbf{D}\mathbf{V}^\top$ with orthonormal $\mathbf{V}$ and diagonal $\mathbf{D}$. Due to the orthogonality of $\mathbf{V}$ we get

$$\tilde{\boldsymbol{\Sigma}} = (1-\gamma)\mathbf{V}\mathbf{D}\mathbf{V}^\top + \gamma\nu\mathbf{I} = (1-\gamma)\mathbf{V}\mathbf{D}\mathbf{V}^\top + \gamma\nu\mathbf{V}\mathbf{I}\mathbf{V}^\top = \mathbf{V}\left((1-\gamma)\mathbf{D} + \gamma\nu\mathbf{I}\right)\mathbf{V}^\top$$

as eigenvalue decomposition of $\tilde{\boldsymbol{\Sigma}}$. That means

- $\tilde{\boldsymbol{\Sigma}}$ and $\hat{\boldsymbol{\Sigma}}$ have the same Eigenvectors (columns of $\mathbf{V}$)
- extreme eigenvalues (large or small) are modified (shrunk or elongated) towards the average ν.
- $\gamma = 0$ yields unregularized LDA, $\gamma = 1$ assumes spherical covariance matrices.

Using LDA with such modified covariance matrix is termed covariance-regularized LDA or LDA with shrinkage. The parameter γ needs to be estimated from training data. This is often done by cross validation, which is a time consuming process. But recently, a method to analytically calculate the optimal shrinkage parameter for certain directions of shrinkage was found ([37]; see also [38] for the first application in BCI). It is quite surprising that an analytic solution exists, since the optimal shrinkage parameter $\gamma^\star$ is defined by minimizing the Frobenius norm $\|\cdot\|_F^2$ between the shrunk covariance matrix and the *unknown* true covariance matrix Σ:

$$\gamma^\star = \mathrm{argmin}_{\gamma \in \mathbb{R}} \|\tilde{\boldsymbol{\Sigma}}(\gamma) - \boldsymbol{\Sigma}\|_F^2.$$

When we denote by $(\mathbf{x}_k)_i$ resp. $(\hat{\mu})_i$ the i-th element of the vector $\mathbf{x}_k$ resp. $\hat{\mu}$ and denote by s_{ij} the element in the i-th row and j-th column of $\hat{\boldsymbol{\Sigma}}$ and define

$$z_{ij}(k) = ((\mathbf{x}_k)_i - (\hat{\mu})_i)\,((\mathbf{x}_k)_j - (\hat{\mu})_j),$$

then the optimal parameter for shrinkage towards identity (as defined by (2)) can be calculated as [39]

$$\gamma^\star = \frac{n}{(n-1)^2} \frac{\sum_{i,j=1}^{d} \mathrm{var}_k(z_{ij}(k))}{\sum_{i \neq j} s_{ij}^2 + \sum_i (s_{ii} - \nu)^2}.$$

3 BBCI Control Using Motor Paradigms

3.1 High Information Transfer Rates

To preserve ecological validity (i.e., the correspondence between intention and control effect), we let the users perform motor tasks for applications like cursor movements. For paralyzed patients the control task is to attempt movements (e.g., left hand or right hand or foot); other participants are instructed to perform kinesthetically imagined movements [40] or quasi-movements [41].

We implemented a 1D cursor control as a test application for the performance of our BBCI system. One of two fields on the left and right edge of the screen was highlighted as target at the beginning of a trial; see Fig. 5. The cursor was initially at the center of the screen and started moving according to the BBCI classifier output

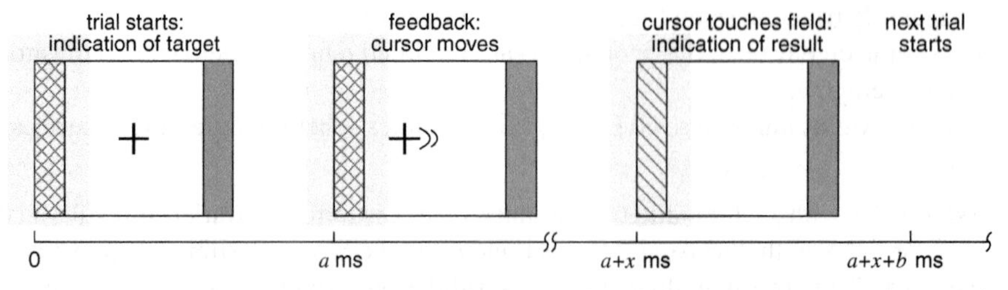

Fig. 5 Course of a feedback trial. The target cue (field with crosshatch) is indicated for a ms, where a is chosen individual according to the capabilities of the user. Then the cursor starts moving according to the BCI classifier output until it touches one of the two fields at the edge of the screen. The duration depends on the performance and is therefore different in each trial (x ms). The touched field is colored green or red according to whether its was the correct target or not (for this black and white reproduction, the field is hatched with *diagonal lines*). After b ms, the next trial starts, where b is chosen individually for each user

about half a second after the indication of the target. The trial ended when the cursor touched one of the two fields. That field was then colored green or red, depending on whether or not it was the correct target. After a short period the next target cue was presented (see [7, 42] for more details).

The aim of our first feedback study was to explore the limits of possible information transfer rates (ITRs) in BCI systems without relying on user training or evoked potentials. The ITR derived in Shannon's information theory can be used to quantify the information content, which is conveyed through a noisy (i.e., error introducing) channel. In BCI context, this leads to

$$\text{bitrate}(p,N) = \left(p\log_2(p) + (1-p)\log_2\left(\frac{1-p}{N-1}\right) + \log_2(N)\right), \tag{3}$$

where p is the accuracy of the user in making decisions between N targets, e.g., in the feedback explained above, $N = 2$ and p is the accuracy of hitting the correct bars. To include the speed of decision into the performance measure, we define

$$\text{ITR [bits/min]} = \frac{\text{\# of decisions}}{\text{duration in minutes}} \cdot \text{bitrate}(p,N). \tag{4}$$

In this form, the ITR takes different average trial durations (i.e., the speed of decisions) and different number of classes into account. Therefore, it is often used as a performance measure of BCI systems [43]. Note that it gives reasonable results only if some assumptions on the distribution of errors are met, see [44].

The participants of the study [7, 14] were six staff members, most of whom had performed feedback with earlier versions of the BBCI system before. (Later, the study was extended by four further volunteers, see [42]). First, the parameters of preprocessing were selected and a classifier was trained based on a calibration individually for each participant. Then, feedback was switched on and further parameters of the feedback were adjusted according to the user's request.

For one participant, no significant discrimination between the mental imagery conditions was found; see [42] for an analysis of that specific case. The other five participants performed eight runs of 25 cursor control trials as explained above. Table 1 shows the performance results as accuracy (percentage of trials in which the user hit the indicated target) and as ITR (see above).

As a test of practical usability, participant *al* operated a simple text entry system based on BBCI cursor control. In a free spelling mode, he spelled three German sentences with a total of 135 characters in 30 min, which is a spelling speed of 4.5 letters per minute. Note that the participant corrected all errors using the deletion symbol. For details, see [45]. Recently, using the novel mental text entry system Hex-o-Spell, which was developed in cooperation with the Human-Computer Interaction Group at the University of Glasgow, the same user achieved a spelling speed of more than 7 letters per minute, cf. [1, 46, 47].

Table 1 Results of a feedback study with six healthy volunteers (identification code in the first column). From the three classes used in the calibration measurement, the two chosen for feedback are indicated in the second column (L: left hand, R: right hand, F: right foot). The accuracies obtained online in cursor control are given in column 3. The average duration ± standard deviation of the feedback trials is provided in column 4 (duration from cue presentation to target hit). Participants are sorted according to feedback accuracy. Columns 5 and 6 report the information transfer rates (ITR) measured in bits per minute as obtained by Shannon's formula, cf. (3). Here, the complete duration of each run was taken into account, i.e., also the inter-trial breaks from target hit to the presentation of the next cue. The column *overall ITR* (oITR) reports the average ITR of all runs (of 25 trials each), while column *peak ITR* (pITR) reports the peak ITR of all runs

id	classes	accuracy [%]	duration [s]	oITR [b/m]	pITR [b/m]
al	LF	98.0 ± 4.3	2.0 ± 0.9	24.4	35.4
ay	LR	95.0 ± 3.3	1.8 ± 0.8	22.6	31.5
av	LF	90.5 ± 10.2	3.5 ± 2.9	9.0	24.5
aa	LR	88.5 ± 8.1	1.5 ± 0.4	17.4	37.1
aw	RF	80.5 ± 5.8	2.6 ± 1.5	5.9	11.0
mean		90.5 ± 7.6	2.3 ± 0.8	15.9	27.9

3.2 Good Performance Without Subject Training

The goal of our second feedback study was to investigate which proportion of naive subjects could successfully use our system in the very first session [48]. The design of this study was similar to the one described above. But here, the subjects were 14 individuals who never performed in a BCI experiment before.

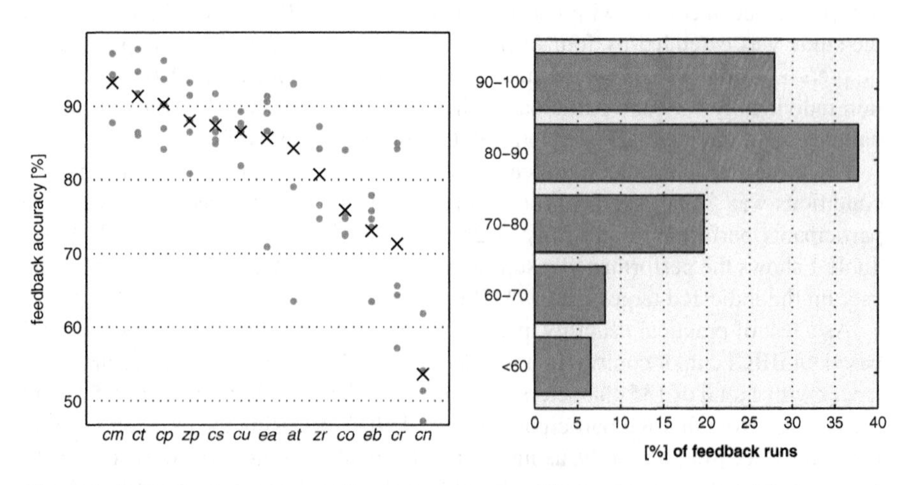

Fig. 6 *Left:* Feedback accuracy of all runs (*gray dots*) and intra-subject averages (*black crosses*). *Right:* Histogram of accuracies obtained in BBCI-controlled cursor movement task in all feedback runs of the study

Furthermore, the parameters of the feedback have been fixed beforehand for all subjects to conservative values.

For one subject, no distinguishable classes were identified. The other 13 subjects performed feedback: 1 near chance level, 3 with 70–80%, 6 with 80–90% and 3 with 90–100% hits. The results of all feedback runs are shown in Fig. 6.

This clearly shows that a machine learning based approach to BCI such as the BBCI is able to let BCI novices perform well from the first session on.

3.3 BCI Illiteracy

One of the biggest challenges in BCI research is to understand and solve the problem of "BCI Illiteracy", which is that a non-negligible portion of users cannot obtain control accurate enough to control applications by BCI. In SMR-based BCI systems illiteracy is encountered regardless of whether a machine learning or an operant conditioning approach is used [49]. The actual rate of illiteracy is difficult to determine, in particular since only a few BCI studies have been published that have a sufficient large number of participants who were not prescreened for being potentially good performers. There rate of illiteracy in SMR-based noninvasive BCI systems can roughly be estimated to be about 15–30% and therefore poses a major obstacle for general broad BCI deployment. Still, very little is known about possible reasons of such failures in BCI control. Note that for BCI systems based on stimulus-related potentials, such as P300 or SSVEP, the rate of nonperformers is lower and some studies do not report any case of failure.

A deeper understanding of this phenomenon requires determining factors that may serve to predict BCI performance, and developing methods to quantify a predictor value from given psychological and/or physiological data. Such predictors may then help to identify strategies for future development of training methods to combat BCI-illiteracy and thereby provide more people with BCI communication.

With respect to SMR-based BCI systems, we found a neurophysiological predictor of BCI performance. It is computed as band-power in physiologically specified frequency bands from only 2 min of recording a "relax with eyes open" condition using two Laplacian channels selectively placed over motor cortex areas. A correlation of $r = 0.53$ between the proposed predictor and BCI feedback performance was obtained on a large data base with $\mathrm{N} = 80$ BCI-naive participants [50, 51].

In a screening study, N=80 subjects performed motor imagery first in a calibration (i.e., without feedback) measurement and then in a feedback measurement in which they could control a 1D cursor application. Coarsely, we observed three categories of subjects: subjects for whom (I) a classifier could be successfully trained and who performed feedback with good accuracy; (II) a classifier could be successfully trained, but feedback did not work well; (III) no classifier with acceptable accuracy could be trained. While subjects of Cat. II had obviously difficulties with the transition from offline to online operation, subjects of Cat. III did not show the expected modulation of sensorimotor rhythms (SMRs): either no SMR idle rhythm was observed over motor areas, or this idle rhythm was not attenuated during motor imagery.

Here, we present preliminary results of a pilot study [52, 53] that investigated whether co-adaptive learning using machine learning techniques could help subjects suffering from BCI illiteracy (i.e., being Cat. II or III) to achieve successful feedback. In this setup, the session immediately started with BCI feedback. In the first 3 runs, a subject-independent classifier pretrained on simple features (band-power in alpha (8–15 Hz) and beta (16–32 Hz) frequency ranges in 3 Laplacian channels at C3, Cz, C4) was used and adapted (covariance matrix and pooled mean [54]) to the subject after each trial. For the subsequent 3 runs, a classifier was trained on a more complex band-power feature in a subject-specific narrow band composed from optimized CSP filters in 6 Laplacian channels. While CSP filters were static, the position of the Laplacians was updated based on a statistical criterion, and the classifier was retrained on the combined CSP plus Laplacians feature in order to provide flexibility with respect to spatial location of modulated brain activity. Finally, for the last 2 runs, a classifier was trained on CSP features, which have been calculated on runs 4–6. The pooled mean of the linear classifier was adapted after each trial [54].

Initially, we verified the novel experimental design with 6 subjects of Cat. I. Here, very good feedback performance was obtained within the first run after 20–40 trials (3–6 min) of adaptation, and further increased in subsequent runs. In the present pilot study, 2 more subjects of Cat. II and 3 of Cat. III took part. All those 5 subjects did not have control in the first three runs, but they were able to gain it when the machine learning based techniques came into play in runs 4–6 (a jump from run 3 to run 4 in Cat. II, and a continuous increase in runs 4 and 5 in Cat. III, see Fig. 7). This level of performance could be kept or even improved in runs 7 and 8 which used unsupervised adaptation. Summarizing, it was demonstrated that subjects categorized as BCI illiterates before could gain BCI control within one session.

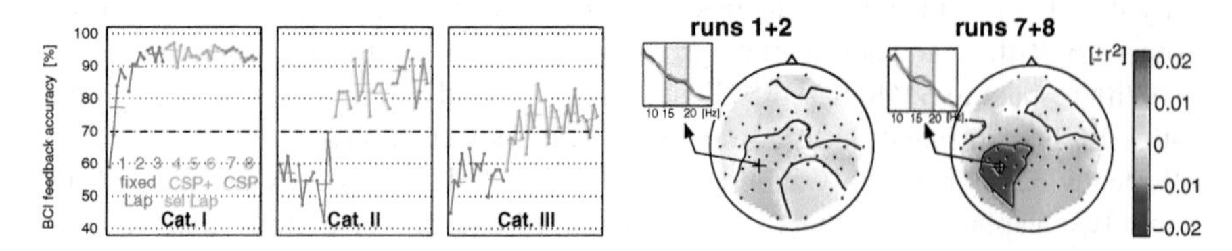

Fig. 7 *Left:* Grand average of feedback performance within each run (*horizontal bars* and *dots* for each group of 20 trials) for subjects of Cat. I (N=6), Cat. II (N=2) and Cat. III (N=3). An accuracy of 70% is assumed to be a threshold required for BCI applications. Note that all runs of one subject have been recorded within one session. *Right:* For one subject of Cat. III, spectra in channel CP3 and scalp topographies of band-power differences (signed r^2-values)[1] between the motor imagery conditions are compared between the beginning (runs 1+2) and the end (runs 7+8) of the experiment

[1] The r^2 value (squared biserial correlation coefficient) is a measure for how much of the variance in one variable is explained by the variance in a second variable, i.e., it is a measure for how good a single feature is in predicting the label.

In particular, one subject who had no SMR idle rhythm in the beginning of the measurement could develop such with our feedback training, see Fig. 7. This important finding gives rise to the development of neurofeedback training procedures that might help to cure BCI illiteracy.

4 Applications of BBCI Technology

Subsequently, we will discuss BBCI applications for rehabilitation (prosthetic control and spelling [1, 2, 46]) and *beyond* (gaming, mental state monitoring [55, 56] etc.). Our view is that the development of BCI to enhance man machine interaction for the healthy will be an important step to broaden and strengthen the future development of neurotechnology.

4.1 Prosthetic Control

Motor-intention based BCI offers the possibility of a direct and intuitive control modality for persons disabled by high-cervical spinal cord injury, i.e., tetraplegics, whose control of all limbs is severely impaired. The advantage of this type of BCI over other interface modalities is that by directly translating movement intention into a command to a prosthesis, the link between cortical activity related to motor control of the arm and physical action is restored, thereby offering a possible rehabilitation function, as well as enhanced motivation factor for daily use. Testing this concept is the main idea driving the Brain2Robot project (see Acknowledgment). However, two important challenges must be fully met before noninvasive, EEG based motor imagery BCIs can be practically used by the disabled.

One such challenge is the cumbersome nature of standard EEG set-up, involving application of gel, limited recording time, and subsequent removal of the set-up, which involves washing the hair. It is unlikely that disabled persons, in need of BCI technology for greater autonomy, would widely adopt such a system. Meanwhile, short of any invasive or minimally invasive recording modality, the only available option is the use of so called “dry” electrodes, which do not require the use of conductive gel or other liquids in such a way that electrode application and removal takes place in a matter of minutes. We have developed such a technology (“dry cap”) and tested it for motor-imagery based BCI [57]. The cap required about 5 min for set-up and exhibited an average of 70% of the information transfer rate achieved for the same subjects with respect to a standard EEG “gel cap”, the difference being most likely attributed to the use of 6 electrodes used in the dry cap vs. 64 electrodes used in the gel cap. Although the locations of the 6 electrodes were chosen judiciously (by analyzing which electrode positions in the gel cap were most important, as expected 3 electrodes over each cortical motor area), some performance degradation was unavoidable and necessary – a full 64 electrode dry cap would also be cumbersome.

Another challenge for EEG-BCI control of prosthetics is inherent safety. This is of paramount importance, whether the prosthetic controlled is an orthosis (a worn mechanical device which augments the function of a set of joints) or a robot (which may move the paralysed arm or be near the body but unattached to it, as in the case of Brain2Robot), or even a neuroprosthesis, i.e. a system that electrically activates muscles in the user's arm or peripheral neurons that innervate these muscles. Specifically, the BCI interface should not output spurious or unintended action commands to the prosthetic device, as these could cause injuries. Even if the probability of injury is low and secondary safety "escape commands" are incorporated, it may (reasonably) cause fear in the otherwise immobile user and therefore discourage him or her from continuing to use the system. Therefore, we have looked at necessary enhancements to commonly used "BCI feedback" control which could incorporate the use of a "rest" or "idle" state, i.e. a continuous output of the classifier that not only outputs a command related to a trained brain state (say, imagination of left hand movement) but a "do nothing" command related to a state in which the user performs daily activities unrelated to motor imagination and in which the prosthetic should do nothing. Thus, we have begun to look at the trade-off between speed of BCI (information transmission rate or ITR) and safety (false positive rate) achievable by incorporating a "control" law, which is a differential equation whose inputs are continuous outputs of the classifer, in our case a quadratic-type classifier, and whose output is the command to the prosthetic [58]. It remains to be seen how much each particular subject, whose "standard" BCI performance varies greatly, must trade reduced speed for increased safety.

A final implicit goal of all BCI research is to improve the maximally achievable ITR for each brain imaging modality. In the case of EEG, the ITR seems to be limited to about 1 decision every 2 s ([42]. The fastest subject performed at an average speed of 1 binary decision every 1.7 s; see also [59]), despite intensive research effort to improve speed. In the case of Brain2Robot, further information about the desired endpoint of arm movement is obtained by 3D tracking of gaze – eye movement and focus are normally intact in the tetraplegic population, and the achievable ITR is sufficient, since it lies in the range of the frequency of discrete reaching movements of the hand. However, competing issues of cognitive load, safety and achievable dexterity can only be assessed by testing BCIs for prosthetic control with the intended user group while paying attention to the level of disability and motor-related EEG patterns in each subject, as both are likely to vary significantly.

4.2 Time-Critical Applications: Prediction of Upcoming Movements

In time-critical control situations, BCI technology might provide early detection of reactive movements based on preparatory signals for the reduction of the time span between the generation of an intention (or reactive movements) and the onset of the intended technical operation (e.g. in driver-assisted measures for vehicle safety). Through detection of particularly early readiness potentials (see Section 1.2), which

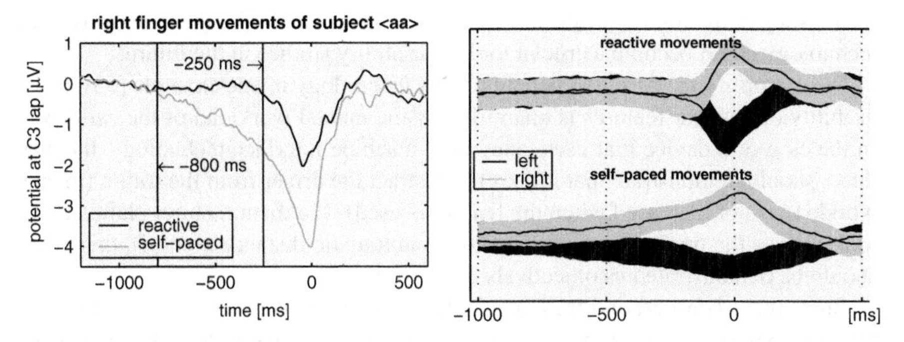

Fig. 8 *Left:* Averaged readiness potential in spontaneous selfpaced (*grey*) and reactive (*dark*) finger movements (with $t = 0$ at key press) for one subject. *Right:* Distribution of the continuous classifier output in both experimental settings

reflect the mental preparation of movements, control actions can be prepared or initiated before the actual movement and thus we intend to decode these signals in a very timely and accurate manner.

In order to explore the prospective value of BCIs for such applications, we conducted a two alternative forced choice experiment (d2-test), in which the subject had to respond as fast as possible with a left or right index finger key press, see [60]. Figure 8 (left) compares the readiness potentials in such reactive finger movements with those in selfpaced finger movements ($t = 0$ for key press). Figure 8 (right) shows the traces of continuous classifier output for reactive (upper subplot) and self-paced (lower subplot) finger movements. As expected, the discrimination between upcoming left vs. right finger movements is better for the self-paced movements at an *early* stage, but is similar towards the time point of key press performance. In particular, 100 ms before the keypress, even for movements in fast reactions, a separation becomes substantial. The discriminability already at this point in time confirms the potential value of BCI technology for time-critical applications. For more details and classification results, we refer the interested reader to [60].

4.3 Neuro Usability

In the development of many new products or in the improvement of existing products, usability studies play an important role. They are performed in order to measure to what degree a product meets the intended purpose with regard to the aspects effectiveness, efficiency and user satisfaction. A further goal is to quantify the joy of use. While effectiveness can be quantified quite objectively, e.g., in terms of task completion, the other aspects are more intricate to assess. Even psychological variables consciously inaccessible to the persons themselves might be involved. Furthermore, in usability studies, it is of interest to perform an effortless continuous acquistion of usability parameters whilst not requiring any action on the side of the subject as this might interfer with the task at hand. For these reasons, BCI technology could become a crucial tool for usability studies in the future.

We exemplify the potential benefit of BCI technology in one example [55]. Here, usability of new car features is quantified by the mental workload of the car driver. In the case of a device that uses fancy man-machine interface technology, the producer should demonstrate that it does not distract the driver from the traffic (mental workload is not increased when the feature is used). If a manufacturer claims that a tool relieves the driver from workload (e.g., automatic distance control), this effect should be demonstrated as objectively as possible.

Since there is no ground truth available on the cognitive workload to which the driver is exposed, we designed a study[2] in which additional workload was induced in a controlled manner. For details, confer [55]. EEG was acquired from 12 male and 5 female subjects while driving on a highway at a speed of 100 km/h (primary task). Second, the subjects had an auditory reaction task: one of two buttons mounted on the left and right index finger had to be hit every 7.5 s according to a given vocal prompt. For the tertiary task, two different conditions have been used. (a) mental calculation; (b) following one of two simultaneously broadcast voice recordings. In an initial calibration phase, the developed BBCI workload detector was adapted to the individual driver. After that, the system was able to predict the cognitive workload of the driver online. This information was used in the test phase to switch off the auditory reaction task when high workload was detected ("mitigation").

As a result of the mitigation strategy, the reaction time in the test phase was on average 100 ms faster than in the (un-mitigated) calibration phase [55]. Since in total the workload during the two phases has been equal, it can be conjectured that the average reactivity was the same. Thus, the difference in reaction times can only be explained by the fact that the workload detector switched off the reaction task during periods of reduced reactivity.

Note that the high intersubject variability, which is a challenge for many BCI applications comes as an advantage here: for neuro-usability studies, top subjects (with respect to the detectability of relevant EEG components) of a study can be selected according to the appropriateness of their brain signals.

Beyond the neuro usability aspect of the study, one could speculate that such devices might be incorporated in future cars in order to reduce distractions (e.g., navigation system is switched off during periods of high workload) to a minimum when the driver's brain is already over-loaded by other demands during potentially hazardous situations.

4.4 Mental State Monitoring

When aiming to optimize the design of user interfaces or, more general, of a work flow, the mental state of a user during the task execution can provide useful information. This information can not only be exploited for the improvement of BCI

[2]This study was performed in cooperation with the Daimler AG. For further information, we refer to [55].

applications, but also for improving industrial production environments, the user interface of cars and for many other applications. Examples of these mental states are the levels of arousal, fatigue, emotion, workload or other variables whose brain activity correlates (at least partially) are amenable to measurement. The improvement of suboptimal user interfaces reduces the number of critical mental states of the operators. Thus it can lead to an increase in production yield, less errors and accidents, and avoids frustration of the users.

Typically, information collected about the mental states of interest is exploited in an offline analysis of the data and leads to a redesign of the task or the interface. In addition, a method for mental state monitoring can be applied online during the execution of a task might be desirable. Traditional methods for capturing mental states and user ratings are questionnaires, video surveillance of the task, or the analysis of errors made by the operator. However, questionnaires are of limited use for precisely assessing the information of interest, as the delivered answers are often distorted by subjectiveness. Questionnaires cannot determine the quantities of interest in real-time (during the execution of the task) but only in retrospect; moreover, they are intrusive because they interfere with the task. Even the monitoring of eye blinks or eye movements only allows for an indirect access to the user's mental state. Although the monitoring of a user's errors is a more direct measure, it detects critical changes of the user state post-hoc only. Neither is the anticipation of an error possible, nor can suitable countermeasures be taken to avoid it.

As a new approach, we propose the use of EEG signals for mental state monitoring and combine it with BBCI classfication methods for data analysis. With this approach, the brain signals of interest can be isolated from background activity as in BCI systems; this combination allows for the non-intrusive evaluation of mental states in real-time and on a single-trial basis such that an online system with feedback can be built.

In a pilot study [56] with four participants we evaluated the use of EEG signals for arousal monitoring. The experimental setting simulates a security surveillance system where the sustained concentration ability of the user in a rather boring task is crucial. As in BCI, the system had to be calibrated to the individual user in order to recognize and predict mental states, correlated with attention, task involvement or a high or low number of errors of the subject respectively.

4.4.1 Experimental Setup for Attention Monitoring

The subject was seated approx. 1 m in front of a computer screen that displayed different stimuli in a forced choice setting. She was asked to respond quickly to stimuli by pressing keys of a keyboard with either the left or right index finger; recording was done with a 128 channel EEG at 100 Hz. The subject had to rate several hundred x-ray images of luggage objects as either dangerous or harmless by a key press after each presentation. The experiment was designed as an oddball paradigm where the number of the harmless objects was much larger than that of the dangerous objects. The terms standard and deviant will subsequently be used for the two conditions. One trial was usually performed within 0.5 s after the cue presentation.

The subject was asked to perform 10 blocks of 200 trials each. Due to the monotonous nature of the task and the long duration of the experiment, the subject was expected to show a fading level of arousal, which impairs concentration and leads to more and more erroneous decisions during later blocks.

For the offline analysis of the collected EEG signals, the following steps were applied. After exclusion of channels with bad impedances, a spatial Laplace filter was applied and the band power features from 8–13 Hz were computed on 2 s windows. The resulting band power values of all channels were concatenated into a final vector. As the subject's correct and erroneous decisions were known, a supervised LDA classifier was trained on the data. The classification error of this procedure was estimated by a cross-validation scheme that left out a whole block of 200 trials during each fold for testing. As the number of folds was determined by the number of experimental blocks, it varied slightly from subject to subject.

4.4.2 Results

The erroneous decisions taken by a subject were recorded and smoothed in order to form a measure for the arousal. This measure is hereafter called the *error index* and reflects the subject's ability to concentrate and fulfill the security task. To enhance the contrast of the discrimination analysis, two thresholds were introduced for the error index and set after visual inspection. Extreme trials outside these thresholds defined two sets of trials with a rather high or low value. The EEG data of the trials were labeled as *sufficiently concentrated* or *insufficiently concentrated* depending on these thresholds for later analysis. Figure 9 shows the error index. The subject did perform nearly error-free during the first blocks, but then showed increasing errors beginning with block 4. However, as the blocks were separated by short breaks, the subject could regain attention at the beginning of each new block at least for a small number of trials. The trials of high and low error index formed the training data for teaching a classifier to discriminate mental states of insufficient arousal based on single trial EEG data.

A so-called Concentration Insufficiency Index (CII) of a block was generated by an LDA classifier that had been trained off-line on the labeled training data of the remaining blocks. The classifier output (CII) of each trial is plotted in Fig. 9 together with the corresponding error index. It can be observed that the calculated CII mirrors the error index for most blocks. More precisely the CII mimics the error increase inside each block and in blocks 3 and 4 it can anticipate the increase of later blocks, i.e. out-of-sample. For those later blocks, the CII reveals that the subject could not recover full arousal during the breaks. Instead, it shows a short-time arousal for the time immediately after a break, but the CII accumulates over time.

The correlation coefficient of both time series with varying temporal delay is shown in the right plot of Fig. 9. The CII inferred by the classifier and the errors that the subject had actually produced correlate strongly. Furthermore, the correlation is high even for predictions that are up to 50 trials into the future.

For a physiological analysis please refer to the original paper [56].

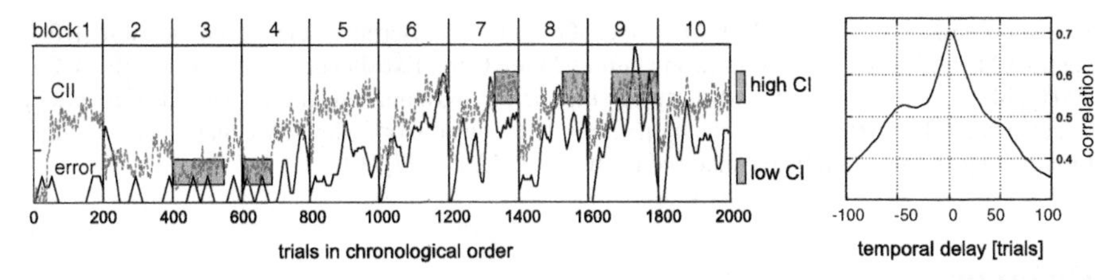

Fig. 9 *Left*: Comparison of the concentration insufficiency index (CII, *dotted curve*) and the error index for the subject. The error index (the true performed errors smoothed over time) reflects the inverse of the arousal of the subject. *Right*: Correlation coefficient between the CII (returned by the classifier) and the true performance for different time shifts. Highest correlation is around a zero time shift, as expected. Note that the CII has an increased correlation with the error even before the error appears

5 Conclusion

The chapter provides a brief overview on the Berlin Brain-Computer Interface. We would like to emphasize that the use of modern machine learning tools – as put forward by the BBCI group – is pivotal for a successful and high ITR operation of a BCI from the first session [48, 42]. Note that, due to space limitations, the chapter can only discuss general principles of signal processing and machine learning for BCI; for details ample references are provided (see also [2]). Our main emphasis was to discuss the wealth of applications of neurotechnology beyond rehabilitation. While BCI is an established tool for opening a communication channel for the severely disabled [61–66], its potential as an instrument for enhancing man-machine interaction is underestimated. The use of BCI technology as a direct channel additional to existing means to communicate opens applications in mental state monitoring [55, 56], gaming [67, 68], virtual environment navigation [69], vehicle safety [55], rapid image viewing [70] and enhanced user modeling. To date, only proofs of concept and first steps have been given that still need to move a long way to innovative products, but the attention monitoring and neuro usability applications outlined in Sections 4.3 and 4.4 already show the usefulness of neurotechnology for monitoring complex cognitive mental states. With our novel technique at hand, we can make direct use of mental state monitoring information to enable Human-Machine Interaction to exhibit adaptive anticipatory behaviour.

To ultimately succeed in these promising applications the BCI field needs to proceed in multiple aspects: (a) improvement of EEG technology beyond gel electrodes and (e.g. [57]) towards cheap and portable devices, (b) understanding of the BCI-illiterates phenomenon, (c) improved and more robust signal processing and machine learning methods, (d) higher ITRs for noninvasive devices and finally (e) the development of compelling industrial applications outside the realm of rehabilitation.

Acknowledgment We are very grateful to Nicole Krämer (TU Berlin) for pointing us to the analytic solution of the optimal shrinkage parameter for regularized linear discriminant analysis, see Section 2.2. The studies were partly supported by the *Bundesministerium für Bildung und*

Forschung (BMBF), Fkz 01IB001A/B, 01GQ0850, by the German Science Foundation (DFG, contract MU 987/3-1), by the European Union's Marie Curie Excellence Team project MEXT-CT-2004-014194, entitled "Brain2Robot" and by their IST Programme under the PASCAL Network of Excellence, ICT-216886. This publication only reflects the authors' views. We thank our coauthors for allowing us to use published material from [7, 48, 52, 55–57, 60].

References

1. B. Blankertz, M. Krauledat, G. Dornhege, J. Williamson, R. Murray-Smith, and K.-R. Müller, A note on brain actuated spelling with the Berlin Brain-Computer Interface, In C. Stephanidis, (Ed.), *Universal access in HCI, Part II, HCII 2007*, ser. LNCS, vol. 4555. Springer, Berlin Heidelberg, pp. 759–768 (2007).
2. G. Dornhege, J. del R. Millán, T. Hinterberger, D. McFarland, and K.-R. Müller (Eds.), *Toward brain-computer interfacing*. MIT Press, Cambridge, MA, (2007).
3. J.R. Wolpaw, N. Birbaumer, D.J. McFarland, G. Pfurtscheller, and T.M. Vaughan, Brain-computer interfaces for communication and control, Clin Neurophysiol, 113(6), 767–791, (2002).
4. B. Allison, E. Wolpaw, and J. Wolpaw, Brain-computer interface systems: progress and prospects. Expert Rev Med Devices, 4(4), 463–474, (2007).
5. G. Pfurtscheller, C. Neuper, and N. Birbaumer, Human Brain-Computer Interface, In A. Riehle and E. Vaadia (Eds.), *Motor cortex in voluntary movements*, CRC Press, New York NY: ch. 14, pp. 367–401, (2005).
6. J. Haynes, K. Sakai, G. Rees, S. Gilbert, and C. Frith, Reading hidden intentions in the human brain, Curr Biol, 17, 323–328, (2007).
7. B. Blankertz, G. Dornhege, S. Lemm, M. Krauledat, G. Curio, and K.-R. Müller, The Berlin Brain-Computer Interface: machine learning based detection of user specific brain states. J Universal Computer Sci, 12(6), 581–607, (2006).
8. T. Elbert, B. Rockstroh, W. Lutzenberger, and N. Birbaumer, Biofeedback of slow cortical potentials. I. Electroencephalogr Clin Neurophysiol, 48, 293–301, (1980).
9. N. Birbaumer, A. Kübler, N. Ghanayim, T. Hinterberger, J. Perelmouter, J. Kaiser, I. Iversen, B. Kotchoubey, N. Neumann, and H. Flor, The thought translation device (TTD) for completly paralyzed patients. IEEE Trans Rehabil Eng, 8(2), 190–193, (June 2000).
10. M. Schreuder, B. Blankertz, and M. Tangermann, A new auditory multi-class brain-computer interface paradigm: Spatial hearing as an informative cue. PLoS ONE, 5(4), p. e9813, (2010).
11. H. H. Kornhuber and L. Deecke, Hirnpotentialänderungen bei Willkürbewegungen und passiven Bewegungen des Menschen: Bereitschaftspotential und reafferente Potentiale. Pflugers Arch, 284, 1–17, (1965).
12. W. Lang, M. Lang, F. Uhl, C. Koska, A. Kornhuber, and L. Deecke, Negative cortical DC shifts preceding and accompanying simultaneous and sequential movements. Exp Brain Res, 74(1), 99–104, (1988).
13. R.Q. Cui, D. Huter, W. Lang, and L. Deecke, Neuroimage of voluntary movement: topography of the Bereitschaftspotential, a 64-channel DC current source density study. Neuroimage, 9(1), 124–134, (1999).
14. B. Blankertz, G. Dornhege, M. Krauledat, K.-R. Müller, V. Kunzmann, F. Losch, and G. Curio, The Berlin Brain-Computer Interface: EEG-based communication without subject training. IEEE Trans Neural Syst Rehabil Eng, 14(2), 147–152, (2006). [Online]. Available: http://dx.doi.org/10.1109/TNSRE.2006.875557. Accessed 14 Sept 2010.
15. B. Blankertz, G. Dornhege, M. Krauledat, V. Kunzmann, F. Losch, G. Curio, and K.-R. Müller, The berlin brain-computer interface: machine-learning based detection of user specific brain states, In G. Dornhege, J. del R. Millán, T. Hinterberger, D. McFarland, and K.-R. Müller (Eds.), *Toward Brain-Computer Interfacing*, MIT, Cambridge, MA, pp. 85–101, (2007).

16. B. Blankertz, G. Dornhege, C. Schäfer, R. Krepki, J. Kohlmorgen, K.-R. Müller, V. Kunzmann, F. Losch, and G. Curio, Boosting bit rates and error detection for the classification of fast-paced motor commands based on single-trial EEG analysis. IEEE Trans Neural Syst Rehabil Eng, 11(2), 127–131, (2003). [Online]. Available: http://dx.doi.org/10.1109/TNSRE.2003.814456. Accessed on 14 Sept 2010.
17. D. Krusienski, G. Schalk, D.J. McFarland, and J. Wolpaw, A mu-rhythm matched filter for continuous control of a brain-computer interface. IEEE Trans Biomed Eng, 54(2), 273–280, (2007).
18. V.V. Nikulin and T. Brismar, Phase synchronization between alpha and beta oscillations in the human electroencephalogram. Neuroscience, 137, 647–657, (2006).
19. G. Pfurtscheller and F.H.L. da Silva, Event-related EEG/MEG synchronization and desynchronization: basic principles. Clin Neurophysiol, 110(11), 1842–1857, (Nov 1999).
20. P.L. Nunez, R. Srinivasan, A.F. Westdorp, R.S. Wijesinghe, D.M. Tucker, R.B. Silberstein, and P.J. Cadusch, EEG coherency I: statistics, reference electrode, volume conduction, Laplacians, cortical imaging, and interpretation at multiple scales. Electroencephalogr Clin Neurophysiol, 103(5), 499–515, (1997).
21. B. Blankertz, R. Tomioka, S. Lemm, M. Kawanabe, and K.-R. Müller, Optimizing spatial filters for robust EEG single-trial analysis. IEEE Signal Process Mag, 25(1), 41–56, (Jan. 2008). [Online]. Available:http://dx.doi.org/10.1109/MSP.2008.4408441. Accessed on 14 Sept 2010.
22. D.J. McFarland, L.M. McCane, S.V. David, and J.R. Wolpaw, Spatial filter selection for EEG-based communication. Electroencephalogr Clin Neurophysiol, 103, 386–394, (1997).
23. N. Hill, T.N. Lal, M. Tangermann, T. Hinterberger, G. Widman, C.E. Elger, B. Schölkopf, and N. Birbaumer, Classifying event-related desynchronization in EEG, ECoG and MEG signals, In G. Dornhege, J. del R. Millán, T. Hinterberger, D. McFarland, and K.-R. Müller (Eds.), *Toward brain-computer interfacing*, MIT, Cambridge, MA, pp. 235–260, (2007).
24. M. Grosse-Wentrup, K. Gramann, and M. Buss, Adaptive spatial filters with predefined region of interest for EEG based brain-computer-interfaces, In B. Schölkopf, J. Platt, and T. Hoffman, (Eds.), *Advances in neural information processing systems 19*, pp. 537–544, (2007).
25. G. Dornhege, M. Krauledat, K.-R. Müller, and B. Blankertz, General signal processing and machine learning tools for BCI, In G. Dornhege, J. del R. Millán, T. Hinterberger, D. McFarland, and K.-R. Müller, (Eds.), *Toward brain-computer interfacing*, MIT, Cambridge, MA, pp. 207–233, (2007).
26. L.C. Parra, C.D. Spence, A.D. Gerson, and P. Sajda, Recipes for the linear analysis of EEG, Neuroimage, 28(2), 326–341, (2005).
27. K. Fukunaga, *Introduction to statistical pattern recognition*, 2nd ed. Academic, San Diego CA, (1990).
28. C. Guger, H. Ramoser, and G. Pfurtscheller, Real-time EEG analysis with subject-specific spatial patterns for a Brain Computer Interface (BCI). IEEE Trans Neural Syst Rehabil Eng, 8(4), 447–456, (2000).
29. H. Ramoser, J. Müller-Gerking, and G. Pfurtscheller, Optimal spatial filtering of single trial EEG during imagined hand movement. IEEE Trans Rehabil Eng, 8(4), 441–446, (2000).
30. S. Lemm, B. Blankertz, G. Curio, and K.-R. Müller, Spatio-spectral filters for improving classification of single trial EEG. IEEE Trans Biomed Eng, 52(9), 1541–1548, (2005). [Online]. Available: http://dx.doi.org/10.1109/TBME.2005.851521. Accessed on 14 Sept 2010.
31. G. Dornhege, B. Blankertz, M. Krauledat, F. Losch, G. Curio, and K.-R. Müller, Optimizing spatio-temporal filters for improving brain-computer interfacing, In Y. Weiss, B. Schölkopf, and J. Platt, (Eds.), *Advances in neural Information Processing Systems (NIPS 05)*, vol. 18. MIT, Cambridge, MA, pp. 315–322, (2006).
32. R. Tomioka, K. Aihara, and K.-R. Müller, Logistic regression for single trial EEG classification, In B. Schölkopf, J. Platt, and T. Hoffman (Eds.), *Advances in neural information processing systems 19*, MIT, Cambridge, MA, pp. 1377–1384, (2007).
33. K.-R. Müller, C.W. Anderson, and G.E. Birch, Linear and non-linear methods for brain-computer interfaces. IEEE Trans Neural Syst Rehabil Eng, 11(2), 165–169, (2003).

34. R.O. Duda, P.E. Hart, and D.G. Stork, *Pattern classification*, 2nd ed. Wiley, New York, (2001).
35. C. Stein, Inadmissibility of the usual estimator for the mean of a multivariate normal distribution. Proceeding of 3rd Berkeley Symposium Mathematical Statistics. Probability 1, 197–206, (1956).
36. J.H. Friedman, Regularized discriminant analysis. J Am Stat Assoc, 84(405), 165–175, (1989).
37. O. Ledoit and M. Wolf, A well-conditioned estimator for large-dimensional covariance matrices. J Multivar Anal, 88, 365–411, (2004).
38. C. Vidaurre, N. Krämer, B. Blankertz, and A. Schlögl, Time domain parameters as a feature for EEG-based Brain Computer Interfaces. Neural Netw, 22, 1313–1319, (2009).
39. J. Schäfer and K. Strimmer, A shrinkage approach to large-scale covariance matrix estimation and implications for functional genomics. Stat Appl Genet Mol Biol, 4, Article32, (2005) [Online]. Available: http://www.bepress.com/sagmb/vol4/iss1/art32/.
40. C. Neuper, R. Scherer, M. Reiner, and G. Pfurtscheller, Imagery of motor actions: Differential effects of kinesthetic and visual-motor mode of imagery in single-trial EEG. Brain Res Cogn Brain Res, 25(3), 668–677, (2005).
41. V.V. Nikulin, F.U. Hohlefeld, A.M. Jacobs, and G. Curio, Quasi-movements: a novel motor-cognitive phenomenon, Neuropsychologia, 46(2), 727–742, (2008). [Online]. Available: http://dx.doi.org/10.1016/j.neuropsychologia.2007.10.008. Accessed on 14 Sept 2010.
42. B. Blankertz, G. Dornhege, M. Krauledat, K.-R. Müller, and G. Curio, The non-invasive Berlin Brain-Computer Interface: fast acquisition of effective performance in untrained subjects. Neuroimage, 37(2), 539–550, (2007). [Online]. Available: http://dx.doi.org/10.1016/j.neuroimage.2007.01.051. Accessed on 14 Sept 2010.
43. J.R. Wolpaw, D.J. McFarland, and T.M. Vaughan, Brain-computer interface research at the Wadsworth Center. IEEE Trans Rehabil Eng, 8(2), 222–226, (2000).
44. A. Schlögl, J. Kronegg, J. Huggins, and S.G. Mason, Evaluation Criteria for BCI Research, In G. Dornhege, J. del R. Millán, T. Hinterberger, D. McFarland, and K.-R. Müller, (Eds.), *Towards Brain-Computer Interfacing*, MIT, Cambridge, MA, pp. 297–312, 2007.
45. G. Dornhege, Increasing information transfer rates for brain-computer interfacing, Ph.D. dissertation, University of Potsdam, (2006).
46. K.-R. Müller and B. Blankertz, Toward noninvasive brain-computer interfaces. IEEE Signal Process Mag, 23(5), 125–128, (Sept 2006).
47. J. Williamson, R. Murray-Smith, B. Blankertz, M. Krauledat, and K.-R. Müller, Designing for uncertain, asymmetric control: Interaction design for brain-computer interfaces. Int J Hum-Comput Stud, 67(10), 827–841, (2009).
48. B. Blankertz, F. Losch, M. Krauledat, G. Dornhege, G. Curio, and K.-R. Müller, The Berlin Brain-Computer Interface: Accurate performance from first-session in BCI-naive subjects. IEEE Trans Biomed Eng, 55(10), 2452–2462, 2008. [Online]. Available: http://dx.doi.org/10.1109/TBME.2008.923152. Accessed on 14 Sept 2010.
49. A. Kübler and K.-R. Müller, An introduction to brain computer interfacing, In G. Dornhege, J. del R. Millán, T. Hinterberger, D. McFarland, and K.-R. Müller, (Eds.), *Toward Brain-Computer Interfacing*, MIT, Cambridge, MA, pp. 1–25, (2007).
50. B. Blankertz, C. Sannelli, S. Halder, E. M. Hammer, A. Kübler, K. R. Müller, G. Curio, and T. Dickhaus, Neurophysiological predictor of SMR-based BCI performance. NeuroImage, 51, 1303–1309, (2010).
51. T. Dickhaus, C. Sannelli, K.-R. Müller, G. Curio, and B. Blankertz, Predicting BCI performance to study BCI illiteracy. BMC Neurosci 2009, 10(Suppl 1), P84, (2009).
52. B. Blankertz and C. Vidaurre, Towards a cure for BCI illiteracy: Machine-learning based co-adaptive learning. BMC Neuroscience 2009, 10, (Suppl 1), P85, (2009).
53. C. Vidaurre and B. Blankertz, Towards a cure for BCI illiteracy, Open Access Brain Topogr, 23, 1303–1309, (2010).
54. C. Vidaurre, A. Schlögl, B. Blankertz, M. Kawanabe, and K.-R. Müller, Unsupervised adaptation of the LDA classifier for Brain-Computer Interfaces, in Proceedings of the 4th International Brain-Computer Interface Workshop and Training Course 2008. Verlag der Technischen Universität Graz, (2008), pp. 122–127.

55. J. Kohlmorgen, G. Dornhege, M. Braun, B. Blankertz, K.-R. Müller, G. Curio, K. Hagemann, A. Bruns, M. Schrauf, and W. Kincses, Improving human performance in a real operating environment through real-time mental workload detection, In G. Dornhege, J. del R. Millán, T. Hinterberger, D. McFarland, and K.-R. Müller, (Eds.), *Toward Brain-Computer Interfacing*, MIT, Cambridge, MA, pp. 409–422, (2007).
56. K.-R. Müller, M. Tangermann, G. Dornhege, M. Krauledat, G. Curio, and B. Blankertz, Machine learning for real-time single-trial EEG-analysis: from brain-computer interfacing to mental state monitoring. J Neurosci Methods, 167(1), 82–90, (2008). [Online]. Available: http://dx.doi.org/10.1016/j.jneumeth.2007.09.022. Accessed on 14 Sept 2010.
57. F. Popescu, S. Fazli, Y. Badower, B. Blankertz, and K.-R. Müller, Single trial classification of motor imagination using 6 dry EEG electrodes, PLoS ONE, 2(7), (2007). [Online]. Available: http://dx.doi.org/10.1371/journal.pone.0000637
58. S. Fazli, M. Danóczy, M. Kawanabe, and F. Popescu, Asynchronous, adaptive BCI using movement imagination training and rest-state inference. IASTED's Proceedings on Artificial Intelligence and Applications 2008. Innsbruck, Austria, ACTA Press Anaheim, CA, USA, (2008), pp. 85–90. [Online]. Available: http://portal.acm.org/citation.cfm?id=1712759.1712777
59. L. Ramsey, M. Tangermann, S. Haufe, and B. Blankertz, Practicing fast-decision BCI using a "goalkeeper" paradigm. BMC Neurosci 2009, 10(Suppl 1), P69, (2009).
60. M. Krauledat, G. Dornhege, B. Blankertz, G. Curio, and K.-R. Müller, The Berlin brain-computer interface for rapid response. Biomed Tech, 49(1), 61–62, (2004).
61. A. Kübler, B. Kotchoubey, J. Kaiser, J. Wolpaw, and N. Birbaumer, Brain-computer communication: Unlocking the locked in. Psychol Bull, 127(3), 358–375, (2001).
62. A. Kübler, F. Nijboer, J. Mellinger, T. M. Vaughan, H. Pawelzik, G. Schalk, D. J. McFarland, N. Birbaumer, and J. R. Wolpaw, Patients with ALS can use sensorimotor rhythms to operate a brain-computer interface. Neurology, 64(10), 1775–1777, (2005).
63. N. Birbaumer and L. Cohen, Brain-computer interfaces: communication and restoration of movement in paralysis. J Physiol, 579, 621–636, (2007).
64. N. Birbaumer, C. Weber, C. Neuper, E. Buch, K. Haapen, and L. Cohen, Physiological regulation of thinking: brain-computer interface (BCI) research. Prog Brain Res, 159, 369–391, (2006).
65. L. Hochberg, M. Serruya, G. Friehs, J. Mukand, M. Saleh, A. Caplan, A. Branner, D. Chen, R. Penn, and J. Donoghue, Neuronal ensemble control of prosthetic devices by a human with tetraplegia. Nature, 442(7099), 164–171, (Jul 2006).
66. J. Conradi, B. Blankertz, M. Tangermann, V. Kunzmann, and G. Curio, Brain-computer interfacing in tetraplegic patients with high spinal cord injury. Int J Bioelectromagnetism, 11, 65–68, (2009).
67. R. Krepki, B. Blankertz, G. Curio, and K.-R. Müller, The Berlin Brain-Computer Interface (BBCI): towards a new communication channel for online control in gaming applications. J Multimedia Tools, 33(1), 73–90, (2007). [Online]. Available: http://dx.doi.org/10.1007/s11042-006-0094-3. Accessed on 14 Sept 2010.
68. R. Krepki, G. Curio, B. Blankertz, and K.-R. Müller, Berlin brain-computer interface - the hci communication channel for discovery. Int J Hum Comp Studies, 65, 460–477, (2007), special Issue on Ambient Intelligence.
69. R. Leeb, F. Lee, C. Keinrath, R. Scherer, H. Bischof, and G. Pfurtscheller, Brain-computer communication: motivation, aim, and impact of exploring a virtual apartment. IEEE Trans Neural Syst Rehabil Eng, 15(4), 473–482, (2007).
70. A. Gerson, L. Parra, and P. Sajda, Cortically coupled computer vision for rapid image search. IEEE Trans Neural Syst Rehabil Eng, 14(2), 174–179, (2006).

Practical Designs of Brain–Computer Interfaces Based on the Modulation of EEG Rhythms

Yijun Wang, Xiaorong Gao, Bo Hong, and Shangkai Gao

1 Introduction

A brain–computer interface (BCI) is a communication channel which does not depend on the brain's normal output pathways of peripheral nerves and muscles [1–3]. It supplies paralyzed patients with a new approach to communicate with the environment. Among various brain monitoring methods employed in current BCI research, electroencephalogram (EEG) is the main interest due to its advantages of low cost, convenient operation and non-invasiveness. In present-day EEG-based BCIs, the following signals have been paid much attention: visual evoked potential (VEP), sensorimotor mu/beta rhythms, P300 evoked potential, slow cortical potential (SCP), and movement-related cortical potential (MRCP). Details about these signals can be found in chapter "Brain Signals for Brain–Computer Interfaces". These systems offer some practical solutions (e.g., cursor movement and word processing) for patients with motor disabilities.

In this chapter, practical designs of several BCIs developed in Tsinghua University will be introduced. First of all, we will propose the paradigm of BCIs based on the modulation of EEG rhythms and challenges confronting practical system designs. In Sect. 2, modulation and demodulation methods of EEG rhythms will be further explained. Furthermore, practical designs of a VEP-based BCI and a motor imagery based BCI will be described in Sect. 3. Finally, Sect. 4 will present some real-life application demos using these practical BCI systems.

1.1 BCIs Based on the Modulation of Brain Rhythms

Many of the current BCI systems are designed based on the modulation of brain rhythms. For example, power modulation of mu/beta rhythms is used in the BCI system based on motor imagery [4]. Besides, phase modulation is another method

S. Gao (✉)
Department of Biomedical Engineering, School of Medicine, Tsinghua University, Beijing, China
e-mail: gsk-dea@tsinghua.edu.cn

B. Graimann et al. (eds.), *Brain–Computer Interfaces*, The Frontiers Collection,
DOI 10.1007/978-3-642-02091-9_8, © Springer-Verlag Berlin Heidelberg 2010

which has been employed in a steady-state visual evoked potential (SSVEP) based BCI [5]. More generally, evoked potentials can be considered to result partially from a reorganization of the phases of the ongoing EEG rhythms [6].

From the viewpoint of psychophysiology, EEG signals are divided into five rhythms in different frequency bands: delta rhythm (0.1–3.5 Hz), theta rhythm (4–7.5 Hz), alpha rhythm (8–13 Hz), beta rhythm (14–30 Hz), and gamma rhythm (>30 Hz) [7]. Although the rhythmic character of EEG has been observed for a long period, many new studies on the mechanisms of brain rhythms emerged after the 1980s. So far, the cellular bases of EEG rhythms are still under investigation. The knowledge of EEG rhythms is limited; however, numerous neurophysiologic studies indicate that brain rhythms can reflect changes of brain states caused by stimuli from the environment or cognitive activities. For example, EEG rhythms can indicate working state or idling state of the functional areas of the cortex. It is known that the alpha rhythm recorded over the visual cortex is considered to be an indicator of activities in the visual cortex. The clear alpha wave while eyes are closed indicates the idling state of the visual cortex, while the block of the alpha rhythm when eyes are open reflects the working state of the visual cortex. Another example is mu rhythm, which can be recorded over the sensorimotor cortex. A significant mu rhythm only exists during the idling state of the sensorimotor cortex. The block of mu rhythm accompanies activation of the sensorimotor cortex.

Self control of brain rhythms serves an important function in brain–computer communication. *Modulation of brain rhythms* is used here to describe the detectable changes of EEG rhythms. BCIs based on the modulation of brain rhythms recognize changes of specific EEG rhythms induced by self-controlled brain activities. In our studies, we focus on EEG rhythms located in the frequency band of alpha and beta rhythms (8–30 Hz). Compared with the low-frequency evoked potentials, these components have several advantages as follows:

(1) Affected less by artifacts. The lower frequency components are always easily affected by the artifacts introduced by electrooculogram (EOG) and electromyogram (EMG). For example, removal of the eye movement artifact is an important procedure in event-related potential (ERP) processing, whereas it can be omitted in analyzing mu/beta rhythms after specific band-pass filtering.
(2) More stable and lasting, and thus with a higher signal-to-noise ratio (SNR). Modulation of these rhythms can last for a relatively long period (e.g., several seconds), while transient evoked potentials only occur within several 100 ms. Therefore, it is more promising to realize single-trial identification for high-frequency brain rhythms. Moreover, due to the phase-locked character of the evoked potentials, strict time precision is necessary for data alignment in the averaging procedure, while analysis of the power-based changes of the high-frequency rhythms permits a lower precision, and thus facilitates online designs of hardware and software.
(3) More applicable signal processing techniques. For wave-based analysis of evoked potentials, analysis is always done in the temporal domain. However, for the high-frequency rhythms, various temporal and frequency methods (e.g.,

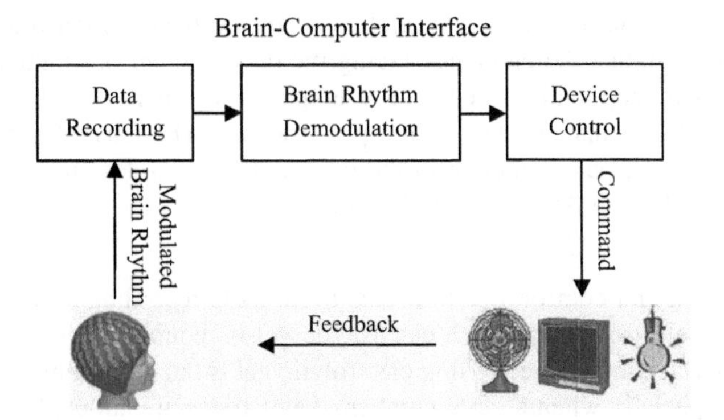

Fig. 1 Diagram of a BCI based on the modulation of brain rhythms

power and coherence analysis methods in the temporal and frequency domains, respectively) can be applied to extract EEG features. Therefore, the BCI based on the high-frequency rhythms provides a better platform for studying signal processing techniques and has a bigger potential in system performance.

Figure 1 is the block diagram of a BCI based on the modulation of brain rhythms. The control intention of the subject is embedded in the modulated brain rhythms through specific approaches, e.g., frequency coding or phase coding. According to the modulation approaches, the signal processing procedure aims to demodulate the brain rhythms and then extract the features, which will be translated into control signals to operate the output device. For instance, the SSVEP-based BCI and the motor imagery based BCI adopt typical approaches of brain rhythm modulation. The SSVEP system uses detection of frequency-coded or phase-coded SSVEPs to determine the gaze or spatial selective attention direction of the subject. Unlike the SSVEP BCI, the motor imagery BCI recognizes spatial distributions of amplitude-modulated mu/beta rhythms corresponding to the motor imagery states of different body parts. The methods for demodulating brain rhythms applied in BCIs will be introduced in the next section.

1.2 Challenges Confronting Practical System Designs

The design and implementation of an online system plays an important role in current BCI research, with the purpose of producing practical devices for real-life clinical application [8]. Compared with offline data analysis, an online BCI has difficulties in fulfilling real-time processing, system practicability and brain–machine co-adaptation. In this chapter, the principle of an online BCI based on the rhythmic modulation of EEG signals will be proposed. In our studies, two rhythmic EEG components corresponding to brain activities from the sensorimotor cortex and the visual cortex, i.e., mu rhythm and SSVEP, have been investigated and employed in constructing different types of online BCI systems.

After many studies carried out to implement and evaluate demonstration systems in laboratory settings, the challenge facing the development of practical BCI systems for real-life applications needs to be emphasized. There is still a long way to go before BCI systems can be put into practical use. The feasibility for practical application is a serious challenge in the current study. To design a practical BCI product, the following issues need to be addressed.

(1) Convenient and comfortable to use. Current EEG recording systems use standard wet electrodes, in which electrolytic gel is required to reduce electrode-skin interface impedance. Using electrolytic gel is uncomfortable and inconvenient, especially when a large number of electrodes are adopted. An electrode cap with a large numbers of electrodes is uncomfortable for users to wear and thus unsuitable for long-term recording. Besides, the preparation for EEG recording takes a long time, thus making BCI operation boring. Moreover, recording hardware with a large amount of channels is quite expensive, so that it is difficult for common users to afford. For these reasons, reducing the number of electrodes in BCI systems is a critical issue for the successful development of clinical applications of BCI technology.
(2) Stable system performance. Considering data recording in unshielded environments with strong electromagnetic interference, employment of an active electrode may be much better than a passive electrode. It can ensure that the recorded signal is insensitive to interference [9]. During system operation, to reduce the dependence on technical assistance, ad hoc functions should be provided in the system to adapt to non-stationarity of the signal caused by changes of electrode impedance or brain state. For example, software should be able to detect bad electrode contact in real time and automatically adjust algorithms to be suitable for the remaining good channels.
(3) Low-cost hardware. Most BCI users belong to the disabled persons' community; therefore, the system can not be popularized if it costs too much, no matter how good its performance is. When considering the following aspects, reducing system costs while maintaining performance might be expected. On one hand, to reduce the cost of commercial EEG equipment, a portable EEG recording system should be designed just to satisfy the requirement of recording the specific brain rhythms. On the other hand, to eliminate the cost of a computer used for signal processing in most current BCIs, a digital signal processor (DSP) should be employed to construct a system without dependency on a computer.

2 Modulation and Demodulation Methods for Brain Rhythms

In rhythm modulation-based BCIs, the input of a BCI system is the modulated brain rhythms with embedded control intentions. Brain rhythm modulation is realized by executing task-related activities, e.g., attending to one of several visual stimuli. Demodulation of brain rhythms can extract the embedded information, which will be converted into a control signal. The brain rhythm modulations could be sorted into the following three classes: power modulation, frequency modulation, and phase modulation. For a signal $s(t)$, its analytical signal $z(t)$ is a complex function defined as:

$$z(t) = s(t) + j\hat{s}(t) = A(t)e^{j\phi(t)} \quad (1)$$

where $\hat{s}(t)$ is the Hilbert transform of $s(t)$, $A(t)$ is the envelope of the signal, and $\phi(t)$ is the instantaneous phase. Changes of the modulated brain rhythms can be reflected by $A(t)$ or $\phi(t)$. In this section, three examples will be introduced to present the modulation and demodulation methods for mu rhythm and SSVEPs. More information about BCI related signal processing methods can be found in chapter "Digital Signal Processing and Machine Learning" of this book.

2.1 Power Modulation/Demodulation of Mu Rhythm

BCI systems based on classifying single-trial EEGs during motor imagery have developed rapidly in recent years [4, 10, 11]. The physiological studies on motor imagery indicate that EEG power differs between different imagined movements in the motor cortex. The event-related power change of brain rhythms in a specific frequency band is well known as event-related desynchronization and synchronization (ERD/ERS) [12]. During motor imagery, mu (8–12 Hz) and beta (18–26 Hz) rhythms display specific areas of ERD/ERS corresponding to each imagery state. Therefore, motor imagery can be performed to implement a BCI based on the power modulation of mu rhythms.

The information coding by power modulation is reflected by $A(t)$ in (1). The features extracted from the modulated mu rhythms for further recognition of the N-class motor imagery states can be defined as:

$$f_i^{\,\mathrm{Power}}(t) = A_i(t), i = 1, 2, ...N \quad (2)$$

The method by means of the analytical signal presents a theoretical description of EEG power modulation and demodulation. In real systems, demodulation is usually realized through estimating power spectral density (PSD) [13]. In online systems, to reduce computational cost, a practical approach is to calculate band-pass power in the temporal domain, i.e., calculating the mean value of the power samples derived from squaring the amplitude samples. Figure 2 shows three single-trial EEG waveforms and PSDs over the left motor cortex (electrode C3), corresponding to imaginations of left/right hand and foot movements for one subject. PSD was estimated by the periodogram algorithm, which can be easily executed by fast Fourier transform (FFT). Compared to the state of left-hand imagination, right-hand imagination induces an obvious ERD (i.e., power decrease), while foot imagination results in a significant ERS characterized as a power increase.

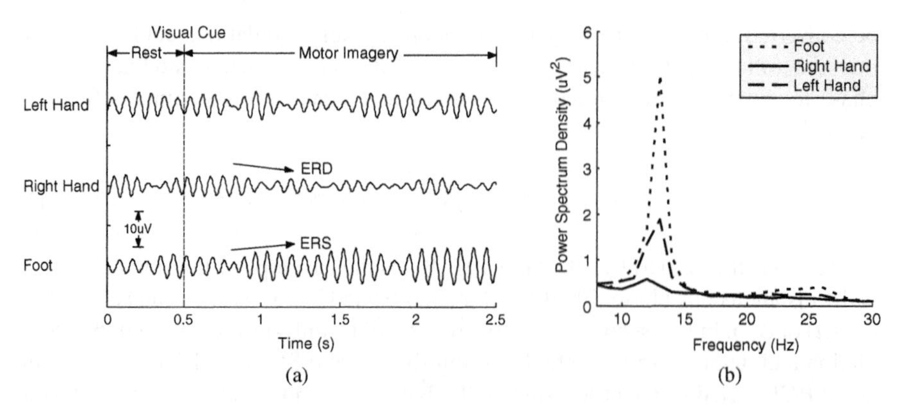

Fig. 2 (**a**) Examples of single-trial waveforms of mu rhythms on electrode C3 during three different motor imagery tasks. Data were band-pass filtered at 10–15 Hz. A visual cue for starting motor imagery appears at 0.5 s, indicated by the dotted line. On electrode C3, ERD and ERS of mu rhythms correspond to imagination of right hand and foot movements, respectively. (**b**) Averaged PSDs on electrode C3 corresponding to three different motor imagery tasks

2.2 Frequency Modulation/Demodulation of SSVEPs

The BCI system based on VEPs has been studied since the 1970s [14]. Studies on the VEP BCI demonstrate convincing robustness of system performance through many laboratory and clinical tests [15–21]. The recognized advantages of this BCI include easy system configuration, little user training, and a high information transfer rate (ITR). VEPs are derived from the brain's response to visual stimulation. SSVEP is a response to a visual stimulus modulated at a frequency higher than 6 Hz. The photic driving response, which is characterized by an increase in amplitude at the stimulus frequency, results in significant fundamental and second harmonics. The amplitude of SSVEP increases enormously as the stimulus is moved closer to the central visual field. Therefore, different SSVEPs can be produced by looking directly at one of a number of frequency-coded stimuli. Frequency modulation is the basic principle of the BCIs using SSVEPs to detect gaze direction.

$s(t)$ in (1) is supposed to be the frequency-coded SSVEP, and the feature used for identifying the fixed visual target is the instantaneous frequency, which can be calculated as:

$$f_i{}^{\text{Frequency}}(t) = \frac{d\phi_i(t)}{dt}, i = 1, 2, ..., N \qquad (3)$$

where $\phi(t)$ is the instantaneous phase, and N is the number of visual targets. Ideally, the instantaneous frequency of the SSVEP should be in accord with the stimulation frequency. In a real system, frequency recognition adopts the approach of detecting the peak value in the power spectrum. The visual target will induce a peak in the amplitude spectrum at the stimulus frequency. For most subjects, the amplitudes in the frequency bands on both sides will be depressed, thus facilitating peak detection.

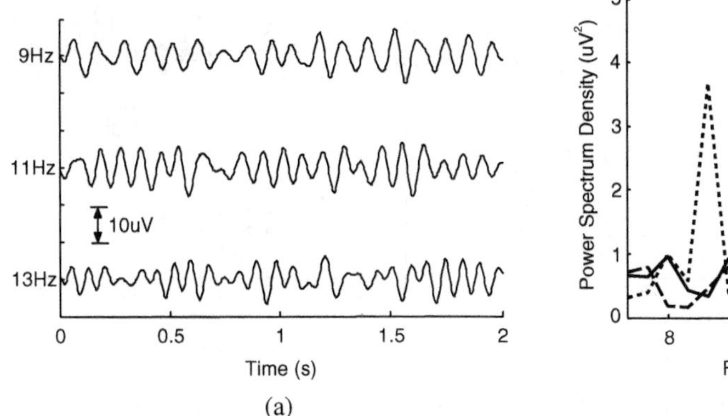

(a)

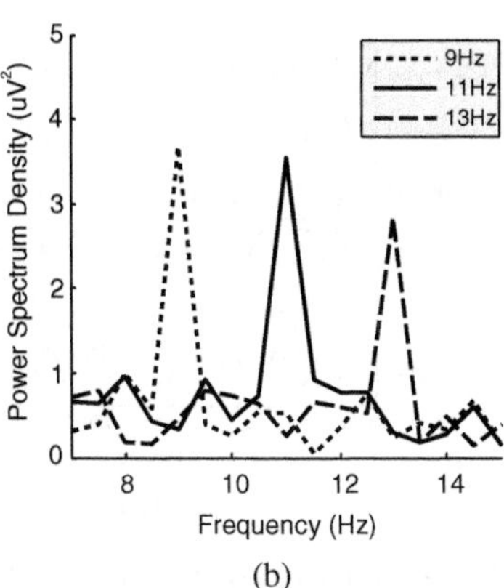

(b)

Fig. 3 (**a**) Examples of single-trial SSVEP waveforms over the occipital region. Data were band-pass filtered at 7–15 Hz. The stimulation frequencies were 9, 11, and 13 Hz. (**b**) PSDs of the three single-trial SSVEPs. The stimulation frequency is clearly shown at the peak value of the PSD

Demodulation of a frequency-coded SSVEP is to search for the peak value in the power spectrum and determine the corresponding frequency. Figure 3 shows three waveforms of frequency-coded SSVEPs (at 9 Hz, 11 Hz, and 13 Hz, respectively) and their corresponding power spectra. The target can be easily identified through peak detection in the power spectrum.

2.3 Phase Modulation/Demodulation of SSVEPs

In the SSVEP-BCI system based on frequency modulation, flickering frequencies of the visual targets are different. In order to ensure a high classification accuracy, a sufficient frequency interval should be kept between two adjacent stimuli and the number of targets will then be restricted. If phase information embedded in SSVEP can be added, the number of flickering targets may be increased and a higher ITR can be expected. An SSVEP BCI based on phase coherent detection was first proposed in [5], and its effectiveness was confirmed. However, only two stimuli with the same frequency but different phases were dealt with in their design, and the advantages of phase detection were not sufficiently shown. Inspired by their work, we tried to further the work by designing a system with stimulating signals of six different phases under the same frequency.

The phase-coded SSVEP can ideally be considered a sine wave at a frequency similar to the stimulation frequency. Suppose $s(t)$ in (1) is the phase-coded SSVEP, the feature used for identifying the fixed visual target is the instantaneous phase:

$$f_i^{\text{Phase}}(t) = \phi_i(t), i = 1, 2, ...N \tag{4}$$

The first step of implementing a phase-encoded SSVEP BCI is its stimuli design. Spots flickering on a computer screen at the same frequency with strictly constant

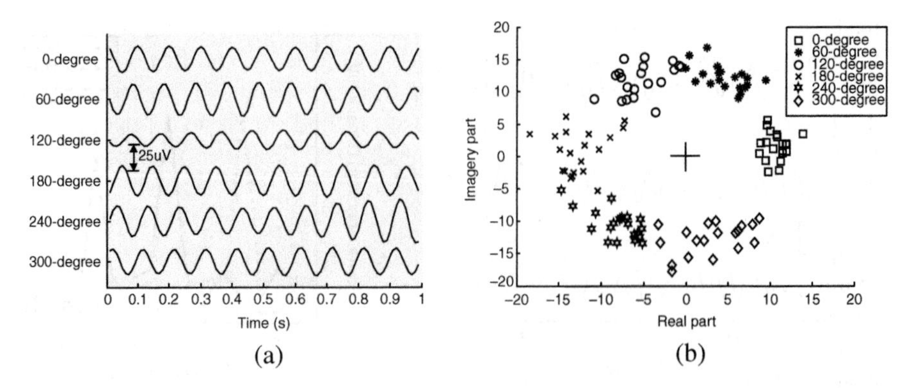

Fig. 4 (**a**) Examples of six phase-coded SSVEP waveforms over the occipital region. The stimulation frequency is 10 Hz. Data were preprocessed through narrow-band filtering at 9–11 Hz. (**b**) Scatter diagram of complex spectrum values at the stimulation frequency. Horizontal and vertical axes correspond to real and imaginary parts of the spectrum values. The cross indicates the origin of the coordinate system

phase differences are required. For example, a computer screen refreshing signal (60 Hz) can be used as a basic clock to produce stable 10 Hz signals. In our phase modulation-based BCI, six visual targets flickering at the same frequency of 10 Hz but with different phases appear on the screen. The flashing moments of the visual targets are staggered by one refreshing period of the screen (1/60 s), and thus produce a phase difference of 60 degrees between two adjacent stimuli (taking six refreshing periods as 360 degrees). During the experiment, the subject was asked to gaze at the six targets, respectively, and the SSVEP was gathered from electrodes located in the occipital region. The initial phases $\phi_{0i}(t)$ of the SSVEPs can be obtained through calculating the angle of the spectrum value at the characteristic frequency simply by the following formula:

$$\phi_{0i} = \text{angle}[\frac{1}{K}\sum_{n=1}^{K} s_i(n)e^{-j2\pi(f_0/f_s)n}], i = 1, 2, ..., N \tag{5}$$

where f_s is the sampling frequency, f_0 is the stimulation frequency, and K is the data length (the number of samples). The six-target phase-coding SSVEP system was tested with a volunteer. The results are shown in Fig. 4. It is clearly shown that the SSVEPs and the stimulating signals are stably phase locked. The responses evoked by stimulating signals with a phase difference of 60 degrees also have a phase difference of approximately 60 degrees.

3 Designs of Practical BCIs

The principles of BCIs based on the modulation of EEG rhythms have been systematically described above. Toward the aim of practical applications, we have made great efforts in facilitating system configuration and improving system performance. According to the practicality issues described in the first section, we focus on the two aspects: parameter optimization and information processing. These designs can significantly reduce system cost and improve system performance. In the present BCI systems, reducing the number of electrodes is important for developing clinical applications of BCI. Therefore, design of electrode layout is a common problem for all the BCI systems. In our studies, the method of bipolar lead was employed. To ensure a stable system performance, appropriate approaches to information processing play important roles. Frequency features of SSVEP harmonics have been investigated in our designs for a frequency-coded system and have shown an increased performance. In the motor imagery BCI, phase synchrony measurement has been employed, providing information in addition to the power feature.

3.1 Designs of a Practical SSVEP-based BCI

The SSVEP BCI based on frequency coding seems to be rather simple in principle, but a number of problems have to be solved during its implementation. Among them, *lead position*, *stimulation frequency*, and *frequency feature extraction* are most important. Due to differences in the subjects' physiological conditions, a preliminary experiment was designed to be carried out for a new user to set the subject-specific optimal parameters. The practicability of the system has been demonstrated by tests in many normal subjects and some patients with spinal cord injury (SCI).

3.1.1 Lead Position

The goal of lead selection is to achieve SSVEPs with a high SNR using the fewest number of electrodes. Only one bipolar lead is chosen as input in our system. The procedure includes two steps: finding a signal electrode and a reference electrode. From physiological knowledge, the electrode giving the strongest SSVEP, which is generally located in the occipital region, is selected as the signal electrode. The reference electrode is searched under the following considerations: its amplitude of the SSVEP should be lower, and its position should lie in the vicinity of the signal electrode so that its noise component is similar to that of the signal electrode. A high SNR can then be gained when the potentials of the two electrodes are subtracted, producing a bipolar signal. Most of the spontaneous background activities are eliminated after the subtraction, while the SSVEP component is retained. Details of this method can be found in [21]. According to our experience, although the selection varies between subjects, once it is selected, it is relatively stable over a period of time. This finding makes this method feasible for practical BCI application. For a new subject, the multi-channel recording only needs to be done once for optimization of the lead position.

3.1.2 Stimulation Frequency

Three problems related to stimulation frequency must be considered carefully. The first one concerns false positives. Generally, the SSVEPs in the alpha region (8–13 Hz) have high amplitudes, which can facilitate frequency detection. However, if the stimulation frequency band overlaps with alpha rhythms, the spontaneous EEG may be likely to satisfy the criteria of peak detection even though the user has not performed any intentional action. To implement an asynchronous BCI that allows the user to operate the system at any moment, avoidance of false positive is absolutely necessary. The second problem is about the efficiency of frequency detection. The criteria for confirming a stimulus frequency is the SNR threshold (SNR is defined as the ratio of EEG power at the stimulation frequency to the mean power of the adjacent frequency bands). For most subjects, background components in the SSVEP are depressed, while the signal amplitude at the stimulus frequency increases enormously. However, for some subjects, the majority of signal energy still lies within the region of background alpha activities. In these circumstances, although the stimulus frequency can be clearly identified, the SNR cannot reach the threshold predefined, and thus the decision of a command can not be made. Due to these reasons, some frequency components in the alpha region should be excluded to avoid the interference of background alpha rhythms effectively. The third problem concerns the bandwidth of usable SSVEPs. Increasing the number of visual targets is an effective approach to increase the ITR. It can be realized through extending the stimulation frequency bandwidth. In our previous study, we demonstrated that the high-frequency SSVEP (>20 Hz) has an SNR similar to the low-frequency SSVEP [22]. By extending stimulation frequency to a wider range, a system with more options can be designed and a higher performance can be expected.

3.1.3 Frequency Feature

Due to the nonlinearity during information transfer in the visual system, strong harmonics may often be found in the SSVEPs. For example, in Fig. 5, the SSVEPs elicited by the 9 and 10 Hz stimulations show characteristic frequency components with peaks at the fundamental and the second harmonic frequencies (18 and 20 Hz, respectively). Müller-Putz et al. investigated the impact of using SSVEP harmonics on the classification result of a four-class SSVEP-based BCI [23]. In their study, the accuracy obtained with combined harmonics (up to the third harmonic) was significantly higher than with only the first harmonic. In our experience, for some subjects, the intensity of the second harmonic may sometimes be even stronger than that of the fundamental component. Thus, analysis of the frequency band should cover the second harmonic and the frequency feature has to be taken as the weighted sum of their powers, namely:

$$P_i = \alpha P_{f_{1i}} + (1-\alpha) P_{f_{2i}}, i = 1, 2, ..., N \tag{6}$$

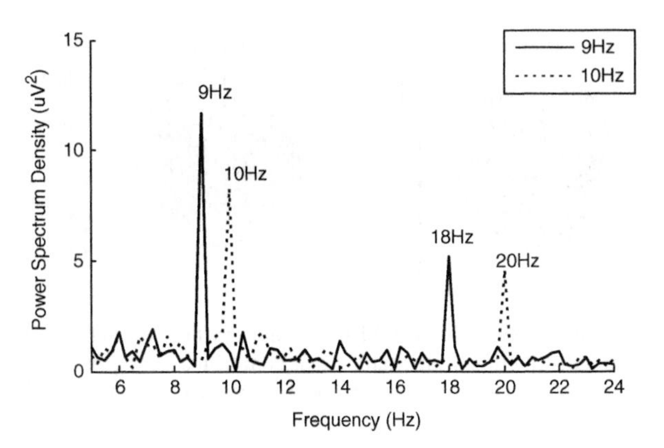

Fig. 5 Power spectra of SSVEPs at the O2 electrode for one subject. The stimulation frequencies were 9 and 10 Hz. Length of the data used for calculating the PSD is 4 s. The spectra show clear peaks at the fundamental and the 2nd harmonic frequencies

where N is the number of targets and $P_{f_{1i}}$, $P_{f_{2i}}$ are respectively the spectrum peak values of fundamental and second harmonics of the ith frequency (i.e., ith target) and α is the optimized weighting factor that varies between subjects. Its empirical value may be taken as:

$$\alpha = \frac{1}{N}\sum_{i=1}^{N} P_{f_{1i}} / \left(P_{f_{1i}} + P_{f_{2i}}\right). \tag{7}$$

3.2 Designs of a Practical Motor Imagery Based BCI

In the study of EEG-based BCI, the system based on imagery movement is another active theme due to its relatively robust performance for communication and intrinsic neurophysiological significance for studying the mechanism of motor imagery [11]. Moreover, it is a totally independent BCI system which is likely to be more useful for completely paralyzed patients than the SSVEP-based BCI. Most of the current motor imagery based BCIs are based on characteristic ERD/ERS spatial distributions corresponding to different motor imagery states. Figure 6 displays characteristic mappings of ERD/ERS for one subject corresponding to three motor imagery states, i.e., imagining movements of left/right hands and foot. Due to the widespread distribution of ERD/ERS, techniques of spatial filtering, e.g, common spatial pattern (CSP), were widely used to obtain a stable system performance. However, due to the limit of electrodes in a practical system, electrode layout has to be carefully considered. With only a small number of electrodes, searching for new features using new information processing methods will contribute significantly to classifying motor imagery states.

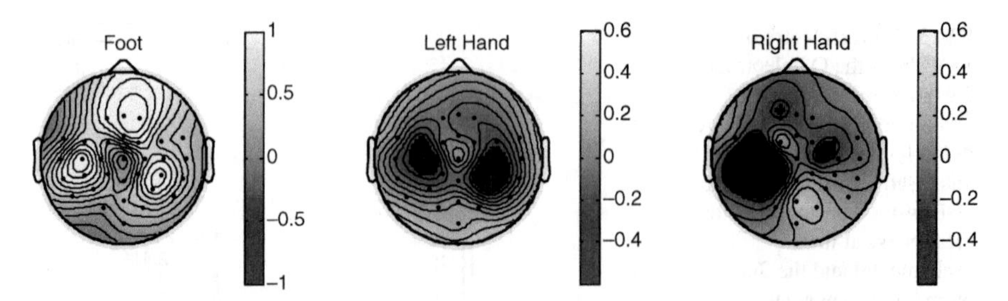

Fig. 6 Mappings of ERD/ERS of mu rhythms during motor imagery. ERD over hand areas has a distribution with contralateral dominance during hand movement imagination. During foot movement imagination, an obvious ERS appears in central and frontal areas

3.2.1 Phase Synchrony Measurement

In recent years, measurement of phase coupling (or phase locking) of EEG or magnetoencephalogram (MEG) has been used for exploring the dynamics of brain networking [24]. The phase-locking value (PLV) measurement has been introduced recently to extract EEG features in BCI research [25, 26].

Given $s_{\mathrm{x}}(t)$ and $s_{\mathrm{y}}(t)$ as the signals at electrodes x and y, and $\phi_{\mathrm{x}}(t)$ and $\phi_{\mathrm{y}}(t)$ as their corresponding instantaneous phases, the instantaneous phase difference between the two signals is defined as $\Delta\phi(t)$. $\Delta\phi(t)$ is a constant when the two signals are perfectly synchronized. In scalp EEG signals with low SNR, the true synchrony is always buried in a considerable background noise; therefore, a statistical criterion has to be provided to quantify the degree of phase locking [24]. A single-trial phase-locking value is defined for each individual trial as:

$$\mathrm{PLV} = \left| \left\langle e^{j\Delta\phi(t)} \right\rangle_t \right| \tag{8}$$

where $\langle\cdot\rangle_t$ is the operator of averaging over time. In the case of completely synchronized signals, $\Delta\phi(t)$ is a constant and PLV is equal to 1. If the signals are unsynchronized, $\Delta\phi(t)$ follows a uniform distribution and PLV approaches 0.

Since supplementary motor area (SMA) and primary motor cortex (M1) areas are considered primary cortical regions involved in the task of motor imagery, we investigated EEG synchrony between these regions (i.e., electrode pairs of FCz-C3, FCz-C4, and C3-C4 shown in Fig. 7). The statistical PLV obtained through averaging over all trials in each class presents a contralateral dominance during hand movement imagery, e.g., PLV of C3-FCz has a higher value during right-hand imagery than that of the left hand. In contrast to C3-FCz and C4-FCz, PLV shows a low synchrony level between C3 and C4 and there exists no significant difference between left- and right-hand imagery. Power features derived from ERD/ERS indicate the brain activities focused on both M1 areas, while synchronization features introduce additional information from the SMA areas. Compared with detecting the power change, the synchrony measure was proved effective to supply additional information of the brain activities in the motor cortex. Therefore, the combination of power and synchrony features is theoretically expected to improve the performance

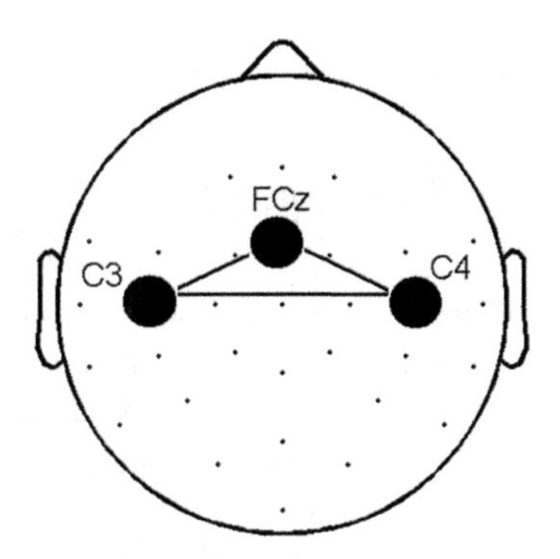

Fig. 7 Placement of electrodes in the motor imagery based BCI. Electrodes C3 and C4 represent lateral hand areas in M1, and electrode FCz indicates the SMA. Anatomical regions of the SMA and M1 areas can be found in [28]

Table 1 Accuracy (±standard deviation, %) of classifying single-trial EEG during imagining movements of left/right hands and foot. The dataset of each subject consists of 360 trials (120 trials per class). Three electrodes, i.e. C3/C4 and FCz, were used for feature extraction

Subject	Synchrony	Power	Synchrony+Power
S1	87.47±2.19	84.83±2.47	91.50±1.98
S2	83.25±1.78	86.64±1.97	90.60±1.79
S3	82.08±2.24	80.64±2.44	87.19±2.12
Mean	84.27	84.04	89.76

of classification. As illustrated in Table 1, the averaged accuracy derived from the synchrony and the power features was 84.27 and 84.04%, respectively, on three subjects (10-fold cross-validation with 120 trials per class for each subject). Feature combination led to an improved performance of 89.76%. The synchrony feature vector consists of two PLVs of two electrode pairs, i.e., FCz-C3, FCz-C4. The power features are the band-pass power on C3 and C4 electrodes. A subject-specific band-pass filter was used to preprocess the EEG data in order to focus on the mu rhythm. PLV and power features were calculated using a time window corresponding to the motor imagery period (0.5–6 s after a visual cue). More details about the experiment paradigm can be found in [27]. For feature combination, classification is applied to the concatenation of the power features and the synchrony features. Linear discriminant analysis (LDA) was used as the classifier. The multi-class classifier was designed in a one-versus-one manner.

3.2.2 Electrode Layout

Since the SMA can be considered zero-phase synchronized with the M1 area displaying ERD, the power difference between M1 areas can be more significant if using FCz as the reference electrode [29]. For example, during left-hand imagination, the subtraction of zero-phase synchronized $S_{\mathrm{FCz}}(t)$ from $S_{\mathrm{C4}}(t)$ results in a much lower power, whereas the power of $S_{\mathrm{C3}}(t)$ changes slightly after the subtraction ($S_{\mathrm{C3}}(t)$, $S_{\mathrm{C4}}(t)$, and $S_{\mathrm{FCz}}(t)$ are ear-referenced EEG signals on C3/C4 and FCz, between which the power difference is not very significant). This idea can be summarized as the following inequality:

$$\left\langle |S_{C4}(t) - S_{FCz}(t)|^2 \right\rangle_t < \left\langle |S_{C4}(t)|^2 \right\rangle_t < \left\langle |S_{C3}(t)|^2 \right\rangle_t \approx \left\langle |S_{C3}(t) - S_{FCz}(t)|^2 \right\rangle_t \quad (9)$$

where $\langle \cdot \rangle_t$ is the operator of averaging over the left-hand imagination period, and the power difference between $S_{C3}(t)$ and $S_{C4}(t)$ is due to ERD. Therefore, the midline FCz electrode is a good reference to extract the power difference between left and right hemispheres, because it contains additional information derived from brain synchronization.

Compared with the approach of concatenating power and synchrony features, this bipolar approach has the advantages of a lower dimension of the features and smaller computational cost. A lower feature dimension can obtain a better generalization ability of the classifier, and smaller computational cost can assure the real-time processing. In the online system, the bipolar approach (i.e. FCz-C3 and FCz-C4) was employed to provide integrated power and synchrony information over the motor cortex areas. Considering the necessity of a practical BCI (e.g., easy electrode preparation, high performance, and low cost), electrode placement with an implicit embedding method is an efficient approach to implement a practical motor imagery based BCI. Recently, a method for optimizing subject specific bipolar electrodes was proposed and demonstrated in [30].

4 Potential Applications

4.1 Communication and Control

BCI research aims to provide a new channel for the motion disabled to communicate with the environment. Up to now, although clinical applications have been involved in some studies, most BCIs have been tested in the laboratory with normal subjects [31]. Practical designs of BCIs proposed in this chapter will benefit spreading real-life applications for the patients. For the BCI community, further investigation of clinical applications should be emphasized. In our studies, online demonstrations have been designed to fulfill some real-life applications. Due to its advantage of a high ITR, the SSVEP BCI has been employed in various applications, e.g., spelling, cursor control, and appliance control. Moreover, an environmental controller has been tested in patients with spinal cord injury [21]. Figure 8 presents the scene of using the SSVEP-based BCI to make a phone call [32]. The system consists of a small EEG recording system, an LED stimulator box including a 12-target number pad and a digitron display, and a laptop for signal processing and command transmission. A simple head-strap with two embedded electrodes was used to record data on one bipolar channel. The user could input a number easily through directing his or her gaze on the target number, and then fulfill the task of making a phone call conveniently with the laptop modem.

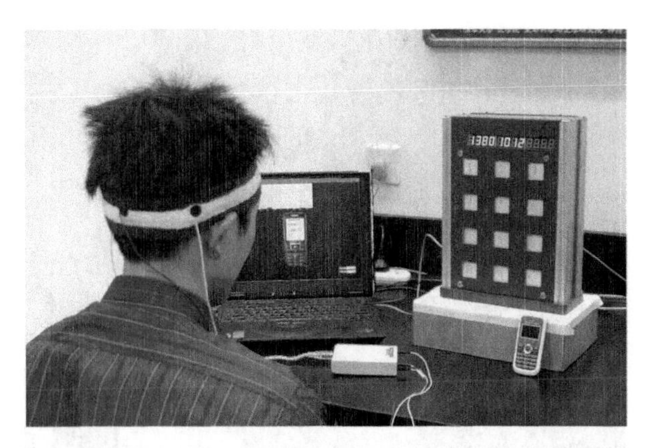

Fig. 8 Application of the SSVEP-based BCI for making a phone call. The system consists of two-lead EEG recording hardware, an LED visual stimulator, and a laptop

4.2 Rehabilitation Training

In addition to applications for communication and control technologies, a new area of BCI application has emerged in rehabilitation research [33] (see also chapter "Brain–Computer Interface in Neurorehabilitation" in this book). It has been proposed that BCIs may have value in neurorehabilitation by reinforcing the use of damaged neural pathways [34]. For example, some studies have demonstrated that motor imagery can help to restore motor function for stroke patients [35–37]. Based on these findings, positive rehabilitation training directed by motor intention should have a better efficiency than conventional passive training. Figure 9 shows such a device used for lower limb rehabilitation. Imagination of foot movement will start the device, while a resting state without foot movement imagination will stop the training. This system still needs further clinical investigation in the future to confirm its superiority.

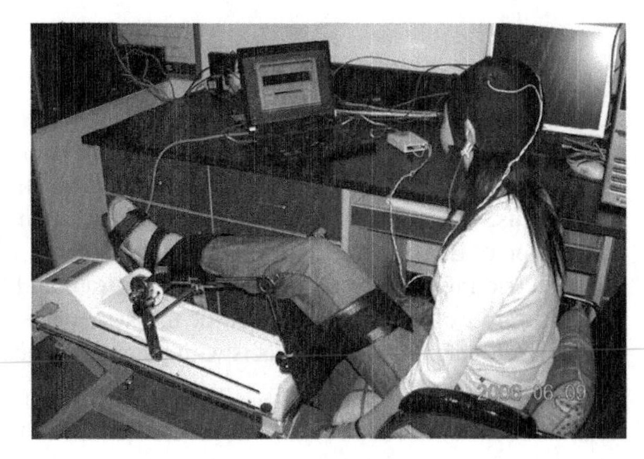

Fig. 9 A system used for lower limb rehabilitation. The motor imagery based BCI is used to control the device. Positive rehabilitation training is performed in accordance with the concurrent motor imagery of foot movement

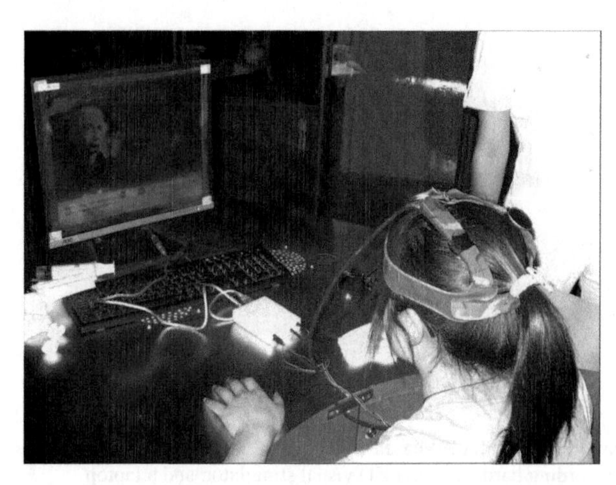

Fig. 10 A player is playing a computer game controlled through a motor imagery based BCI. This system is available for visitors to the Zhengzhou Science and Technology Center in China

4.3 Computer Games

BCI technology was first proposed to benefit users belonging to the disabled persons' community. However, it also has potential applications for healthy users. With numerous players, computer gaming is a potential area to employ the BCI technique [38, 39]. Through integrating BCI to achieve additional mental control, a computer game will be more appealing. Recently, BCI technology has been used in the electronic games industry to make it possible for games to be controlled and influenced by the player's mind, e.g., Emotiv and NeuroSky systems [40, 41]. In Fig. 10, a player is absorbed in playing a game using a portable motor imagery based BCI developed in our lab. In the near future, besides the gaming industry, applications of BCI will also span various potential industries, such as neuroeconomics research and neurofeedback therapy.

5 Conclusion

Most of current BCI studies are still at the stage of laboratory demonstrations. Here we described the challenges in changing BCIs from demos to practically applicable systems. Our work on designs and implementations of the BCIs based on the modulation of EEG rhythms showed that by adequately considering parameter optimization and information processing, system cost could be greatly decreased while system usability could be improved at the same time. These efforts will benefit future development of BCI products with potential applications in various fields.

Acknowledgments This project is supported by the National Natural Science Foundation of China (30630022) and the Science and Technology Ministry of China under Grant 2006BAI03A17.

References

1. J.R. Wolpaw, N. Birbaumer, D.J. McFarland, G. Pfurtscheller, and T.M. Vaughan, Brain-computer interfaces for communication and control. Clin Neurophysiol, 113, 6, 767–791, (2002).
2. M.A. Lebedev and M.A.L. Nicolelis, Brain-machine interfaces: past, present and future. Trends Neurosci, 29(9), 536–546, (2006).
3. N. Birbaumer, Brain-computer-interface research: Coming of age. Clin Neurophysiol, 117(3), 479–483, (2006).
4. G. Pfurtscheller and C. Neuper, Motor imagery and direct brain-computer communication. Proc IEEE, 89(7), 1123–1134, (2001).
5. T. Kluge and M. Hartmann, Phase coherent detection of steady-state evoked potentials: experimental results and application to brain-computer interfaces. Proceedings of 3rd International IEEE EMBS Neural Engineering Conference, Kohala Coast, Hawaii, USA, pp. 425–429, 2–5 May, (2007).
6. S. Makeig, M. Westerfield, T.P. Jung, S. Enghoff, J. Townsend, E. Courchesne, and T.J. Sejnowski, Dynamic brain sources of visual evoked responses. Science, 295(5555), 690–694, (2002).
7. E. Niedermeyer and F.H. Lopes da Silva, *Electroencephalography: Basic principles, clinical applications and related fields,* Williams and Wilkins, Baltimore, MD, (1999).
8. Y. Wang, X. Gao, B. Hong, C. Jia, and S. Gao, Brain-computer interfaces based on visual evoked potentials: Feasibility of practical system designs. IEEE EMB Mag, 27(5), 64–71, (2008).
9. A.C. MettingVanRijn, A.P. Kuiper, T.E. Dankers, and C.A. Grimbergen, Low-cost active electrode improves the resolution in biopotential recordings. Proceedings of 18th International IEEE EMBS Conference, Amsterdam, Netherlands, pp. 101–102, 31 Oct–3 Nov, (1996).
10. J.R. Wolpaw and D.J. McFarland, Control of a two-dimensional movement signal by a noninvasive brain-computer interface in humans. Proc Natl Acad Sci USA, 101(51), 17849–17854, (2004).
11. B. Blankertz, K.R. Muller, D.J. Krusienski, G. Schalk, J.R. Wolpaw, A. Schlogl, G. Pfurtscheller, J.D.R. Millan, M. Schroder, and N. Birbaumer, The BCI competition III: Validating alternative approaches to actual BCI problems. IEEE Trans Neural Syst Rehab Eng, 14(2), 153–159, (2006).
12. G. Pfurtscheller and F.H. Lopes da Silva, Event-related EEG/MEG synchronization and desynchronization: basic principles., Clin Neurophysiol, 110(11), 1842–1857, (1999).
13. D.J. McFarland, C.W. Anderson, K.R. Muller, A. Schlogl, and D.J. Krusienski, BCI Meeting 2005 – Workshop on BCI signal processing: Feature extraction and translation. IEEE Trans Neural Syst Rehab Eng, 14(2), 135–138, (2006).
14. J.J. Vidal, Real-time detection of brain events in EEG. Proc. IEEE, 65(5), 633–641, (1977).
15. E.E. Sutter, The brain response interface: communication through visually-induced electrical brain response. J Microcomput Appl, 15(1), 31–45, (1992).
16. M. Middendorf, G. McMillan, G. Calhoun, and K.S. Jones, Brain-computer interfaces based on the steady-state visual-evoked response. IEEE Trans Rehabil Eng, 8(2), 211–214, (2000).
17. M. Cheng, X.R. Gao, S.G. Gao, and D.F. Xu, Design and implementation of a brain-computer interface with high transfer rates. IEEE Trans Biomed Eng, 49(10), 1181–1186, (2002).
18. X. Gao, D. Xu, M. Cheng, and S. Gao, A BCI-based environmental controller for the motion-disabled. IEEE Trans Neural Syst Rehabil Eng, 11(2), 137–140, (2003).
19. B. Allison, D. McFarland, G. Schalk, S. Zheng, M. Jackson, and J. Wolpaw, Towards an independent brain–computer interface using steady state visual evoked potentials. Clin Neurophysiol, 119(2), 399–408, (2007).
20. F. Guo, B. Hong, X. Gao, and S. Gao, A brain–computer interface using motion-onset visual evoked potential. J Neural Eng, 5(4), 477–485, (2008).
21. Y. Wang, R. Wang, X. Gao, B. Hong, and S. Gao, A practical VEP-based brain-computer interface. IEEE Trans Neural Syst Rehabil Eng, 14(2), 234–239, (2006).

22. Y. Wang, R. Wang, X. Gao, and S. Gao, Brain-computer interface based on the high frequency steady-state visual evoked potential. Proceedings of 1st International NIC Conference, Wuhan, China, pp. 37–39, 26–28 May, (2005).
23. G.R. Müller-Putz, R. Scherer, C. Brauneis, and C. Pfurtscheller, Steady-state visual evoked potential (SSVEP)-based communication: impact of harmonic frequency components. J Neural Eng, 2(4), 123–130, (2005).
24. J.P. Lachaux, E. Rodriguez, J. Martinerie, and F.J. Varela, Measuring phase synchrony in brain signals. Hum Brain Mapp, 8(4), 194–208, (1999).
25. E. Gysels and P. Celka, Phase synchronization for the recognition of mental tasks in a brain-computer interface. IEEE Trans Neural Syst Rehabil Eng, 12(4), 406–415, (2004).
26. Y. Wang, B. Hong, X. Gao, and S. Gao, Phase synchrony measurement in motor cortex for classifying single-trial EEG during motor imagery. Proceedings of 28th International IEEE EMBS Conference, New York, USA, pp. 75–78, 30 Aug–3 Sept, (2006).
27. Y. Wang, B. Hong, X. Gao, and S. Gao, Implementation of a brain-computer interface based on three states of motor imagery. Proceedings of 29th International IEEE EMBS Conference, Lyon, France, pp. 5059–5062, 23–26 Aug, (2007).
28. M.F. Bear, B.W. Connors, and M.A. Paradiso, *Neuroscience: exploring the brain*, Lippincott Williams and Wilkins, Baltimore, MD, (2001).
29. Y. Wang, B. Hong, X. Gao, and S. Gao, Design of electrode layout for motor imagery based brain-computer interface. Electron Lett, 43(10), 557–558, (2007).
30. B. Lou, B. Hong, X. Gao, and S. Gao, Bipolar electrode selection for a motor imagery based brain-computer interface. J Neural Eng, 5(3), 342–349, (2008).
31. S.G. Mason, A. Bashashati, M. Fatourechi, K.F. Navarro, and G.E. Birch, A comprehensive survey of brain interface technology designs. Ann Biomed Eng, 35(2), 137–169, (2007).
32. C. Jia, H. Xu, B. Hong, X. Gao, and S. Gao, A human computer interface using SSVEP-based BCI technology. Lect Notes Comput Sci, 4565, 113–119, (2007).
33. E. Buch, C. Weber, L.G. Cohen, C. Braun, M.A. Dimyan, T. Ard, J. Mellinger, A. Caria, S. Soekadar, A, Fourkas, and N. Birbaumer, Think to move: a neuromagnetic brain-computer interface (BCI) system for chronic stroke. Stroke, 39(3), 910–917, (2008).
34. A. Kubler, V.K. Mushahwar, L.R. Hochberg, and J.P. Donoghue, BCI Meeting 2005 – Workshop on clinical issues and applications. IEEE Trans Neural Syst Rehabil Eng, 14(2), 131–134, (2006).
35. S. de Vries and T. Mulder, Motor imagery and stroke rehabilitation: A critical discussion. J Rehabil Med, 39(1), 5–13, (2007).
36. D. Ertelt, S. Small, A. Solodkin, C. Dettmers, A. McNamara, F. Binkofski, and G. Buccino, Action observation has a positive impact on rehabilitation of motor deficits after stroke. NeuroImage, 36(suppl 2), T164–T173, (2007).
37. M. Iacoboni and J.C. Mazziotta, Mirror neuron system: Basic findings and clinical applications. Ann Neurol, 62(3), 213–218, (2007).
38. J.A. Pineda, D.S. Silverman, A. Vankov, and J. Hestenes, Learning to control brain rhythms: making a brain-computer interface possible. IEEE Trans Neural Syst Rehabil Eng, 11(2), 181–184, (2003).
39. E.C. Lalor, S.P. Kelly, C. Finucane, R. Burke, R. Smith, R.B. Reilly, and G. McDarby, Steady-state VEP-based brain-computer interface control in an immersive 3D gaming environment. EURASIP J Appl Signal Process, 19, 3156–3164, (2005).
40. Emotiv headset (2010). Emotiv – brain computer interface technology. Website of Emotiv Systems Inc., http://www.emotiv.com. Accessed 14 Sep 2010.
41. NeuroSky mindset (2010). Website of Neurosky Inc., http://www.neurosky.com. Accessed 14 Sep 2010.

Brain–Computer Interface in Neurorehabilitation

Niels Birbaumer and Paul Sauseng

1 Introduction

Brain–computer interfaces (BCI) are using brain signals to drive an external device outside the brain without activation of motor output channels. Different types of BCIs were described during the last 10 years with an exponentially increasing number of publications devoted to Brain–computer interface research [32]. Most of the literature describes mathematical algorithms capable of translating brain signals online and real-time into signals for an external device, mainly computers or peripheral prostheses or orthoses. Publications describing applications to human disorders are relatively sparse with very few controlled studies available on the effectiveness of BCIs in treatment and rehabilitation (for a summary see [8, 20]).

Historically, BCIs for human research developed mainly out of the neurofeedback literature devoted to operant conditioning and biofeedback of different types of brain signals to control seizures [16], to treat attention deficit disorder [34], stroke and other neuro-psychiatric disorders ([28]; for a review see [31]). Brain–computer interface research departed from neurofeedback by applying advanced mathematical classification algorithms and using basic animal research with direct recording of spike activity in the cortical network to regulate devices outside the brain.

Brain–computer interface research in animals and humans used many different brain signals: in animals spike trains from multiple microelectrode recordings (for a summary see Schwartz and Andrasik (2003) and chapter "Intracortical BCIs: A Brief History of Neural Timing" in this book), but also in humans. One study [14] reported "Neural ensemble control of prosthetic devices by a human with tetraplegia". In human research mainly non-invasive EEG recordings using sensory motor rhythm [18], slow cortical potentials [3] and the P300 event-related evoked brain potential [13] were described.

N. Birbaumer (✉)
Institute of Medical Psychology and Behavioral Neurobiology, University of Tübingen, Tübingen, Germany
e-mail: niels.birbaumer@uni-tuebingen.de

B. Graimann et al. (eds.), *Brain–Computer Interfaces*, The Frontiers Collection,
DOI 10.1007/978-3-642-02091-9_9,

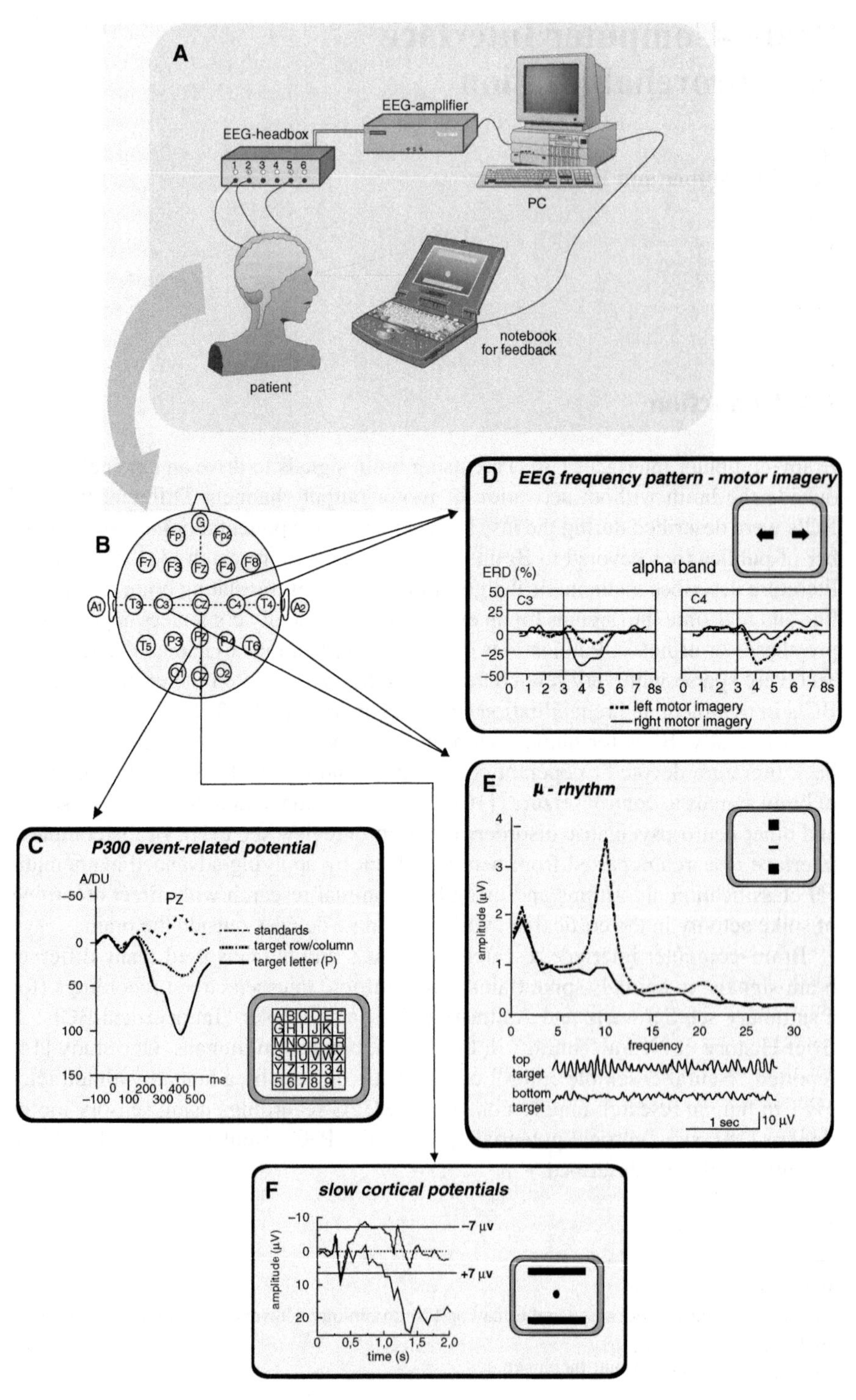

Fig. 1 Basic principle of a typical EEG BCI. (**a**) and (**b**) show the setup of a BCI. (**c–f**) Examples of EEG signals used for BCI

Figure 1 shows the most frequently used EEG BCI. More recently, a BCI system based on magnetoencephalography was tested [8] and applied to stroke rehabilitation. On an experimental basis near-infrared spectroscopy (NIRS) and functional magnetic resonance-BCIs were introduced by our group [33, 10]. Of particular interest is the use of signals ranging in spatial resolution between single spikes and local field potentials (LFP) and the EEG, namely the electrocorticogramm (ECoG). Electrical signals recorded subdurally or epidurally provide a better spatial resolution with the same exquisite time resolution as EEG but avoids the smearing and filtering of the EEG signals. In addition, it allows recording and training of high frequency components particularly in the gamma range (from 30 to 100 Hz). First attempts to classify movement directions from the electrocorticogramm were particularly successful (see [2, 22, 35, 38]; see also chapters "BCIs Based on Signals from Between the Brain and Skull" and "A Simple, Spectral-change Based, Electrocorticographic Brain–Computer Interface" in this book). At present, brain–computer interface research in neurorehabilitation is dominated by interesting basic science approaches holding great promises for human applications but very modest clinical use (see [7]). As in the tradition of clinical applications of neurofeedback, the field is lacking large-scale clinical studies with appropriate controls proving the superior efficiency of BCI systems over classical rehabilitation attempts.

2 Basic Research

Animal research stimulated to a large extent the interest and enthusiasm in BCI research. Nicolelis [26, 27] summarized these attempts: using more than 100 electrodes in the motor cortex or in the parietal cortex projecting into the motor system monkeys were trained to move a prosthesis mounted outside their body or even in another laboratory using spike sequences; monkeys were trained over long time periods to move prostheses with spike frequencies and were rewarded when the peripheral orthoses performed an aiming or reaching movement usually grapping a food reward. Even more impressive were studies from the laboratory of Eberhard Fetz [15] with operant conditioning of spike sequences. Monkeys were able to produce many different types of single cell activity, rhythmic and non-rhythmic, and used the trained spike sequences for manipulating external machinery. Of particular interest was the demonstration of a neurochip in which the monkey learned to operate an electronic implant with action potentials recorded on one electrode to trigger electrical stimuli delivered to another location on the cortex. After several days of training, the output produced from the recording site shifted to resemble the output from the corresponding stimulation site, consistent with a potentiation of synaptic connections between the artificially synchronized neurons. The monkeys themselves produced long-term motor cortex plasticity by activating the electronic neuronal implant. These types of Hebbian implants may turn out to be of fundamental importance for neurorehabilitation after brain damage using the brain–computer interface to permanently change cortical plasticity in areas damaged or suppressed by pathophysiological activity.

3 Brain–Computer Interfaces for Communication in Complete Paralysis

The author's laboratory was the first to apply an EEG–brain–computer interface to communication in the completely paralyzed [4]. This research was based on earlier work from our laboratory showing amazing and consistent and long-lasting control of brain activity in intractable epilepsy [16]. Some of these patients learned to increase and decrease the amplitude of their slow cortical potentials (SCP) with a very high success rate (from 90 to 100%) and they were able to keep that skill stable over more than a year without any intermediate training. If voluntary control over one's own brain activity is possible after relatively short training periods ranging from 20 to 50 sessions as demonstrated in these studies (see [4]), selection of letters from a computer menu with slow cortical potential control seems to be possible. The first two patients reported in [5] learned over several weeks and months sufficient control of their slow brain potentials and were able to select letters and write words with an average speed of one letter per minute by using their brain potentials only. These patients suffered from amyotrophic lateral sclerosis (ALS) in an advanced stage, having only unreliable motor twitches of eyes or mouth muscles left for communication. Both patients were artificially ventilated and fed but had some limited and unreliable motor control. After this first report in the last 10 years, 37 patients with amyotrophic lateral sclerosis at different stage of their disease were trained in our laboratory to use the BCI to select letters and words from a spelling menu in a PC. Comparing the performance of patients with a different degree of physical restrictions from moderate paralysis to the locked-in state, we showed that there is no significant difference in BCI performance in the different stages of the disease, indicating that even patients with locked-in syndrome are able to learn an EEG-based brain–computer interface and use it for communication. These studies, however, also showed that the remaining 7 patients suffering from a complete locked-in state without any remaining muscle twitch or any other motor control who started BCI training *after* entering the completely locked-in state were unable to learn voluntary brain control or voluntary control over any other bodily function: limited success was reported for a Ph-value based communication device using the Ph-value of saliva during mental imagery as a basis for "yes" and "no" communication [37]. Saliva, skin conductance, EMG and cardiovascular functions are kept constant or are severely disturbed in end-stage amyotrophic lateral sclerosis and therefore provide no reliable basis for communication devices. Thus, brain–computer interfaces need an already established learned voluntary control of the brain activity before the patient enters the complete locked-in state where no interaction with the outside environment is possible any more. On the basis of these data, the first author developed the hypothesis which was later termed "goal directed thought extinction hypothesis": all directed, output-oriented thoughts and imagery extinguishes in the complete locked-in state because no reliable contingencies (response reward sequences) exist in the environment of a completely locked-in patient. Any particular thoughts related to a particular outcome ("I would like to be turned around", "I would like my saliva to be sucked out of my throat", "I would like to see my friend" etc.) are not followed by the anticipated or desired consequence. Therefore, extinction takes place within the first weeks or months of the complete locked-in state. Such a negative learning process can only be abolished and voluntary, operant control can be reinstated if reliable contingencies re-occur. Without the knowledge of the patients' goal-directed thoughts or its electrophysiological antecedents a reinstatement seems impossible. We therefore recommend early initiation of BCI training before entering the complete locked-in state.

These negative results for the completely locked-in and the "thought extinction hypothesis" illustrate the failure of operant conditioning of autonomic responses in the long-term curarized rat studied by Neal Miller and his students at Rockefeller University during the 60s and 70s. Miller [25] reported successful operant conditioning of different types of autonomic signals in the curarized rat. He proposed that autonomic function is under voluntary (cortical) control comparable to motor responses. The term "autonomic" for vegetative system responsivity does not seem to be the appropriate term for body functions which are essential under voluntary control. Miller's argumentation opened the door for the learning treatment and biofeedback of many bodily functions and disorders such as heart disease, high blood pressure, diseases of the vascular system such as arrythmias and disorders of the gastrointestinal system. But the replication of these experiments on the curarized (treated with curare to relax the skeletal muscles) rat turned out to be impossible [12] and the consequent clinical applications in humans in training patients with hypertension or gastrointestinal disorders were also largely unsuccessful. The reason for the failure to replicate the curarized rat experiments remained obscure. The extinction-of-thought hypothesis [7, 8] tries to explain both the failure of operant brain conditioning in completely paralyzed locked-in patients and the failure to train long-term curarized artificially respirated rats to increase or decrease autonomic functions. In two of our completely locked-in patients, we tried to improve the signal-to-noise ratio by surgical implantation of an electrode grid subdurally and tried to train the patient's electrocorticogramm in order to reinstate communication. At the time of implantation, one patient was already completely locked-in for more than a year. After implantation, no successful communication was possible and no voluntary operant control of any brain signal was possible, supporting the extinction-of-thought hypothesis. A second patient with minimal eye control left was implanted recently in our laboratory with a 120 electrode grid epidurally and was able to communicate by using electrocorticographic oscillations from 3 to 40 Hz. This patient had some motor contingencies still left at the time of implantation and therefore the general extinction of goal directed behavioral responses was not complete, allowing the patient to reinstate voluntary control and social communication after complete paralysis.

At present with such a small number of cases any definite recommendation is premature but we strongly recommend invasive recording of brain activity if EEG-based BCI fails and there are still some behavior-environment contingencies left.

The application of brain–computer interfaces in disorders with severe restriction of communication such as autism, some vegetative state patients and probably in patients with large brain damages, such as hydrancephaly (a very rare disorder in which large parts of the brain are substituted by cerebrospinal fluid) seems to hold some promise for future development. Particularly in the vegetative state and minimal responsive state after brain damage, the application of BCI in cases with relatively intact cognitive function measured with evoked cognitive brain potentials is an important indication for future development of BCI in neurorehabilitation (see [17]).

4 Brain–Computer Interfaces in Stroke and Spinal Cord Lesions

Millions of people suffer from motor disorders where intact movement-related areas of the brain can not generate movements because of damage to the spinal cord, muscles or the primary motor output fibres of the cortex. First clinically relevant attempts to by-pass the lesion with a brain–computer interface were reported by the group of Pfurtscheller [29, 14], and in stroke by [9]). Pfurtscheller reported a patient who was able to learn grasping movements, even to pick up a glass and bring it to the mouth by using sensory motor rhythm (SMR) control of the contralateral motor cortex and electrical stimulation devices attached to the muscle or nerve of the paralyzed hand. Hochberg et al. implanted a 100 electrode grid in the primary motor cortex of a tetraplegic man and trained the patient to use spike sequences classified with simple linear discriminate analysis to move a prosthetic hand. No functional movement, however, was possible with this invasively implanted device. Buch et al. from our laboratory and the National Institutes of Health, National Institute of Neurological Disorders and Stroke developed a non-invasive brain–computer interface for chronic stroke patients using magnetoencephalography [9]. Figure 2 demonstrates the design of the stroke BCI: chronic stroke patients with no residual movement 1 year after the incident do not respond to any type of rehabilitation; their prognosis for improvement is extremely bad. Chronic stroke with residual movement profits from physical restraint therapy developed by Taub (see [39]) where the healthy non-paralyzed limb is fixated with a sling on the body forcing the patient to use the paralyzed hand for daily activities over a period of about 2–3 weeks. Learned non-use, largely responsible for maladaptive brain reorganization and persistent paralysis of the limb, is responsible for the lasting paralysis. Movement restraint therapy, however, can not be applied to stroke patients without any residual movement capacity because no successful contingencies of the motor response and the reward are possible. Therefore, brain–computer interfaces can be used to by-pass the lesion usually subcortically and drive a peripheral device or peripheral muscles or nerves with motor activity generating brain activity.

In the Buch et al. study, 10 patients with chronic stroke without residual movement were trained to increase and decrease sensory motor rhythm (8–15 Hz) or its harmonic (around 20 Hz) from the ipsilesional hemisphere. MEG with 250 channels distributed over the whole cortical surface was used to drive a hand orthosis fixed to the paralyzed hand as seen in Fig. 2. Voluntary increase of sensorimotor rhythm amplitude opened the hand and voluntary decrease of sensorimotor rhythm from

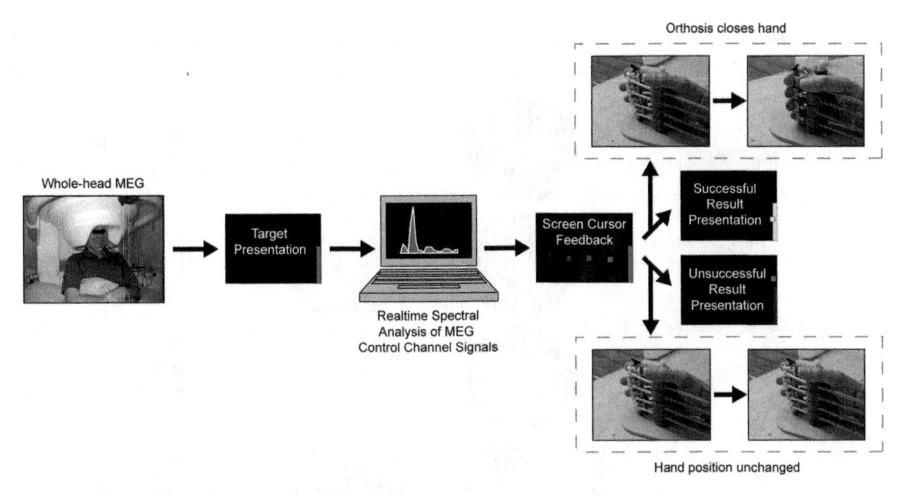

Fig. 2 Trial description for BCI training. Whole-head MEG data (153 or 275-channels) was continuously recorded throughout each training block. At the initiation of each trial, one of two targets (*top-right* or *bottom-right edge of screen*) appeared on a projection screen positioned in front of the subject. Subsequently, a screen cursor would appear at the left edge of the screen, and begin moving towards the right edge at a fixed rate. A computer performed spectral analysis on epochs of data collected from a pre-selected subset of the sensor array (3–4 control sensors). The change in power estimated within a specific spectral band was transformed into the vertical position of the screen cursor feedback projected onto the screen. At the conclusion of the trial, if the subject was successful in deflecting the cursor upwards (net increase in spectral power over the trial period) or downwards (net decrease in spectral power over the trial period) to contact the target, two simultaneous reinforcement events occurred. The cursor and target on the visual feedback display changed colors from red to yellow. At the same time, the orthosis initiated a change in hand posture (opening or closing of hand). If the cursor did not successfully contact the target, no orthosis action was initiated

a group of sensors located over the motor strip closed the hand. Patients received visual feedback of their brain activity on a video screen and at the same time observed and felt the opening and closing of their own hand (proprioceptive perception was clinically assessed in all patients and absent in most of them) by the orthosis device.

Figure 3 demonstrates the effects of the training in 8 different patients.

This study demonstrated for the first time with a reasonable number of cases and in a highly controlled fashion that patients with complete paralysis after chronic stroke are able to move their paralyzed hand with an orthotic device. Movement without that orthotic device was not possible despite some indications of cortical reorganization after training. In chronic stroke, reorganizational processes of the intact hemisphere seem to block adaptive reorganization of the ipsilesional hemisphere. On the other hand, rehabilitation practices profit from activation of both hands and therefore simultaneous brain activation of both hemispheres. It is unclear whether future research and neurorehabilitation of stroke should train patients to move their paralyzed hand exclusively from the ipsilesional intact brain parts or

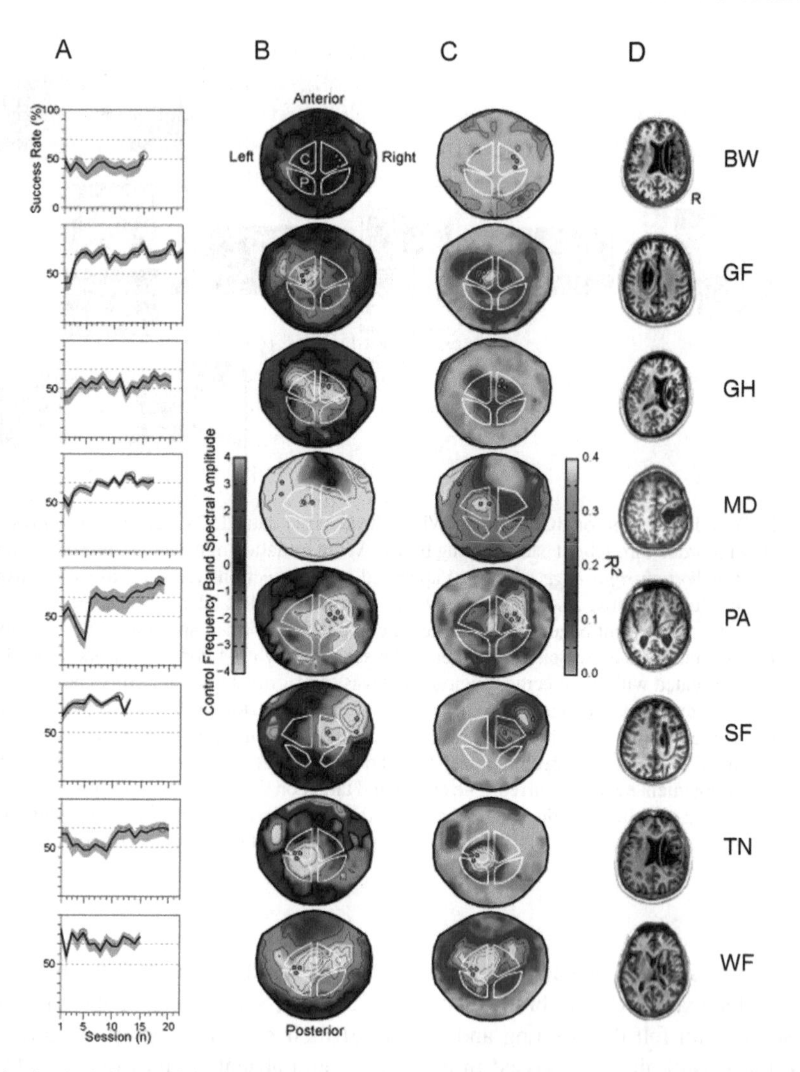

Fig. 3 On the left side of the figure, (**a**) the performance of each patient is depicted demonstrating significant learning in most patients except one. Within 20 sessions of training (1–2 h each), most patients were able to open and close their completely paralyzed hand with the help of the orthosis and their brain activity in more than 70–80% of the trials. Column; (**b**) displays a flat map of the spectral amplitude differences across the MEG array between both target conditions (increase SMR rhythm or decrease SMR rhythm). The sensor locations used to produce feedback and control of the orthosis are highlighted by green filled circles. Column (**c**) of Fig. 3 displays a statistical map (*r*-square) of the correlation of the SMR rhythm amplitude across the MEG array with target location. Column (**d**) displays MR scans obtained for each subject. The red circles highlight the location of each patient's lesion

if simultaneous activation of both hemispheres should be allowed for the activation of the contralesional hand. Whether cortical reorganization after extended BCI training will allow the reinstatement of peripheral control remains an open question. By-passing the lesion by re-growth of new or non-used axonal connections in the primary motor output path seems highly unlikely. However, a contribution of fibres from the contralesional hemisphere reaching the contralesional hand to activate the paralysed hand is theoretically possible. In those cases, a generalisation of training from a peripheral orthosis or any other rehabilitative device to the real world condition seems to be possible. In most cases, however, patients will depend on the activation of electrical stimulation of the peripheral muscles or an orthosis as provided in the study of Buch et al. Some patients will also profit from an invasive approach using epicortical electrodes or microelectrodes implanted in the ipsilesional intact motor cortex, stimulating peripheral nerves or peripheral muscles with voluntary generated brain activity. A new study on healthy subjects using magnetoencephalography from our laboratory (see [35]) has shown that with a non-invasive device such as the MEG with a better spatial resolution than the EEG directional movements of the hand can be classified from magnetic fields online from one single sensor over the motor cortex. These studies demonstrate the potential of non-invasive recordings but do not exclude invasive approaches for a subgroup of patients non-responding to the non-invasive devices. Also, for real life use, invasive internalised BCI systems seem to function better because the range of brain activity usable for peripheral devices is much larger and artefacts from movements do not affect implanted electrodes. The discussion whether invasive or non-invasive BCI devices should be used is superfluous: only a few cases select the invasive approaches, and it is an empirical and not a question of opinion which of the two methods under which conditions provide the better results.

5 The "Emotional" BCI

All reported BCI systems use cortical signals to drive an external device or a computer. EEG, MEG and ECoG as well as near infrared spectroscopy do not allow the use of subcortical brain activity. Many neurological and psychiatric and psychological disorders, however, are caused by pathophysiological changes in subcortical nuclei or by disturbed connectivity and connectivity dynamics between cortical and subcortical and subcortical-cortical areas of the brain. Therefore, particularly for emotional disorders, caused by subcortical alterations brain–computer interfaces using limbic or paralimbic areas are highly desirable. The only non-invasive approach possible for operant conditioning of subcortical brain activity in humans is functional magnetic resonance imaging (fMRI). The recording of blood-flow and conditioning of blood-flow with positron emission tomography (PET) does not constitute a viable alternative because the time delay between the neuronal response, its neurochemical consequences and the external reward is variable and too long to be used for successive learning. That is not the case in functional magnetic resonance imaging, where special gradients and echo-planar imaging allows on-line feedback of the BOLD response (BOLD – blood oxygen level dependent) with a delay of 3s from the neuronal response to the hemodynamic change (see [10, 23, 36]). The laboratory of the author reported the first successful and well-controlled studies of subcortical fMRI-BCI systems (see [36] for a summary). Figure 4 shows the effect of operant conditioning of the anterior insula within 3 training sessions, each session lasting 10 min. Subjects received feedback of their BOLD response in the Region-of-interest with a red or blue arrow, red arrow indicated increase of BOLD in the respective area relative to baseline, the blue arrow pointing downwards indicated a decrease of BOLD referred to baseline. Before and after training subjects were presented with a selection of emotional pictures from the International Affective Picture System (IAPS [21]) presenting negative and neutral emotional slides. As shown in Fig. 4 subjects achieved a surprising degree of control over their brain activity in an area strongly connected to paralimbic area, being one of phylogenetically oldest "cortical" areas generating mainly negative emotional states.

Figure 4 also shows that, after training, subjects indicated a specific increase in aversion to negative emotional slides only. Neutral slides were not affected by the training of BOLD increase in the anterior insula, proving the anatomical specificity of the effect. The rest of the brain was controlled for concomitant increases or decreases and it was demonstrated that no general arousal or general increase or decrease of brain activity is responsible for the behavioral effects. In addition, two control groups one with inconsistent feedback and another with instructions to emotional imagery only did not show a learned increase in BOLD in the anterior insula nor behavioral emotional valence specific effects on emotion.

Conditioning of subcortical areas such as the anterior cingulate and the amygdala were also reported. Ongoing studies investigating the possibility of increasing and decreasing *dynamic connectivity* between different brain areas by presenting feedback only if increased connectivity between the selected areas is produced voluntarily are promising. Most behavioral responses depend not on a single brain area but on the collaboration or disconnection between particular brain areas; behavioural effects should be much larger for connectivity training than for the training of single brain areas alone.

Two clinical applications of the fMRI-BCI were reported: DeCharms et al. [11] showed effects of anterior cingulate training on chronic pain and our laboratory showed that criminal psychopaths are able to regulate their underactivated anterior insula response (see [6]). Whether this has a lasting effect on the behavioral outcome of criminal psychopaths remains to be demonstrated. However, the direction of research is highly promising and studies on the conditioning of depression relevant areas and on schizophrenia are on the way in our laboratory.

The remarkable ease and speed of voluntary control of vascular brain response such as BOLD suggests a superior learning of instrumental control for these response categories. EEG, spike trains, and electrocorticographic activity all need extended training for voluntary control. Vascular responses seem to be easier to regulate, probably because the brain receives feedback of the dynamic status of the vascular system, and these visceral perceptions allow for a better regulation of a non-motor response which can not be detected by the brain in neuroelectric responses.

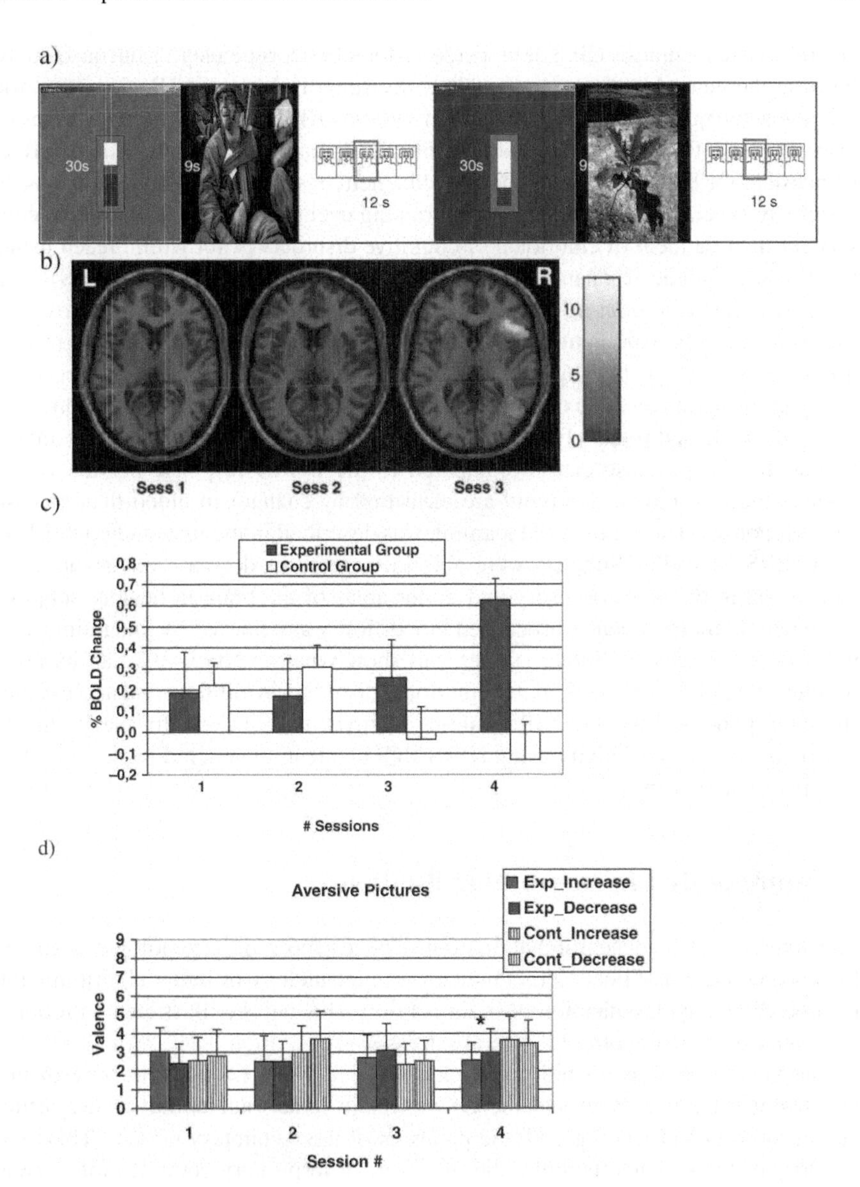

Fig. 4 (**a**) Experimental design. A single run consisted of a 30s increase or decrease block followed by a 9s picture presentation block, that in turn was followed by a 12s rating block. During rating blocks, participants were shown the Self-Assessment Manikin, SAM[32], which allow them to evaluate emotional valence and arousal. (**b**) Random effects analysis on the experimental group confirmed an increased BOLD-magnitude in the right anterior insular cortex over the course of the experiment. (**c**) % BOLD increase in the anterior insula averaged in the experimental group and control group (sham feedback) across the training sessions. (**d**) Valence ratings for aversive pictures for the experimental (Exp) and control group (Cont). During the last training session, aversive pictures presented after the increase condition were rated as significantly more negative (lower valence) than after the decrease condition

Neuroelectric responses can not be perceived neither consciously or unconsciously because the central nervous system does not seem to have specific receptors for its own activity such as peripheral organ systems. Brain perception analogous to visceral perception (see [1]) is not possible for neuroelectric activity but seems to be possible for brain vascular responses. Magnetic resonance imaging scanners are extremely expensive and routine clinical training over longer periods of times necessary for the treatment of emotional or cognitive disorders is not within reach using functional magnetic resonance imaging. *Near Infrared Spectroscopy* (NIRS) may serve as a cheap and non-invasive alternative to fMRI. Near Infrared Spectroscopy uses infrared light from light sources attached to the scalp and measures the reflection or absorption of that light by the cortical tissue, which is largely dependent on the oxygenation and deoxygenation of cortical blood-flow. Devices are commercially available and relatively cheap, and a multitude of channels can be recorded. Essentially the response can be compared to the BOLD response insofar as the consequences of neuronal activity are measured by changes in blood-flow or vascular responses. Therefore, rapid learning was described in the first study published on a NIRS-BCI [33]. Subjects were able to increase or decrease oxygenation of their blood in the somatosensory and motor areas of the brain in healthy subjects using mainly motor imagery. Localised blood-flow was achieved by imagining contralateral hand activity. Future studies will show whether NIRS-BCI can be used for clinical application, particularly emotional disorders in children and adolescents should respond positively to NIRS training. A first controlled trial for the treatment of fronto-central connectivity using NIRS-BCI in attention deficit disorder is on the way in our laboratory.

6 Future of BCI in Neurorehabilitation

The future of BCI in neurorehabilitation depends more on psychological, sociological and social political factors than on new technology or better algorithms for the decoding and classification of brain activity. This will be illustrated with brain communication in amyotrophic lateral sclerosis:

Despite the obvious visibility and success of BCI in ALS patients, 95% of the patients at least in Europe and the US (fewer in Israel) decide not to use artificial respiration and feeding with the paralysis of the respiratory system. This vast majority of the patients therefore die of unknown respiratory complications under unknown circumstances. Countries allowing assisted suicide or euthanasia such as the Netherlands, Belgium, Oregon, Australia and others report even larger death rates before artificial respiration than countries more restrictive on assisted death practices such as Germany and Israel. Controlled studies on large populations of ALS patients have shown [19] that quality of life even in the advanced stages of ALS is comparable to healthy subjects and emotional status is even better (see [24]). Despite these data, no reduction in death rates and no increase in artificial respiration in end-stage ALS are detectable. The great majority of those patients who decide for life and go under artificial respiration have no brain–computer interface available.

Most of them will end in a completely locked-in state where no communication with the outside world is possible. Life expectancy in artificially respirated completely paralyzed patients can be high and exceeds the average life expectancy of 5 years in ALS for many other years. Therefore, every ALS patient who decides for life should be equipped and trained with a brain–computer interface early. Insurance companies are reluctant to pay for the personal and technical expertise necessary to learn brain control. An easy-to-use and easy-to-handle affordable BCI system for brain communication is not commercially available. Obviously, expected profit is low and the industry has no interest in marketing a device for a disorder affecting only a small percentage of the population comparable to the lack of interest of the pharmaceutical industry to develop a drug treatment for rare diseases or for diseases affecting developing countries. We therefore propose a state funded and state run national information campaign affecting end-of-life decisions in these chronic neurological disorders and forcing insurance companies to pay for BCI-related expenses.

The situation is much brighter in the case of motor restoration in stroke and high spinal cord lesions (see [30]). The large number of cases promises great profit and alternative treatments are not within reach of years. The first stroke-related brain–computer interface (connected with central EEG) commercially available is presently built in Israel by Motorika (Cesarea). It combines a neurorehabilitative robotic device which can be connected non-invasively with the brain of the patient and allowing the patient even in complete paralysis to run the device directly with voluntary impulses from his brain and voluntary decisions and planning for particular goal-directed movements. Also, research on invasive devices for stroke rehabilitation will be performed during the next years. Whether they will result in profitable easy-to-use economic devices for chronic stroke and other forms of paralysis remains to be seen.

Acknowledgement Supported by the Deutsche Forschungsgemeinschaft (DFG).

References

1. G. Adam, *Visceral perception*, Plenum Press, New York, (1998).
2. T. Ball, E. Demandt, I. Mutschler, E. Neitzl, C. Mehring, K. Vogt, A. Aertsen, A. Schulze-Bonhage, Movement related activity in the high gamma-range of the human EEG. NeuroImage, 41, 302–310, (2008).
3. N. Birbaumer, T. Elbert, A. Canavan, and B. Rockstroh, Slow potentials of the cerebral cortex and behavior. Physiol Rev, 70, 1–41, (1990).
4. N. Birbaumer, Slow cortical potentials: Plasticity, operant control, and behavioral effects. The Neuroscientist, 5(2), 74–78, (1999).
5. N. Birbaumer, N. Ghanayim, T. Hinterberger, I. Iversen, B. Kotchoubey, A. Kubler, J. Perelmouter, E. Taub, and H. Flor A spelling device for the paralyzed. Nature, 398, 297–298, (1999).
6. N. Birbaumer, R. Veit, M. Lotze, M. Erb, C. Hermann, W. Grodd, H. Flor, Deficient fear conditioning in psychopathy: A functional magnetic resonance imaging study. Arch Gen Psychiatry, 62, 799–805, (2005).
7. N. Birbaumer, Brain–Computer–Interface Research: Coming of Age. Clin Neurophysiol, 117, 479–483, (2006).

8. N. Birbaumer and L. Cohen, Brain–Computer-Interfaces (BCI): Communication and Restoration of Movement in Paralysis. J Physiol, 579(3), 621–636, (2007).
9. E. Buch, C. Weber, L.G. Cohen, C. Braun, M. Dimyan, T. Ard, J. Mellinger, A. Caria, S. Soekadar, N. Birbaumer, Think to move: a neuromagnetic Brain–Computer interface (BCI) system for chronic stroke. Stroke, 39, 910–917, (2008).
10. A. Caria, R. Veit, R. Sitaram, M. Lotze, N. Weiskopf, W. Grodd, and N. Birbaumer, (2007). Regulation of anterior insular cortex activity using real-time fMRI. NeuroImage, 35, 1238–1246.
11. R.C. DeCharms, F. Maeda, G.H. Glover, D. Ludlow, J.M. Pauly, D. Soneji, J.D. Gabrieli, and S.C. Mackey, Control over brain activation and pain learned by using real-time functional MRI. Proc Natl Acad Sci, 102(51), 18626–18631, (2005).
12. B.R. Dworkin, and N.E. Miller, Failure to replicate visceral learning in the acute curarized rat preparation. Behav Neurosci, 100, 299–314, (1986).
13. L.A. Farwell and E. Donchin, Talking off the top of your head: toward a mental prosthesis utilizing event-related brain potentials. Electroencephalogr Clin Neurophysiol, 70, 510–523, (1988).
14. L.R. Hochberg, M.D. Serruya, G.M. Friehs, J.A. Mukand, M. Saleh, A.H. Caplan, A. Branner, D. Chen, R.D. Penn, and J.P. Donoghue, Neural ensemble control of prosthetic devices by a human with tetraplegia. Nature, 442, 164–171, (2006).
15. A. Jackson, J. Mavoori, E. Fetz, Long-term motor cortex plasticity induced by an electronic neural implant. Nature, 444, 56–60, (2006).
16. B. Kotchoubey, U. Strehl, C. Uhlmann, S. Holzapfel, M. König, W. Fröscher, V. Blankenhorn, and N. Birbaumer Modification of slow cortical potentials in patients with refractory epilepsy: a controlled outcome study. Epilepsia, 42(3), 406–416, (2001).
17. B. Kotchoubey, A. Kübler, U. Strehl, H. Flor, and N. Birbaumer, Can humans perceive their brain states? Conscious Cogn, 11, 98–113, (2002).
18. A. Kübler, B. Kotchoubey, J. Kaiser, J. Wolpaw, and N. Birbaumer, Brain–Computer communication: unlocking the locked-in. Psychol Bull, 127(3), 358–375, (2001).
19. A. Kübler, S. Winter, A.C. Ludolph, M. Hautzinger, and N. Birbaumer, Severity of depressive symptoms and quality of life in patients with amyotrophic lateral sclerosis. Neurorehabil Neural Repair, 19(3), 182–193, (2005).
20. A. Kübler and N. Birbaumer, Brain–Computer Interfaces and communication in paralysis: Extinction of goal directed thinking in completely paralysed patients. Clin Neurophysiol, 119, 2658–2666, (2008).
21. P. Lang, M. Bradley, and B. Cuthbert, *International Affective Picture System (IAPS)*. The Center for Research in Psychophysiology, University of Florida, Gainesville, Fl, (1999).
22. E.C. Leuthardt, K. Miller, G. Schalk, R.N. Rao, and J.G. Ojemann, Electrocorticography-Based Brain Computer Interface – The Seattle Experience. IEEE Trans Neur Sys Rehab Eng, 14, 194–198, (2006).
23. N. Logothetis, J. Pauls, M. Augath, T. Trinath, and A. Oeltermann, Neurophysiological investigation of the basis of the fMRI signal. Nature, 412, 150–157, (2001).
24. D. Lulé, V. Diekmann, S. Anders, J. Kassubek, A. Kübler, A.C. Ludolph, and N. Birbaumer, Brain responses to emotional stimuli in patients with amyotrophic lateral sclerosis (ALS). J Neurol, 254(4), 519–527, (2007).
25. N. Miller, Learning of visceral and glandular responses. Science, 163, 434–445, (1969).
26. M.A.L. Nicolelis, Actions from thoughts, Nature, 409, 403–407, (2001).
27. M.A. Nicolelis, Brain-machine interfaces to restore motor function and probe neural circuits. Nat Rev Neurosci, 4(5), 417–422, (2003).
28. L.M. Oberman, V.S. Ramachandran, and J.A. Pineda, Modulation of mu suppression in children with autism spectrum disorders in response to familiar or unfamiliar stimuli: the mirror neuron hypothesis. Neuropsychologia, 46, 1558–65, (2008).
29. G. Pfurtscheller, C. Neuper, and N. Birbaumer, Human Brain–Computer Interface (BCI). In Alexa Riehle & Eilon Vaadia (Eds.), *Motor cortex in voluntary movements. A distributed system for distributed functions*, CRC Press, Boca Raton, FL, pp. 367–401, (2005).

30. G. Pfurtscheller, G. Müller-Putz, R. Scherer, and C. Neuper, Rehabilitation with Brain–Computer Interface systems. IEEE Comput Sci, 41, 58–65, (2008).
31. M. Schwartz, F. Andrasik, (Eds.), *Biofeedback: A Practitioner's Guide*, 3rd edn, Guilford Press, New York, (2003).
32. G. Schalk, Brain–Computer symbiosis. J Neural Eng, 5, 1–15, (2008).
33. R. Sitaram, H. Zhang, C. Guan, M. Thulasidas, Y. Hoshi, A. Ishikawa, K. Shimizu, and N. Birbaumer, Temporal classification of multi-channel near-infrared spectroscopy signals of motor imagery for developing a Brain–Computer interface. NeuroImage, 34, 1416–1427, (2007).
34. U. Strehl, U. Leins, G. Goth, C. Klinger, T. Hinterberger, and N. Birbaumer, Self-regulation of Slow Cortical Potentials – A new treatment for children with Attention-Deficit/Hyperactivity Disorder. Pediatrics, 118(5), 1530–1540, (2006).
35. S. Waldert, H. Preissl, E. Demandt, C. Braun, N. Birbaumer, A. Aertsen, and C. Mehring, Hand movement direction decoded from MEG and EEG. J Neurosci, 28, 1000–1008, (2008).
36. N. Weiskopf, F. Scharnowsi, R. Veit, R. Goebel, N. Birbaumer, and K. Mathiak, Self-regulation of local brain activity using real-time functional magnetic resonance imaging (fMRI). J Physiol, Paris, 98, 357–373, (2005).
37. B. Wilhelm, M. Jordan, and N. Birbaumer, Communication in locked-in syndrome: effects of imagery on salivary pH. Neurology, 67, 534–535, (2006).
38. J.A. Wilson, E.A. Felton, P.C. Garell, G. Schalk, and J.C. Williams, ECoG factors underlying multimodal control of a Brain–Computer interface. IEEE Trans Neur Sys Rehab Eng, 14, 246–250, (2006).
39. S.L. Wolf, et al., Effect of constraint-induced movement therapy on upper extremity function 3 to 9 months after stroke: The EXCITE randomized clinical trial. Jama, 296(17), 2095–104, (2006).

Non Invasive BCIs for Neuroprostheses Control of the Paralysed Hand

Gernot R. Müller-Putz, Reinhold Scherer, Gert Pfurtscheller, and Rüdiger Rupp

1 Introduction

About 300,000 people in Europe alone suffer from a spinal cord injury (SCI), with 11,000 new injuries per year [20]. SCI is caused primarily by traffic and work accidents, and an increasing percentage of the total population also develops SCI from diseases like infections or tumors. About 70% of SCI cases occur in men. 40% are tetraplegic patients with paralyses not only of the lower extremities (and hence restrictions in standing and walking) but also of the upper extremities, which makes it difficult or impossible for them to grasp.

1.1 Spinal Cord Injury

SCI results in deficits of sensory, motor and autonomous functions, with tremendous consequences for the patients. The total loss of grasp function resulting from a complete or nearly complete lesion of the cervical spinal cord leads to an all-day, life-long dependency on outside help, and thus represents a tremendous reduction in the patients' quality of life [1]. Any improvement of lost or limited functions is highly desirable, not only from the patients' point of view, but also for economic reasons [19]. Since tetraplegic patients are often young persons due to sport and diving accidents, modern rehabilitation medicine aims at restoration of the individual functional deficits.

1.2 Neuroprostheses for the Upper Extremity

Today, the only way to permanently restore restricted or lost functions to a certain extent in case of missing surgical options [4] is the application of Functional

G.R. Müller-Putz (✉)
Laboratory of Brain-Computer Interfaces, Institute for Knowledge Discovery, Graz University of Technology, Krenngasse 37, 8010 Graz, Austria
e-mail: gernot.mueller@tugraz.at

B. Graimann et al. (eds.), *Brain–Computer Interfaces*, The Frontiers Collection,
DOI 10.1007/978-3-642-02091-9_10, © Springer-Verlag Berlin Heidelberg 2010

Electrical Stimulation (FES). The stimulation devices and systems that are used for this purpose are called neuroprostheses [27].

A restoration of motor functions (e.g., grasping) by using neuroprostheses is possible if the peripheral nerves connecting the neurons of the central nervous system to the muscles are still intact [26]. By placing surface electrodes near the motor point of the muscle and applying short (< 1 ms) constant-current pulses, the potential of the nerve membrane is depolarized and the elicited action potential leads to a contraction of the innervated muscle fibers, somewhat similar to natural muscle contractions. There are some differences between the physiological and the artificial activation of nerves: physiologically, small and thin motor fibers of a nerve, which innervate fatigue resistant muscle fibers, are activated first. As people grasp more strongly, more and more fibers with larger diameters are recruited, ultimately including muscle fibers that are strong, but fatigue rapidly. Applying artificial stimulation pulses, this activation pattern is reversed. The current pulses lead to action potentials in the fibers with large diameter first, and by further increasing the current, the medium sized and thin fibers also get stimulated [29].

This "inverse recruitment" leads to a premature fatiguing of electrically stimulated muscles. Additionally, the fact that every stimulation pulse activates the same nerve fibers all at once further increases muscle fatigue. Muscle fatigue is especially problematic with higher stimulation frequencies for tetanic contractions. A frequency around 35 Hz leads to a strong tonus, however the muscle gets tired earlier than in the case of e.g., 20 Hz. Therefore, the pulse repetition rate has to be carefully chosen in relation to the desired muscle strength and the necessary contraction duration.

Some neuroprostheses for the upper extremity are based on surface electrodes for external stimulation of muscles of the forearm and hand. Examples are the commercially available (N200, Riddenderk, Netherlands) System [5] and other more sophisticated research prototypes [12, 31]. To overcome the limitations of surface stimulation electrodes concerning selectivity, reproducibility and practicability, an implantable neuroprostheses (the Freehand® system, Neurocontrol, Valley View, OH, USA) was developed, where electrodes, cables and the stimulator reside permanently under the skin [7]. This neuroprosthesis was proven effective in functional restoration and user acceptance [21]. All FES systems for the upper extremity can only be used by patients with preserved voluntary shoulder and elbow function, which is the case in patients with an injury of the spinal cord below C5. The reason for this limitation is that the current systems are only able to restore grasp function of the hand, and hence patients must have enough active control movements for independent use of the system.

Until now, only two groups have dealt with the problem of restitution of elbow and shoulder movement. Memberg and colleagues [13] used an extended Freehand system to achieve elbow extension, which is typically not apparent in patients with a lesion at cervical level C5. In the mid 80 s, Handa's group [6] developed a system based on intramuscular electrodes for restoration of shoulder function in hemiparesis and SCI. Both systems represent exclusive FES systems, which stimulate the appropriate muscle groups not only for dynamic movements but also for maintaining a static posture. Due to the weight of the upper limb and the non-physiologic synchronous activation of the paralyzed muscles, these systems are difficult to use for most activities throughout the day.

One of the main problems in functional restoration of the grasping and reaching function in tetraplegic patients is the occurrence of a combined lesion of the central and peripheral nervous structures. In almost one third of the tetraplegic patients, an ulnar denervation occurs due to damage of the motor neurons in the cervical spinal cord [2]. Flaccid and denervated muscles can not be used for functional restoration by electrical stimulation where action potentials on the nerve are elicited. Much higher currents are necessary to directly stimulate the muscle, which can damage the skin. Here, further research is necessary to evaluate the possibility of transferring results obtained by direct muscle stimulation at the lower extremities [8].

In principle, all types of grasp neuroprostheses are controlled with external control units, such as a shoulder position sensor (see Fig. 1).

Today, residual movements not directly related to the grasping process are usually used to control the neuroprostheses. In the highest spinal cord injuries, not enough functions are preserved for control, which so far hampered the development of neuroprostheses for patients with a loss of not only hand and finger but also elbow and shoulder function.

Kirsch [10] presents a new concept of using implanted facial and neck muscles sensors to control an implanted neuroprosthesis for the upper extremity. The system itself is designed to consist of two stimulation units providing 12 stimulation

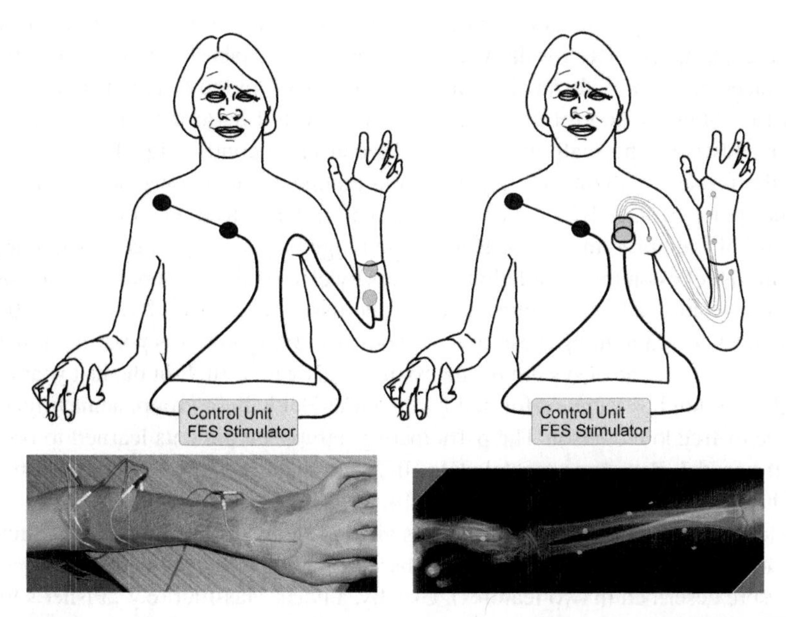

Fig. 1 Grasp neuroprostheses. *Left*: Neuroprosthesis using surface electrodes. *Right*: Neuroprosthesis with implanted electrodes (e.g., Freehand(R)). Both systems are controlled with a shoulder position sensor fixed externally on the contralateral shoulder

channels each. They will be implanted to control simple basic movements such as eating and grooming. Control signals will be obtained from implanted face and neck electromyographic (EMG) sensors and from an additional position sensor placed on the head of the user. Additionally, two external sensors are fixed at the forearm and the upper arm to provide the actual position of the arm. A part of this concept was already realized and implanted in nine arms in seven C5/C6 SCI individuals. The study described in [9] showed that it is possible to control the neuroprostheses for grasp-release function with EMG signals from strong, voluntarily activated muscles with electrical stimulation in nearby muscles.

Rupp [28] presented a completely non invasive system to control grasp neuroprostheses based on either surface or implanted electrodes. With this system, it is possible to measure the voluntary EMG activity of very weak, partly paralysed muscles that are directly involved in, but do not efficiently contribute to, grasp function. Due to a special filter design, the system can detect nearby applied stimulation pulses, remove the stimulation artefacts, and use the residual voluntary EMG-activity to control the stimulation of the same muscle in the sense of "muscle force amplification".

2 Brain-Computer Interface for Control of Grasping Neuroprostheses

To overcome the problems of limited degrees of freedom for control or controllers that are not appropriate for daily activities outside the laboratory, Brain-Computer Interfaces might provide an alternative control option in the future. The ideal solution for voluntary control of a neuroprosthesis would be to directly record motor commands from the scalp and transfer the converted control signals to the neuroprosthesis itself, realizing a technical bypass around the interrupted nerve fiber tracts in the spinal cord. A BCI in general is based on the measurement of the electrical activity of the brain, in case of EEG in the range of μV [32]. In contrast, a neuroprosthesis relies on the stimulation of nerves by electrical current pulses in the range of up to 40 mA, which assumes an electrode-tissue resistance of 1 kΩ. One of the challenges for combining these two methods is proving that it is possible to realize an artefact free control system for neuroprosthesis with a BCI. In the last years, two single case studies were performed by the Graz-Heidelberg group, achieving a one degree of freedom control. The participating tetraplegic patients learned to operate a self-paced 1-class (one mental state) BCI and thereby control a neuroprosthesis, and hence control their grasp function [14, 15, 23].

The basic idea of a self-paced brain switch is shown in Fig. 2. In the beginning, patients were trained to control the cue-based BCI with two types of motor imagery (MI, here described in two features). Usually, a linear classifier (e.g., Fisher's linear discriminant analysis, LDA) fits a separation hyper plane in a way, e.g., to maximise the distance of the means between the two classes (Fig. 2a). This requires analyzing the classifier output time series, often as presented in Fig. 2b. One class (class 1)

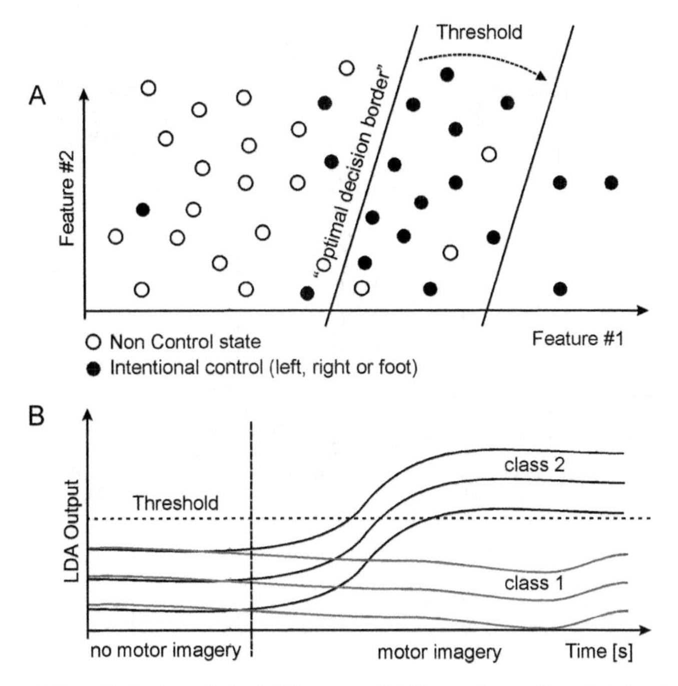

Fig. 2 General idea of a brain-switch. (**a**) Features of different classes (e.g., left hand, right hand or feet MI) are usually separated by an optimal decision border. Introducing an additional threshold a switch function can be designed. (**b**) LDA output of two types of MI during no MI and MI

does not really change from a period without MI to a period with MI. However, the other class (class 2) does. A significant change can be seen. It is assumed that the first case (class 1) is a very general case. This means that the classifier would also select this class when no MI is performed. Therefore, this describes the non-control state. Introducing an additional threshold into the class of the other MI pattern (class 2), a switch function can be designed (control state). A control signal is only triggered when the MI is recognized clearly enough to exceed the threshold. For the BCI application (see Fig. 3), the grasp is divided into distinct phases (Fig. 3c, d), e.g., hand open, hand close, hand open, hand relax (stimulation off). Whenever a trigger signal is induced, the users' hand is stimulated so that the current phase is subsequently switched to the next grasp phase. If the end of the grasp sequence is reached, a new cycle is started.

2.1 Patients

The first patient enrolled in the study, TS, is a 32 year old man who became tetraplegic because of a traumatic spinal cord injury in April 1998. He has a complete (ASIA A, described by the American Spinal Cord Injury Association, [30]) motor and sensory paralysis at the level of the 5th cervical spinal vertebra. Volitional

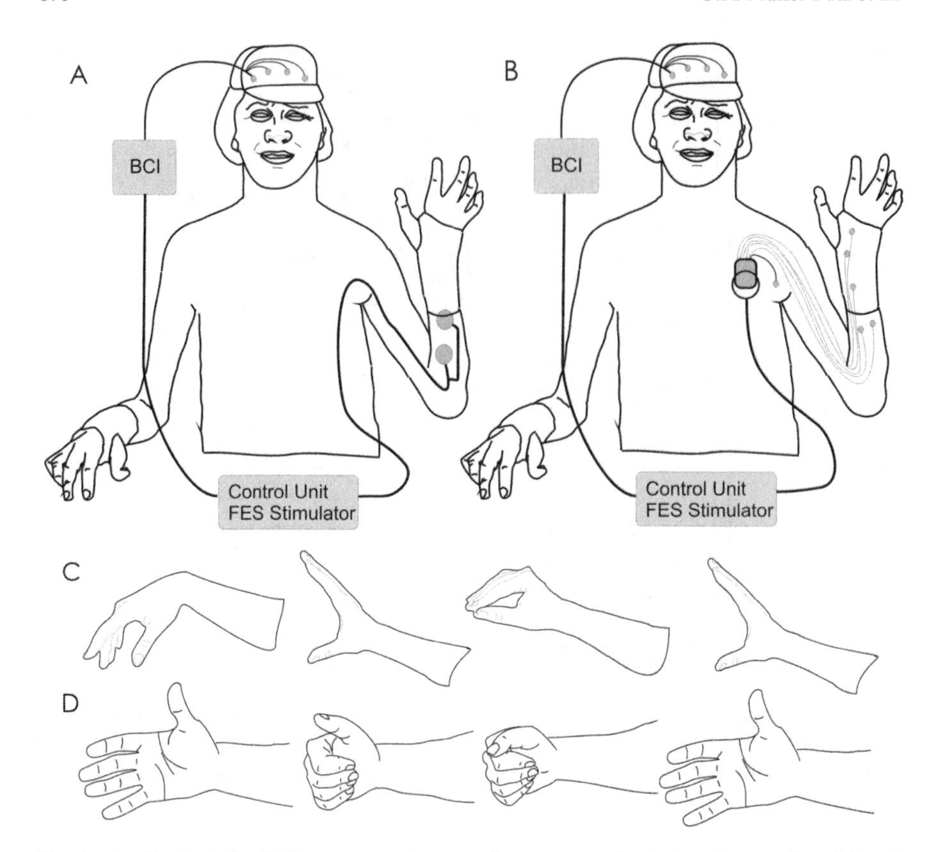

Fig. 3 Application of a BCI as a control system for neuroprostheses based on surface (**a**) and implanted (**b**) electrodes, respectively. (**c**) Grasp pattern for palmar grasp. (**d**) Grasp pattern for lateral grasp

muscle function is preserved in both shoulders and in his left biceps muscle for active elbow flexion, as well as very weak extension. He has no active hand and finger function. As a preparation for the neuroprosthesis he performed a stimulation training program using surface electrodes for both shoulders and the left arm/hand (lasting about 10 months) with increasing stimulation frequency and stimulation time per day until he achieved a strong and fatigue resistant contraction of the paralyzed muscles of the left (non-dominant) forearm and hand. The stimulation device used for this training was then configured per software as a grasp neuroprosthesis by implementing the stimulation patterns of three distinct grasp phases (Microstim, Krauth & Timmermann, Hamburg, Germany).

The second patient, a 42 year old man called HK, has a neurological status like TS, with a complete (ASIA A) motor and sensory paralysis at the 5th cervical spinal vertebra because of a car accident in December 1998. His volitional muscle activation is restricted to both shoulders and bilateral elbow flexion with no active movements of his hands or fingers. He underwent a muscle conditioning program of the paralysed and atrophied muscles of the forearm and hand at the Orthopaedic University Hospital II in Heidelberg starting at January 2000. The Freehand(R) neuroprosthesis was implanted in September 2000 in his right arm and hand, which was his dominant hand prior to the injury. After the rehabilitation program, he gained a substantial functional benefit in performing many activities of everyday life.

2.2 EEG Recording and Signal Processing

In general, the EEG was bipolarly recorded from positions 2.5 cm anterior and posterior to C3, Cz and C4 (overlying the sensorimotor areas of right hand, left hand and feet) according the international 10–20 electrode system using gold-electrodes. In the final experiments, EEG was recorded from the vertex; channel Cz (foot area) in patient TS, and from positions around Cz and C4 (left hand area) in patient HK. In both cases, the ground electrode was placed on the forehead. The EEG-signals were amplified (sensitivity was 50 μV) between 0.5 and 30 Hz with a bipolar EEG-amplifier (Raich, Graz, Austria, and g.tec, Guger Technologies, Graz, Austria), notch filter (50 Hz) on, and sampled with 125/250 Hz.

Logarithmic band power feature time series were used as input for both experiments. For identifying reactive frequency bands, time-frequency maps were calculated. These maps provide data about significant power decrease (event-related desynchronization, ERD) or increase (event-related synchronization, ERS) in predefined frequency bands related to a reference period within a frequency range of interest (for more details see Chapter 3). Usually, these relative power changes are plotted over the whole trial time and result in so-called ERD/S maps [3].

For both experiments, band power was estimated by band pass filtering (Butterworth IIR filter with order 5, individual cut-off frequency) of the raw EEG, squaring and averaging (moving average) samples over a 1-s period. The logarithm was applied to the band power values, which are generally not normally distributed. The logarithmic band power features were classified using LDA.

LDA projects features on a line so that samples belonging to the same class form compact clusters (Fig. 2a). At the same time, the distance between the different clusters is maximized to enhance discrimination. The weights of the LDA were calculated for different time points starting at second 0 until the end of the trial in steps of 0.5 or 0.25 s. Applying a 10-times 10-fold cross validation statistic the classification accuracy is estimated to avoid over fitting (more details about signal processing can be found in Chapter 17). The weight vector of the time point with the best accuracy was then used for further experiments.

2.3 Setup Procedures for BCI Control

As a first step, both patients went through the standard cue-based or synchronous BCI training to identify reactive frequency bands during hand or foot movement imagination [24]. This means that cues appearing randomly on a screen indicated the imagery task that had to be performed by the patients for about 4 s in each trial. A minimum of 160 trials (one session) were used to identify suitable frequency bands with the help of ERD/S maps. Logarithmic band power features were then calculated from reactive bands and used for LDA classifier setup. The classifier at time of maximum accuracy was then used for further training.

2.3.1 BCI-Training of Patient TS Using a Neuroprosthesis with Surface Electrodes

In 1999, patient TS learned to control the cue-based BCI during a very intensive training period lasting more than 4 months. The training started with left hand vs. right hand imagination. These MI tasks were changed to to right/left hand vs. idling or right hand vs. left foot because of insufficient accuracy. Finally, his performance (session 65) was between 90 and 100 % using right hand and foot motor imaginations [22]. Because of the prominent ERS during foot MI, the paradigm was changed to operate asynchronously. Figure 4a presents the average of the LDA output during an asynchronous neuroprosthetic control experiment. A threshold (TH) was implemented for the foot class (around 0.5). Whenever the LDA output exceeded this TH (at second 3) a switch function was triggered [23]. The corresponding features of the 2 bipolar channels C3 and Cz (α=10–12 Hz, β=15–19 Hz) are shown in Fig. 4b. It can clearly be seen that the LDA output (Fig. 4a) depends mainly on the β-band power of channel Cz. This suggested that the system could be simplified by implementing a TH-comparator of the band power [15].

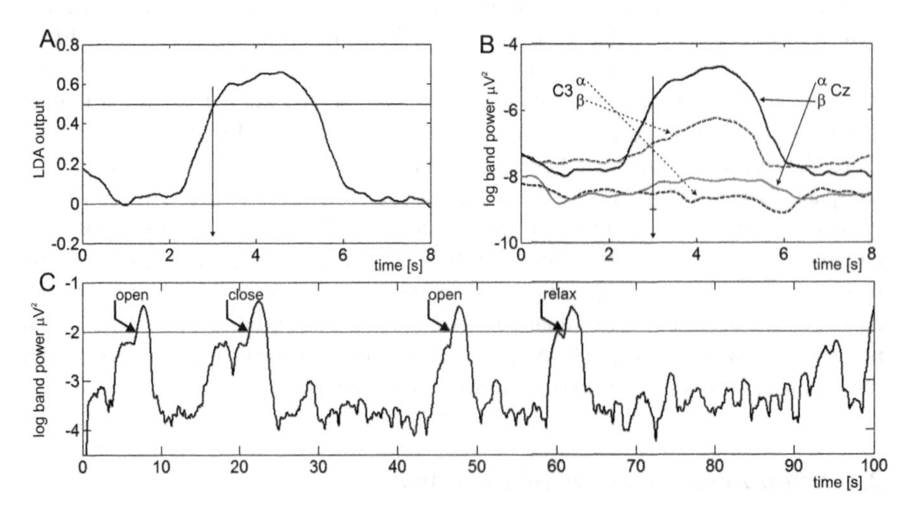

Fig. 4 (**a**) Average LDA outputs during grasp phase switching. (**b**) Corresponding logarithmic band power features (α and β for C3 and Cz) to A. (**c**) Logarithmic β-band power for Cz during one grasp sequence: hand opens, fingers close (around a drinking glass), hand opens again, and stimulation stops, so that corresponding muscles relax (modified from [18])

An example of such a system is given in Fig. 4c. In this case, the TH was computed over a time period of 240 s by performing two "idle" runs without any imagination. The band power was then extracted (15–19 Hz band) and the mean (x) and standard deviation (sd) calculated. Here, TH was set to TH $= x + 3 \cdot sd$. To give the user enough time to changing his mental state, a refractory phase of 5 s was implemented. In this time period, the classification output could not trigger another grasp phase. The first two pictures of Fig. 7 presents two shots during drinking.

2.3.2 BCI-Training of Patient HK Using an Implanted Neuroprosthesis

For practical reasons, training was performed over 3 days at the patient's home. At first, the patient was asked to imagine different feet and left hand movements to determine which movements required the least concentration. After this pre-screening, a cue-guided screening session was performed, where he was asked to imagine feet and left hand movements 160 times. Applying time-frequency analyses, the ERD/S maps were calculated. From these results, the most reactive frequency bands (14–16 and 18–22 Hz) were selected, and a classifier was set up. With this classifier, online feedback training was performed, which consisted of totally 25 training runs using the Basket paradigm [11]. The task in this Basket experiment was to move a ball, falling with constant speed (falling duration was 3 s) from the top of the screen, towards the indicated target (basket) to the left or to the rigth at the bottom of the screen. Four runs (consisting of 40 trials each) with the best accuracy were then used to adapt the frequency bands (resulting in 12–14 Hz and 18–22 Hz) and calculate a new classifier (offline accuracy was 71 %). This classifier was then used in an asynchronous paradigm for unguided training. Because of a significant ERD, left hand MI was used for switching. This decision was supported by the fact that, during the non-control state, foot activity was detected, so the classifier had a bias to the foot class. Therefore, the output of the classifier was compared with a threshold implemented in the class representing left hand movement imaginations (compare the brain switch scheme given in Fig. 2). Whenever the classifier output exceeded this threshold for a dwell time of 1 s, a switching signal was generated. Consecutively a refractory period of 3 s was implemented so that the movement control was stable and the probability of false positives was reduced. A 180-s run without any MI showed no switch-action. After the BCI-training paradigm, the classifier output of the BCI was coupled with the Freehand(R) system. In Fig. 5a the averaged LDA output during the evaluation experiment is presented. The predefined threshold in this case 0.7 was exceeded 1 s prior the real switching action (dwell time) at second 0. Fig. 5b shows the corresponding band power features [14].

2.4 Interferences of Electrical Stimulation with the BCI

A grasp neuroprosthesis produces short electrical impulses with up to 40 mA (= 40 V assuming 1 kΩ electrode-skin resistance) in a frequency range from 16 to 35 Hz. These strong stimulation pulses lead to interference in the μV-amplitude

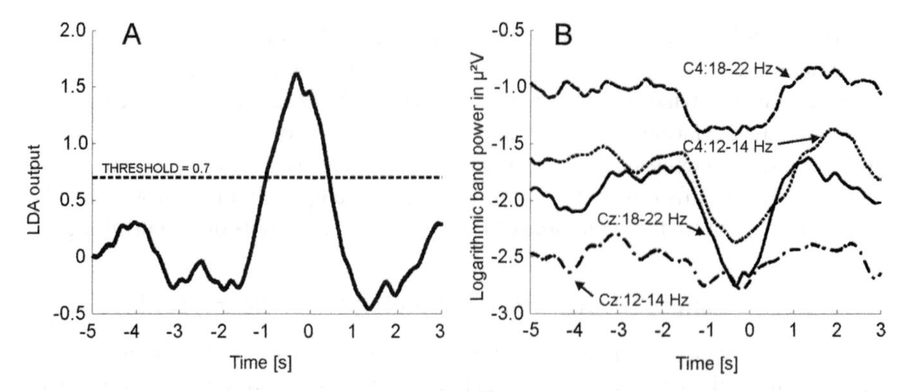

Fig. 5 (**a**) Average LDA output for the classifier used during the grasp test (16 triggered grasp phases). (**b**) Corresponding averaged logarithmic band power features from electrode positions Cz and C4 during the grasp test (modified from [17])

EEG. In the case of the neuroprosthesis with surface electrodes, the EEG was recorded in a bipolar manner so that synchronized interference could effectively be suppressed. Since every stimulation channel had a bipolar design, the influence on EEG recordings was minimized. Compared to the use of surface electrodes there was a much higher degree of interferences apparent in the EEG using the Freehand system.

This stimulation device consists of eight monopolar stimulation electrodes and only one common anode, which is the stimulator implanted in the chest. To avoid contamination by stimulation artifacts, spectral regions with these interferences were not included in the frequency bands chosen for band power calculation. When the analyzed frequency bands of the BCI and stimulation frequency of stimulation device overlap, the repetition rate of the stimulation pulses can be shifted towards a frequency outside of the reactive EEG bands. Techniques such as Laplacian EEG derivations or regression methods with additional artifact recording channels would inevitably lead to more electrodes and therefore be a severe limitation for the application of a EEG-based brain-switch in everyday life (Fig. 6).

2.5 Evaluation of the Overall Performance of the BCI Controlled Neuroprostheses

For evaluation of the whole system performance in patient HK, a part of the internationally accepted grasp-and-release test [33] was carried out. The task was to repetitively move one paperweight from one place to another within a time interval of 3 min (see Fig. 7 picture 3 and 4). For the completion of one grasp sequence, 3 classified left hand MI were necessary. During this 3-min interval, the paperweight was moved 5 times with a sixth grasp sequence initiated. The mean duration of the 16 performed switches was 10.7 s ± 8.3 s [14].

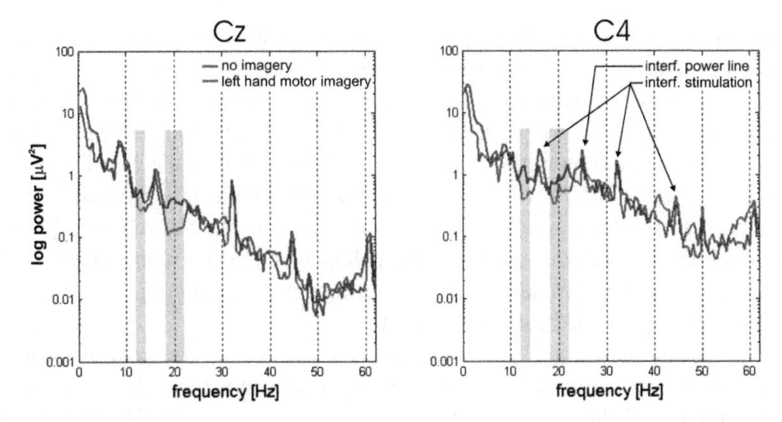

Fig. 6 Spectra of channel Cz and C4 during the grasp-release test. Gray bars indicate the frequency bands used for classification. The stimulation frequency as well as the power line interference can be seen (modified from [14])

Fig. 7 Pictures of both patients during BCI control of their individual neuroprosthesis

3 Conclusion

It has been shown that brain-controlled neuroprosthetic devices do not only exist in science fiction movies but have become reality. Although overall system performance is only moderate, our results have proven that BCI systems can work together with systems for functional electrical stimulation and thus open new possibilities for restoration of grasp function in high cervical spinal cord injured people. Still, many open questions must answered before patients can be supplied with BCI controlled neuroprosthetic devices on a routinely clinical basis. One major restricition of current BCI systems is that they provide only very few degrees for control and only digital control signals. This means that the brain-switch described above can be used to open or close fingers, or switch from one grasp phase to the next. Also, no way to control the muscle strength or stimulation duration in an analogue matter has yet been implemented. However, for patients who are not able to control their neuroprosthetic devices with additional movements (e.g., shoulder or arm muscles), BCI control may be a real alternative for the future.

BCI applications are still mainly performed in the laboratory environment. Small wireless EEG amplifiers, together with a form of an electrode helmet (or something similar with dry electrodes (e.g., [25]), must be developed for domestic or public use. This problem is addressed already from some research groups in Europe, which are confident that these systems can be provided to patients in the near future. In the meantime the current FES methods for restoration of the grasp function have to be extended to restoration of elbow (first attempts are already reported [16]) and shoulder function to take advantage of the enhanced control possibilities of the BCI systems. This issue is also addressed at the moment by combining an motor driven orthosis with electrical stimulation methods.

In the studies described above the direct influence of the neuroprosthesis to EEG recording is discussed. A new and important aspect for the future will be the investigation of the plasticity of involved cortical networks. The use of a neuroprosthesis for grasp/elbow movement restoration offers realistic feedback to the individual. Patterns of imagined movements used for BCI control then compete with patterns from observing the user's own body movement. These processes must be investigated and used to improve BCI control.

Acknowledgments The authors would like to acknowledge the motovation and patience of the patients during the intensive days of trainning. This work was partly supported by Wings for Life - The spinal cord research foundation, Lorenz Böhler Gesellschaft, "Allgemeine Unfallversicherungsanstalt - AUVA", EU COST BM0601 Neuromath, and Land Steiermark.

References

1. K.D. Anderson, Targeting recovery: priorities of the spinal cord-injured population. J Neurotrauma, 21(10), 1371–1383, (Oct 2004).
2. V. Dietz and A. Curt, Neurological aspects of spinal-cord repair: promises and challenges. Lancet Neurol, 5(8), 688–694, (Aug 2006).
3. B. Graimann, J.E. Huggins, S.P. Levine, and G. Pfurtscheller, Visualization of significant ERD/ERS patterns multichannel EEG and ECoG data. Clin Neurophysiol, 113, 43–47, (2002).
4. V. Hentz and C. Le Clercq, *Surgical Rehabilitation of the Upper Limb in Tetraplegia*. W.B. Saunders Ltd., London, UK, (2002).
5. M.J. Ijzermann, T.S. Stoffers, M.A. Klatte, G. Snoeck, J.H. Vorsteveld, and R.H. Nathan, The NESS Handmaster Orthosis: restoration of hand function in C5 and stroke patients by means of electrical stimulation. J Rehabil Sci, 9, 86–89, (1996).
6. J. Kameyama, Y. Handa, N. Hoshimiya, and M. Sakurai, Restoration of shoulder movement in quadriplegic and hemiplegic patients by functional electrical stimulation using percutaneous multiple electrodes. Tohoku J Exp Med, 187(4), 329–337, (Apr 1999).
7. M.W. Keith and H. Hoyen, Indications and future directions for upper limb neuroprostheses in tetraplegic patients: a review. Hand Clin, 18(3), 519–528, (2002).
8. H. Kern, K. Rossini, U. Carraro, W. Mayr, M. Vogelauer, U. Hoellwarth, and C. Hofer, Muscle biopsies show that fes of denervated muscles reverses human muscle degeneration from permanent spinal motoneuron lesion. J Rehabil Res Dev, 42(3 Suppl 1), 43–53, (2005).
9. K.L. Kilgore, R.L. Hart, F.W. Montague A.M. Bryden, M.W. Keith, H.A. Hoyen, C.J. Sams, and P.H. Peckham, An implanted myoelectrically-controlled neuroprosthesis for upper extremity function in spinal cord injury. Proceedings Of the 2006 IEEE Engineering in Medicine and Biology 28th Annual Conference, San Francisco, CA, pp. 1630–1633, (2006).

10. R. Kirsch, Development of a neuroprosthesis for restoring arm and hand function via functional electrical stimulation following high cervical spinal cord injury. Proceedings Of the 2005 IEEE Engineering in Medicine and Biology 27th Annual Conference, Shanghai, China, pp. 4142–4144, (2005).
11. G. Krausz, R. Scherer, G. Korisek, and G. Pfurtscheller, Critical decision-speed and information transfer in the "graz brain-computer interface". Appl Psychophysiol Biofeedback, 28, 233–241, (2003).
12. S. Mangold, T. Keller, A. Curt, and V. Dietz, Transcutaneous functional electrical stimulation for grasping in subjects with cervical spinal cord injury. Spinal Cord, 43(1), 1–13, (Jan 2005).
13. W.D. Memberg, P.E. Crago, and M.W. Keith, Restoration of elbow extension via functional electrical stimulation in individuals with tetraplegia. J Rehabil Res Dev, 40(6), 477–486, (2003).
14. G.R. Müller-Putz, R. Scherer, G. Pfurtscheller, and R. Rupp, EEG-based neuroprosthesis control: a step towards clinical practice. Neurosci Lett, 382, 169–174, (2005).
15. G.R. Müller-Putz, New concepts in brain-computer communication: use of steady-state somatosensory evoked potentials, user training by telesupport and control of functional electrical stimulation. PhD thesis, Graz University of Technology, (2004).
16. G.R. Müller-Putz, R. Scherer, and G. Pfurtscheller, Control of a two-axis artificial limb by means of a pulse width modulated brain-switch. Challenges for assistive Technology - AAATE '07, San Sebastian, Spain, pp. 888–892, (2007).
17. G.R. Müller-Putz, R. Scherer, G. Pfurtscheller, and R. Rupp, Brain-computer interfaces for control of neuroprostheses: from synchronous to asynchronous mode of operation. Biomed Tech, 51, 57–63, (2006).
18. C. Neuper, G.R. Müller-Putz, R. Scherer, and G. Pfurtscheller, Motor imagery and EEG-based control of spelling devices and neuroprostheses. Prog Brain Res, 159, 393–409, (2006).
19. NSCISC. Nscisc: 2006 annual statistical report. Last access September 2010 (2006). http://main.uab.edu.
20. M. Ouzky. Towards concerted efforts for treating and curing spinal cord injury, last access September 2009, (2002). http://assembly.coe.int/Mainf.asp?link=/Documents/WorkingDocs/Doc02/EDOC9401.htm.
21. P.H. Peckham, M.W. Keith, K.L. Kilgore, J.H. Grill, K.S. Wuolle, G.B. Thrope, P. Gorman, J. Hobby, M.J. Mulcahey, S. Carroll, V.R. Hentz, and A. Wiegner, Efficacy of an implanted neuroprosthesis for restoring hand grasp in tetraplegia: a multicenter study. Arch Phys Med Rehabil, 82, 1380–1388, (2001).
22. G. Pfurtscheller, C. Guger, G. Müller, G. Krausz, and C. Neuper, Brain oscillations control hand orthosis in a tetraplegic. Neurosci Lett, 292, 211–214, (2000).
23. G. Pfurtscheller, G.R. Müller, J. Pfurtscheller, H.J. Gerner, and R. Rupp, "Thought" – control of functional electrical stimulation to restore handgrasp in a patient with tetraplegia. Neurosci Lett, 351, 33–36, (2003).
24. G. Pfurtscheller and C. Neuper, Motor imagery and direct brain-computer communication. Proc IEEE, 89, 1123–1134, (2001).
25. F. Popescu, S. Fazli, Y. Badower, B. Blankertz, and K.-R. Müller, Single trial classification of motor imagination using 6 dry EEG electrodes. PLoS ONE, 2(7), e637, (2007).
26. J.P. Reilly and H. Antoni, *Electrical Stimulation and Electropathology*. Cambridge University Press, Cambridge, UK, (1997).
27. R. Rupp and H.J. Gerner, Neuroprosthetics of the upper extremity–clinical application in spinal cord injury and challenges for the future. Acta Neurochir Suppl, 97(Pt 1), 419–426, (2007).
28. R. Rupp, Die motorische Rehabilitation von Querschnittgelähmten mittels Elektrostimulation Ein integratives Konzept für die Kontrolle von Therapie und funktioneller Restitution. PhD thesis, University of Karlsruhe, (2008).
29. T. Stieglitz, Diameter-dependent excitation of peripheral nerve fibers by mulitpolar electrodes during electrical stimulation. Expert Rev Med Devices, 2, 149–152, (2005).

30. L.S. Stover, D.F. Apple, W.H. Donovan, and J.F. Ditunno. Standards für neurologische und funktionelle klassifikationen von rückenmarksverletzungen. Technical report, American Spinal Injury Association, New York NY, (1992).
31. R. Thorsen, R. Spadone, and M. Ferrarin. A pilot study of myoelectrically controlled FES of upper extremity. IEEE Trans Neural Syst Rehabil Eng, 9(2), 161–168, (Jun 2001).
32. J.R. Wolpaw, N. Birbaumer, D.J. McFarland, G. Pfurtscheller, and T.M. Vaughan, Brain-computer interfaces for communication and controls. Clin Neurophysiol, 113, 767–791, (2002).
33. K.S. Wuolle, C.L. Van Doren, G.B. Thrope, M.W. Keith, and P.H. Peckham, Development of a quantitative hand grasp and release test for patients with tetraplegia using a hand neuroprosthesis. J Hand Surg [Am], 19, 209–218, (1994).

Brain–Computer Interfaces for Communication and Control in Locked-in Patients

Femke Nijboer and Ursula Broermann

If you really want to help somebody, first you must find out where he is. This is the secret of caring. If you cannot do that, it is only an illusion if you think you can help another human being. Helping somebody implies you understanding more than he does, but first of all you must understand what he understands.[1]

1 Introduction

Most Brain–Computer Interface (BCI) research aims at helping people who are severely paralyzed to regain control over their environment and to communicate with their social environment. There has been a tremendous increase in BCI research the last years, which might lead to the belief that we are close to a commercially available BCI applications to patients. However, studies with users from the future target group (those who are indeed paralyzed) are still outnumbered by studies on technical aspects of BCI applications and studies with healthy young participants. This might explain why the number of patients who use a BCI in daily life, without experts from a BCI group being present, can be counted on one hand.

In this chapter we will focus on the feasibility and flaws of BCIs for locked-in and complete locked-in patients (the difference between these conditions will be explained in paragraph 2). Thus, we will speak a lot about problems BCI researchers face when testing a BCI or implementing a BCI at the home of patients. With this, we hope to stimulate further studies with paralyzed patients. We believe that patients

F. Nijboer (✉)
Institute of Medical Psychology and Behavioral Neurobiology, Eberhard Karls University of Tübingen, Tübingen, Germany; Human-Media Interaction, University of Twente, Enschede, The Netherlands
e-mail: femke.nijboer@utwente.nl

[1] Sören Kierkegaard: The point of view from my work as an author, 39.

B. Graimann et al. (eds.), *Brain–Computer Interfaces*, The Frontiers Collection,
DOI 10.1007/978-3-642-02091-9_11, © Springer-Verlag Berlin Heidelberg 2010

become BCI experts themselves during BCI training and provide useful insights into the usability of different BCI systems. Thus, it only seemed logical that for this chapter a BCI expert and a BCI user should work together to describe the experiences with BCI applications in the field.

First, we will describe the user group that might benefit from BCI applications and explain the difference between the locked-in syndrome and the completely locked-in syndrome (Sect. 2). These terms are often wrongfully used by BCI researchers and medical staff, and it is important to clarify them. Second, we review the studies that have shown that Brain–Computer interfaces might provide a tool to communicate or to control the environment for locked-in patients.

Third, in a unique interview, Ursula Broermann comments on multiple questions the reader might have in his or her head after reading the description of the disease amyotrophic lateral sclerosis (Sect. 3). She explains her daily life, what is important in her life, and how technical tools improve her quality of life significantly. These personal words from a BCI user are even more impressive when one considers that she had to select every single letter by carefully manoeuvring a cursor over a virtual keyboard by a lower lip-controlled joystick. It took her several months to write the text provided in this chapter.

In Sect. 4, we describe a typical BCI training process with a patient on the verge of being locked-in in a very non-scientific yet illustrative way. Through this description we can deduce simple requirements BCI applications should meet to become suitable for home use.

2 Locked-in the Body and Lock-Out of Society

The future user group of BCI applications consists mainly of people with neurodegenerative diseases like amyotrophic lateral sclerosis (ALS), also known as Lou Gehrig's disease. ALS is a fatal motor neuron disease of unknown etiology and cure. ALS is a neurodegenerative disorder of large motor neurons of the cerebral cortex, brain stem, and spinal cord that results in progressive paralysis and wasting of muscles [1]. ALS has an incidence of 2/100,000 and a prevalence of 6–8/100,000 [2]. Survival is limited by respiratory insufficiency. Most patients die within 3–5 years after onset of the disease [1], unless they choose life-sustaining treatment [3].

As the disease progresses, people get more and more paralyzed. The first symptoms most people initially experience include weakness in arms or legs, after which the paralysis spreads to other extremities and finally also neck and head areas. This form of ALS is called spinal ALS. On contrary, bulbar ALS starts with symptoms of weakness and paralysis in neck and mouth regions and then spreads to other extremities.

The choice (written down in a living will) to accept or decline life-sustaining treatment, such as artificial nutrition and artificial ventilation, is probably the most difficult choice a patient has to make during his disease progress. In fact, most people find it so difficult that they do not make a decision at all and decisions are made by caregivers or medical staff in case of emergency. Buchardi [4] found that ALS patients estimate their actual and anticipated quality of life as the most important criteria for their prospective decision regarding life-sustaining treatments. They refuse life-sustaining treatments if they expect that quality of life is low.

Maintaining and improving the quality of life of chronically ill patients is the major task for most health related professions. Since BCI may maintain or even restore communication, it is critical to inform patients about this possibility once the BCI is commercially available and to continue research on optimising BCI and implementing it in daily life.

Unfortunately, patients are not sufficiently informed about the course of the disease and the treatment options (for example against pain), particularly in the end-stage of the illness [5, 6] to such an extent that, in Germany (for example) only about 1/3 of neurological centres offer invasive ventilation for patients with motoneuron disease [6]. Neglecting to inform patients about all palliative measures, including artificial ventilation and communication methods, is a violation of fundamental ethical rules and a lack of respect for autonomy.

Most people, whose lives are defined as not worth living by a "healthy" society, state that they do not wish to die [7]. Nevertheless, studies from neurological centres in Germany show that only 4.5–7.4% of ALS patients receive invasive artificial respiration, while the rest die from respiratory insufficiency [8, 9]. Patients' attitudes towards life-sustaining treatment and towards artificial ventilation are influenced by the physician and family members' opinions [10]. How physicians and relatives perceive patient's quality of life shapes their attitude toward life-sustaining treatment for the patients. Unfortunately, most physicians, caregivers and family members (significant others) assume that quality of life of ALS patients is poor [11]. Empirical data on quality of life in ALS patients show instead that quality of life does not necessarily depend on the physical situation and that it can be maintained despite physical decline [12–15]. A study from Kübler and colleagues [12] even showed that ventilated people have a quality of life comparable to non-ventilated people.

Although major depression occurs in only about 9–11% of ALS patients, a certain degree of depressive symptoms occur in higher percentage of patients, and treatment may be indicated not only in full blown depression [16]. However, the psychological aspects of the disease are often neglected, although the degree of depressive symptoms is negatively correlated with quality of life [11, 13, 14, 17] and psychological distress shortens the survival rate [11, 18]. Thus, it should be of high priority to identify patients with depression.

In a study by Hayashi and Oppenheimer [3], more than half of 70 patients who did choose for artificial ventilation (in this case a tracheotomy positive pressure ventilation) survived 5 years or more *after* respiratory failure and 24% even survived 10 years or more. Thus, artificial ventilation may prolong the life of the patients significantly. Some of these patients may enter the so-called locked-in state (LIS). LIS patients are almost completely paralyzed, with residual voluntary control over few muscles, such as eye movement, eye blinks, or twitches with the lip [3, 19]. The locked-in state might also be a result from traumatic brain-injury, hypoxia, stroke, encephalitis, a tumor or the chronic Guillain-Barré syndrome.

Patients in the LIS state may communicate by moving their eyes or blinking. Most patients and their caregivers use assistive communication aids which can rely on only one remaining muscle (single switch). For example, patients control a virtual keyboard on a computer with one thumb movement or control a cursor on a screen with a joystick moved by the lower lip; a selection can be made (similar to left mouse click) by blinking with the eye lid. Another useful system constitutes of a letter matrix (see Fig. 1) which is often memorized by the patients by heart. The caregiver serves as an interlocutor and slowly reads out loud the numbers of the rows and the patient blinks with his eye when the row containing his desired letter has been read. Then, the caregiver reads out loud the letters in this row until the patient blinks again and a letter is selected. Jean Dominique Bauby, a French journalist who entered the locked-in state after a stroke, also communicated in such a ways, although his caregiver slowly recited the whole alphabet. His book "The Diving Bell and the Butterfly" [20] gives a glimpse of what it is like to be locked-in and how Bauby dealt with his condition. One quote from his book describes very precise how coping with this physical situation may be like making a conscious decision: "I decided to stop pitying myself. Other than my eye, two things aren't paralyzed, my imagination and my memory."

Fig. 1 A letter matrix used by many locked-in patients. A caregiver reads out loud the row numbers 1–5. The patient signals a "yes" with his remaining muscle to select a row. The caregiver then reads out loud the letters in that row until the patient selects a letter

1	A	B	C	D	E	F
2	G	H	I	J	K	L
3	M	N	O	P	Q	R
4	S	T	U	V	W	X
5	Y	Z	Ä	Ö	Ü	–

However, a further stage in the ALS disease can lead to an even worse physical situation. Some patients may enter the complete locked-in state (CLIS) and are then totally immobile [3, 21]. These people are unable to communicate at all, because even eye movements become unreliable and are finally lost altogether.

Let us imagine for a second that you are bedridden, speechless and immobile, without any possibility to express your thoughts and feelings. Imagine furthermore that this imprisonment is not temporary, but lasts for days, weeks, months and even years – in fact for the rest of your life. A healthy brain is locked into a paralyzed body and locked out of society by lack of communication. Brain–Computer Interfaces may be particularly useful for these LIS and CLIS patients and bridge the gap between inner and outer world.

3 BCI Applications for Locked-in Patients

Traditionally, BCI research has focused on restoring or maintaining communication, which is reported as the most important thing to the quality of life of locked-in patients [22]. In 1999, Birbaumer and colleagues reported the first 2 locked-in patients using a BCI to communicate messages [23–25]. One of these patients was able to use the BCI system without experts from the lab being present for private communication with his friends, such that he wrote and printed letters with a BCI [26]. A caregiver was only needed to apply the electrodes to the head of the patient.

Recently, Vaughan and colleagues [27] reported how a locked-in patient used a BCI on daily basis to communicate messages. Daily cap placement and supervision of the BCI was performed by a caregiver and data was daily uploaded to the lab via the internet. Further studies have also proven the feasibility of different BCIs in ALS patients, but they did not result in a clinical application for the patients who were tested. In chapter "Brain Signals for Brain–Computer Interfaces" of this book, we learned that commonly used brain signals are slow cortical potentials (SCPs) [23, 28, 29], sensorimotor rhythms (SMR) [30–33] or event-related potentials [34–36]. Nijboer and colleagues investigated ERP-BCI and SMR-BCI performance of participants with ALS in a within subject-comparison, which means that participants serves in each experimental condition. The ERP-BCI yielded a higher information transfer rate, and only 1 session was needed for most patients to use the BCI for communication. In contrast, the self-regulation of SMR for BCI use seemed to be more difficult for ALS patients than for healthy controls [30, 45] and required extensive training of the patients. Similarly, the self-regulation of SCPs requires extensive training [24]. Thus, the best initial approach for the purpose of communication with a locked-in patient seems to be to implement an ERP-BCI.

However, to our knowledge, there are no *complete* locked-in patients who can use a BCI with *any* of those signals (SCP, SMR or ERP) to communicate. Kübler and Birbaumer [19, 37] reviewed the relation between physical impairment and BCI performance of 35 patients who were trained with SCP-, SMR- or P300-BCIs or more than one approach. These patients were trained over the past 10 years in our institute and constitute a unique patient sample for the BCI community. Twenty-nine patients were diagnosed with ALS and six had other severe neurological disorders.

Kübler and Birbaumer [19] found a strong relation between BCI performance and physical impairment. Seven patients were in the complete locked-in state (CLIS) with no communication possible. When these CLIS patients were not included in the analysis, the relationship between physical impairment and BCI performance disappeared. Basic communication (yes/no) was not restored in any of the CLIS patients with a BCI. Hill and colleagues [38] also did not succeed in obtaining classifiable signals from complete locked-in patients. Thus, BCIs can be used by severely paralyzed patients, but its feasibility has not yet been proven in CLIS patients.

A variety of causes might explain the difficulty of making a BCI application work for a CLIS patient. A first issue arises when considering that BCI use based on the self-regulation of SMR or SCP is a skill that needs to be learned and maintained [39]. This skill is learned through operant conditioning. This term simply means that the chance an organism will perform a particular kind of behaviour increases when it is rewarded for it, and decreases when it is punished or not rewarded for it. A good example of operant conditioning in our own life can be observed when our email program produces its typical sound for incoming email, we rush to the inbox, because we have been rewarded with the sight of new emails in the past and we expect to be rewarded again. Our behaviour, the checking of the inbox, will vanish after we've heard the typical sound many times without having the reward of a new email. When behaviour vanishes, because no reward for the behaviour is given, psychologists call this extinction. The same principle can now be applied to Brain–computer interface use. When the user produces brain signals that result in a successful manipulation of the environment (reward), the likelihood that the user will produce the brain signal again in future increases.

In 2006, Birbaumer [21] suggested why learning through operant conditioning might be very difficult, if not impossible, for complete locked-in patients (see also chapter "Brain–Computer Interface in Neurorehabilitation" in this book). Birbaumer remembered a series of studies on rats [40] paralyzed with curare. The rats were artificially ventilated and fed and could not affect their environment at all, which made it impossible to them to establish a link between their behaviour and the reward from outside. It was not possible for these rats to learn through operant conditioning. Birbaumer hypothesized that when CLIS patients can not affect their environment for extended periods of time, the lack of reward from the environment might cause extinction of the ability to voluntarily produce brain signals. If this hypothesis is correct, it supports the idea that BCI training should begin before loss of muscle control, or as soon as possible afterwards [38]. It should be noted that no patients have proceeded with BCI training from the locked-in state into the complete locked-in state. Maybe a LIS patient who is able to control a BCI is very capable of transferring this ability in the complete locked-in state.

Other ideas have been put forward as to why CLIS patients seem unable to learn BCI control. Hill and colleagues [38] proposed that it might be very difficult for a long term paralyzed patient to imagine movements as is required for SMR self-regulation or that it is impossible to self-regulate SMR, because of physiological changes due to the disease. Also, general cognitive deficits, lack of attention, fatigue, lack of motivation or depression have been mentioned as possible factors that hamper BCI learning [38, 41].

Neumann [42, 43] and Neumann and Kübler [44] describe how attention, motivation and mood can affect BCI performance of ALS patients. Nijboer and colleagues also found that mood and motivation influences BCI performance in healthy subjects [45] and in ALS patients (data is being prepared for publication). It is impossible to ask CLIS patients how motivated they are to train with the BCI, but data from communicating ALS patients show that ALS patients are generally more motivated to perform well with BCI than healthy study participants. Personal experience with ALS patients and healthy subjects is compatible with this conclusion. ALS patients who participated in our studies during the past few years were highly intrinsically motivated, because they see the BCI as a final option when communication with the muscles is no longer possible. ALS patients were also very willing to participate in our studies, despite extensive training demands, discomfort of putting the electrode cap on and physical impairments due to their disease. Healthy subjects were often very enthusiastic about BCI studies, but seemed more extrinsically motivated by money and comparison to performance of other participants.

Until now, we have mentioned many user-related issues that might hamper BCI performance. However, also BCI-related issues that might affect a user's ability to control a BCI. Many CLIS patients have compromised vision, and may not be able to use a visually-based BCI. Thus, BCIs on the basis of other sensory modalities need to be explored. Usually the auditory system is not compromised in these patients. First results with auditory BCIs [36, 45–47] and vibrotactile BCIs [48–50] have been promising, although these BCIs have not been tested on LIS or CLIS patients.

To conclude, until now only few locked-in patients have used BCI applications to communicate and no CLIS patients could use the system. More BCI studies should focus on those users who are on the verge of the complete locked-in state. It would take only one CLIS patient who can control a BCI to disprove the hypothesis of Birbaumer [21]. Furthermore, we hope to see more non-visual BCI applications to anticipate the vision problems of the user in the locked-in state.

4 Experiences of a BCI User

Dr. Ursula Broermann is a 50 year-old pharmacist, who enjoys being in nature with her husband, reading, hearing classical music and who loves children very much. She lives in a beautiful apartment in the black forest, which is decorated with skill and love for aesthetics. She knows everything about food, cooking and herbs and she constantly redesigns her own garden. She often receives friends and never lets them leave with an empty stomach. Yet, she cannot open the front door to let her friends in. Nor can she shout a loud welcome salute to them when they enter the room. Ursula Broermann is in the locked-in state.

As if her car accident in 1985 (after which she was in a wheelchair) wasn't enough, she got the disease amyotrophic lateral sclerosis (ALS). She is now quadriplegic, and more and more facial muscles are getting paralyzed. In 2004, she received artificial nutrition and respiration so she could survive. When a journalist once asked her why she chose to be artificially respirated, she replied with a BCI: "Life is always beautiful, exciting and valuable". Since 2007, she has been what we would call locked-in. For communication she moves a cursor with a joystick attached to her lower lip over the computer screen and selects letters from a virtual keyboard. When asked if she would help write the chapter she quickly indicted a "yes" (by blinking one time). The following section is written by her over the course of several months (partly translated from German to English by the first author keeping intact the original syntax and verbiage and partly written by Dr. Broermann in English):

Awake – Locked in! Locked out? – Living with ALS

In the beginning of 2003 I got diagnosed with "suspicion of ALS" (it is a motoneuron disease and also called the Lou Gehrig's disease). The time began, when I got artificially respired only during the nights[2]. Since 2004 I have tracheal respiration and artificial nutrition - otherwise there weren't big changes to my life.

Since my car accident in 1985, I have been sitting in a wheelchair. It happened two years after my wedding with my very beloved and best man in the world, and right in the middle of both our graduation. Leg amputation and paralysis of bladder and colon, as well as brachial plexus injury to my right hand side made me very dependent on other people's help.

I was often asked when I first realized the specific symptoms of ALS. Because of my earlier accident, I can't exactly tell. Both, my mother and her brother (and probably their mother, that means my grandma), were diagnosed with ALS before they died. At the end of the year 2003 it was quite evident that I had problems in moving and breathing. After intensive neurological examinations in the hospital, I got the diagnosis: ALS, a very rare inheritable form. Although there is no clear evidence that the criteria according to El Escorial[3] were confirmed, neither at present time nor in the past. There is always a big difference between the theory and the practical life. Well, who knows, whether itś for the good.

Only because of our graduation my husband and me, at that time abstained from having babies. During the time I had to stay in the hospital, my uncle died of ALS (1985). The turn of events (the consequences of my accident and my chance of 50% to come down with ALS) we decided with a heavy heart to do without bearing children everlasting. After all children do not end in themselves – far be it from me to deny it, that again and again I quarrel with our decision about children.

My day usually starts with the night shift- or morning shift: inhalation, measuring blood sugar level, injecting insulin, temperature, blood pressure, and oxygen and CO2 level measurement and so on. Then: laxating[4], washing, getting dressed. Then I am carried from the bed to the wheelchair with the help of a lift attached to the ceiling. I am in the wheelchair until the evening; most of the time, I am working. My wheelchair is a real high-tech device. It is equipped with lots of adjustment possibilities for the seat and some functioning environmental control options. It is always exciting to see if one of those functions stops working. I dare you to say a wheelchair doesn't have a soul. . . . . . .

The following things represent quality of life for me:

1. Love, Family and Friends

I am privileged with my husband, my caring family and a lot of really reliable friends at my side, who still have the force, when I do not have any left. My husband

---

[2]Most ALS patients at some point will use a non-invasive respiration mask, which can be placed over the mouth and nose. Most patients use this during the night to avoid fatigue in daytime.

[3]The El Escorial consists of a set of medical criteria used by physicians that classify patients with amyotrophic lateral sclerosis into categories reflecting different levels of diagnostic certainty. Common diagnosis are "probable ALS" (as in the case of Dr. Broermann), whereas hardly any patient receives the diagnose "definite ALS". Diagnosing ALS can take a long time (years!). This uncertainty is not only difficult for the patient and his/her caregivers but also for the physician.

[4]Patients with ALS are often constipated. Caregivers often need to help initiate defecation. Sphincter control is often one of the last muscles over which ALS patients lose voluntary control and might even be used for signalling yes/no [51].

and a good deal of friends have a technical comprehension which considerably excels mine. When I look back I am surprised at how much commitment it takes for every little success and how sometimes that commitment is without success. It helps incredibly to be able to jointly walk difficult roads. I cannot express how grateful I am for that, also when the ALS drives us again and again at our limits, and, sometimes also over it.
My husband supports my - at any one time - remaining functions with everything that can be done technically. In this way he was able to allow me control over my laptop and the operation of my wheelchair until this moment. I am very happy about that, but even happier for the love he lets me perceive every day!
2. Mobility
Handicapped accessible public transport is a really good thing, especially for people like me, who otherwise have to rely on special disability transportation services. The transport schedules of these transportation services do not allow spontaneous trips. Usually you have to book many days if not weeks in advance and there is always a real possibility that they cancel your travel plans. With the public transportation you are also considering the environment and, not unimportant, your wallet.
Not being able to head somewhere spontaneously with my own car, but having to wait due time after advance reservation to let someone drive you, takes a bit of getting used to. However, I don't want to be ungrateful. At least I still get around.
3. Acceptance
The personal environment has an unimaginable influence on the psyche (and with that also directly and indirectly on the physique) of people. In individuals who are physically dependent on others' assistance, these effects are roughly proportional to the degree of their dependence. In this environment, a strong social pressure finally aimed at the death of the corresponding person can develop within a very short time. Notably, when the social environment is characterized by a positive attitude, and when the patient can feel this attitude, suicidal thoughts and the wish for a physician-assistant suicide are very rarely expressed.
Unfortunately, the public opinion about the physically dependent does not improve, and even politicians dispense negative attitude toward such individuals. Thus, the environment of severely ill people could already be seen as problematic.
I have a negative approach against the theme 'living will', because of the way it is currently formulated. I think it's inhumane and inadequate to die of hunger and thirst in our alleged civil and social society. In addition, living wills are often written in times of good physical health. For some people the opinion might change once they are really affected by the disease.
This is also true for the right to live of unborn alleged or actual physically or mentally disabled children. Nowadays one has to excuse and justify oneself for having a disabled child. Friends of mine (both physicians) were expecting a baby with the Down syndrome (trisomy 21). In the delivery room they were told by a doctor who 'meant well': "Something like that this is not necessary anymore these days!"
However, when you ask adults who were impaired from birth about this theme, you will almost always get an answer like: 'Fortunately, they didn't have the extensive

screening tests for pregnant women back then. I enjoy living very much, even though I have disabilities!' This applies accordingly for me too.
4. Communication
My phonetic is a mystery to my fellow men, because I use a cannula with 8 little windows for speaking[5]*. The BCI training with the University of Tübingen continues as usual. The goal is to control a computer with thought. Slow waves, μ rhythm and P300 are the methods that I tried. Finally, I stuck to the P300 method. One sees on a computer screen the whole alphabet as well as special characters or special functions. According to an algorithm, to me unknown, the characters in the rows and columns light up irregularly one after the other, but in every sequence equally often. This happens after one has been put an EEG-cap on, which electrodes are filled with contact gel and which is connected to the laptop with a cable. My task is to concentrate myself on a single letter and when the computer freaks at the university did their job very well, I am lucky. In the rare occasion that I make a mistake they programmed a backspace key, which I use in the same way as choosing a letter. The only disadvantage is that, after every training session, I always have to wash my hair.*

5 BCI Training with Patients

In this section we describe an exemplary BCI session of patient J. and the valuable lessons we have learned from that concerning requirements for BCI systems. J. is a 39-year old man, who was diagnosed with ALS 3 years before the session we will describe. He is in a wheelchair and has great difficulties speaking and swallowing. He is not artificially ventilated, although he has a non-invasive ventilation mask during the night. He says he has not decided yet if he will accept a tracheotomy.

After arriving at his house, we normally first drink coffee and get cookies, which is a pleasant side-effect of our job. We talk about his sons and about motor bikes (J's life motto is "live to ride, ride to live"). Sometimes we have to ask him to repeat himself because we cannot understand him. J. has to go to the toilet before training starts and we call for his wife to help him get to the bathroom. This might take 15 min so we have time to set up our system. *The first requirement for any person supervising a BCI system for a patient is that he or she must have patience.*

The system consists of a computer screen for the patient, an amplifier to amplify the brain signals, and a laptop which contains the BCI system. In contrast to working in the lab, where a system is generally not moved around, working with patients at home implies that we assemble and disassemble our system many times per week and transport it through all weather and traffic conditions. Also if we would leave

[5]With a tracheotomy air no longer runs through the larynx. An add-on speech cannula that uses incoming air instead of outgoing air can enable a person to continue to produce speech. In this case, Mrs. Broermann says her "phonetic" is a mystery to her fellow men, because it's difficult to understand her whispery and raspy voice.

the BCI at the patient's bedside damage cannot be avoided. Caregivers perform their duties and handle medical equipment in close vicinity of the patient. While pulling the bed sheets for example, one can also easily pull some cables of the electrodes. Or, while washing the patient's hair after training (most patients stay in bed while this happens!), water might easily drip on the amplifier. *Thus, the second requirement of a BCI system for patients' homes is that it is robust.*

When J. has returned, we start adjusting the computer screen so that it is on his eye height. When patients are in a wheelchair they often sit higher than people on a normal chair and cannot always move their head such that they can see the screen. When the neck is also paralyzed, the head of a patient is often attached to the back of the wheel chair to prevent it from dropping down (something that happened with J. during run 7; see protocol in Fig. 3). Fortunately, J. has many board game boxes lying around and we use them to put under the screen. At the house of Ursula Broermann, we use encyclopaedias (see Fig. 2). For some patients who are bed-ridden, it would be useful to attach the computer screen above their bed facing down. *The third requirement of a BCI system at the home of patients is that it is compact, so that you can carry it around easily and put it on strange places like game board boxes.* In addition, the cables attached to the monitor, the electrode cap and the computer should be long, because one never knows where you can put the system. The direct space around a typical ALS patient is already filled with many technical devices.

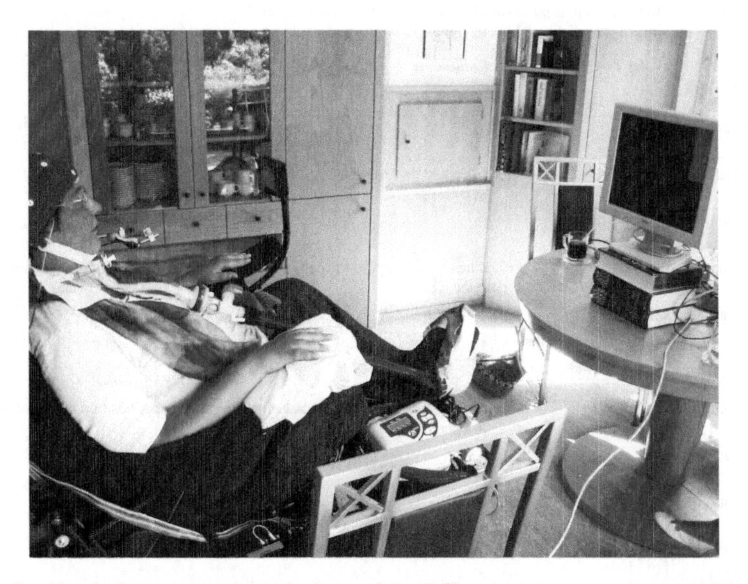

Fig. 2 Dr. Ursula Broermann sitting in front of the BCI computer screen (supported by encyclopaedias). On the *left* one can see the patient in a wheel chair with the electrode cap on. The wheelchair is slightly tilted to the back to stabilize the head into the chair. This way, the head does not need to be supported with a head band, which would influence the signal quality. Beside the wheelchair, in the *middle* of the picture, on can see the artificial ventilation machine. The amplifier is sitting on a chair besides the patient (no visible in this picture)

Now that the BCI system is set up, we can place the electrode cap with 16 electrodes on the head of J. Because a good contact between the electrode surface and the skin needs to be established, we need to fill the electrodes with electrode gel. This procedure in itself is relatively easy and takes no longer than 5–10 min for an experienced EEG researcher. However, it creates a big problem for any BCI user who is severely paralyzed, because it is very cumbersome to wash the hair after training, especially when the patient is in a wheel chair or in a bed. *A future requirement of BCI systems would therefore be the use of electrodes which do not need to be filled with electrode gel*. First results with so-called dry electrodes [52] have been promising.

After we have shown J. his brain signals and jokingly told him we could still detect a brain, we are finally ready to start the BCI session. In Fig. 3 the protocol of session 4 can be found. During this session we tested him with a P300 BCI. We instructed J. to "write" the sentence "Franz jagt im komplett verwahrlosten Taxi quer durch Bayern" (translation: "Franz hurries in a completely shabby taxi across Bavaria").

SubjectCode: Patient J.
Date: 10.03.2005
Instructors: Tamara & Femke
Session number: 4
Parameterfile: 3Speller_DAS.prm
Cap: Blue; 16 channels
Comments: Referenced to right mastoid; grounded with left mastoid; all impedances below 5 kOhm; signals looking good

Run Number (and time)	Word To Copy	Comments
1 (10:44 am)	Franz	
2 (10:47 am)	jagt	Telephone is ringing
3 (10:50 am)	im	
4 (10:55 am)	komplett	When he was trying to spell the letter ‚L' he had fasciculations that were visible in the EEG.
5 (11:03 am)	verwahr	
6 (11:10 am)	losten	Again fasciculations
7 (11:15 am)	taxi	We suspended this run, because he cannot keep his head up anymore. His head dropped down!
7 (11:40 am)	taxi	
8 (11:44 am)	quer	
9 (11:48 am)	durch	Yawning ; patient reported not sleeping much the night before
10 (11:52 am)	Bayern	

Fig. 3 Training protocol from session 4 with patient J. The header contains information on technical aspects of this measurement and how the signals were acquired. The first row specifies the run number and the time the run started. Then, the Word-To-Copy column specifies which word J. had to copy in each of the ten runs. In the final column there is space to comment on the run

The training that follows is unproblematic during the first 3 runs (only one phone call during run 2). Then, in run 4 the brain signals are confounded by electrical activity coming from fasciculations in J's neck. Fasciculations are spontaneous, irregularly discharging motor unit potentials with visible muscle twitches [53] and constitute a universal feature in ALS patients. They are not painful to the patient, but they create noise in the brain signals, which makes it more difficult for the BCI to detect the real signals, and thus decreases the communication performance.

ALS patients may also have difficulties swallowing, and they might have to simultaneously deal with increased saliva production. Patients may thus have to swallow with great effort during BCI use. The electrical noise arising from these movements in the neck again affects the signal quality. These confounding factors, referred to as artefacts by BCI researchers, are usually not seen in healthy subjects, who can sit perfectly still during BCI experiments. These concerns again underscore the need for more BCI studies with target user groups. *Thus, a fifth requirement for home BCI systems is that they filter out artefacts during BCI use.* Also here, first results are promising [54].

Because we can not do anything against the fasciculations and they do not bother J. personally, we decide continue the session until, in run seven, something unexpected happens. J's head falls forward onto his chest and he is not able to lift it up anymore. Lately, we and his family have noticed his neck muscles getting weaker, but this is the first time we see the effect of this new disease progress. Although we are very taken aback by this event, J. laughs at the view of our nervous faces and insists we find a solution for his "loose" head to prevent it from dropping again. We look around in the room and see a bandana lying around. It's very colourful, decorated with flowers and probably belongs to J.'s wife. We find a way to put it around his forehead (Rambo style) and tie it to the wheel chair. After we've made sure this system is stable we repeat the suspended run 7 and continue the rest of the experiment.

During run 9 we write down in the protocol that J. is yawing constantly. He already informed us that he did not sleep much the night before. Fatigue is a common symptom of ALS patients [55, 56] and may affect BCI performance [42]. *An ideal BCI system would be able to recognize mental states within the user and adapt to them.* For example, when the user falls asleep the system might switch to standby mode by itself.

We motivate J. to give his last effort for the last word of the session and he's able to finish without falling asleep. After we've packed up the system, and washed and dried J.'s hair, we thank him for his contribution and head home. Another session is planned for later in the week.

6 Conclusion

To summarize, BCI systems should be robust and compact and have electrodes that are easy to attach and to remove, without cumbersome hair washing. Furthermore, BCI systems should be able to detect artefacts and different mental states and adapt to them. Finally, we would like to add that BCI system should not be very expensive. Although some patients have successfully convinced their health insurance of the necessity of communication, we anticipate that some patients will have to pay for a BCI system themselves. ALS patients are already faced with enormous costs, because of non-complying health insurance agencies, which regularly cause depressive symptoms in. We urge the BCI community to find cheap solutions.

The BCI community should also consider ethical issues related to their work [57, 58]. Perhaps the most important issue involves not giving patients and their caregivers too much hope. Due to media coverage and exaggeration of successes, we get about one phone call every 2 weeks by a patient or caregiver who wants to order a BCI system. These people are very disappointed when I have to tell them that there is no commercial BCI available and that using a BCI system currently involves our presence or at least supervision over a long period of time.

Finally, we would like to respond to the ethical concerns raised by a neurologist in 2006 [59], who is worried that the development of Brain–Computer Interfaces for communication might compel physicians and patients to consider life-sustaining treatments more often, and that patients are therefore at higher risk of becoming locked-in at some point. He argues that the implications of BCIs and life-sustaining treatments might thus inflict heavy burdens on the quality of life of the patients and the emotional and financial state of the caregivers. For these reasons he asks: "even when you *can* communicate with a locked-in patient, *should* you do it"?

First, as we have seen, the assumption that quality of life in the locked-in state is necessarily low is factually wrong. Second, even if quality of life in the locked-in state was low and caregivers are financially and emotionally burdened (which is in fact often true), this does not plead for a discontinuation of BCI development nor for the rejection of life-sustaining treatments. Low quality of life in patients, financial problems and emotional burdens plead for better palliative care, more support for caregivers and more money from the health care, not for "allowing" patients to die. The question that we pose is: "When you can communicate with a locked-in patient, why should you *not* do it?".

Acknowledgements BCI research for and with BCI users is not possible without the joint effort of many dedicated people. People who added to the content of this chapter are: mr. JK, mrs. LK, mrs. KR, mr. WW, mr. RS, mr. HM, professor Andrea Kübler, professor Niels Birbaumer, professor Boris Kotchoubey, Jürgen Mellinger, Tamara Matuz, Sebastian Halder, Ursula Mochty, Boris Kleber, Sonja Kleih, Carolin Ruf, Jeroen Lakerveld, Adrian Furdea, Nicola Neumann, Slavica von Hartlieb, Barbara Wilhelm, Dorothée Lulé, Thilo Hinterberger, Miguel Jordan, Seung Soo Lee, Tilman Gaber, Janna Münzinger, Eva Maria Hammer, Sonja Häcker, Emily Mugler. Also thanks for useful comments on this chapter to Brendan Allison, Stefan Carmien and Ulrich Hoffmann. A special thanks to mr. GR, mr. HC, mr. HPS and Dr. Hannelore Pawelzik.

Dedication On the 9th of June 2008, 2 months after completion of the first draft of this chapter, Dr. Ursula Broermann passed away. And in her memory her husband and I decided to dedicate this chapter to her with words from "the little prince", the book from Saint-Exupéry, that she loved so much [60]: "It is only with the heart that one can see rightly; what is essential is invisible to the eye."

References

1. M. Cudkowicz, M. Qureshi, and J. Shefner, Measures and markers in amyotrophic lateral sclerosis. NeuroRx, 1(2), 273–283, (2004).
2. B.R. Brooks, Clinical epidemiology of amyotrophic lateral sclerosis. Neurol Clin, 14(2), 399–420, (1996).
3. H. Hayashi and E.A. Oppenheimer, ALS patients on TPPV: Totally locked-in state, neurologic findings and ethical implications. Neurology, 61(1), 135–137, (2003).
4. N. Buchardi, O. Rauprich, and J. Vollmann, Patienten Selbstbestimmung und Patientenverfügungen aus der Sicht von Patienten mit amyotropher Lateralsklerose: Eine qualitative empirische Studie. Ethik der Medizin, 15, 7–21, (2004).
5. P.B. Bascom and S.W. Tolle, Responding to requests for physician-assisted suicide: These are uncharted waters for both of us. JAMA, 288(1), 91–98, (2002).
6. F.J. Erbguth, Ethische und juristische Aspekte der intensicmedizinischen Behandlung bei chronisch-progredienten neuromuskulären Erkrankungen. Intensivmedizin, 40, 464–657, (2003).
7. A. Kübler, C. Weber and N. Birbaumer, Locked-in – freigegeben für den Tod. Wenn nur Denken und Fühlen bleiben – Neuroethik des Eingeschlossenseins. Zeitschrift für Medizinische Ethik, 52, 57–70, (2006).
8. G.D. Borasio, Discontinuing ventilation of patients with amyotrophic lateral sclerosis. Medical, legal and ethical aspects. Medizinische Klinik, 91(2), 51–52, (1996).
9. C. Neudert, D. Oliver, M. Wasner, G.D. Borasio, The course of the terminal phase in patients with amyotrophic lateral sclerosis. J Neurol, 248(7), 612–616, (2001).
10. A.H. Moss, Home ventilation for amyotrophic lateral sclerosis patients: outcomes, costs, and patient, family, and physician attitudes. Neurology, 43(2), 438–443, (1993).
11. E.R. McDonald, S.A. Wiedenfield, A. Hillel, C.L. Carpenter, R.A. Walter, Survival in amyotrophic lateral sclerosis. The role of psychological factors. Arch Neurol 51(1), 17–23, (1994).
12. A. Kübler, S. Winters, A.C. Ludolph, M. Hautzinger, N. Birbaumer, Severity of depressive symptoms and quality of life in patients with amyotrophic lateral sclerosis. Neurorehabil Neural Repair, 19, 1–12, (2005).
13. Z. Simmons, B.A. Bremer, R.A. Robbins, S.M. Walsh, S. Fischer, Quality of life in ALS depends on factors other than strength and physical function. Neurology, 55(3), 388–392, (2000).
14. R.A. Robbins, Z. Simmons, B.A. Bremer, S.M. Walsh, S. Fischer, Quality of life in ALS is maintained as physical function declines. Neurology, 56(4), 442–444, (2001).
15. A. Chio, G.A., Montuschi, A. Calvo, N. Di Vito, P. Ghiglione, R. Mutani, A cross sectional study on determinants of quality of life in ALS. J Neurol Neurosurg Psychiatr, 75(11), 1597–601, (2004).
16. A. Kurt, F. Nijboer, T. Matuz, A. Kübler, Depression and anxiety in individuals with amyotrophic lateral sclerosis: epidemiology and management. CNS Drugs, 21(4), 279–291, (2007).
17. A. Kübler, S. Winter, J. Kaiser, N. Birbaumer, M. Hautzinger, Das ALS-Depressionsinventar (ADI): Ein Fragebogen zur Messung von Depression bei degenerativen neurologischen Erkrankungen (amyotrophe Lateralsklerose). Zeit KL Psych Psychoth, 34(1), 19–26, (2005).
18. C. Paillisse, L. Lacomblez, M. Dib, G. Bensimon, S. Garcia-Acosta, V. Meininger, Prognostic factors for survival in amyotrophic lateral sclerosis patients treated with riluzole. Amyotroph Lateral Scler Other Motor Neuron Disord, 6(1), 37–44, (2005).
19. A. Kubler, and N. Birbaumer (2008) Brain-computer interfaces and communication in paralysis: Extinction of goal directed thinking in completely paralysed patients? Clin Neurophysiol, 119, 2658–2666.
20. J.D. Bauby, *Le scaphandre et le papillon*. Editions Robert Laffont, Paris, (1997).
21. N. Birbaumer, Breaking the silence: Brain-computer interfaces (BCI) for communication and motor control. Psychophysiology, 43(6), 517–532, (2006).

22. J.R. Bach, Amyotrophic lateral sclerosis – communication status and survival with ventilatory support. Am J Phys Med Rehabil, 72(6), 343–349, (1993).
23. N. Birbaumer, N. Ghanayim, T. Hinterberger, I. Iversen, B. Kotchoubey, A. Kübler, J. Perelmouter, E. Taub, H. Flor, A spelling device for the paralysed. Nature, 398(6725), 297–298, (1999).
24. A. Kübler, N. Neumann, J. Kaiser, B. Kotchoubey, T. Hinterberger, and N. Birbaumer, Brain-computer communication: Self-regulation of slow cortical potentials for verbal communication. Arch Phys Med Rehabil, 82, 1533–1539, (2001).
25. A. Kübler, N. Neumann, J. Kaiser, B. Kotchoubey, T. Hinterberger, N. Birbaumer, Brain-computer communication: Self-regulation of slow cortical potentials for verbal communication. Arch Phys Med Rehabil, 82(11), 1533–1539, (2001).
26. N. Neumann, A. Kübler, J. Kaiser, T. Hinterberger, N. Birbaumer, Conscious perception of brain states: Mental strategies for brain-computer communication. Neuropsychologia, 41(8), 1028–1036, (2003).
27. T.M. Vaughan, D.J. McFarland, G. Schalk, W.A. Sarnacki, D.J. Krusienski, E.W. Sellers, J.R. Wolpaw, The Wadsworth BCI Research and Development Program: At home with BCI. IEEE Trans Neural Syst Rehabil Eng, 14(2), 229–233, (2006).
28. T. Hinterberger, S. Schmidt, N. Neumann., J. Mellinger, B. Blankertz, G. Curio, N. Birbaumer, Brain-computer communication and slow cortical potentials. IEEE Trans Biomed Eng, 51(6), 1011–1018, (2004).
29. A. Kübler, B. Kotchoubey, T. Hinterbebrg, N. Ghanayim, J. Perelmouter, M. Schauer, C. Fritsch, E. Taub, N. Birbaumer, The thought translation device: A neurophysiological approach to communication in total motor paralysis. Exp Brain Res, 124, 223–232, (1999).
30. A. Kübler, F. Nijboer, K. Mellinger, T.M. Vaughan, H. Pawelzik, G. Schalk, D.J. McFarland, N. Birbaumer, J.R. Wolpaw, Patients with ALS can use sensorimotor rhythms to operate a brain-computer interface. Neurology, 64(10), 1775–1777, (2005).
31. C. Neuper, R. Scherer, M. Reiner, G. Pfurtscheller, Imagery of motor actions: differential effects of kinesthetic and visual-motor mode of imagery in single-trial EEG. Brain Res Cogn Brain Res, 25(3), 668–677, (2005).
32. G. Pfurtscheller, B. Graimann, J.E. Huggins, S.P. Levine, Brain-computer communication based on the dynamics of brain oscillations. Suppl Clin Neurophysiol, 57, 583–591, (2004).
33. D.J. McFarland, G.W. Neat, R.F. Read, J.R. Wolpaw, An EEG-based method for graded cursor control. Psychobiology, 21(1), 77–81, (1993).
34. F. Nijboer, E.W. Sellers, J. Mellinger, T. Matuz, S. Halder, U. Mochty, M.A. Jordan, D.J. Krusienski, J.R. Wolpaw, N. Birbaumer, A. Kübler, P300-based brain-computer interface (BCI) performance in people with ALS. Clin Neurophys, 119(8), 1909–1916, (2008).
35. U. Hoffmann, J.M. Vesin, T. Ebrahimi, K. Diserens, An efficient P300-based brain-computer interface for disabled subjects. J Neurosci Methods, 167(1), 115–25, (2008).
36. E.W. Sellers, A. Kubler, and E. Donchin, Brain-computer interface research at the University of South Florida Cognitive Psychophysiology Laboratory: the P300 Speller. IEEE Trans Neural Syst Rehabil Eng, 14, 221–224, (2006).
37. A. Kübler, F. Nijboer, and N. Birbaumer, Brain-computer interfaces for communication and motor control – Perspectives on clinical applications. In G. Dornhege, et al. (Eds.), *Toward brain-computer interfacing*, The MIT Press, Cambridge, MA, Pp. 373–392, (2007).
38. N.J. Hill, T.N. Lal, M. Schoeder, T. Hinterberger, B. Wilhelm, F. Nijboer, U. Mochty, G. Widman, C. Elger, B. Schoelkopf, A. Kübler, N. Birbaumer, Classifying EEG and ECoG signals without subject training for fast BCI implementation: Comparison of nonparalyzed and completely paralyzed subjects. IEEE Trans Neural Syst Rehabil Eng, 14(2), 183–186, (2006).
39. J.R. Wolpaw, N. Birbaumer, D.J. McFarland, G. Pfurtscheller, T.M. Vaughan, Brain-computer interfaces for communication and control. Clin Neurophysiol, 113(6), 767–791, (2002).
40. B.R. Dworkin, and N.E. Miller, Failure to replicate visceral learning in the acute curarized rat preparation. Behav Neurosci, 100(3), 299–314, (1986).

41. E.A. Curran and M.J. Stokes: Learning to control brain activity: A review of the production and control of EEG components for driving brain-computer interface (BCI) systems. Brain Cogn, 51(3), 326–336, (2003).
42. N. Neumann, Gehirn-Computer-Kommunikation, Einflussfaktoren der Selbstregulation langsamer kortikaler Hirnpotentiale, in Fakultär für Sozial- und Verhaltenswissenschaften., Eberhard-Karlsuniversität Tübingen: Tübingen, (2001).
43. N. Neumann and N. Birbaumer, Predictors of successful self control during brain-computer communication. J Neurol Neurosurg Psychiatr, 74(8), 1117–1121, (2003).
44. N. Neumann and A. Kübler, Training locked-in patients: A challenge for the use of brain-computer interfaces. IEEE Trans Neural Syst Rehabil Eng, 11(2), 169–172, (2003).
45. F. Nijboer, A. Furdea, I. Gunst, J. Mellinger, D.J. McFarland, N. Birbaumer, A. Kübler, An auditory brain-computer interface (BCI). J Neurosci Methods, 167(1), 43–50, (2008).
46. T. Hinterberger, N. Neumann, M. Pham, A. Kübler, A. Grether, N. Hofmayer, B. Wilhelm, H. Flor, N. Birbaumer, A multimodal brain-based feedback and communication system. Exp Brain Res, 154(4), 521–526, (2004).
47. A. Furdea, S. Halder, D.J. Krusienski, D. Bross, F. Nijboer et al., An auditory oddball (P300) spelling system for brain-computer interfaces. Psychophysiology, 46, 617–625, (2009).
48. A. Chatterjee, V. Aggarwal, A. Ramos, S. Acharya, N.V. Thakor, A brain-computer interface with vibrotactile biofeedback for haptic information. J Neuroeng Rehabil, 4, 40, (2007).
49. F. Cincotti, L. Kauhanen, F. Aloise, T. Palomäki, N. Caporusso, P. Jylänki, D. Mattia, F. Babiloni, G. Vanacker, M. Nuttin, M.G. Marciani, J.R. Millan, Vibrotactile feedback for brain-computer interface operation. Comput Intell Neurosci, 2007 (2007), Article048937, doi:10.1155/2007/48937.
50 G.R. Müller-Putz, R. Scherer, C. Neuper, G. Pfurtscheller, Steady-state somatosensory evoked potentials: Suitable brain signals for Brain-computer interfaces? IEEE Trans Neural Syst Rehabil Eng, 14(1), 30–37.
51. T. Hinterberger and colleagues, Assessment of cognitive function and communication ability in a completely locked-in patient. Neurology, 64(7), 307–1308, (2005).
52. F. Popescu, S. Fazli, Y. Badower, N. Blankertz, K.R. Müller, Single trial classification of motor imagination using 6 dry EEG electrodes. PLoS ONE, 2(7), e637, (2007).
53. F.J. Mateen, E.J. Sorenson, J.R. Daube Strength, physical activity, and fasciculations in patients with ALS. Amyotroph Lateral Scler, 9, 120–121, (2008).
54. S. Halder, M. Bensch, J. Mellinger, M. Bogdan, A. Kübler, N. Birbaumer, W. Rosenstiel, Online artifact removal for brain-computer interfaces using support vector machines and blind source separation. Comput Intell Neurosci, Article82069, (2007), doi:10.1155/2007/82069.
55. J.S. Lou, A. Reeves, T. benice, G. Sexton, Fatigue and depression are associated with poor quality of life in ALS. Neurology, 60(1), 122–123, (2003).
56. C. Ramirez, M.E. Piemonte, D. Callegaro, H.C. Da Silva, Fatigue in amyotrophic lateral sclerosis: Frequency and associated factors. Amyotroph Lateral Scler, 9, 75–80, (2007).
57. A. Kübler, V.K. Mushahwar, L.R. Hochberg, J.P. Donoghue, BCI Meeting 2005 – Workshop on clinical issues and applications. IEEE Trans Neural Syst Rehabil Eng, 14(2), 131–134, (2006).
58. J.R. Wolpaw, G.E. Loeb, B.Z. Allison, E. Donchin, O. Feix do Nascimento, W.J. Heetderks, F. Nijboer, W.G. Shain, J.N. Turner, BCI Meeting 2005 – Workshop on signals and recording methods. IEEE Trans Neural Syst Rehabil Eng, 14(2), 138–141, (2006).
59. L.H. Phillips, Communicating with the "locked-in" patient – because you can do it, should you? Neurology, 67, 380–381, (2006).
60. A. Saint-Exupéry, *Le petit prince (R. Howard, (trans.), 1st edn.),* Mariner Books, New York, (1943).

Intracortical BCIs: A Brief History of Neural Timing

Dawn M. Taylor and Michael E. Stetner

1 Introduction

In this chapter, we will explore the option of using neural activity recorded from tiny arrays of hair-thin microelectrodes inserted a few millimeters into the brain itself. These tiny electrodes are small and sensitive enough to detect the firing activity of individual neurons. The ability to record individual neurons is unique to recording technologies that penetrate the brain. These microelectrodes are also small enough that many hundreds of them can be implanted in the brain at one time without displacing much tissue. Therefore, the activity patterns of hundreds or even thousands of individual neurons could potentially be detected and used for brain–computer interfacing (BCI) applications.

Having access to hundreds of individual neurons opens up the possibility of controlling very sophisticated devices directly with the brain. For example, you could assign 88 individual neurons to control the 88 individual keys on a digital piano. Theoretically, an individual could play the piano by firing the associated neurons at the appropriate times. However, our ability to play Mozart directly from the brain is still far off in the future (other than using a BCI to turn on the radio and select your favorite classical station). There still are many technical challenges to overcome before we can make use of all the potential information that can be extracted from intracortical microelectrodes.

2 Why Penetrate the Brain?

To illustrate the practical differences between non-invasive and invasive brain recording technologies, we will expand on a metaphor often used to explain how extracortical recordings can provide useful information about brain activity without recording individual neurons, i.e. "You don't have to measure the velocity of every

D.M. Taylor (✉)
Dept of Neurosciences, The Cleveland Clinic, Cleveland, OH 44195, USA
e-mail: dxt42@case.edu

B. Graimann et al. (eds.), *Brain–Computer Interfaces*, The Frontiers Collection,
DOI 10.1007/978-3-642-02091-9_12, © Springer-Verlag Berlin Heidelberg 2010

molecule in a room to determine the temperature in the room". A very true statement! You could simply put a thermometer in the room or even infer the temperature inside the room by placing a thermometer on the outside of the door. Similarly, brain recording technologies that record the "electrical field potentials" from the surface of the brain, or even from the scalp, can be used to determine the general activity level (a.k.a. "temperature") of a given part of the brain without having to measure the activity of each individual neuron. Many useful BCIs can be controlled just by measuring general activity levels in specific parts of the brain.

Expanding the metaphor further, let us suppose you had sensitive thermometers on the doors of every room in a school building. Based on the temperature of each room, you could make inferences about when the students in certain classrooms were taking a nap versus when they were actively engaged in discussions versus when they were all using their computers. Similarly, by placing multiple recording electrodes over different parts of the brain that control different functions, such as hand, arm, or leg moments, you can deduce when a paralyzed person is trying to move their right hand versus their left hand versus their feet. Again, you don't need to measure each individual neuron to know when a body part is being moved or is at rest. The electrical signals recorded on different electrodes outside the brain can tell you which body parts are moving or resting.

Now let's suppose you placed thermometers at each desk within each classroom; you might be able to infer even more details about which students are sitting quietly and paying attention versus which students are fidgeting and expending excess energy. If you know what topics each student is interested in, you may even be able to infer what subject is being discussed in class. Similarly, electrode manufactures are making smaller and smaller grids of brain surface electrodes that can detect the activity level of even smaller, more-defined regions of cortex – still without penetrating the brain's surface. Complex activity patterns across these arrays of brain surface electrodes could tell us quite a lot about a person's attempted activities, such as attempting to make one hand grasp pattern over another.

So if we can use smaller and smaller electrodes outside the brain to detect the activity levels in increasingly more focused locations of the brain, why would we need to penetrate the brain with microelectrodes? The metaphor, "You don't have to measure the velocity of each individual molecule to determine the temperature of the room" has another key component representing the differences between intracortical and non-penetrating brain recordings – temporal resolution. Most thermometers are designed to represent the energy of the molecules within a given area averaged over the recent past. But, average energy is not the only useful measure that can be extracted from the room. If you measured the velocity of each air molecule in the room over time, you would see patterns generated by the sound pressure waves traveling out of each person's mouth as he/she speaks. If you measured the velocity of these air molecules with enough spatial and temporal resolution, you could use that information to determine what each person was saying, and not simply which person is actively using energy and which person is at rest.

Similarly, the implanted microelectrode recording systems can detect the exact firing patterns of the individual neurons within the brain. These detailed firing patterns are rich with complex information, such as how much you need to activate each specific muscle to make a given movement or the detailed position, velocity, and acceleration of the limb about each joint. This level of detail about what a person is thinking or trying to do, so far, has eluded researchers using non-invasive recording methods.

Finally, to carry this metaphor even further, let us compare the technical challenges with measuring the room's temperature versus detecting the velocity of each individual molecule. It would be impractical to install sensors at every possible location within the room for the purpose of measuring the velocity of every molecule. There would be no room left for the people or for the molecules of air to travel. Similarly, you cannot put enough microelectrodes in the brain to detect the activity of every neuron in a given area. The electrodes would take up all the space leaving no room for the neurons. Microelectrode arrays sample only a very small fraction of the available neurons. And, although complex information can be extracted from this sparse sampling of neurons, this information is inherently just an estimate of the true information encoded in the detailed firing patterns of one's brain. In the next section, we will describe the imperfect process by which the firing patterns of individual neurons are recorded using intracortical microelectrode technologies. Then, in the rest of this chapter, we will discuss how researchers have been able to use even a single neuron to a few hundreds of neurons to very effectively control different BCI systems.

3 Neurons, Electricity, and Spikes

As you are reading this sentence, the neurons in your brain are receiving signals from photoreceptors in your eyes that tell the brain where light is falling on your retina. This information gets passed from the eyes to the brain via electrical impulses called "action potentials" that travel along neurons running from the eyes to the brain's visual processing center. These electrical impulses then get transmitted to many other neurons in the brain through cell-to-cell connections called synapses. Each neuron in the brain often receives synaptic inputs from many different neurons and also passes on its own synaptic signals to many different neurons. Your brain is made up of about a hundred billion neurons that synapse together to form complex networks with specific branching and connection patterns. Amazingly, these networks can transform the electrical impulses, generated by the light and dark patterns falling on your retina, into electrical impulses that encode your symbolic understanding of words on the page, and (hopefully) into electrical impulse patterns that encode your understanding of the more abstract concepts this book is trying to get across.

At the heart of perceiving light and dark, or understanding a word, or abstractly thinking about how we think – is the action potential. The action potential is an electrical phenomenon specific to neurons and muscle cells. These electrically active cells naturally have a voltage difference across their cell membrane due to different concentrations of ions inside and outside the cell. Special channels within the cell membrane can be triggered to briefly allow certain ions to pass through the membrane resulting in a transient change in the voltage across the membrane – an action potential. These action potentials are often called "spikes" as they can be detected as a spike in the measured voltage lasting only about a millisecond.

Although the action potential itself is the fundamental unit of neural activity, our perceptions, thoughts, and actions emerge from how action potentials form and travel through the complex network of neurons that make up our nervous systems. Action potentials travel within and between neurons in a very specific unidirectional manner. Each neuron is made up of four parts (depicted in Fig. 1) – the dendrites, the cell body, the axon, and the synaptic terminal. Neural signals flow through each neuron in that order.

1) The *dendrites* are thin branching structures that can receive signals or synaptic inputs from many other neurons or from sensory receptors. Some synaptic inputs will work to excite or increase the probability that an action potential will be generated in the receiving neuron. Synaptic inputs from other neurons will inhibit or reduce the likelihood of an action potential taking place.
2) The large *cell body* or *soma* contains the structures common to all cells, such as a nucleus as well as the machinery needed to make proteins and to process other molecules essential for the cell's survival. Most importantly, the cell

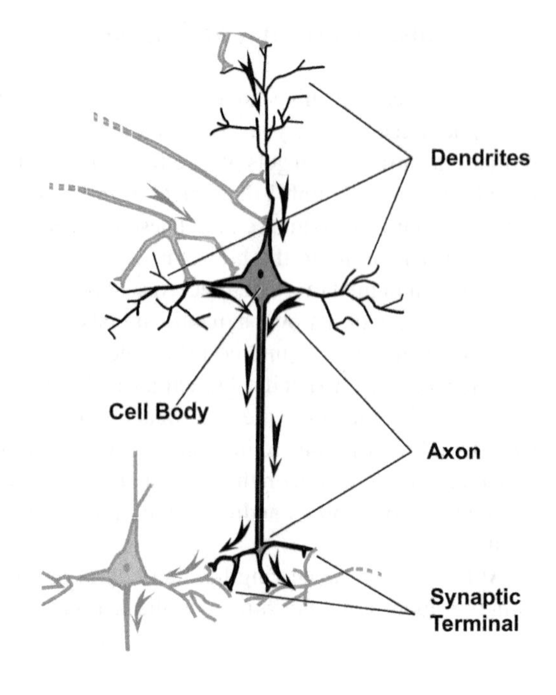

Fig. 1 The four main parts of a neuron. The dendrites receive synaptic inputs ("yes" or "no" votes) from upstream neurons. The large cell body or soma combines the synaptic inputs and initiates an *action potential* if enough "yes" votes are received. The action potential travels down the axon (sometimes long distances) to the synaptic terminal. The action potential triggers neurotransmitters to be releases from the synaptic terminal. These neurotransmitters become the "yes" or "no" votes to the dendrites of the next neurons in the network. Arrows indicate the direction in which the electrical signals travel in a neuron

body (specifically a part called the axon hillock) is where action potentials are first generated. To take an anthropomorphic view, each of the other neurons that synapse on the cell's dendrites gets to "vote" on whether or not an action potential should be generated by the cell body. If the neuron received more excitatory inputs, or "yes" votes, than inhibitory inputs, or "no" votes, then an action potential will be initiated in the cell body.

3) The *axon* is a thin (and sometimes very long) extension from the cell body. The axon carries the electrical impulses (i.e. action potentials) from one part of the nervous system to another. When you wiggle your toes, axons over a meter long carry action potentials from the cell bodies in your spinal cord to the muscles in your foot.
4) The *synaptic terminals* form the output end of each neuron. These synaptic terminals connect with other neurons or, in some cases, directly with muscles. When an action potential traveling along an axon reaches these synaptic terminals, the neural signal is transmitted to the next set of cells in the network by the release of chemical signals from the synaptic terminals called neurotransmitters. These neurotransmitters are the "synaptic outputs" of the cell and also make up the "synaptic inputs" to the dendrites of the next neuron in the processing stream. As discussed in #1 above, these synaptic signals will either vote to excite or inhibit the generation of an action potential in the next neuron down the line.

Most neurons receive synaptic inputs to their dendrites from many different cells. Each neuron, in turn, usually synapses with many other neurons. These complex wiring patterns in the brain form the basis of who we are, how we move, and how we think. Measuring the overall electrical energy patterns at different locations outside of the brain can tell us what parts of this network are active at any given time (i.e. taking the temperature of the room). Measuring the exact firing patterns of a relatively small number of neurons sampled from the brain can provide imperfect clues to much more detailed information about our thoughts and actions.

4 The Road to Imperfection

Intracortical recordings are imperfect, not only because they detect the firing patterns of only a small part of a very complex network of neurons, but also because the process of detecting the firing patterns is itself imperfect. When we stick an array of electrodes a couple millimeters into the cortex, the sensors are surrounded by many neurons, each neuron generating its own sequences of action potentials. Each intracortical microelectrode simply measures the voltage at a given point in the brain. Action potentials in any of the nearby neurons will cause a spike in the voltage measured by the electrode. If a spike in voltage is detected, you know a neuron just fired – but which neuron? Many hundreds of neurons can be within "ear shot" of the electrode. Larger neurons have more ions flowing in and out during an action potential. Therefore neurons with large cell bodies generate larger spikes in the measured voltage than smaller neurons. However, the size of the spike is also affected by how far away the neuron is from the electrode. The size of the voltage spike drops off rapidly for neurons that are farther away. The end result is that these microelectrodes often pick up a lot of small overlapping spikes from hundreds or even a thousand different neurons located within a few microns from the recording site. However, each microelectrode can also pick up larger spikes from just a few very close neurons that have relatively large cell bodies.

The many small overlapping spikes from all the neurons in the vicinity of the electrode mix together to form a general background activity level that can reflect the overall amount of activation within very localized sections of the brain (much like taking the local "temperature" in a region only a couple hundred microns in diameter). This background noise picked up by the electrodes is often called "multiunit activity" or "hash". It can be very useful as a BCI control signal just like the electric field potentials measured outside the brain can be a useful measure of activity level (a.k.a. temperature) over much larger regions of the brain.

The real fun begins when you look at those few large nearby neurons whose action potentials result in larger spikes in voltage that stand out above this background level. Instead of just conveying overall activity levels, these individual neural firing patterns can convey much more unique information about what a person is sensing, thinking, or trying to do. For example, one neuron may increase its firing rate when you are about to move your little finger to the right. Another neuron may increase its firing rate when you are watching someone pinch their index finger and thumb together. Still another neuron may only fire if you actually pinch your own index finger and thumb together. So what happens when spikes from all these neurons are picked up by the same electrode? Unless your BCI system can sort out which spikes belonged to which neurons, the BCI decoder would not be able to tell if you are moving your little finger or just watching someone make a pinch.

Signal processing methods have been developed to extract out which spikes belong to which neuron using complicated and inherently flawed methodologies. This process of extracting the activity of different neurons is depicted in Fig. 2 and summarized here. Spikes generated from different neurons often have subtle differences in the size and shape of their waveforms. Spikes can be "sorted" or designated as originating from one neuron or another based on fine differences between their waveforms. This spike sorting process is imperfect because sometimes the spike waveforms from two neurons are too similar to tell apart, or there is too much noise or variability in the voltage signal to identify consistent shapes in the waveforms generated from each neuron.

In order to even see the fine details necessary for sorting each spike's waveforms, these voltage signals have to be recorded at a resolution two orders of magnitude higher than the recording resolution normally used with non-penetrating brain electrodes. Therefore, larger, more sophisticated processors are needed to extract all the useful information that is available from arrays of implanted microelectrodes. However, for wheelchair-mobile individuals, reducing the hardware and power requirements is essential for practical use. For mobile applications, several

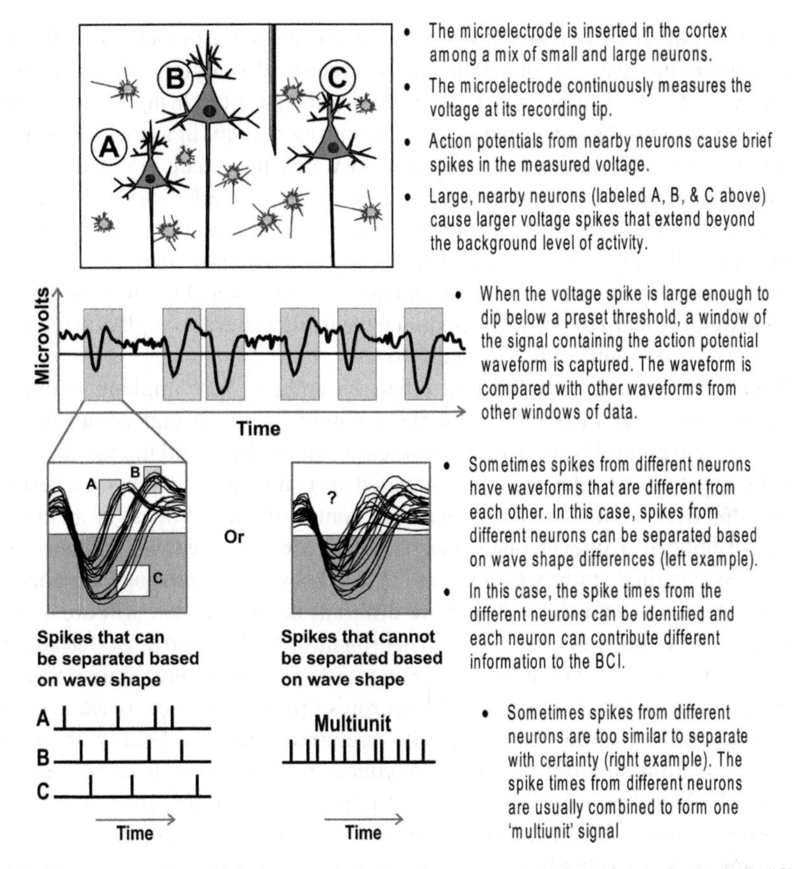

Fig. 2 The process by which intracortical signals are processed to determine when different neurons are firing

companies and research groups are developing simplified spike sorting systems that are either less accurate or don't worry about spike sorting at all and just lump all large spikes on an electrode together. Fortunately, different neurons that are picked up on the same electrode often convey somewhat similar information. Therefore, much useful information can still be extracted even without spike sorting.

5 A Brief History of Intracortical BCIs

The digital revolution has made the field of BCIs possible by enabling multiple channels of brain signals to be digitized and interpreted by computers in real time. Impressively, some researchers were already developing creative BCIs with simple analog technology before computers were readily available. In the late sixties and early seventies Dr. Eberhard Fetz and his colleagues implanted monkeys with microelectrodes in the part of the brain that controls movements [1, 2]. These researchers designed analog circuits that transformed the voltage spikes detected by the electrodes into the movement of a needle on a voltmeter. The circuits were designed so that the needle would move when the monkey increased the firing rate of individually recorded neurons sometimes while decreasing activity in associated muscles. Most importantly, the animal was allowed to watch the needle's movements and was rewarded when he moved the needle past a certain mark. The animal quickly learned to control the firing activity of the neurons as assigned in order to move the voltmeter needle and get a reward. These early studies show the power of visual feedback, and they demonstrated that monkeys (and presumably humans) have the ability to quickly learn to willfully modulate the firing patterns of individual neurons for the purpose of controlling external devices.

Along side this initial BCI work, other researchers were implanting intracortical microelectrodes in brains of monkeys simply to try to understand how our nervous system controls movement. These early studies have laid the foundation for many BCI systems that decode one's intended arm and hand movement in real time and use that desired movement command to control the movements of a computer cursor, a robotic arm, or even one's own paralyzed arm activated via implanted stimulators. One seminal study was done in the 1980s by Dr. Apostolos Georgopoulos and his colleagues who showed that the different neurons in the arm areas of the brain are "directionally tuned" [3]. Each neuron has what's called a "preferred direction". Each neuron will fire at its maximum rate as you are about to move your arm in that neuron's preferred direction. The neuron's firing rate will decrease as the arm movements start deviating from that cell's preferred direction. Finally, the neuron will fire at its minimum rate as you move your arm opposite the neuron's preferred direction. Different neurons have different preferred directions. By looking at the firing rates of many different directionally-tuned neurons, we can deduce how the arm is moving fairly reliably [4].

Unfortunately, each neuron's firing patterns are not perfectly linked to one's movement intent. Neurons have a substantial amount of variability in their firing patterns and each one only provides a noisy estimate of intended movement. In 1988, Georgopoulos and Massey conducted a study to determine how accurately movement direction could be predicted using different numbers of motor cortex neurons [5]. They recorded individual neurons in monkeys from the area of the brain that controls arm movements. They recorded these neurons while the monkeys made a sequence of "center-out" movements with a manipulandum, which is a device that tracks hand position as one moves a handle along a tabletop. In these experiments, the table top had a ring of equally-spaced target lights and one light in the center of this ring that indicated the starting hand position. The monkey would start by holding the manipulandum over the center light. When one of the radial target lights would come on, the animal would move the manipulandum outward as fast as it could to the radial target to get a reward[1]. The researchers recorded many different

---

[1]Because the researchers imposed time limits on hitting the targets, the subjects tried to move quickly to the targets and did not always hit the target accurately.

individual neurons with different preferred directions as these animals repeated the same sequence of movements over many days.

Once these data were collected, Georgopoulos and his colleagues then compared how much information one could get about the animal's intended target based on either the actual hand trajectories themselves or from the firing rates of different numbers of neurons recorded during those same movements (see Fig. 3). Amazingly, they found that once they combined the activity of only 50 or more neurons, they could extract *more accurate* information about the intended target than they could deduce from the most accurate hand trajectories generated by any subject[2].

The implications of this study for BCIs are enormous. For example, this study would suggest that you should be able to move your computer cursor to targets more accurately by controlling the cursor directly with 50+ neurons than by moving a mouse with your hand. So, does this mean we don't really need the many millions of neurons that our brain has for controlling movements? ...Yes and no. We do need all those neurons to be able to effortlessly control all the joints of the body over a wide range of limb configurations and movement speeds and under a variety of environmental conditions. However, for specific simplified tasks, such as moving to a small number of fixed targets, intracortical signals may provide a more efficient target selection option than using your own hand to select targets with a mouse.

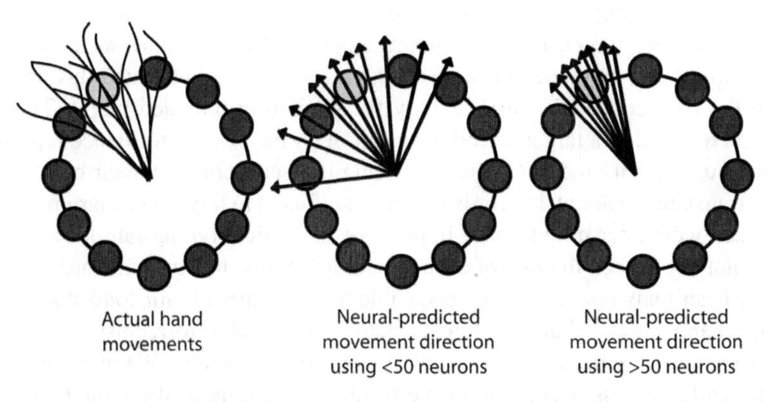

Fig. 3 Diagram (not actual data) illustrating hand movements versus neural-predicted movement directions compared in the Georgopoulos and Massey study [5]. Left diagram shows hand paths toward a highlighted target in a center-out task. Note that the subjects were required to make rapid movements and not all movements went precisely to the intended target. Middle and right diagrams illustrate direction vectors generated using neural data collected while the subjects made movements to the highlighted targets. Middle diagram illustrates direction vectors that might be generated when less than 50 neurons were used. Right diagram illustrates direction vectors that might be generated when more than 50 neurons were used

[2]The researchers also had human subjects do the center-out movement task but without simultaneously recording neural signals. The 50+ neurons could predict the intended target better than using the actual trajectories from any monkey *or* human subject.

Many useful BCI systems consist of a limited number of targets at fixed locations (e.g. typing programs or icon selection menus for the severely paralyzed). Therefore, these systems are particularly well suited to efficiently use intracortical signals; the neural activity can be decoded into specific target locations directly without needing the user to generate (or the BCI system to decipher) the detailed muscle activations and joint configurations that would be needed to physically move one's arm to these same fixed targets.

Dr. Krishna Shenoy and colleagues have recently demonstrated this concept of rapid efficient selection from a small fixed number of targets using intracortical signals [6]. These researchers trained monkeys in a center-out reaching task that included a delay and a "go" cue. The animals had to first hold their hand at a center start location; a radial target would then light up. However, the animal was trained to wait to move to the target until a "go" cue was given after some delay. These researchers were able to fairly reliably predict the target to which the animal was about to move using only the neural activity generated while the animal was still waiting for the cue to move. The researchers then implemented a real-time neural decoder that selected the target directly based on neural activity generated before the animal started to move. Decoding of the full arm movement path was not needed to predict and select the desired target.

In any BCI, there is usually a trade off between speed and accuracy. In the Shenoy study, the longer the BCI system collected the monkey's neural activity while the animal was planning its move, the more accurately the decoder could predict and select the desired target. However, longer neural collection times mean fewer target selections can be made per minute. These researchers optimized this time over which the BCI collected neural activity for predicting intended target. This optimization resulted in a target selection system with the best performance reported in the literature at that time – 6.5 bits of information per second[3] in their best monkey, which translates to about 15 words per minute if used in a typing program.

Fifteen words per minute is an impressive theoretical typing rate for a BCI, but it does not yet exceed the rate of even a mediocre typist using a standard keyboard. So why hasn't any research group been able to show direct brain control surpassing the old manual way of doing things? Many labs are able to record from over 50 neural signals at a time. Based on the Georgopoulos study, 50+ neurons should be enough to surpass normal motor performance in some of these target selection tasks. The answer lies in the way in which the neural signals must be collected for real-time BCI applications. In the Georgopoulos study, the animal's head had a chamber affixed over an opening in the skull. The chamber lid was opened up each day and a new electrode temporarily inserted into the cortex and moved slowly down until a good directionally-tuned neuron was recorded well above the background firing activity of all the other cells. Under these conditions, one can very accurately

[3]Bits per second is a universal measure of information transfer rate that allows one to compare BCI performance between different studies and research groups as well as compare BCIs with other assistive technologies.

detect the firing patterns of the individual neuron being recorded. This recording process was repeated each day for many different neurons as the monkey repeated the same movement to the different targets over and over again. The researchers then combined the perfectly-recorded firing patterns of the different neurons to try to predict which target the animal was moving to when those neurons were recorded.

Unfortunately, current permanently-implanted, multi-channel, intracortical recording systems are usually not adjustable once they are implanted. We do not have the luxury of fine tuning the individual location of each electrode to optimally place the recording tip where it can best detect the firing pattern of individual neurons. After implantation, you are stuck with whatever signals you can get from the initial implant location. Depending on how far away the recording sites are from the large cell bodies, you may or may not be able to isolate the firing patterns of individual neurons. Some recording sites may only detect background activity made up of the sum of many hundreds of small or distant neurons. Some channels may be near enough to a large neuronal cell body to detect unique, large-amplitude, voltage spikes from that specific neuron. However, many recording sites often are located among several neurons whose spike waveforms are hard to differentiate from each other or from the background noise. Therefore, some of the specific information encoded by the action potentials of different neurons gets mixed together at the detection stage. This mixing of the firing patterns makes decoding somewhat less accurate than what has been shown with acutely recorded neurons using repositionable electrodes.

6 The Holy Grail: Continuous Natural Movement Control

In spite of getting some "mixed signals" when using permanently implanted intracortical recording arrays, very useful information can easily and reliably be decoded from chronically-implanted intracortical microelectrodes. The Shenoy study demonstrated the reliability of intracortical recordings for accurately choosing between a small fixed number of targets (between 2 and 16 targets in this case). These targets in the lab could represent letters in a typing program or menu choices for controlling any number of assistive devices when applied in the real world. Many BCIs, including most extracortical BCIs, are being developed as discrete selection devices. However, intracortical signals contain all the fine details about one's limb movements, and, therefore, have the potential for being used to predict and replicate one's continuous natural limb movements.

Virtually all aspects of arm and hand movement have been decoded from the neural firing patterns of individual neurons recorded in the cortex (e.g. reach direction [3, 4], speed [6, 7], position [8–10], force [11, 12], joint kinematics [13], muscle activation [14], etc.). This level of detail has never been attained from extracortically recorded signals. If we were to accurately decode one's intended arm and hand movements from these intracortical signals, we could theoretically use this information to accurately drive a realistic prosthetic limb or a virtual limb, and the person wouldn't be able to tell the difference. The artificial limb would respond just as they intend their own limb to move.

It is unclear how many individual neurons are needed to accurately decode full arm and hand movements over a continuous range of possible joint configurations and muscle forces. Most studies so far try to predict some subset of movement characteristics, such as predicting two or three dimensional hand position in space and ignoring other details of the hand and arm configuration. Figure 4 compares actual hand trajectories and brain-controlled movement trajectories from a study by Taylor et al. where a monkey used intracortical signals to move a virtual cursor to targets in 3D space [15]. This figure illustrates the importance of visual feedback and the brain's ability to adapt to the BCI. Figure 4 also illustrates the benefit of concurrently adapting the decoding function to track and encourage learning and beneficial changes in the brain signals.

Part (a) of Fig. 4 shows the monkey's actual hand trajectories as it made center-out movements in 3D space (split into two 2D plots for easier viewing on the printed page). Part (b) plots trajectories decoded from the intracortical signals recorded while the animal made the hand movements in part (a). These trajectories were generated after the animal completed the experiment, so the animal did not have any real-time visual feedback of where the neural trajectories were going. Part (c) shows what happened when the animal did have visual feedback of the brain-controlled trajectories in real time. Here the animal used its intracortical signals to directly control the 3D movements of the cursor to the targets. In this case, the monkey modified its brain signals as needed to correct for errors and steer the cursor to the targets. Part (d) shows what happened when the experimenters also adapted the decoding function to track and make use of learning induced changes in the animal's brain signals. Adapting the BCI decoder to the brain as the brain adapts to the decoder resulted in substantial improvements in movement control. Many neural signals that did not convey much movement information during normal arm movements became very good at conveying intended movement with practice. Regularly updating the decoding function enabled the decoder to make use of these improvements in the animal's firing patterns.

Dr. Schwartz and colleagues have moved further toward generating whole arm and hand control with intracortical signals. They trained a monkey to use a brain-controlled robotic arm to retrieve food and bring it back to its mouth. This self-feeding included 3D control of the robot's hand position as well as opening and closing of the hand itself for control of four independent dimensions or actions [16].

The United States government's Department of Defense is currently sponsoring research investigating use of intracortical signals to control very sophisticated prosthetic limbs to benefit soldiers who have lost limbs in the war efforts in the Middle East. The goal of this funding is to develop realistic whole-arm-and-hand prosthetic limbs with brain signals controlling 22 independent actions or dimensions including control of individual finger movements. Researchers involved in this effort so far have been able to decode individual finger movements from the intracortical signals recorded in monkeys [17]. This government funding effort is intended to fast-track

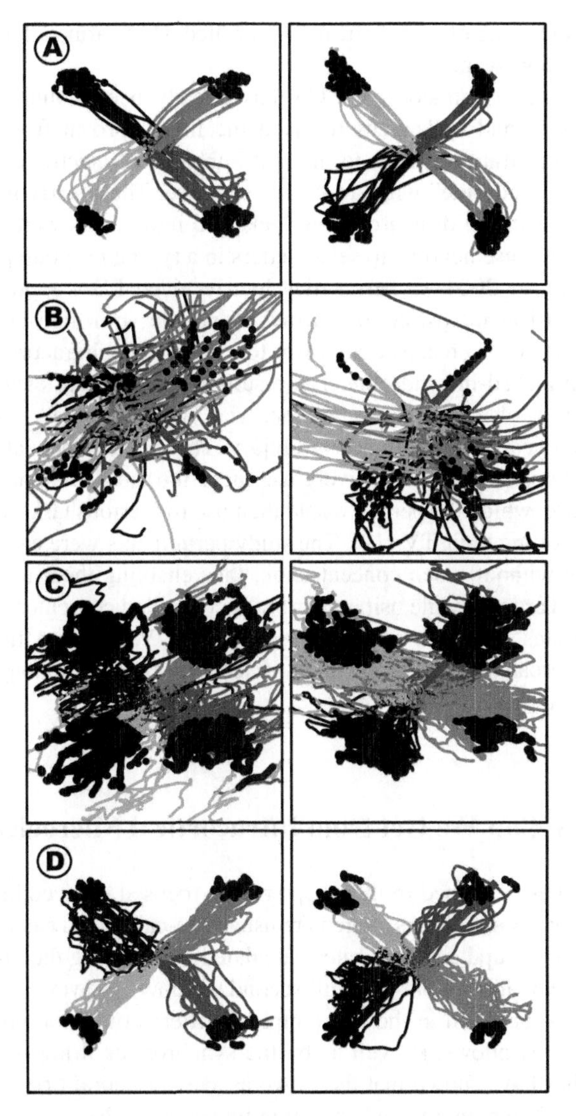

Fig. 4 Trajectories from an eight-target 3D virtual center-out task comparing actual hand movements and various brain-control scenarios. Trajectories go from a center start position out to one of eight radial targets located at the corners of an imaginary cube. Trajectories to eight targets are split into two plots of four for easier 2D viewing. Line shading indicates intended target. Black dots indicate when the intended target was hit. Thicker solid lines are drawn connecting the center start position to the eight outer target locations. (**a**) Actual hand trajectories generated when the monkey moved the virtual cursor under "hand control". (**b**) Trajectories predicted off line from the neural activity recorded during the hand movements shown in part a. (**c**) Trajectories generated with the cursor under real-time brain control using a fixed (i.e. non-adaptive) decoding function. (**d**) Trajectories generated with the cursor under real-time brain control using an adaptive decoding function that was regularly adjusted to make use of learning-induced changes in the neural coding. In a–c, the animal's arm was free to move. In D, both of the animal's arms were held still

eventual human testing of intracortically-controlled whole arm and hand prosthetics within the next few years.

Human testing of intracortical BCIs has already been going on since 1998. Dr. Phil Kennedy and colleagues received the first approval from the US Food and Drug Administration (FDA) to implant intracortical electrodes in individuals with "locked-in syndrome" who are unable to move. These early implant systems recorded only a handful of neurons, but locked-in individuals were able to use the firing patterns of these neurons to select letters in a typing program [18].

In the Spring of 2004, Cyberkinetics Inc. received FDA approval to test 96-channel intracortical microelectrode arrays in people with high-level spinal cord injuries. They later also received approval to evaluate this intracortical technology in people with locked-in syndrome. The results of their initial testing showed that the firing activity of motor cortex neurons was still modulated with thoughts of movement even years after a spinal cord injury. These researchers also showed that the study participants could effectively use these intracortical signals to control a computer mouse, which the person would then use for various tasks such as sending emails or controlling their TV [19]. The study participants were easily able to control the mouse without much concentration, thus enabling them to talk freely and move their head around while using their brain-controlled computer mouse. The US National Institutes of Health and the Department of Veteran's Affairs continue to strongly fund research efforts to develop intracortical BCI technologies for severely paralyzed individuals.

7 What Else Can We Get from Intracortical Microelectrodes?

So far, the studies described in this chapter have focused on decoding information from the firing *rates* of neurons, which are usually re-calculated each time the brain-controlled device is updated (e.g. rates calculated by counting the number of times a neuron fires over about a 30–100 millisecond window). However, useful information may also be encoded in the fine temporal patterns of action potentials within those short time windows, as well as by the synchronous firing between pairs of neurons. Studies have shown that these two aspects of neural firing are modulated independently while planning and executing movements [20].

Over the course of planning and executing a movement, the timing of spikes across neurons changes back and forth between being synchronized (i.e. spikes from different neurons occur at the same time more often than would be expected by chance) and being independent (i.e. two neurons fire at the same time about as often as you would expect by chance). What these changes in synchronous firing represent or encode is a much debated topic in neuroscience. A study by Oram and colleagues suggested that the changes in spike timing synchronization do not encode any directional information that isn't already encoded in the firing rates [21]. Instead, spikes become synchronized when a movement is being planned. The level of synchrony is highest at the end of movement planning just before the arm starts to move [22].

While synchronous firing does not appear to be as useful as firing rates for determining movement direction, synchronous firings might be used to provide a more reliable "go" signal to turn on a device and/or initiate a movement.

When large groups of cortical neurons start to fire synchronously in time rather than firing independently, the combined synaptic currents resulting from this synchronous firing activity cause voltage changes in the tissue that are large enough to be recorded on the surface of the brain and even on the surface of the scalp. These extracortical voltage changes are detected by the EEGs and ECoG recordings talked about throughout this book. Note the voltage change due to an individual action potential lasts only about a millisecond. For action potentials from individual neurons to sum together and generate detectable voltage changes on the scalp surface, large groups of neurons would have to synchronize their firing to well within a millisecond of each other. Given the inherent variability in neuronal activation delays, this precision of synchrony across large numbers of neurons is not common. However, voltage changes due to neurotransmitter release and synaptic activation (i.e. the "yes" and "no" votes passed on from one neuron to the next) can last in the tens of milliseconds. Therefore, if large groups of cortical neurons are receiving "yes" votes within several milliseconds of each other, the voltage changes from all of these neurons will overlap in time and sum together to generate detectable voltage changes that can be seen as far away as the scalp surface.

If only small clusters of cells receive synchronous synaptic inputs, the resulting, more-focal voltage changes may still be detectable extracortically by using high-resolution grids of small, closely-spaced electrodes placed on the brain surface. With today's microfabrication technology, many labs are making custom subdural micro-grid electrode arrays with recording sites of about a millimeter or less [23]. These custom arrays are able to detect synchronous neural activity within small clusters of cortical neurons localized under the small recording contacts.

Intracortical microelectrodes that are normally used to detect action potentials also can detect these slower voltage changes resulting from the synaptic currents of even smaller, more-focal clusters of synchronous neurons. These very localized slow voltage changes are termed "local field potentials" when recorded on intracortical microelectrodes. These local field potentials are part of the continuum of field potential recordings that range from the lowest spatial resolution (when recorded by large EEG electrodes on the scalp surface) to the highest resolution (when recorded within the cortex using arrays of microelectrodes a few tens of microns in diameter).

Like scalp or brain-surface recordings, these local field potentials, or "LFPs", can be analyzed by looking for changes in the power in different frequency bands or for changes in the natural time course of the signal itself. Local field potentials have been implicated in aspects of behavior ranging from movement direction to attention. Rickert and colleagues have shown that higher frequency bands (60–200 Hz) can be directionally tuned just like the firing rates of individual neurons [24]. Also, some frequency bands have been shown to be modulated with higher-level cognitive processes. For example, Donoghue and colleagues have shown that oscillations in the gamma range (20–80 Hz) are stronger during periods of focused attention [25]. At movement onset, the local field potential has a characteristic negative peak followed by a positive peak, and the amplitude of these peaks can vary with movement direction. Work by Merhing and colleagues suggest that intended targets can be estimated just as well from these LFP amplitudes as from the firing rates of individual neurons in monkeys performing a center-out task [26].

Information useful for BCI control has been extracted in many forms from intracortical microelectrodes (firing rates, spike synchronization, local field potentials, etc.). All forms can be recorded from the same intracortical electrodes, and, if you have the computational power, all can be analyzed and used simultaneously. Complex detailed information about our activities and behavior can be gleaned from the signals recorded on these tiny microelectrode arrays. Current studies with intracortical electrodes show much promise for both rapid target selection and generation of signals that can be used to control full arm and hand movements. Because intracortical electrodes are so small, many hundreds to even thousands of microelectrodes can potentially be implanted in the cortex at one time. We have only begun to scratch the surface of the complex information that will likely be decoded and used for BCI control from intracortical electrodes in the future.

References

1. E.E. Fetz, Operant conditioning of cortical unit activity. Science, 163(870), 955–958, (1969)
2. E.E. Fetz, D.V. Finocchio, Operant conditioning of specific patterns of neural and muscular activity. Science, 174(7), 431–435, (1971)
3. A.B. Schwartz, R.E. Kettner, A.P. Georgopoulos, Primate motor cortex and free arm movements to visual targets in three-dimensional space. I. Relations between single cell discharge and direction of movement. J Neurosci, 8(8), 2913–2927, (1988)
4. A.P. Georgopoulos, R.E. Kettner, A.B. Schwartz, Primate motor cortex and free arm movements to visual targets in three-dimensional space II: coding the direction of movement by a neural population. J Neurosci, 8(8), 2928–2937, (1988)
5. A.P. Georgopoulos, J.T. Massey, Cognitive spatial-motor processes. 2. Information transmitted by the direction of two-dimensional arm movements and by neuronal populations in primate motor cortex and area 5. Exp Brain Res, 69(2), 315–326, (1988)
6. G. Santhanam, S.I. Ryu, et al., A high-performance brain-computer interface. Nature, 442(7099):195–198, (2006)
7. D.W. Moran, A.B. Schwartz, Motor cortical representation of speed and direction during reaching. J Neurophysiol, 82(5), 2676–2692, (1999).
8. L. Paninski, M.R. Fellows, et al., Spatiotemporal tuning of motor cortical neurons for hand position and velocity. J Neurophysiol, 91(1), 515–532, (2004)
9. R.E. Kettner, A.B. Schwartz, A.P. Georgopoulos, Primate motor cortex and free arm movements to visual targets in three-dimensional space. III. Positional gradients and population coding of movement direction from various movement origins. J Neurosci, 8(8), 2938–2947, (1988)
10. R. Caminiti, P.B. Johnson, A. Urbano, Making arm movements within different parts of space: dynamic aspects in the primary motor cortex. J Neurosci, 10(7), 2039–2058, (1990)
11. E.V. Evarts, Relation of pyramidal tract activity to force exerted during voluntary movement. J Neurophysiol, 31(1), 14–27, (1968)
12. J. Ashe, Force and the motor cortex. Behav Brain Res, 86(1), 1–15, (1997)
13. Q.G. Fu, D. Flament, et al., Temporal coding of movement kinematics in the discharge of primate primary motor and premotor neurons. J Neurophysiol, 73(2), 2259–2263, (1995)

14. M.M. Morrow, L.E. Miller, Prediction of muscle activity by populations of sequentially recorded primary motor cortex neurons. J Neurophysiol, 89(4), 2279–2288, (2003)
15. D.M. Taylor, S.I. Helms Tillery, A.B. Schwartz, Direct cortical control of 3D neuroprosthetic devices. Science, 296(5574), 1829–1832, (2002)
16. M. Velliste, S. Perel, et al., Cortical control of a prosthetic arm for self-feeding. Nature, 453(7198), 1098–1101, (2008)
17. V. Aggarwal, S. Acharya, et al., Asynchronous decoding of dexterous finger movements using M1 neurons. IEEE Trans Neural Syst Rehabil Eng, 16(1), 3–4, (2008)
18. P.R. Kennedy, R.A. Bakay, et al., Direct control of a computer from the human central nervous system. IEEE Trans Rehabil Eng, 8(2), 198–202, (2000)
19. L.R. Hochberg, M.D. Serruya, et al., Neuronal ensemble control of prosthetic devices by a human with tetraplegia. Nature, 442(7099), 164–171, (2006)
20. A. Riehle, S. Grun, et al., Spike Synchronization and Rate Modulation Differentially Involved in Motor Cortical Function. Science, 278(5345), 1950–1953, (1997)
21. M.W. Oram, N.G. Hatsopoulos, et al., Excess synchrony in motor cortical neurons provides redundant direction information with that from coarse temporal measures. J Neurophysiol, 86(4), 1700–1716, (2001)
22. A. Riehle, F. Grammont, et al., Dynamical changes and temporal precision of synchronized spiking activity in monkey motor cortex during movement preparation. J Physiol, Paris, 94(5–6), 569–582, (2000)
23. A. Kim, J.A. Wilson, J.C. Williams, A cortical recording platform utilizing microECoG electrode arrays. Conf Proc, IEEE Eng Med Biol Soc, 5353–5357, (2007)
24. J. Rickert, S.C. Oliveira, et al., Encoding of movement direction in different frequency ranges of motor cortical local field potentials. J Neurosci, 25(39), 8815–8824, (2005)
25. J.P. Donoghue, J.N. Sanes, et al., Neural Discharge and Local Field Potential Oscillations in Primate Motor Cortex during Voluntary Movements. J Neurophysiol, 79(1), 159–173, (1998)
26. C. Mehring, Inference of hand movements from local field potentials in monkey motor cortex. Nat Neurosci, 6(12), 1253–1254, (2003)

BCIs Based on Signals from Between the Brain and Skull

Jane E. Huggins

1 Introduction

This chapter provides an introduction to electrocorticogram (ECoG) as a signal source for brain–computer interfaces (BCIs). I first define ECoG, examine its advantages and disadvantages, and outline factors affecting successful ECoG experiments for BCI. Past and present BCI projects that utilize ECoG and have published results through early 2008 are then summarized. My own ECoG work with the University of Michigan Direct Brain Interface project is described in detail, as the first and (at the time of writing) longest running targeted exploration of ECoG for BCI. The well established ECoG research at the University of Washington is described only briefly, since Chapter "A Simple, Spectral-change Based, Electrocorticographic Brain–Computer Interface" in this volume provides a first-hand description. This chapter concludes with a few thoughts on the growth of BCI research utilizing ECoG and potential future applications of BCI methods developed for ECoG.

2 Electrocorticogram: Signals from Between the Brain and Skull

The commands for a BCI must come from the collection of individual cells that make up the brain. Observation and interpretation of the activity of each of the billions of individual cells is obviously impossible with current technology and unlikely to become possible in the near future. BCI researchers must therefore choose either to observe activity from a small percentage of single cells or observe the field potentials from groups of cells whose activity has been combined by the filtering affect of brain tissue and structures such as the brain's protective membranes, the skull, and the scalp. Field potentials can be observed using electrodes in a wide

J.E. Huggins (✉)
University of Michigan, Ann Arbor, MI, USA
e-mail: janeh@umich.edu

B. Graimann et al. (eds.), *Brain–Computer Interfaces*, The Frontiers Collection,
DOI 10.1007/978-3-642-02091-9_13, © Springer-Verlag Berlin Heidelberg 2010

range of sizes from micro-electrodes intended to record single cells (e.g. [1, 2]) to the more common electroencephalography (EEG) electrodes placed on the scalp (see Fig. 1). As the distance from the signal source (the individual cells) increases, the strength of the filtering effect also increases. Observing the brain activity from the scalp (EEG) can be compared to listening to a discussion from outside the room. The intervening scalp and bone hide most subtle variations and make the signals of interest difficult to isolate. To avoid this effect, it is advantageous to at least get inside the room (inside the skull), even if you remain at a distance from individual speakers.

ECoG electrodes allow observation of brain activity from within the skull, since they are surgically implanted directly on the surface of the brain. However, they do not penetrate the brain itself. ECoG electrodes can be arranged in grids or strips (see Fig. 2), providing different amounts of cortical coverage. ECoG grids are placed through a craniotomy, in which a window is cut in the skull to allow placement of the grid on the surface of the brain and then the bone is put back in place. ECoG strips can be placed through a burrhole, a hole drilled in the skull through which the narrow strips of electrodes can be slipped into place. ECoG electrodes are placed subdurally, (i.e. under the membranes covering the brain), although epidural placements (above these membranes, but still inside the skull) are sometimes done simultaneously. In this chapter, the term ECoG will be used to refer to signals from both subdural and epidural electrodes, with placements solely involving epidural electrodes specifically noted. ECoG electrodes are commonly 4 mm diameter disks arranged in either strips or grids with a 10 mm center-to-center distance [e.g. [3–6]]. Smaller, higher density ECoG electrodes are also used [e.g. [7]] and the optimal size of ECoG electrodes for BCIs remains an open question. Platinum or stainless-steel ECoG electrodes are mounted in a flexible silicone substrate from which a cable carries the signals through the skin and connects to the EEG recording equipment. Standard EEG recording equipment is used for clinical monitoring, although some units having settings for ECoG.

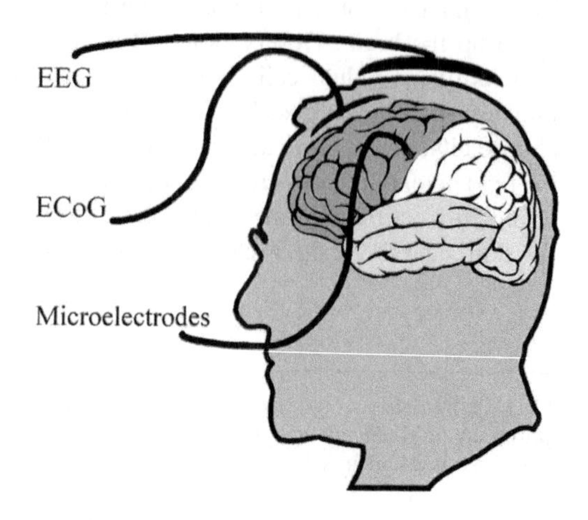

Fig. 1 Electrode location options for BCI operation

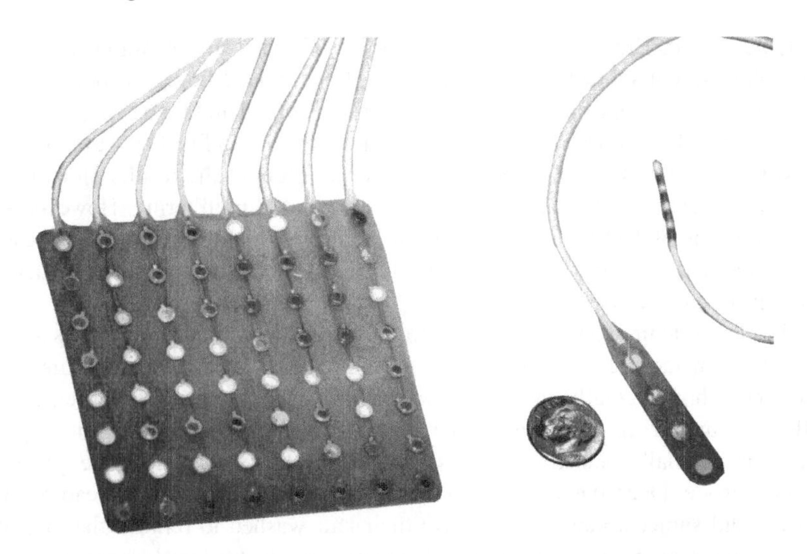

Fig. 2 Grid and strip arrangements of ECoG electrodes with a US dime for size comparison

3 Advantages of ECoG

ECoG provides advantages over both EEG and microelectrode recordings as a BCI input option.

3.1 Advantages of ECoG Versus EEG

Signal quality is perhaps the most obvious advantage of ECoG in comparison to EEG. By placing electrodes inside the skull, ECoG avoids the filtering effects of the skull and several layers of tissue. Further, these same tissue layers essentially insulate ECoG from electrical noise sources such as muscle activity (electromyogram (EMG)) and eye movement (electrooculogram (EOG)) resulting in reduced vulnerability to these potential signal contaminates [8]. ECoG also provides greater localization of the origin of the signals [9] with spatial resolution on the order of millimeters for ECoG versus centimeters for EEG [4, 7]. ECoG has a broader signal bandwidth than EEG, having useful frequencies in the 0–200 Hz range [4] with a recent report of useful information in the 300 Hz−6 kHz frequency range [5]. In EEG, the higher frequencies are blocked by the lowpass filter effect of the scalp and intervening tissues, which limits the useful frequency range of EEG to about 0–40 Hz [4]. In addition to a wider available frequency band, ECoG provides higher amplitude signals with values in the range of 50–100 microVolts compared to 10–20 microVolts for EEG [4].

As an initial comparison of the utility of EEG and ECoG, the University of Michigan Direct Brain Interface (UM-DBI) project and the Graz BCI group performed a classification experiment [10] on data segments from either event or idle (rest) periods in EEG and in ECoG that were recorded under similar paradigms but with different subjects. Approximately 150 self-paced finger movements were performed by 6 normal subjects while EEG was recorded and by 6 epilepsy surgery patients while ECoG was recorded. Classification results on ECoG always exceeded classification results on EEG. Spatial prefiltering brought the results on EEG to a level equivalent to that found on ECoG without spatial prefiltering. However, spatial prefiltering of the ECoG achieved a similar improvement in classification results. Thus, in all cases, ECoG proved superior to EEG for classification of brain activity as event or idle periods.

The electrode implantation required for ECoG, although seen by some as a drawback, may in fact be another advantage of ECoG because the use of implanted electrodes should greatly minimize the set-up and maintenance requirements of a BCI. Setup for an EEG-based BCI requires an electrode cap (or the application of individually placed electrodes) and the application of electrode gel under each electrode. Once the subject is done using the BCI, the electrode cap must be washed and subjects may want to have their hair washed to remove the electrode gel. While research on electrodes that do not require electrode gel is underway [e.g. [11]], developing electrodes capable of high signal quality without using electrode gel is a serious technical challenge and would still require daily electrode placement with inherent small location variations. Setup for an ECoG-based BCI would require initial surgical implantation of electrodes, ideally with a wireless transmitter. After recovery, however, setup would likely be reduced to the time required to don a simple headset and maintenance to charging a battery pack. Further, instead of the appearance of an EEG electrode cap (which resembles a swim cap studded with electrodes), the BCI could be a relatively low visibility device, with a possible appearance similar to a behind-the-ear hearing aid. So, from the user's perspective, as well as from the algorithm development perspective, ECoG provides many advantages over EEG.

In summary, the advantages of ECoG over EEG include:

- Improved signal quality
  - Insulation from electrical noise sources
  - Improved resolution for localization of signal origin
  - Broader signal bandwidth
  - Larger signal amplitude
- Reduced setup time
- Reduced maintenance
- Consistent electrode placement
- Improved cosmesis

3.2 Advantages over Microelectrodes

ECoG may also have advantages over recordings from microelectrodes. Microelectrodes that are intended to record the activity from single cells are implanted inside both the skull and the brain itself. Recordings from microelectrodes therefore share with ECoG the same insulation from potential EMG and EOG contamination. However, these recordings may actually provide too great a degree of localization. While single cells are certainly involved in the generation of movements and other cognitive tasks, these actions are produced by the cooperative brain activity from multiple cells. A specific cell may not be equally active every time an action is performed, and individual cells may be part of only one portion of an action or imagery task. Interpreting the purpose of brain activity from the microelectrode recording therefore requires integrating the results from multiple cell recordings. Although good BCI performance has been achieved with as few as 10 cells [e.g. [12]], it is possible that microelectrodes could provide *too close* a view of the brain activity. One might compare the recording of brain activity related to a specific task with recording a performance by a large musical ensemble. You can either record individual singers separately with microphones placed close to them, or you can place your microphones some distance from the ensemble and record the blend of voices coming from the entire ensemble. If you are unable to record all the individual singers, your sample of singers may inadvertently omit a key component of the performance, such as the soprano section. While both recording individual singers and recording the entire ensemble can allow you to hear the whole piece of music, the equipment necessary to place a microphone by each singer is daunting for a large ensemble and the computation required to recombine the individual signals to recreate the blended performance of the ensemble may be both intensive and unnecessary. ECoG provides recordings at some distance from the ensemble of neurons involved in a specific brain activity.

When ECoG-based BCI research was first initiated, microelectrode recordings in human subjects were not available for clinical use. Even now, human microelectrode recordings for BCI research are only used by a few pioneering laboratories (e.g. [1, 13, 14]) and the ethical considerations and safeguards necessary for this work are daunting. Further, microelectrodes have long been subject to issues of stability, both of electrode location and electrical recording performance [6, 15, 16]. The large size of ECoG electrodes and the backing material in which they are mounted provide resistance to movements and increase the opportunity to anchor ECoG electrodes. Along with the large electrode size (in comparison to microelectrodes), this makes electrode movement a negligible concern when using ECoG electrodes. The placement of ECoG electrodes on the surface of the brain, without penetrating the brain itself, has caused ECoG to be referred to as "minimally invasive." The appropriateness of this term is debatable, considering that subdural implantation of ECoG electrodes does require penetrating not only the skull, but the membranes covering the brain. However, ECoG does not require penetration of the brain itself, and therefore (as mentioned in [4]) could have less potential for cellular damage and cellular reactions than microelectrodes.

As a practical implementation issue, Wilson et al. [7] points out that the sampling frequencies used for ECoG are low in comparison to the sampling frequencies (25–50 kHz) for spike trains recorded with microelectrodes, indicating that ECoG would require less bandwidth for wireless transmission of recorded data from a fully implanted BCI. While the recent report of useful event-related ECoG in the

300–6 kHz range [5] could reduce this difference, when multiplied by the number of individual electrodes whose data is required for a functional BCI, it remains a distinct technical advantage.

In summary, the advantages of ECoG over microelectrodes include:

- Ensemble activity recording
- Stable electrode placement
- Preservation of brain tissue
- Lower sampling rates

3.3 Everything Affects the Brain

Despite the advantages of ECoG over EEG and microelectrodes, it should be noted that wherever brain activity is recorded from, it will be impossible to avoid recording irrelevant brain activity, which must be ignored during the signal analysis. While ECoG is less susceptible to noise from muscle artifact or electrical noise in the environment, it is still subject to the extraneous signals from the other activities that the brain is involved in. Indeed, any list of factors that can be expected to affect brain activity quickly becomes lengthy, since any sensory experience that the subjects have will affect activity in some part of the brain, as may the movement of any part of their body, their past history, their expectations and attitude toward the experiment and even their perception of the experimenter's reaction to their performance. So, while using ECoG may simplify the signal analysis challenges of detecting brain activity related to a particular task, the detection is still a significant challenge.

4 Disadvantages of ECoG

The advantages of ECoG also come with distinct disadvantages. A primary disadvantage is limited patient access and limited control of experimental setup. Subjects for ECoG experiments cannot be recruited off the street in the same way as subjects for EEG experiments. ECoG is only available through clinical programs such as those that use ECoG for pre-surgical monitoring for epilepsy surgery. Epilepsy surgery involves the removal of the portion of the brain where seizures begin. While removing this area of the brain can potentially allow someone to be seizure-free, any surgery that involves removing part of the brain must be carefully planned. Epilepsy surgery planning therefore frequently involves the temporary placement of ECoG electrodes for observing seizure onset and for mapping the function of specific areas of the brain. This placement of ECoG electrodes for clinical purposes also provides an opportunity for ECoG-based BCI experiments with minimal additional risk to the patient. While this is an excellent opportunity, it also results in serious limitations on experimental design. BCI researchers working with ECoG typically have no control over the placement of the ECoG electrodes. While they could exclude subjects based on unfavorable electrode locations, they generally do not have the ability to add additional electrodes or to influence the locations at which the clinically placed electrodes are located. Thus, ECoG-based BCI studies usually include large numbers of recordings from electrodes placed over areas that are suboptimal for the task being studied.

Additionally, not all eligible subjects will agree to participate in research. Some subjects are excited by the opportunity to "give back" to medical research by participating in research that will hopefully help others as they are themselves being helped. However, other subjects are tired, distracted, in pain (they have just undergone brain surgery after all), apprehensive, or simply sick of medical procedures and thankful to have something to which they can say "no." This further limits the number of subjects available for ECoG research.

Further, working with a patient population in a clinical environment severely limits time with the subjects. Patient care activities take priority over research and medical personnel frequently exercise the right to interact with the patient, sometimes in the middle of an experiment. Further, the patient's time is scheduled for clinically important tasks such as taking medication, monitoring vital signs, cortical stimulation mapping, and procedures such as CT scans and personal activities such as family visits. Other factors that may affect the brain activity are also out of the researcher's control. Subjects may be on a variety of seizure medications, or they may have reduced their normal seizure medications in an effort to precipitate a seizure so that its location of origin can be identified. Other measures, such as sleep deprivation or extra exercise, may also be ongoing in an effort to precipitate a seizure. Finally, a patient may have had a seizure recently, which can dramatically affect concentration and fatigue in a manner unique to that individual. Of course, a subject may also have a seizure during an experiment, which could be only a minor interruption, or could bring the experimental session to an end, depending on the manifestation of the seizure. These issues highlight the generally understood, though perhaps seldom stated, fact that the brains of subjects who have ECoG electrodes placed for epilepsy surgery (or any other clinical condition) are not in fact "normal" examples of the human brain. While the brain of a person who needs a BCI may also not be "normal" (depending on the reason the BCI is needed), there is no evidence that the subjects who currently provide access to ECoG are good models for people who need a BCI; they are simply the only models available.

In addition to practical issues of experimental control, there are practical technical challenges. Since the ECoG electrodes have been placed for clinical purposes, monitoring using these electrodes cannot be interrupted during the experiment. While digital recording of brain activity is now standard during epilepsy monitoring, the ability to provide custom analysis and BCI operation with a clinical recording setup is not. So, any experiment that involves more than passively recording subjects' brain activity during known actions may require a separate machine to record and process the ECoG for BCI operation. Using this machine to temporarily replace the clinical recording machine for the duration of the experimental setup would interrupt standard patient care and would therefore require close involvement of all clinical epilepsy monitoring personnel and an unusually flexible patient care regime, although it has been reported [5]. A more common recording setup (see Fig. 3)

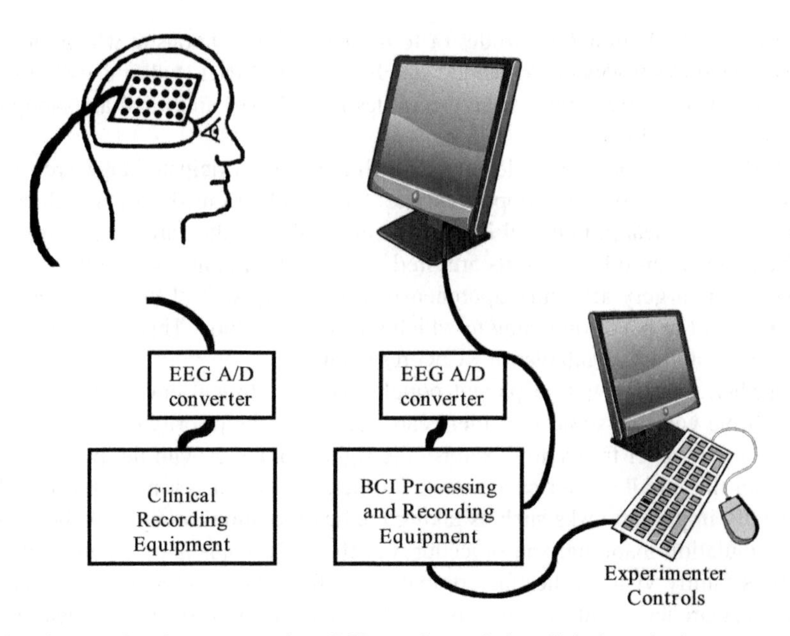

Fig. 3 Diagram of equipment setup for a BCI experiment during clinical monitoring

therefore includes splitting the analog signals from the ECoG electrodes prior to digitization so that they can be sent both to the clinical recording equipment and to the BCI. This setup must be carefully tested according to hospital regulations for electrical safety. While signal splitting can increase the electrical noise apparent in the ECoG signals, it has been successfully used by several researchers [3, 17]. The sheer quantity of electrical devices that are used in the hospital presents an additional technical challenge that the high quality of clinical recording equipment can only partly overcome. Sources of electrical noise in a hospital room can range from auto-cycling pressure cuffs on a patient's legs that are intended to reduce post-operative complications, to medical equipment being used in the next room (or occasionally by the patient sharing a semi-private room), to other electrical devices such as the hospital bed itself. Vulnerability to electrical noise may be increased by the location of the electrodes used as the ground and recording reference. These important electrodes may be ECoG electrodes, bone screw electrodes or scalp electrodes. When a scalp electrode is used as the reference, vulnerability to widespread electrical noise can easily result.

5 Successful ECoG-Based BCI Research

The use of ECoG for BCI research has advantages of signal quality, spatial resolution, stability, and (eventually) ease of BCI setup. ECoG can be obtained from large areas of the cortex, and ECoG shows event-related signal changes in both the time domain and the frequency domain (see Fig. 4). However, special attention must be paid to the technical challenges of working in an electrically noisy hospital environment within the limited time subjects have available during the constant demands of treatment in a working medical facility. An ECoG-based BCI project should pay special attention to these concerns from the earliest planning stages. Equipment selection and interfaces with clinical equipment should be designed for ease of setup, minimal disruption of clinical routine, and robustness to and identification of electrical contamination on either the ECoG or reference electrode signals. Feedback paradigms and experimental instructions should be quick to explain and easy to understand. Experiments should be designed to be compatible with the limited time available, the unpredictable nature of subject recruitment, and the limited control over basic experimental parameters such as electrode location. While the technical challenges of ECoG recording can usually be conquered by engineering solutions, the challenges of scheduling, access, and interruptions will be better addressed through the active involvement of the medical personnel on the research team. Ideally, these research team members should be excited about the research and have a specific set of research questions of particular interest to them. Since the medical team members may be the logical ones to recruit subjects into the experiment, and act as a liaison between the BCI research team and the clinical personnel, they can have a make-or-break effect on the total project success. Working as a team, a multidisciplinary research group can successfully utilize the opportunity provided by clinical implantation of ECoG electrodes for a variety of medically necessary treatment regimes to conduct strong BCI research.

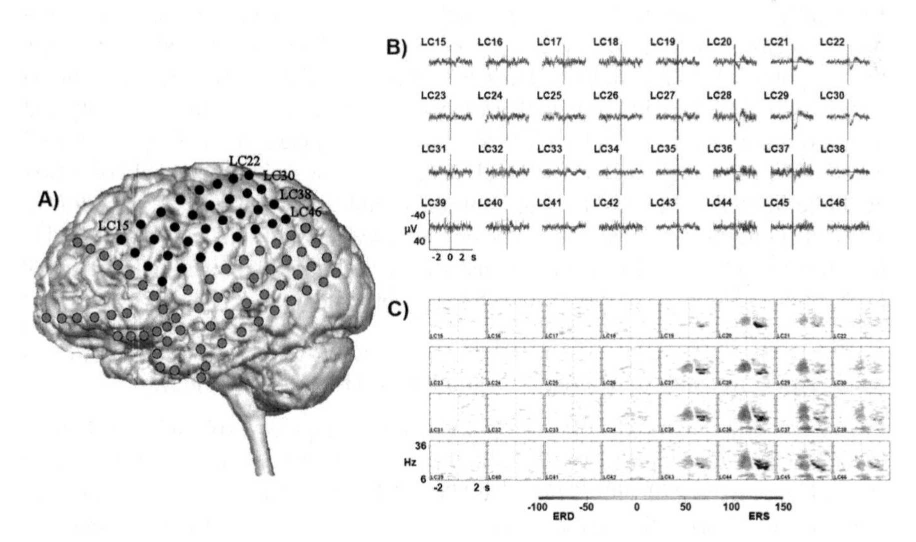

Fig. 4 Event-related potentials and ERD/ERS maps of movement-related ECoG for a subject performing right middle finger extension. (**a**) Electrode locations. (**b**) Averaged ECoG for the black electrodes in (a). (**c**) ERD/ERS for the black electrodes

6 Past and Present ECoG Research for BCI

Although human ECoG has been used for scientific study of brain function and neurological disorders since at least the 1960s [e.g. [18]], its use for BCI research was almost nonexistent prior to the 1990s. The earliest BCI researchers either worked with animals to develop tools for recording the activity of individual neurons (e.g., [16, 19]) or worked with human EEG (e.g.[20, 21]).

6.1 ECoG Animal Research

The first ECoG studies for BCI development were reported in 1972 by Brindley and Craggs [22, 23] who studied the signals from epidural [22] electrode arrays implanted over the baboon motor cortex [22, 23]. ECoG was bandpass filtered to 80–250 Hz and the mean square of the signal calculated to produce a characteristic signal shape preceding specific movements that was localized over arm or leg motor cortex. A simple threshold detector was applied to this feature, generating prediction accuracy as high as 90%.

6.2 Human ECoG Studies

6.2.1 Smith-Kettlewell Eye Research Institute

The first human BCI work using ECoG was reported in 1989 by Sutter [8, 24]. Epidurally implanted electrodes over visual cortex provided the input to a cortically-based eyegaze tracking system. The system was developed and tested using visual evoked potentials (VEP) in EEG. However, implanted electrodes were used when a subject with ALS had difficulty with artifact due to uncontrolled muscle activity and signal variations due to variable electrode placement by caregivers. Subjects viewed a grid of 64 flickering blocks. The block at the center of the subject's field of vision could be identified because the lag between stimulus and VEP onset was known. While this interface did not claim to be operable without physical movement, and is therefore not a "true" BCI, it did demonstrate the use of large intracranial electrodes as the signal source for an assistive technology interface.

6.2.2 The University of Michigan – Ann Arbor (Levine and Huggins)

Levine and Huggins' UM-DBI project was the first targeted exploration of human ECoG for BCI development [3, 25, 26]. The UM-DBI project recorded from their first subject on the 27th July 1994. The UM-DBI project has focused on the detection of actions in a continuous segment of ECoG data during which subjects repeated actions at their own pace. ECoG related to actual movement has been used almost exclusively to allow good documentation of the time at which the subject chose to do the action.

Subjects and Actions

Subjects who participated in the UM-DBI project studies were patients in either the University of Michigan Epilepsy Surgery Program or the Henry Ford Hospital Comprehensive Epilepsy Program and had signed consent forms approved by the appropriate institutional review board. Early subjects of the UM-DBI project performed a battery of discrete movements while ECoG was recorded for off-line analysis [3, 27]. Later subjects performed only a few actions and then participated in feedback experiments [28], described below. Actions were sometimes modified to target unique areas over which the individual subject's electrodes were placed or to accommodate a subject's physical impairments. The most common actions were a pinch movement, extension of the middle finger, tongue protrusion, and lip protrusion (a pouting motion). The actions ankle flexion and saying a phoneme ("pah" or "tah") were also used with some frequency. Subjects repeated a single action approximately 50 times over a period of several minutes. The time at which each repetition of the action was performed was documented with recorded EMG (or another measure of physical performance [3]). For each subject, the ECoG related to a particular action was visualized using triggered averaging and/or spectral analysis of the event-related desynchronization (ERD) and event-related synchronization (ERS) [29] (see Fig. 4). This analysis showed the localized nature of the ECoG related to particular actions and also revealed features that could be used for BCI detection analysis.

Training and Testing Data and Performance Markup

Detection experiments were generally done on ECoG from single electrodes (e.g. [26, 30]) with only an initial exploration of the beneficial effects of using combinations of channels [31]. The first half of the repetitions of an action were used for training the detection method and the second half of the repetitions were used for testing. Detection method testing produced an activation/no activation decision for each sample in the several-minute duration of the test data. An activation was considered to be valid (a "hit") if it occurred within a specified activation acceptance window around one of the actions as shown by EMG recording. False activations were defined as activations that occurred outside this acceptance window. The dimensions of the acceptance window used for individual experiments are reported below along with the results.

The HF-Difference

The UM-DBI project introduced a performance measure called the HF-difference for evaluation of BCI performance. The HF-difference is the difference between the hit percentage and a false activation percentage, with both of these measures calculated from a user's perspective. A user's perception of how well a BCI detected their commands would be based on how many times they tried to use the BCI. Therefore, the hit percentage is the percentage of actions performed by the subject that were correctly detected by the BCI. So, if the subject performed 25 actions, but only 19 of them were detected by the BCI, the hit percentage would be 76%. The false activation percentage is the percentage of the detections the BCI produced that were false (the calculation is the same as that of the false discovery rate [32] used in statistics). So, if the BCI produced 20 activations, but one was incorrect, the false activation percentage would be 5%. Note that the calculation of the false activation percentage is different from the typical calculation for a false positive percentage, where the denominator is the number of data samples that should be classified as no-activation. This large denominator can result in extremely low false positive rates, despite performance that has an unacceptable amount of false positives from the user's perspective. By using the number of detections as the denominator, the false activation rate should better reflect the user's perception of BCI performance. Finally, the HF-difference is the simple difference between the hit percentage and the false activation percentage. So, for the example where the subject performed 25 actions, the BCI produced 20 detections, and 19 of the detections corresponded to the actions, we have 76% hits, 5% false activations, and an HF-difference of 71%.

Detection Results

The first reports from the UM-DBI project of the practicality of ECoG for BCI operation used a cross-correlation template matching method (CCTM) in off-line analysis and occurred in the late 1990s [3, 26, 33, 34]. CCTM detected event-related potentials with accuracy greater than 90% and false activation rates less than 10% for 5 of 17 subjects (giving HF-differences above 90) using an acceptance window of 1 s before to 0.25 s after each action [27]. However, the CCTM method had an unacceptable delay due to the use of a template extending beyond the trigger. A partnership with the BCI group at the Technical University of Graz, in Austria led to further off-line analyses. An adaptive autoregressive method for detecting ERD/ERS was tested on ECoG from 3 subjects with each subject performing the actions finger extension, pinch, tongue protrusion and lip protrusion in a separate dataset for a total of 12 subject/action combinations. The adaptive autoregressive method produced HF-differences above 90% for all three subjects and for 7 of the 12 subject/action combinations [35]. A wavelet packet analysis method found HF-differences above 90% for 8 of 21 subject/action combinations with perfect detection (HF-difference = 100) for 4 subject/action combinations with an acceptance window of 0.25 s before and 1 s after each action [30]. Model-based detection method development continued at the University of Michigan, resulting in two versions of a quadratic detector based on a two-covariance model [36, 37]. The basic quadratic detector produced HF-differences greater than 90% for 5 of 20 subject/action combinations from 4 of 10 subjects with an acceptance window of 0.5 s before and 1 s after each action [37]. The changepoint quadratic detector produced HF-differences greater than 90% for 7 of 20 subject/action combinations from 5 of 10 subjects with the same acceptance window [37].

Feedback Experiments

The UM-DBI project has also performed real-time on-line feedback experiments with epilepsy surgery subjects [28]. Subjects performed one of the self-paced actions described above while the ECoG averages were viewed in real-time. When an electrode was found that recorded brain activity related to a particular action, the initially recorded ECoG was used to setup the feedback system. Under different protocols [28, 38], feedback training encouraged subjects to change their brain activity to improve BCI performance. When subjects were given feedback on the signal-to-noise ratio (SNR) [38], three of six subjects showed dramatic improvements in the SNR of their ECoG, with one subject showing corresponding improvement in off-line BCI accuracy from 79% hits and 22% false positives to 100% hits and 0% false positives. When subjects were given feedback on the correlation values used by the CCTM detection method, [28], one of three subjects showed improvements in online BCI accuracy from 90% hits with 44% false positives to 90% hits with 10% false positives. Note that on-line accuracy calculations differ due to timing constraints of the feedback system (see [28]).

fMRI Studies

One of the significant restrictions of current ECoG research is that researchers lack control over the placement of the electrodes. However, even if it were possible to choose electrode locations solely for research purposes, we do not yet have an a priori method for selecting optimal electrode locations. The UM-DBI project is investigating whether fMRI could be used to select locations for an ECoG-based BCI [39]. Accurate prediction of electrode location is an important issue, since implantation of an electrode strip through a burrhole would entail less risk than opening a large window in the skull to place an electrode grid. Some subjects who were part of the ECoG studies returned after epilepsy surgery to participate in an fMRI study while doing the same actions they performed during ECoG recording. Figure 5 shows the results from a subject who performed a pinch action during both ECoG and fMRI sessions. BCI detection of this action produced high HF-differences over several centimeters of the ECoG grid. Visual comparison of these areas of good detection with areas that were active on the fMRI during the pinch action show good general agreement. However, some good ECoG locations do not correspond to active fMRI areas, and a numerical comparison of individual locations shows some instances that could be problematic for using fMRI to select ECoG placement.

6.2.3 The University of Washington in St. Louis

In 2004, Leuthardt et al. reported the application of the frequency analysis algorithms in BCI2000 [40], developed as part of Wolpaw's EEG-based work, to ECoG [4]. Chapter "A Simple, Spectral-change Based, Electrocorticographic

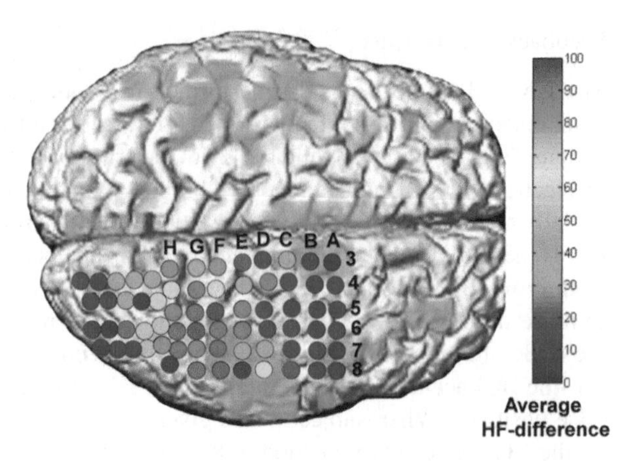

Fig. 5 ECoG detection and fMRI activation for a subject performing palmar pinch. Electrode location color indicates average HF-difference. Green regions were active on fMRI

Brain–Computer Interface" by Miller and Ojemann in this volume describes this work in detail, so only a brief description is given here. Working with epilepsy surgery patients over a 3–8 day period, Leuthardt et al. demonstrated the first BCI operation of cursor control using ECoG. Sensorimotor rhythms in ECoG related to specific motor tasks and motor imagery were mapped in real-time to configure the BCI. With a series of feedback sessions, four subjects were then trained to perform ECoG-based cursor control [4, 41]. Subjects controlled the vertical cursor speed while automatic horizontal movement limited trial duration to 2.1–6.8 s. Subjects were successful at a binary selection task in 74–100% of trials. This work showed that ECoG sensorimotor rhythms in the mu (8–12 Hz), beta (18–26 Hz), and gamma (> 30 Hz) frequency ranges could be used for cursor control after minimal training (in comparison with that required for EEG). This group has also shown that ECoG can be used to track the trajectory of arm movements [4, 42].

6.2.4 University of Wisconsin – Madison

Williams and Garrell at the University of Wisconsin have also started ECoG experiments using BCI2000 [7, 17] that used both auditory and motor imagery tasks for real-time cursor control to hit multiple targets. Subjects participate in 2–7 training sessions of 45 min in duration. Subjects first participate in a screening session during which they perform different auditory and motor imagery tasks which are analyzed off-line to identify tasks and electrodes with significantly different frequency components compared to a rest condition. These electrodes are then utilized in the on-line feedback experiments. Subjects control the vertical position of a cursor moving across the screen at a fixed rate, but the duration of a trial is not reported. Subjects attempt to guide the cursor to hit one of 2, 3, 4, 6, or 8 targets. Subjects were able to achieve significant accuracy in only a few sessions with typical performance on the 2-target task of about 70% of trials [17]. One subject achieved 100% accuracy at a 3 target task, and the same subject had 80, 69 and 84% accuracy on the 4, 6, and 8 target tasks respectively [17].

6.2.5 Tuebingen, Germany

Birbaumer's well-established EEG-based BCI work has led to an initial ECoG-based BCI study [43]. This work used off-line analysis of ECoG and compared it to off-line analysis of EEG recorded from different subjects under a slightly different paradigm. The algorithm performance on ECoG was comparable to that found for EEG results, although the maximum results for EEG were better than those for ECoG. However, the number of trials, sampling rates, and applied filtering (among other parameters) were different, making conclusions difficult. Of greater interest is a single ECoG recording experiment done with a completely paralyzed subject as part of a larger study comparing BCI detection algorithm function on brain activity from healthy subjects and subjects with complete paralysis [44]. The subject gave informed consent for surgical implantation of ECoG electrodes by using voluntarily control of mouth pH to communicate [44, 45]. However, the BCI detection algorithms in this study did not produce results greater than chance for any of the subjects who were completely paralyzed (4 with EEG electrodes and 1 with ECoG). See the Chapters "Brain–Computer Interface in Neurorehabilitation" and "Brain–Computer Interfaces for Communication and Control in Locked-in Patients" in this volume for further discussion of this issue.

6.2.6 University Hospital of Utrecht

Ramsey et al. [46] included ECoG analysis as a minor component (1 subject) of their fMRI study of working memory as a potential BCI input. They looked at the ECoG near the onset of a working memory load and reported the increased activity that occurred in averaged ECoG. Although they were able to show agreement between the fMRI and ECoG modalities, they did not do actual BCI detection experiments either online or offline. Further, the necessary use of working memory in any task for which BCI operation of technology was desired makes this signal source appear to be only practical for detecting concentration or perhaps intent to perform a task. As such, working memory detection could be used to gate the output of another BCI detection method, blocking control signals from the other detection method unless working memory was also engaged. While this might improve BCI function during no-control periods, by requiring concentration during BCI operation, it may also limit BCI use for recreational tasks such as low intensity channel surfing. Further, it could falsely activate BCI function during composition tasks when the user was not yet ready to start BCI typing.

6.2.7 The University of Michigan – Ann Arbor (Kipke)

In a recent study using microelectrodes in rats performing a motor learning task in Kipke's Neural Engineering Laboratory, ECoG from bone-screw electrodes was analyzed in addition to neural spike activity and local field potentials from microelectrodes [2]. This work found ECoG signals that occurred only during the learning period. Comparison of ECoG to local field potentials recorded from microelectrodes at different cortical depths revealed that ECoG includes brain activity evident at different cortical levels, including some frequencies that were only apparent on the deepest cortical levels.

6.2.8 University of Florida – Gainesville

A 2008 report applied principles developed using microelectrode recordings to ECoG recorded at 12 kHz during reaching and pointing tasks by 2 subjects [5]. On average, the correlation coefficients between actual movement trajectory and predicted trajectory were between 0.33 and 0.48 for different frequency bands, position coordinates and subjects. Different frequency bands were found to better correlate with different parts of the movement trajectory with higher frequencies (300 Hz–6 kHz) showing the best correlation with the y-coordinate. Specific instances of high correlation (above 0.90) indicate that specific tasks are better represented than others. This exploration of ECoG frequencies far outside those used for most studies has revealed a hitherto unknown information source and presents exciting opportunities for future work.

6.2.9 Albert-Ludwigs-University, Freiburg, Germany

Another 2008 report also applies BCI methods developed using microelectrode recordings to ECoG during arm movements [6]. Using low frequency components (below about 10 Hz) from 4-electrode subsets of the available electrodes, an average correlation coefficient of 0.43 was found for subjects with ECoG from electrodes over motor cortex containing minimal epileptic artifact. Higher frequency components were also found to correlate with movement trajectory, but their inclusion did not improve the overall prediction accuracy. The authors point out many areas for future work, including recording even broader frequency bands, the use of denser grids of smaller electrode, and provision of real-time feedback to the subjects.

7 Discussion

ECoG is now increasingly included in BCI research. Many BCI research groups pursuing other approaches are starting to incorporate or explore ECoG. The degree to which ECoG-based BCI research becomes a permanent part of the work of these research groups has yet to be seen. Of course, ECoG research is also ongoing for both basic research on brain function and for applications such as epilepsy surgery. A recent paper on real-time functional mapping of brain areas for epilepsy surgery [47] included both background and discussion on potential BCI applications of their work, perhaps indicating a generally increased awareness and acceptance of BCI applications.

While ECoG provides advantages over EEG and microelectrode signal sources, the knowledge developed through the pursuit of ECoG-based BCIs should be applicable to other signal types. Signal processing methods that apply to ECoG may also be useful for EEG. Further, methods developed and tested using ECoG may be applicable to smaller electrodes recording local field potentials, and perhaps eventually to microelectrodes. As shown by the ECoG work in the last decade, methods from both EEG and single cell recordings have now been successfully applied to ECoG signals. Therefore, it is reasonable to suppose that methods developed using ECoG could also be applied to other types of recorded brain activity.

References

1. P.R. Kennedy, M.T. Kirby, M.M. Moore, B. King, A. Mallory, Computer control using human intracortical local field potentials. IEEE Trans Neural Syst Rehabil Eng, 12, 339–344, (2004).
2. T.C. Marzullo, J.R. Dudley, C.R. Miller, L. Trejo, D.R. Kipke, Spikes, local field potentials, and electrocorticogram characterization during motor learning in rats for brain machine interface tasks. Conf Proc IEEE Eng Med Biol Soc, 1, 429–431, (2005).
3. S.P. Levine, J.E. Huggins, S.L. BeMent, et al., Identification of electrocorticogram patterns as the basis for a direct brain interface. J Clin Neurophysiol, 16, 439–447, (1999).
4. E.C. Leuthardt, G. Schalk, J.R. Wolpaw, J.G. Ojemann, D.W. Moran, A brain-computer interface using electrocorticographic signals in humans. J Neural Eng, 1, 63–71, (2004).
5. J.C. Sanchez, A. Gunduz, P.R. Carney, J.C. Principe, Extraction and localization of mesoscopic motor control signals for human ECoG neuroprosthetics. J Neurosci Methods, 167, 63–81, (2008).
6. T. Pistohl, T. Ball, A. Schulze-Bonhage, A. Aertsen, C. Mehring, Prediction of arm movement trajectories from ECoG-recordings in humans. J Neurosci Methods, 167, 105–114, (2008).
7. J.A. Wilson, E.A. Felton, P.C. Garell, G. Schalk, J.C. Williams, ECoG factors underlying multimodal control of a brain-computer interface. IEEE Trans Neural Syst Rehabil Eng, 14, 246–250, (2006).
8. E.E. Sutter, The brain response interface: Communication through visually-induced electrical brain responses. J Microcomput Appl, 15, 31–45, (1992).
9. V. Salanova, H.H. Morris 3rd, P.C. Van Ness, H. Luders, D. Dinner, E. Wyllie, Comparison of scalp electroencephalogram with subdural electrocorticogram recordings and functional mapping in frontal lobe epilepsy. Arch Neurol, 50, 294–299, (1993).
10. B. Graimann, G. Townsend, J.E. Huggins, S.P. Levine, and G. Pfurtscheller, A Comparison between using ECoG and EEG for direct brain communication. IFMBE proceedings EMBEC05 3rd European medical & biological engineering conference, IFMBE European Conference on Biomedical Engineering, vol. 11, 2005, Prague, Czech Republic, CD.
11. F. Popescu, S. Fazli, Y. Badower, B. Blankertz, K.R. Muller, Single trial classification of motor imagination using 6 dry EEG electrodes. PLoS ONE, 2, e637, (2007).
12. W.J. Heetderks, A.B. Schwartz, Command-control signals from the neural activity of motor cortical cells: Joy-stick control. Proc RESNA '95, 15, 664–666, (1995).
13. P.R. Kennedy, R.A. Bakay, M.M. Moore, K. Adams, J. Goldwaithe, Direct control of a computer from the human central nervous system. IEEE Trans Rehabil Eng, 8, 198–202, (2000).
14. L.R. Hochberg, M.D. Serruya, G.M. Friehs, et al., Neuronal ensemble control of prosthetic devices by a human with tetraplegia.see comment. Nature, 442, 164–171, (2006).
15. E. Margalit, J.D. Weiland, R.E. Clatterbuck, et al., Visual and electrical evoked response recorded from subdural electrodes implanted above the visual cortex in normal dogs under two methods of anesthesia. J Neurosci Methods, 123, 129–137, (2003).
16. E.M. Schmidt, Single neuron recording from motor cortex as a possible source of signals for control of external devices. Ann Biomed Eng, 8, 339–349, (1980).
17. E.A. Felton, J.A. Wilson, J.C. Williams, P.C. Garell, Electrocorticographically controlled brain-computer interfaces using motor and sensory imagery in patients with temporary subdural electrode implants. Report of four cases. J Neurosurg, 106, 495–500, (2007).

18. J.D. Frost Jr, Comparison of intracellular potentials and ECoG activity in isolated cerebral cortex. Electroencephalogr Clin Neurophysiol, 23, 89–90, (1967).
19. P.R. Kennedy, The cone electrode: A long-term electrode that records from neurites grown onto its recording surface. J Neurosci Methods, 29, 181–193, (1989).
20. J.J. Vidal, Toward direct brain-computer communication. Annu Rev Biophys Bioeng, 2, 157–180, (1973).
21. J.R. Wolpaw, D.J. McFarland, A.T. Cacace, Preliminary studies for a direct brain-to-computer parallel interface. In: IBM Technical Symposium. Behav Res Meth, 11–20, (1986).
22. G.S. Brindley, M.D. Craggs, The electrical activity in the motor cortex that accompanies voluntary movement. J Physiol, 223, 28P–29P, (1972).
23. M.D. Craggs, Cortical control of motor prostheses: Using the cord-transected baboon as the primate model for human paraplegia. Adv Neurol, 10, 91–101, (1975).
24. E.E. Sutter, B. Pevehouse, N. Barbaro, Intra-Cranial electrodes for communication and environmental control. Technol Persons w Disabil, 4, 11–20, (1989).
25. J.E. Huggins, S.P. Levine, R. Kushwaha, S. BeMent, L.A. Schuh, D.A. Ross, Identification of cortical signal patterns related to human tongue protrusion. Proc RESNA '95, 15, 670–672, (1995).
26. J.E. Huggins, S.P. Levine, S.L. BeMent, et al., Detection of event-related potentials for development of a direct brain interface. J Clin Neurophysiol, 16, 448–455, (1999).
27. S.P. Levine, J.E. Huggins, S.L. BeMent, et al., A direct brain interface based on event-related potentials. IEEE Trans Rehabil Eng, 8, 180–185, (2000).
28. J. Vaideeswaran, J.E. Huggins, S.P. Levine, R.K. Kushwaha, S.L. BeMent, D.N. Minecan, L.A. Schuh, O. Sagher, Feedback experiments to improve the detection of event-related potentials in electrocorticogram signals. Proc RESNA 2003, 23, Electronic publication, (2003).
29. B. Graimann, J.E. Huggins, S.P. Levine, G. Pfurtscheller, Visualization of significant ERD/ERS patterns in multichannel EEG and ECoG data, Clin. Neurophysiol. 113, 43–47, (2002).
30. B. Graimann, J.E. Huggins, S.P. Levine, G. Pfurtscheller, Toward a direct brain interface based on human subdural recordings and wavelet-packet analysis. IEEE Trans Biomed Eng, 51, 954–962, (2004).
31. U.H. Balbale, J.E. Huggins, S.L. BeMent, S.P. Levine, Multi-channel analysis of human event-related cortical potentials for the development of a direct brain interface. Engineering in Medicine and Biology, 1999. 21st Annual Conference and the 1999 Annual Fall Meeting of the Biomedical Engineering Society. BMES/EMBS Conference, 1999. Proceedings of the First Joint 1 447 vol.1, Atlanta, GA, 13–16 Oct (1999).
32. Y. Benjamini, Y. Hochberg, Controlling the false discovery rate: A practical and powerful approach to muiltiple testing. J. R. Statist Soc, 57, 289–300, (1995).
33. S.P. Levine, J.E. Huggins, S. BeMent, L.A. Schuh, R. Kushwaha, D.A. Ross, M.M. Rohde, Intracranial detection of movement-related potentials for operation of a direct brain interface. Proc IEEE EMB Conf 2(7), 1–3, (1996).
34. J.E. Huggins, S.P. Levine, S.L. BeMent, R.K. Kushwaha, L.A. Schuh, M.M. Rohde, Detection of event-related potentials as the basis for a direct brain interface. Proc RESNA '96, 16, 489–491, (1996).
35. B. Graimann, J.E. Huggins, A. Schlogl, S.P. Levine, G. Pfurtscheller, Detection of movement-related desynchronization patterns in ongoing single-channel electrocorticogram. IEEE Trans Neural Syst Rehabil Eng, 11, 276–281, (2003).
36. J.A. Fessler, S.Y. Chun, J.E. Huggins, S.P. Levine, Detection of event-related spectral changes in electrocorticograms. Neural Eng, 2005. Conf Proc 2nd Int'l IEEE EMBS Conf, 269–272, Arlington, VA, 16–19 Mar (2005). doi:10.1109/CNE.2005.1419609
37. J.E. Huggins, V. Solo, S.Y. Chun, et al., Electrocorticogram event detection using methods based on a two-covariance model, (unpublished).
38. M.M. Rohde, Voluntary control of cortical event related potentials, Ph.D. Dissertation, University of Michigan, Published Ann Arbor, MI, (2000).

39. J.E. Huggins, R.C. Welsh, V. Swaminathan, et al., fMRI for the prediction of direct brain interface recording location. Biomed Technik, 49, 5–10, (2004).
40. G. Schalk, D.J. McFarland, T. Hinterberger, N. Birbaumer, J.R. Wolpaw, BCI2000: A general-purpose brain-computer interface (BCI) system. IEEE Trans Biomed Eng, 51, 1034–1043, (2004).
41. E.C. Leuthardt, K.J. Miller, G. Schalk, R.P. Rao, J.G. Ojemann, Electrocorticography-based brain computer interface – the seattle experience. IEEE Trans Neural Syst Rehabil Eng, 14, 194–198, (2006).
42. G. Schalk, J. Kubanek, K.J. Miller, et al., Decoding two-dimensional movement trajectories using electrocorticographic signals in humans. J Neural Eng, 4, 264–275, (2007).
43. T.N. Lal, T. Hinterberger, G. Widman, M. Schroder, N.J. Hill, W. Rosenstiel, C.E. Elger, B. Scholkopf, N. Birbaumer, Methods towards invasive human brain computer interfaces. In: K. Saul, Y. Weiss, and L. Bottou (Eds.), *Advances in neural information processing systems*, MIT Press, Cambridge, MA, pp. 737–744, (2005).
44. N.J. Hill, T.N. Lal, M. Schroder, et al., Classifying EEG and ECoG signals without subject training for fast BCI implementation: Comparison of nonparalyzed and completely paralyzed subjects. IEEE Trans Neural Syst Rehabil Eng, 14, 183–186, (2006).
45. B. Wilhelm, M. Jordan, N. Birbaumer, Communication in locked-in syndrome: Effects of imagery on salivary pH. Neurology, 67, 534–535, (2006).
46. N.F. Ramsey, M.P. van de Heuvel, K.H. Kho, F.S.S. Leijten, Towards human BCI applications based on cognitive brain systems: An investigation of neural signals recorded from the dorsolateral prefrontal cortex. IEEE Trans Neural Syst Rehabil Eng, 14, 214–217, (2006).
47. J.P. Lachaux, K. Jerbi, O. Bertrand, et al., A blueprint for real-time functional mapping via human intracranial recordings. PLoS ONE, 2, e1094, (2007).

A Simple, Spectral-Change Based, Electrocorticographic Brain–Computer Interface

Kai J. Miller and Jeffrey G. Ojemann

1 Introduction

A brain–computer interface (BCI) requires a strong, reliable signal for effective implementation. A wide range of real-time electrical signals have been used for BCI, ranging from scalp recorded electroencephalography (EEG) (see, for example, [1, 2]) to single neuron recordings (see, for example, [3, 4]. Electrocorticography (ECoG) is an intermediate measure, and refers to the recordings obtained directly from the surface of the brain [5]. Like EEG, ECoG represents a population measure, the electrical potential that results from the sum of the local field potentials resulting from 100,000 s of neurons under a given electrode. However, ECoG is a stronger signal and is not susceptible to the artifacts from skin and muscle activity that can plague EEG recordings. ECoG and EEG also differ in that the phenomena they measure encompass fundamentally different scales. Because ECoG electrodes lie on the cortical surface, and because the dipole fields [7] that produce the cortical potentials fall off rapidly ($V(r) \sim r^{-2}$), the ECoG fundamentally reflects more local processes.

Currently, ECoG takes place in the context of clinical recording for the treatment of epilepsy. After implantation, patients recover in the hospital while they wait to have a seizure. Often, that requires a week or longer of observation, during which time patients may chose to participate in experiments relevant to using ECoG to drive BCI. Recently, researchers have used the spectral changes on the cortical surface of these patients to provide feedback, creating robust BCIs, allowing individuals to control a cursor on a computer screen in a matter of minutes [8–12]. This chapter discusses the important elements in the construction of these ECoG based BCIs: signal acquisition, feature selection, feedback, and learning.

K.J. Miller (✉)
Department of Physics, Neurobiology, and Behavior, University of Washington, Seattle, WA 98195, USA
e-mail: kjmiller@u.washington.edu

B. Graimann et al. (eds.), *Brain–Computer Interfaces*, The Frontiers Collection,
DOI 10.1007/978-3-642-02091-9_14,

2 Signal Acquisition

ECoG is available from frequently performed procedures in patients suffering from medically intractable epilepsy (Fig. 1). Such patients undergo elective placement of electrodes on the surface of the brain when the seizure localization is not evident from non-invasive studies. During an inter-operative waiting period, patients stay in the hospital and are monitored for seizure activity. When they have a seizure, neurologists retrace the potential recordings from each electrode, isolating the one in which electrical evidence of seizure first appeared. The cortex beneath this electrode site is then resected. These electrodes are also placed, in some situations, to localize function such as movement or language prior to a neurosurgical resection. The same electrodes can record ECoG and stimulate to evoke disruption in the function of the underlying cortex. The implanted electrode arrays are typically only those which would be placed for diagnostic clinical purposes. Most often, these are ~2.5 mm in diameter have a spacing of 1-cm from center-to-center (Fig. 2). While this is somewhat coarse, it is fine enough to resolve individual finger representation and may be sufficient to extract many independent control signals simultaneously [13]. At some institutions, preliminary results are coming out using smaller electrodes and higher resolution arrays [14], and it may become apparent that finer resolution grids identify independent function and intention better than the current clinical standard. Intra-operative photographs, showing the arrays in-situ can be useful for identifying which electrodes are on gyri, sulci, and vasculature, and also which are near known cortical landmarks.

There are several necessary components of the electrocorticographic BCI experimental setting, as illustrated in Fig. 1. **(a)** *The experimenter.* While this element may seem trivial or an afterthought, the individual who interacts with the subject, in the clinical setting, must have an agreeable disposition. There are several reasons for this. The first is that the subjects are patients in an extremely tenuous position, and it is important to encourage and reinforce them with genuine compassion. Not

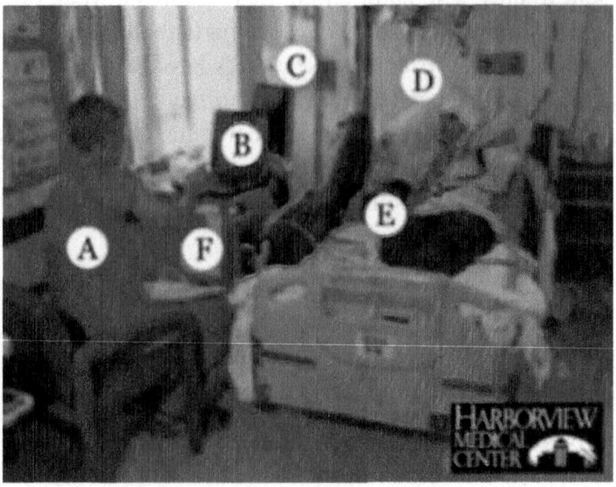

Fig. 1 The necessary components of the electrocorticographic BCI experimental setting (see text for details). (**a**) The experimenter; (**b**) A central computer; (**d**) The subject; (**e**) Signal splitters; (**f**) Amplifiers

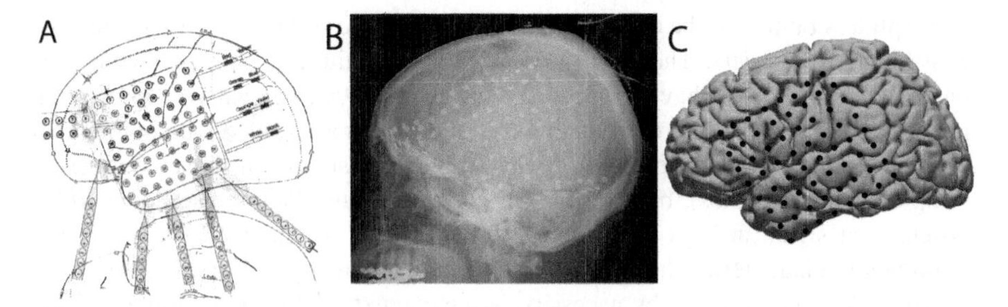

Fig. 2 *Necessary elements for co-registration of electrodes and plotting of data on template cortices.* (**a**) Clinical schematic; (**b**) Diagnostic imaging; (**c**) Cortical electrode position reconstruction

only is this a kind thing to do, but it makes the difference between 10 min and 10 h of experimental recording and participation. The second is that the hospital environment requires constant interaction with physicians, nurses, and technicians, and all of these individuals have responsibilities that take priority over the experimental process at any time. It is important to cultivate and maintain a sympathetic relationship with these individuals. The last reason is that the hospital room is not a controlled environment. There is non-stationary contamination, a clinical recording system to be managed in parallel, and constant interruption from a myriad of sources. The researcher must be able to maintain an even disposition and be able to constantly troubleshoot. (**b**) *A central computer.* This computer will be responsible for recording and processing the streaming amplified potentials from the electrode array, translating the processed signal into a control signal, and displaying the control signal using an interface program. The computer must have a large amount of memory, to buffer the incoming data stream, a fast processor to perform signal processing in real-time and adequate hardware to present interface stimuli with precision. Therefore, it is important to have as powerful a system as possible, while remaining compact enough to be part of a portable system that can easily be brought in and out of the hospital room. An important element not shown in the picture is the software which reads the incoming datastream, computes the power spectral density changes, and uses these changes to dynamically change the visual display of an interface paradigm. We use the BCI2000 program [15] to do all of these things simultaneously (see Chapter "Using BCI2000 in BCI Research" for details about BCI2000). (**c**) *A second monitor.* It is a good idea to have a second monitor for stimulus presentation. It should be compact with good resolution. (**d**) *The subject.* It is important to make sure that the subject is in a comfortable, relaxed position, not just to be nice, but also because an uncomfortable subject will have extraneous sensorimotor phenomena in the cortex and also will not be able to focus on the task. (**e**) *Signal splitters.* If a second set of amplifiers (experimental or clinical) is being used in parallel with the clinical ones used for video monitoring, the signal will be split after leaving the scalp, and before the clinical amplifier jack-box. The ground must be split as well, and be common between both amplifiers, or else there is the potential for current to be passed between the two grounds. Several clinical systems have splitters built in to the clinical wire ribbons, and these should be used whenever possible. **(f)** *Amplifiers*. These will vary widely by institution, and, also depending on the institution, will have to have, for instance, FDA approval (USA) or a CE marking (EU) (the process of obtaining this approval is associated with both higher cost and lower quality amplifiers). Many amplifier systems will have constrained sample rates (A/D rates), built in filtering properties, and large noise floors which obscure the signal at high frequencies. Regardless of which system is used, it is important to characterize the amplifiers independently using a function generator.

The ECoG recording is, by necessity, in the context of clinical amplification and recording, so the experimental recording must take place in the context of clinical amplification with commercially available amplifiers (eg, XLTEK, Synamps, Guger Technologies, Grass). Most clinically relevant EEG findings are detected visually and classically the information explored is between 3 and 40 Hz, so the settings on the clinical amplifiers may be adequate to obtain clinical information, but not for research purposes. Recent advances have suggested that faster frequencies may be clinically relevant so many newer systems include higher sampling rate (at least 1 kHz) as an option to allow for measurement of signals of 200 Hz or higher, but this varies by institution, and the clinical recording settings will vary even within institutions, depending upon the clinical and technical staff managing the patient. Experimentalists must obtain either the clinically amplified signal, or split the signal and amplify it separately. Using the clinical signal has the advantage that less hardware is involved, and that there are no potential complications because of the dual-amplification process. Such complications include artifact/noise introduction from one system to the other, currents between separate grounds if the two do not share a common ground. Splitting the signal has the advantage that the experimenter can use higher fidelity amplifiers and set the amplification parameters at will, rather than having to use the clinical parameters, which typically sample at a lower frequency than one would like, and often have built in filtering properties which limit the usable frequency range. The ground chosen, which must be the same as the clinical ground to avoid complication, will typically be from the surface of the scalp. Most amplifiers will have a built in choice of reference, which each electrode in the array will be measured with respect to. These may also be from the scalp, as they often are clinically, or they may be from an intra-cranial electrode.

The experimenter will often find it useful to re-reference the electrode array in one of several ways. Each electrode may be re-referenced with respect to a single electrode from within the array, chosen because it is relatively "dormant," each may be re-referenced to a global linear combination of electrodes from the entire array, or each may be referenced to one or more nearest neighbors. Re-referencing with respect to a single electrode is useful when the one in the experimental/clinical montage is sub-optimal (noisy, varies with task, etc), but it means that the experimenter has introduced an assumption about which electrode is, in fact, appropriate. The simplest global referencing is a common average re-reference: the average of all electrodes is subtracted from each electrode. The advantage of this is that it is generic (unbiased, not tied to an assumption), and it will get rid of common-mode phenomena. One must be careful that there are not any electrodes that are broken, or have extremely large contamination, or every electrode will be contaminated by the re-referencing process. Local re-referencing may also be performed, such as subtracting the average of nearest-neighbors (Laplacian), which ensures that the potential changes seen in any electrode are spatially localized. One may also re-reference in a pair-wise fashion, producing bipolar channels which are extremely local, but phenomena cannot be tied to a specific electrode from the pair. This re-referencing can also be interpreted as applying a spatial filter. Please see Chapter "Digital Signal Processing and Machine Learning" for details about spatial filters.

In order to appropriately understand both the experimental context and the connection between the structure of the brain and signal processing findings, it is necessary to co-register electrode locations to the brain surface. The simplest method is to use x-rays, and plot data to template cortices (as illustrated in Fig. 2). A clinical schematic will typically be obtained from the surgeon. The position of each electrode may then be correlated with potential recordings from each amplifier channel. Different diagnostic imaging may be obtained from the course of the clinical care, or through specially obtained high-fidelity experimental imaging. The level and quality of this may be highly variable across time and institutions, from x-ray only, to high-fidelity pre-operative magnetic resonance imaging (MRI) and post-operative fine-cut computed tomography (CT). The clinical schemata and diagnostic imaging may be used in concert to estimate electrode positions, recreate cortical locations, and plot activity and analyses. The most simple method for doing this, using x-rays, is the freely-available LOC package [16], although there is the promise of more sophisticated methodology for doing this, when higher fidelity diagnostic imaging is obtained.

Choosing a sampling frequency is important – there is often a trade-off between signal fidelity and the practical issues of manageable data sizes and hardware limitations. Whatever sampling rate is chosen, one should be sure to have high signal fidelity up to at least 150 Hz. This means that the sampling rate should be above 300 Hz, because of a law called the Nyquist Law (aka Nyquist Theorem), which says that you must record data at (at least) twice the frequency of the highest wave you wish to measure. The sampling rate may have to be higher if the amplifiers used have built in filtering properties. The reason for this is that there is a behavioral split in the power spectrum (see Fig. 3) which can be as high as 60 Hz [17]. In order to capture the spatially focal high frequency change, one must have large bandwidth above this behavioral split. Some characteristic properties of motor and imagery associated spectra are shown in Fig. 3. There is a decrease in the power at lower frequencies with activity, and an increase in the power at higher frequencies [6, 18–20]. The intersection in the spectrum is dubbed the "primary junction" (J_0). A recent study involving hand and tongue movement [17] found that, for hand movement, $J_0 = 48 +/- 9$ Hz (mean+/– SD) (range 32–57 Hz), and, for tongue movement, $J_0 = 40 +/- 8$ Hz (range 26–48 Hz). Rather than this indicating two phenomena, a "desynchronization" at low frequencies, and a "synchronization" at high frequencies, as some have proposed [19, 21], this might instead reflect the superposition

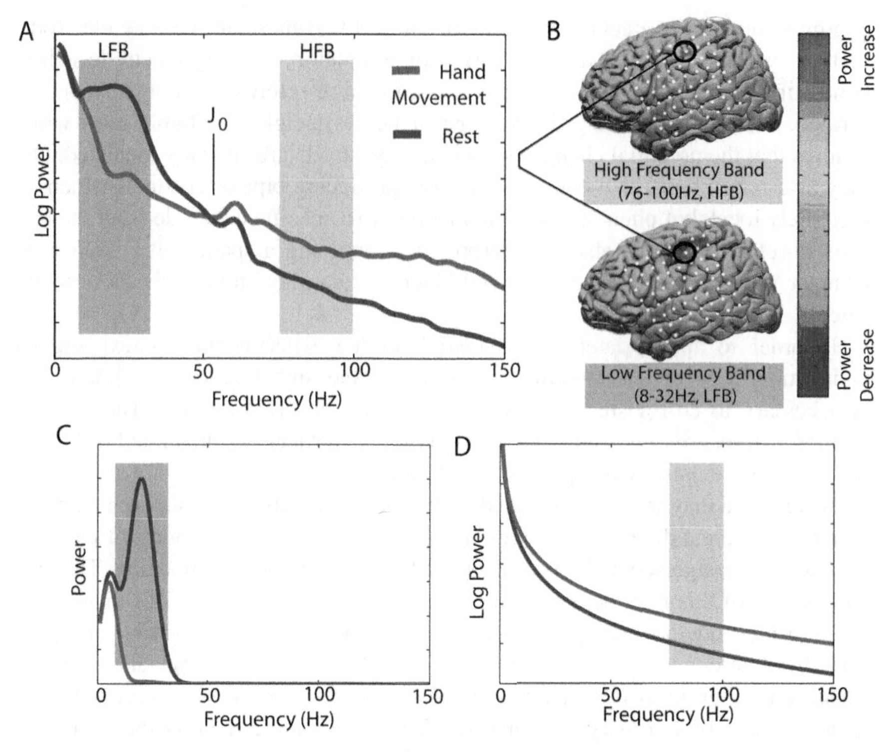

Fig. 3 Characteristic changes in the power spectrum with activity [6]. (**a**) Example of a characteristic spectral change with movement; (**b**) Demonstration of different spatial extent of changes in these high and low frequency ranges on the cortical surface. These changes are related to (**c**) Decoherence of discrete peaks in the power spectrum with movement (ERD), and (**d**) Power-law like broadband power spectrum that shifts upward with movement. The "behavioral split", J_0, where the peaked phenomena and the broadband phenomenon intersect represents a natural partition between features

of the phenomena [6, 13], described in Fig. 3c,d, that produce an intersection (J_0) in the power spectrum in the classic gamma range. Choices for feedback features should explicitly avoid J_0. If one examines a low frequency range (8–32 Hz, LFB), and a high frequency range (76–100 Hz, HFB), one finds that the spatial distribution of the LFB/HFB change is broad/narrow, and corresponds to a decrease/increase in power. These have been demonstrated reflect the classic event-related desynchronization at lower frequencies [6, 20, 22], and shifts in a broadband, power-law like, process. This broadband change is most easily observed at high frequencies because it is masked by peaked ERD at low frequencies (Fig. 3). Recent findings have hypothesized and demonstrated that this power-law like process and the ERD may be decoupled from each other and that the power-law like process [6, 13, 23] may be used as an extremely good correlate of local activity, with extremely high (10–15 ms) temporal precision [13].

3 Feature Selection

In order to implement a BCI paradigm, a specific signal feature must be chosen. This will need to be a feature that can be determined in a computationally rapid fashion. Second, the feature must be translated into a specific output. The choice of signal feature should be an empiric one. There are two complementary approaches to choosing a BCI feature. One approach is to start with a strictly defined task, such as hand movement, and look for a particular feature at the signal change associated with this task. Then, the most reliable signal is identified and used to run a BCI. Another approach is to choose a signal that is less well characterized behaviorally and then, over time, to allow the subject to learn to control the feature by exploiting feedback, and then control the BCI. In a dramatic example of the latter, it was found that the spike rate from an arbitrary neuron that grew into a glass cone could be trained to run BCI [4], without necessary a priori knowledge about the preferred behavioral tuning of the given neuron.

The most straightforward approach is a motor imagery-based, strictly defined, task-related change for feature control. In order to identify appropriate simple features to couple to device control, a set of screening tasks is performed. In these screening tasks, the subject is cued to move or imagine (kinesthetically) moving a given body part for several seconds and then cued to rest for several seconds [6]. Repetitive movement has been found to be useful in generating robust change because cortical activity during tonic contraction is quickly attenuated [19, 20]. Different movement types should be interleaved, so that the subject does not anticipate the onset of each movement cue. Of course, there are multiple forms of motor imagery. One can imagine what the movement looks like, one can imagine what the movement feels like, and one can imagine the action of making the muscular contractions which produce the movement (kinesthetic) [24]. It was demonstrated by Neuper, et al. [24] that kinesthetic imagery produces the most robust cortical spectral change, and, accordingly, we and others have used kinesthetic motor imagery as the paired modality for device control. In order to establish that the control signal is truly imagery, experimenters should exclude, by surface EMG and other methods, subtle motor movement as the underlying source of the spectral change. In the screening task, thirty to forty such movement/imagery cues for each movement/imagery type should be recorded in order to obtain robust statistics (electrode-frequency band shifts with significance of order $p<<0.01$, after Bonferroni correction, are obtained with 30–40 cues). The power spectral density during each of these cue and rest periods can be calculated for each electrode. “Feature maps” of statistically significant changes at each frequency, in each electrode, can be calculated by comparing rest and movement cues. A simple prescription for doing this is shown step-wise in Fig. 4. Each electrode is treated independently, although generally it is a good idea to re-reference each electrode by subtracting the average time series, canceling the effect of the reference from the time series of each electrode (common average re-referencing) [6]. The potential is measured during a sequence of movements performed following visual cues. Here, the cues are for hand and tongue movement in a screening task, with rest interleaved, but the process can also be used for online

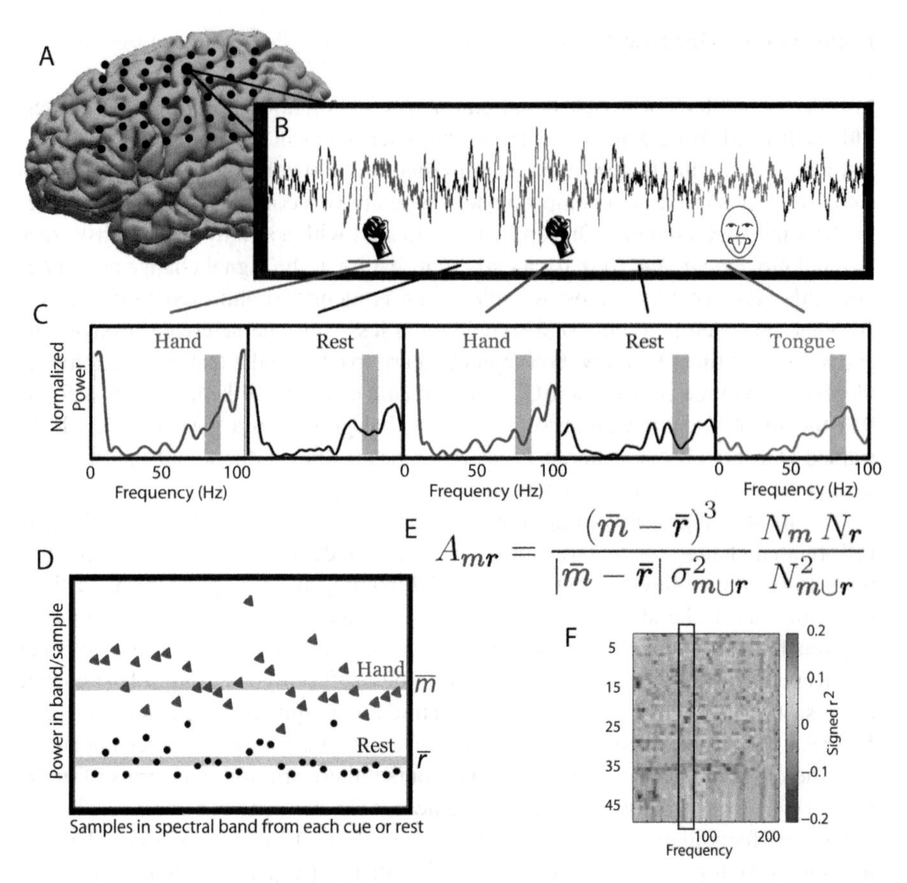

Fig. 4 Identifying areas of activity during simple tasks using cross-correlation: Each electrode is treated independently (**a**), and the potential is measured during a sequence of cues (**b**). The normalized power spectral density (PSD) is then calculated for each cue (or target) period (**c**). Samples of band-limited normalized power (**d**) are used to calculate activation, A_{mr} (**e**). "Feature maps" (**f**) can tell us about which electrode-frequency range combinations discriminate between cues

data, comparing periods where different targets were on the screen. The normalized power spectral density (PSD) can be calculated for each cue (or target) period. This is done by calculating the raw power spectral density for each period, and then dividing each by the average. A sample of the PSD of a specific range for each cue is used to compare cues of different types (one is shown here as the shaded bar in the back of each). Figure 4 shows examples of band-limited PSDs for cues of different types to illustrate the difference in the distribution of hand movement samples and rest samples. Activation (A_{mr}), quantified using the square of the cross-correlation (r^2), is signed to reflect whether the significance represents an increase or a decrease in the band-limited PSD for movement with respect to rest (or, in the case of a target task, whether one target represents an increase or decrease in power with respect to the other target). What this metric tells us is how much of the variation in the joint data set $\sigma^2_{m\cup r}$ can be accounted for by the fact that sub-distributions of movement (m) and rest (r) periods might have different means, $\bar{m}$ and $\bar{r}$ (N_m, N_r, and are the number of samples of type m, r, respectively, and $N_{m\cup r} = N_m + N_r$). "Feature maps" (Fig. 4f) can tell us about which electrode-frequency combinations discriminate between cues. We can calculate A_{mr} for each electrode, frequency band combination, to create feature maps of discriminative potential. When performed on a screening task with actual movements (overt) or imagined movements (covert), we can identify specific electrode-frequency power features as candidates for feedback.

Several previous findings [6, 18–20] have shown a consistent decrease in power at low frequencies, and a characteristic increase in power at high frequencies during movement when compared with rest (Fig. 3). These changes in the cortical spectrum may be decoupled into distinct phenomena [13]. At low frequencies, there is a band limited spectral peak which decreases with activity, consistent with event-related desynchronization (ERD). At high frequencies, a broad, power-law like increase in power may be observed, which is highly correlated with very local cortical activity. This functional change has been denoted the "χ-band" or "χ-index" when explicitly targeting this broad spectral feature [13, 25], and "high-γ" when seen as a band-specific increase in power at high frequencies [19, 26, 27]. As shown in Fig. 4, it is often convenient to choose a low frequency band to capture ERD changes (α/β) in the classic EEG range, and a high frequency band to capture broad spectral changes [6]. Because this high frequency change is more specific for local cortical activity, they are often the best choice for BCI control signals [28].

As with EEG, one feature-driven approach is to look across channels and frequency bands to obtain reliable features for feedback [2, 8–10]. This can be done manually, selecting spectral features from intuitive cortical areas, such as sensorimotor cortex. It can also be done using naïve, blind-source deconvolution and machine learning techniques [21, 29–34]. Sophisticated recombination techniques, optimized for an offline screening task, face the potential confound that the resulting mapping is not intuitive for subject control, particularly if the distribution of cortical spectral change is different for screening than it is for feedback studies (which it can be). Simple features, in contrast, may be employing only a fraction of the potential signal for the feedback task, and also may suffer because the simple feature chosen is not the best out of a family of potential simple features. Our approach, which we describe in detail here, has been to keep feature selection as simple and straightforward as possible, and then examine how brain dynamics change, with feedback, over time. Future strategies will have to take both approaches into account.

Adaptive feature techniques, which dynamically change the parameterization between brain signal and feedback, represent another approach, where the machine iteratively "learns" the signals that the subject is attempting to use during the BCI task. The potential advantage of such adaptive techniques is that they might be robust against non-stationarity in the distribution of cortical spectral change and compensate for shifts in the signal. The disadvantage is that a subject may be trying to adapt the signal at least as fast as the machine algorithm, and we have at times found a negative feedback between these processes. If the brain dynamics are trying to converge on given parameterization, and the adaptive algorithm is continuously changing the parameterization, there is the potential that no stable interface can be converged upon. Recent ECoG studies, however, have demonstrated that, for a simple ECoG BCI, stable control can be maintained for many consecutive days without any change in the feedback feature parameters [35].

4 Feedback

The feedback process is illustrated in Fig. 5. The potential, $V(t)$, is measured from each electrode with respect to a ground, and amplified. The ground may come from the scalp, or from an internal ECoG electrode, but is dictated by the primary clinical system because of the necessity for a common ground between the two amplifiers. These potentials are then converted to an estimate of the power, in a specific range, at each point in time $P(t) = f(V(t))$. The function $f(t)$ will first use a window in time to calculate a power spectrum – a longer window will produce a better estimate of the spectrum, a shorter window will allow for a higher effective bit-rate of the BCI, and the optimal window size will represent a tradeoff between the two (we typically use a quarter-second window). The power $P(t)$ will be chosen from a confined frequency range from the spectrum. It is important that the window length be significantly longer than the time period of the lowest frequency, so that the lowest frequency used may be estimated with some accuracy. Papers that are submitted, and forget to do this, are rejected (really). The spectral estimation method used may be a Fourier transform technique, or may be a parameterized estimation technique, like the autoregressive model method. Parameterized techniques have the potential advantage of better spectral estimation using shorter time windows, but the drawback that it can be hard to know how many parameters to use (model order), and that different model orders can produce widely different results. We couple the power,

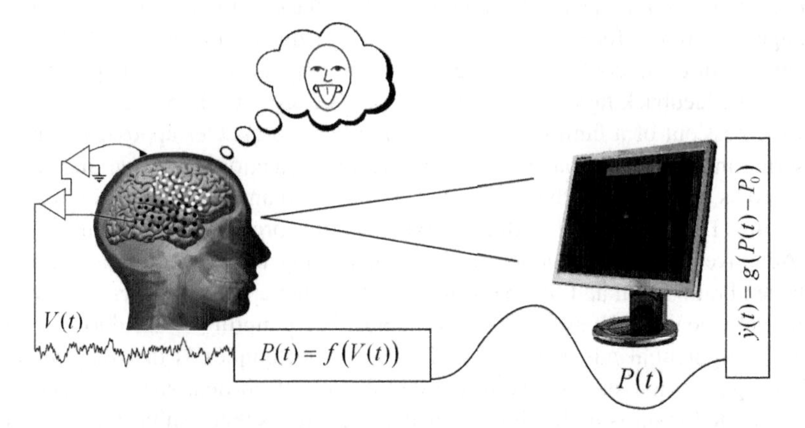

Fig. 5 Schematic representation of the closed-loop brain–computer interface that we use

$P(t)$, to the movement of a cursor in a given direction, according to the simple linear relation $\dot{y} = g(P(t) - P_0)$, where P_0, is a power level somewhere between movement imagery and rest, determined from a screening task (as in Fig. 4). This difference is then coupled to the movement of a cursor, with speed determined by the parameter g, which can be adjusted in the beginning of the task to be comfortable to the subject. The subject views changes in the cursor trajectory, and the subject then modifies the nature of the imagery to cause the cursor to move toward their target choice. After 8–10 min, in the 1-D task, subjects will abstract away from the imagery and are able to imagine the movement of the cursor itself and produce appropriate spectral changes.

The power in this identified feature, P_0, is then linked to the velocity, $\dot{y}$, of a cursor on the computer screen, using the simple translation algorithm $\dot{y} = g(P(t) - P_0)$, where P_0 is a power level somewhere between movement imagery and rest, and g is a velocity parameter. This difference is then coupled to the move-

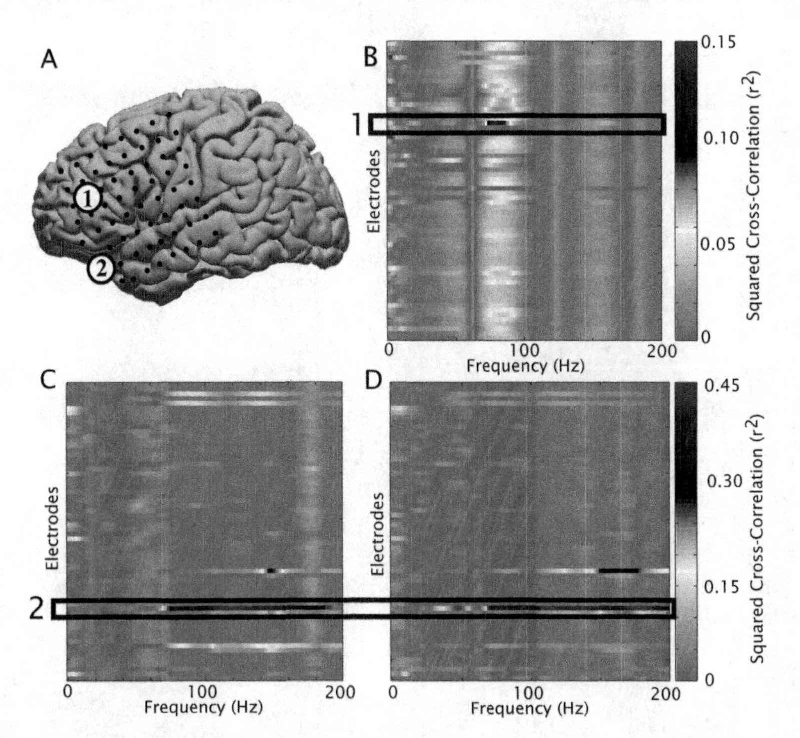

Fig. 6 Feature reassessment after feedback during a speech imagery task. (**a**) Two different feedback sites. An initial feedback site was chosen at site "1", and, upon reassessment, a secondary feedback site was chosen at site "2". (**b**) The feature map from a speech imagery screen identified a frontal site for control (labeled "1" in (a)), which was coupled to a cursor control task. The subject was not able to accurately control the cursor in the task, obtaining only 45% correct. (**c**) A feature map from this unsuccessful feedback run demonstrated that, while there was no significant difference at the feedback site, there was at a different site (labeled "2" in (A)). (**d**) The feature map following feature reassessment. The subject was able to rapidly attain 100% target accuracy with the new feature, and the most significant change was in the reassessed electrode

ment of a cursor, with speed determined by the parameter g, which can be adjusted in the beginning of the task to be comfortable to the subject. The subject is instructed to adjust the position of the cursor so that it hits a target (Fig. 5). Modification of the imagery behavior is allowed, even encouraged, to maximize target accuracy.

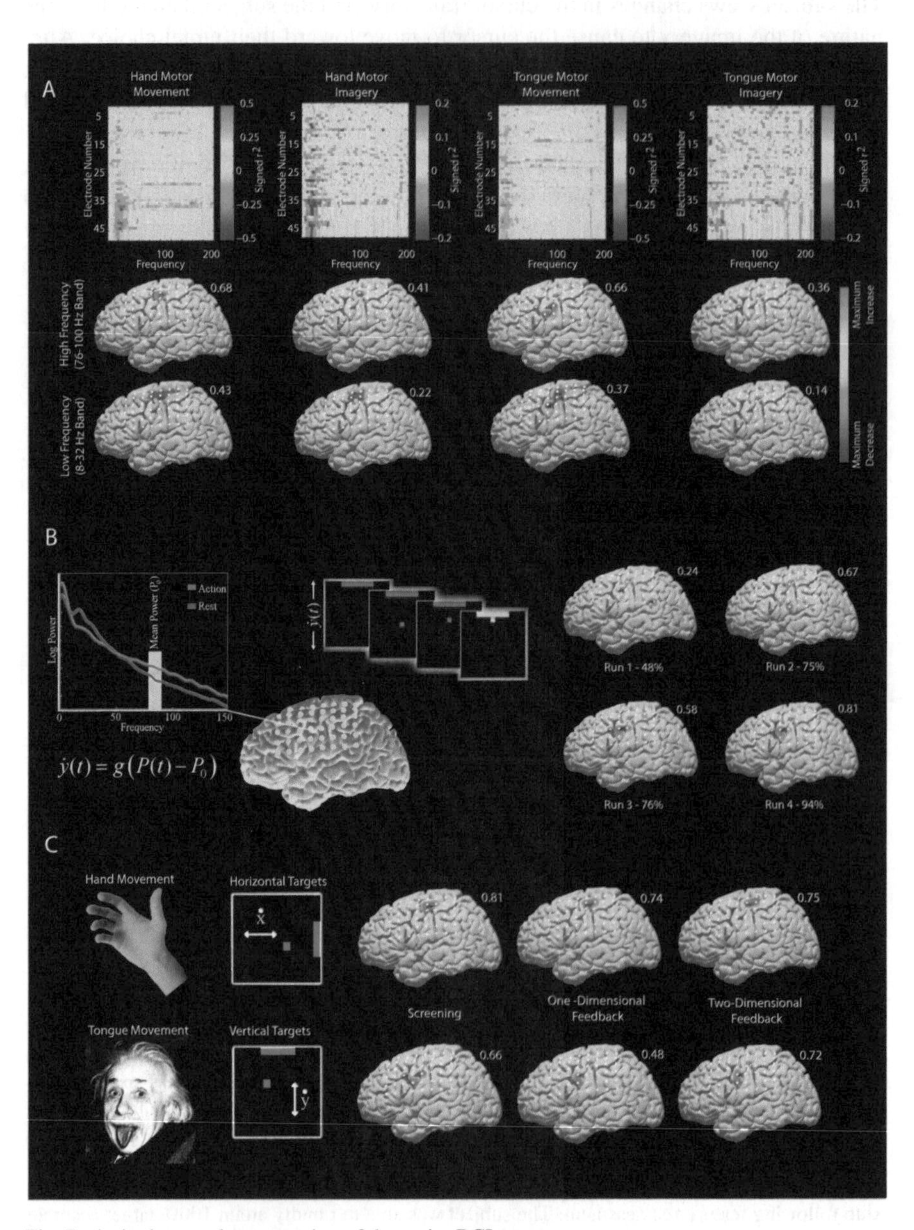

Fig. 7 A single case demonstration of the entire BCI process

Interestingly, even with the most robust signal features, once the behavior is linked to a different task (i.e., the subject is now aiming for a target not just performing the given movement), the electrocorticographic signal may change. We have used the concept of "re-screening" (illustrated in Fig. 6) to make use of the fact that a given signal at a given electrode may be subtly or even dramatically different when the task condition is changed to include a BCI component.

For applications in patients with certain neurologic impairments that impair movement, overt motor activity will, of course, not be accessible for a BCI device. Therefore, attempting to drive BCI with other features is of particular interest. The methods discussed here employ imagined movement and imagined speech to generate screening features. The areas of cortex engaged by, for example, imagined hand movement are remarkably similar to those involved in overt hand movement, though of much weaker strength (Fig. 7a). Because of the weaker signal, initial control can be more difficult with an imagined task, however, the presence of cursor feedback reliably produces an enhancement of the ECoG signal resulting in improved performance. This remarkable ability to enhance the signal has been reported in different imagined motor tasks and silent speech [36].

In most circumstances, accuracy increases over time, usually within a few trials (Fig. 7b, see [8, 10]). BCI2000 allows for a recursive tuning of the weights of a given feature so that the program learns, based on the signal of correct and incorrect trials, what the ideal translation between feature and cursor position should be. However, subjects show a robust learning on their own – the brain is able to learn, as with any new motor task (riding a bike, etc) to subconsciously modify activity based on feedback. Taken to an extreme, this concept has allowed BCI to occur even when the behavior of the cortex underlying the chosen electrode is less well defined [12].

Though originally the cursor control is explicitly linked to the behavior, the subject is free to explore mental states that achieve better cursor control. Anecdotally, successful control often is associated with a wide range of experiential strategies. Some control evolves to being achieved with a throat sensation or, in the most striking example of behavioral plasticity, the subject simply imagines the cursor to move in the preferred direction, no longer using the originally prescribed behavior as an intermediate, as happened with the subject described in the case study at the end of this chapter.

5 Learning

In this particular paradigm, stability of signal is hard to determine given the primary purpose of the implants is clinical and the implant is removed once the seizure focus is determined. However, we have had occasion to assess the stability over multiple days in a 5 day, repeated testing region. Since the recording devices are on the brain, not fixed to it, one could imagine day-to-day variations in the signal. In fact, the use of ECoG appears to rely upon a population signal that is not overly sensitive to daily fluctuations [35] and the same features can be reused with minimal re-training required to achieve accurate cursor control.

In a one dimensional experiment, the learning process is rapid (3–24 min, see [8]). Though not always highly focal, we have observed cortical activity that increases with improved performance (Fig. 7b). Anecdotally, this is associated with a psychological "de-linking" of the screening behavior (e.g., hand movement imagery) and the cursor control, where the cursor gains some kind of abstract, tool-like, representation with the subject. They simply will the cursor to move to the left or the right, and there is an increase in power in the appropriate feature. Recent results (an example is illustrated in Fig. 7c) have demonstrated efficacy using ECoG for two-dimensional control [10]. This requires two independent signals to drive the two degrees (typically up-down and left-right cursor position, as shown in Fig. 7c). The motor system allows for some intuitive, separable features, using, for example, hand and mouth, which are distinct when *high-frequency* ECoG features are used [6, 36, 37]. Despite success, simultaneous 2D control has been less robust than for one dimension. One reason for this is that it requires a good deal more mental coordination by the subject. For example, in this simple form of a 2D target task BCI, there is fundamental uncertainty in the null condition. When increased power in one channel drives the cursor up and a lack of power down, with increased power in another channel driving the cursor to the left, then weak power in both channels could be ambiguous when targeting a down or right sided target. It takes practice to coordinate the two null directions. Even if the signals are not overlapping during the screening for the two different tasks, when combined to achieve two-dimensional cursor control, they may have some initial interdependence. Through feedback, however, these can be segregated by some individuals [37, 38].

6 Case Study

We conclude this chapter with a step-by-step example of successful implementation of electrocorticographic BCI.

Background: The patient is an 18 year old right handed woman with intractable seizures related to a left frontal lesion. Electrodes were placed to map seizures and localize speech function prior to resective surgery, and the array covered speech areas and primary, peri-central, sensorimotor cortex. This excellent coverage of motor cortex made her an ideal candidate for multiple degree of freedom cursor control. Signals were split outside of the patient, with a common ground and reference to the clinical system. The experiments were conducted in three stages spanning two sessions in the same day. The subject was positioned comfortably in her hospital bed throughout the experimental process, and the experimental setting is illustrated in Fig. 5. The electrode localization and cortical plotting were performed using the LOC package [16]. The three steps that the experiment was constructed upon were cue-based screening for feature identification, independent one-dimensional cursor-to-target feedback, and combined two-dimensional cursor-to-target feedback.

Figure 7a, Screening for feature identification: In response to 3 s cues, she repetitively moved either her hand or her tongue in repetition, resting for 3 s in between each cue. There were 30 cues each for hand and tongue, and they were interleaved in random order to reduce the influence of anticipation. Following the 6 min of overt movement, an identical task was performed, except that rather than actually move the hand and tongue, she imagined moving her hand and tongue. She was explicitly instructed to imagine the kinetics of the movement, not just the sensation (this kinesthetic imagery was found to produce the most robust signal change in a study by Neuper et al. [24]). A comparison of movement and rest periods (as shown in Fig. 4) produced feature maps for movement and imagery, shown in the top row of (a). These reveal that changes during imagery mimic those of overt movement, but are less intense (note that the color bar scales are different for movement than imagery). Cortical projections of the imagined and real changes for both high (76–100 Hz) and low (8–32 Hz) frequency changes reveal that the spatial distributions are similar, but the imagery associated change is more focal [36]. The number to the top right of each cortical projection indicates the maximum absolute value of the activation, since each brain is scaled to the maximum. Electrodes' weights with statistically insignificant change were not projected. These types of projections can be a very useful sanity check to verify that the locations of features identified on a feature map make sense anatomically. A similar screening for repetition of the word "move" was performed, to select a feature for the imagery-based feedback shown in part (b).

Figure 7b, One-dimensional cursor feedback: In the second stage, we provided the subject with feedback based upon movement and speech imagery. The imagery was coupled to cursor movement, as detailed in Fig. 5. Figure inset (b) demonstrates the feature chosen, an 80–95 Hz frequency range from an electrode in primary mouth motor area. The mean power, P_0, lay between the mean power during speech imagery and rest. If the mean power, during the task, was above this level (obtained by speech imagery) then the cursor would move up, and if it was below, the cursor would move down, according to the relation $\dot{y} = g(P(t) - P_0)$. The speed parameter, g, was determined online prior to the first experimental run, such that it was "reasonable" i.e. the cursor velocity changes on roughly the same timescale as the electrode-frequency range feature. In theory, this could be estimated offline prior to the first run by examining the variation in the power of the feature during the screening task, but in practice, the rapid adjustment parameter built into the BCI2000 program [15] is a much easier way to get a comfortable speed parameter. The right-most portion of (b) demonstrates the activation during the 4 (~2 min) trials of imagery based feedback. Each cortical projection represents the activation between upper targets (speech imagery) and lower targets (rest); because the feature chosen was at a high frequency, above the intersection in the power spectrum (shown in Fig. 4), the power in the feature increases with activity. She rapidly learned to control the cursor, with the signal becoming both more pronounced and more focused in the feedback area. After the third run, she reported having gone from the coupled imagery (imagined speech) to just thinking about the cursor moving up and down. Again, only significant changes were plotted to the template cortex. The activity in the most ventral posterior electrode (bottom right) hints at activity in Wernicke's area, and persists throughout (it appears to be less, but only because the primary areas are becoming more pronounced), and the overall scale (to the top right) is increasing. Similar one dimensional tasks were performed for tongue and hand. The hand movement was coupled to left-right cursor movement, in preparation for the combination of the two into a two dimensional task.

Figure 7c, Two-Dimensional Cursor Feedback: In the last stage of the experiment, two one dimensional control signals are combined into a single cursor to target task. If the two signals are independent, as is the case here, then the transition between robust control in two one-dimensional tasks and robust control in one two-dimensional task is straightforward. The combination of hand and tongue linked features is good (as in (c)), because they are well demarcated on the precentral gyrus, but the pair chosen in any particular instance will be dependent on the coverage of the electrode array. The example shown in (c), with the frequency range 80–90 Hz demonstrates how robust, screened, features (left) can be used for robust control in one-dimensional tasks (center). The electrode for up-down control in both the one- and two-dimensional tasks was from the classic tongue area. The electrode for left-right control in both the one- and two-dimensional tasks was from the classic hand area. The one- and two-dimensional control tasks were successful (100% Left/Right 1D, 97% Up/Down 1D, and 84% 2D).

7 Conclusion

ECoG provides robust signals that can be used in a BCI system. Using different kinds of motor imagery, subjects can volitionally control the cortical spectrum in multiple brain areas simultaneously. The experimenter can identify salient brain areas and spectral ranges using a cue based screening task. These different kinds of imagery can be coupled to the movement of a cursor on a screen in a feedback process. By coupling them first separately, and later in concert, the subject can learn to control multiple degrees of freedom simultaneously in a cursor based task.

Acknowledgements Special thanks to Gerwin Schalk for his consistent availability and insight. The patients and staff at Harborview Medical Center contributed invaluably of their time and enthusiasm. Author support includes NSF 0130705 and NIH NS07144.

References

1. J.R. Wolpaw, N. Birbaumer, D.J. McFarland, G. Pfurtscheller, and T.M. Vaughan, Brain-computer interfaces for communication and control. Clin Neurophysiol, 113(6), 767–791, (2002).
2. J.R. Wolpaw and D.J. McFarland, Control of a two-dimensional movement signal by a noninvasive brain-computer interface in humans. Proc Natl Acad Sci USA, 101(51), 17849–17854, (2004).
3. L.R. Hochberg, et al., Neuronal ensemble control of prosthetic devices by a human with tetraplegia. Nature, 442(7099), 164–171, (2006).
4. P.R. Kennedy and R.A. Bakay, Restoration of neural output from a paralyzed patient by a direct brain connection. Neuroreport, 9(8), 1707–1711, (1998).

5. J.G. Ojemann, E.C. Leuthardt, and K.J. Miller, Brain-machine interface: Restoring neurological function through bioengineering. Clin Neurosurg, 54(28), 134–136, (2007).
6. K.J. Miller, et al., Spectral changes in cortical surface potentials during motor movement. J Neurosci, 27(9), 2424–2432, (2007).
7. P.L. Nunez and B.A. Cutillo, *Neocortical dynamics and human EEG rhythms,* Oxford University Press, New York, pp. xii, 708 p., (1995).
8. E.C. Leuthardt, K.J. Miller, G. Schalk, R.P. Rao, and J.G. Ojemann, Electrocorticography-based brain computer interface – the Seattle experience. IEEE Trans Neural Syst Rehabil Eng, 14(2), 194–198, (2006).
9. E.C. Leuthardt, G. Schalk, J.R. Wolpaw, J.G. Ojemann, and D.W. Moran, A brain-computer interface using electrocorticographic signals in humans. J Neural Eng, 1(2), 63–71, (2004).
10. G. Schalk, et al., Two-dimensional movement control using electrocorticographic signals in humans. J Neural Eng, 5(1), 75–84, (2008).
11. E.A. Felton, J.A. Wilson, J.C. Williams, and P.C. Garell, Electrocorticographically controlled brain-computer interfaces using motor and sensory imagery in patients with temporary subdural electrode implants. Report of four cases. J Neurosurg, 106(3), 495–500, (2007).
12. J.A. Wilson, E.A. Felton, P.C. Garell, G. Schalk, and J.C. Williams, ECoG factors underlying multimodal control of a brain-computer interface. IEEE Trans Neural Syst Rehabil Eng, 14(2), 246–250, (2006).
13. K.J. Miller, S. Zanos, E.E. Fetz, M. den Nijs, and J.G. Ojemann, Decoupling the cortical power spectrum reveals real-time representation of individual finger movements in humans. J Neurosci, 29(10), 3132, (2009).
14. T. Blakely, K.J. Miller, R.P.N. Rao, M.D. Holmes, and J.G. Ojemann, *Localization and classification of phonemes using high spatial resolution electrocorticography (ECoG) grids.* Proc IEEE Eng Med Biol Soc, 4964–4967, (2008).
15. G. Schalk, D.J. McFarland, T. Hinterberger, N. Birbaumer, and J.R. Wolpaw, BCI2000: A general-purpose brain-computer interface (BCI) system. IEEE Trans Biomed Eng, 51(6), 1034–1043, (2004).
16. K.J. Miller, et al., Cortical electrode localization from X-rays and simple mapping for electrocorticographic research: The "location on cortex" (LOC) package for MATLAB. J Neurosci Methods, 162(1–2), 303–308, (2007).
17. K.J. Miller, et al., Beyond the gamma band: The role of high-frequency features in movement classification. IEEE Trans Biomed Eng, 55(5), 1634–1637, (2008).
18. F. Aoki, E.E. Fetz, L. Shupe, E. Lettich, and G.A. Ojemann, Increased gamma-range activity in human sensorimotor cortex during performance of visuomotor tasks. Clin Neurophysiol, 110(3), 524–537, (1999).
19. N.E. Crone, D.L. Miglioretti, B. Gordon, and R.P. Lesser, Functional mapping of human sensorimotor cortex with electrocorticographic spectral analysis. II. Event-related synchronization in the gamma band. Brain, 121(Pt 12), 2301–2315, (1998).
20. N.E. Crone, et al., Functional mapping of human sensorimotor cortex with electrocorticographic spectral analysis. I. Alpha and beta event-related desynchronization. Brain, 121(Pt 12), 2271–2299, (1998).
21. G. Pfurtscheller, B. Graimann, J.E. Huggins, S.P. Levine, and L.A. Schuh, Spatiotemporal patterns of beta desynchronization and gamma synchronization in corticographic data during self-paced movement. Clin Neurophysiol, 114(7), 1226–1236, (2003).
22. G. Pfurtscheller, *Event-related desynchronization (erd) and event related synchronization (ERS),* Williams and Wilkins, Baltimore, MD, pp. 958–967, (1999).
23. K.J. Miller, L.B. Sorensen, J.G. Ojemann, M. den Nijs: Power-law scaling in the brain surface electric potential. PLOS Comput Biol., 5(12), e1000609, (2009, Dec).
24. C. Neuper, R. Scherer, M. Reiner, and G. Pfurtscheller, Imagery of motor actions: Differential effects of kinesthetic and visual-motor mode of imagery in single-trial EEG. Brain Res, 25(3), 668–677, (2005).
25. K.J. Miller, et al., Real-time functional brain mapping using electrocorticography. NeuroImage, 37(2), 504–507, (2007).

26. A. Brovelli, J.P. Lachaux, P. Kahane, and D. Boussaoud, High gamma frequency oscillatory activity dissociates attention from intention in the human premotor cortex. NeuroImage, 28(1), 154–164, (2005).
27. R.T. Canolty, et al., High gamma power is phase-locked to theta oscillations in human neocortex. Science, 313(5793), 1626–1628, (2006).
28. K.J. Miller, et al., Beyond the gamma band: The role of high-frequency features in movement classification. IEEE Trans Biomed Eng, 55(5), 1634, (2008).
29. B. Blankertz, G. Dornhege, M. Krauledat, K.R. Muller, and G. Curio, The non-invasive Berlin Brain-Computer Interface: Fast acquisition of effective performance in untrained subjects. Neuroimage, 37(2), 539–550, (2007).
30. S. Lemm, B. Blankertz, G. Curio, and K.R. Muller, Spatio-spectral filters for improving the classification of single trial EEG. IEEE Trans Biomed Eng, 52(9), 1541–1548, (2005).
31. K.R. Muller, et al., Machine learning for real-time single-trial EEG-analysis: From brain-computer interfacing to mental state monitoring. J Neurosci Methods, 167(1), 82–90, (2008).
32. P. Shenoy, K.J. Miller, J.G. Ojemann, and R.P. Rao, Generalized features for electrocorticographic BCIs. IEEE Trans Biomed Eng, 55(1), 273–280, (2008).
33. R. Scherer, B. Graimann, J.E. Huggins, S.P. Levine, and G. Pfurtscheller, Frequency component selection for an ECoG-based brain-computer interface. Biomed Tech (Berl), 48(1–2), 31–36, (2003).
34. B. Graimann, J.E. Huggins, S.P. Levine, and G. Pfurtscheller, Visualization of significant ERD/ERS patterns in multichannel EEG and ECoG data. Clin Neurophysiol, 113(1), 43–47, (2002).
35. T. Blakeley, K.J. Miller, S. Zanos, R.P.N. Rao, J.G. Ojemann: Robust long term control of an electrocorticographic brain computer interface with fixed parameters. Neurosurg Focus, 27(1), E13, (2009, Jul).
36. K.J. Miller, G.S. Schalk, E.E. Fetz, M. den Nijs, J.G. Ojemann, and R.P.N. Rao: Cortical activity during motor movement, motor imagery, and imagery-based online feedback. Proc Natl Acad Sci, 107(9), 4430–4435, (2010, Mar).
37. G. Schalk, et al., Two-dimensional movement control using electrocorticographic signals in humans. J Neural Eng, 5(1), 75–84, (2008).
38. K.J. Miller, et al., Three cases of feature correlation in an electrocorticographic BCI. IEEE Eng Med Biol Soc, 5318–5321, (2008).

Using BCI2000 in BCI Research

Jürgen Mellinger and Gerwin Schalk

1 Introduction

BCI2000 is a general-purpose system for brain–computer interface (BCI) research. It can also be used for data acquisition, stimulus presentation, and brain monitoring applications [18, 27]. The mission of the BCI2000 project is to facilitate research and applications in these areas. BCI2000 has been in development since 2000 in a collaboration between the Wadsworth Center of the New York State Department of Health in Albany, New York, and the Institute of Medical Psychology and Behavioral Neurobiology at the University of Tübingen, Germany. Many other individuals at different institutions world-wide have contributed to this project.

BCI2000 has already had a substantial impact on BCI research. As of mid-2010, BCI2000 has been acquired by more than 350 laboratories around the world. The original article that described the BCI2000 system [27] has been cited more than 200 times, and was recently awarded the Best Paper Award by *IEEE Transactions on Biomedical Engineering*. Furthermore, a review of the literature revealed that BCI2000 has been used in more than 50 peer-reviewed publications. Many of these papers set new directions in BCI research, which include: the first online brain–computer interfaces using magnetoencephalographic (MEG) signals [19] or electrocorticographic (ECoG) signals [9, 13, 14, 28, 33]; the first application of BCI technology to functional restoration in patients with chronic stroke [4, 34]; the demonstration that non-invasive BCI systems can support multidimensional cursor movements [16, 17, 36]; the first real-time BCI use of high-resolution EEG techniques [6]; the use of BCI techniques to control assistive technologies [7]; control of a humanoid robot by a noninvasive BCI [2]; and the first demonstrations that people severely paralyzed by amyotrophic lateral sclerosis (ALS) can use BCIs based on sensorimotor rhythms or P300 evoked potentials [11, 22, 32]. In addition, several studies have used BCI2000 for purposes other than BCI research, e.g., the first large-scale motor mapping studies using ECoG signals [12, 21]; real-time

J. Mellinger (✉)
Institute of Medical Psychology and Behavioral Neurobiology, University of Tübingen, Tübingen, Germany
e-mail: juergen.mellinger@uni-tuebingen.de

B. Graimann et al. (eds.), *Brain–Computer Interfaces*, The Frontiers Collection, DOI 10.1007/978-3-642-02091-9_15,

mapping of cortical function using ECoG [3, 20, 24, 26]; the optimization of BCI signal processing routines [5, 23, 37]; the demonstration that two-dimensional hand movement trajectories can be decoded from ECoG signals [25]; and evaluation of steady-state visual evoked potentials (SSVEP) for the BCI purpose [1]. A number of these studies were done across different laboratories, thereby taking advantage of the common exchange of data and experimental paradigms supported by BCI2000.

In summary, BCI2000 is a critical tool for challenging one of the greatest problems with BCIs: the need for considerable time and energy to set up and configure a BCI system. Much BCI research still involves laboratory demonstrations of highly specialized and mutually incompatible BCI systems, or only isolated components of a BCI such as a new signal processing mechanism. By providing a straightforward, easy, usable platform with strong support, BCI2000 has strongly catalyzed the emerging transition to flexible, practical, and clinically relevant BCI systems. This transition is essential to making BCIs practical to severely disabled and other users.

As indicated by the above descriptions of its capacities, its utility for many different aspects of BCI research, the success of its wide dissemination, and its growing prominence in the scientific literature, BCI2000 is fast becoming, or perhaps has already become, a standard software platform for BCI research and development. BCI2000 is available free of charge with full documentation and complete source code at http://www.bci2000.org for research and educational purposes. See Sect. 5 for further information about dissemination and availability.

In this book chapter, we will introduce the reader to the BCI2000 system and give an overview of its design, capabilities, and use. We will also present a number of scenarios that are typical of applications in BCI research, and discuss how BCI2000 may be used to implement them. While all these scenarios are different, we show how each of them benefits from the use of the system.

BCI2000 facilitates the implementation of different BCI systems and other psychophysiological experiments by substantially reducing labor and cost. It does this mainly by providing a number of capable BCI and stimulus presentation paradigms that can either be configured by the investigator or adapted by a software engineer. Because BCI2000 is based on a system model that can describe any BCI system, because its underlying framework is highly generic, and because BCI2000 does not require third-party components, its use is most beneficial in large collaborative research programs with many concurrent and different experiments in different locations. BCI2000 provides these benefits through the following features:

1.1 Proven Components

BCI2000 is provided with fully documented components that have proven to work robustly in many different BCI and other experiments. By its modular design, BCI2000 tries to avoid redundancies in code, and re-uses modules and code in multiple contexts rather than reduplicating it. One of the advantages of this approach is that it maximizes the effect of errors that may exist in any single component, greatly increasing the likelihood to detect such errors early.

1.2 Documentation

We provide hundreds of pages of documentation that encompass user, technical, and programming references that cover all major components. In addition, we include user tutorials that provide an introduction to general BCI operation using sensorimotor rhythms (i.e., mu/beta rhythms) and P300 evoked potentials. This documentation is maintained as a wiki information system on doc.bci2000.org, and is deployed with BCI2000 as HTML help pages. Additionally, these help pages are linked to the BCI2000 user interface, providing an instant reference.

1.3 Adaptability

Because of its flexible and modular design, BCI2000 allows combination of existing software components with relatively little effort. This ranges from simple re-combination of modules (e.g., using the same signal processing algorithm with different feedback paradigms), to modification of the kind and order of signal processing components, and to modification of existing components or creation of new components.

1.4 Access

BCI2000 comes with full source code. Thus, the user may adapt it to individual needs. In most cases, modifying an existing system is easier than creating a new one from scratch.

1.5 Deployment

BCI2000 facilitates deployment to, and collaboration amongst, multiple sites. It is very easy to install because it comes as a simple directory tree, and does not require any specific installation process other than copying this directory tree. Installation and operation is even possible with limited user rights. Aside from hardware drivers necessary for the data acquisition device, BCI2000 does not depend on third-party components. Collaboration is facilitated by generic data and parameter files. Thus, experimental paradigms can be designed in one location, tested in another, and the resulting data files can be easily interpreted.

In the following section, we will first give an overview of BCI2000's basic design and components, and then move on to discuss research scenarios that may profit from the use of BCI2000.

2 BCI2000 Design

2.1 System Model

The BCI2000 system model [27] was designed such that it can accommodate any kind of BCIs. This model is similar to that proposed in [15]. Any BCI consists of three main components: a *data acquisition* component that records signals from the brain; a *signal processing algorithm* that extracts signal features that represent the user's intent and that translates them into commands for an output device; and a *user application* component that performs these commands, e.g., for letter selection. These components of any BCI system are described in e.g. [35] and in chapters "Brain Signals for Brain-Computer Interfaces" and "Brain–Computer Interfaces: A Gentle Introduction in this book.

In BCI2000, these three components correspond to three "core modules": A *Source Module,* which acquires and stores data, a *Signal Processing Module,* and a *User Application Module* (Fig. 1). These core modules are realized as independent executables; they exchange data using a network-capable protocol. There is a Source Module for each type of data acquisition hardware, a Signal Processing Module for each type of signal processing algorithm, and a User Application Module for each type of feedback or output device. These modules may be re-combined by choosing a different set of executables when starting up BCI2000. Typically, this is done by executing different startup scripts in order to run BCI2000.

As an example, conducting sensorimotor rhythm (SMR) feedback experiments using the g.tec^{TM} g.USBamp acquisition system will involve the g.USBamp Source module (for data acquisition and storage), the ARSignalProcessing module (for spatial filtering, autoregressive spectral estimation, linear classification, and signal normalization), and the CursorTask module (to give visual feedback to the user in form of cursor movement).

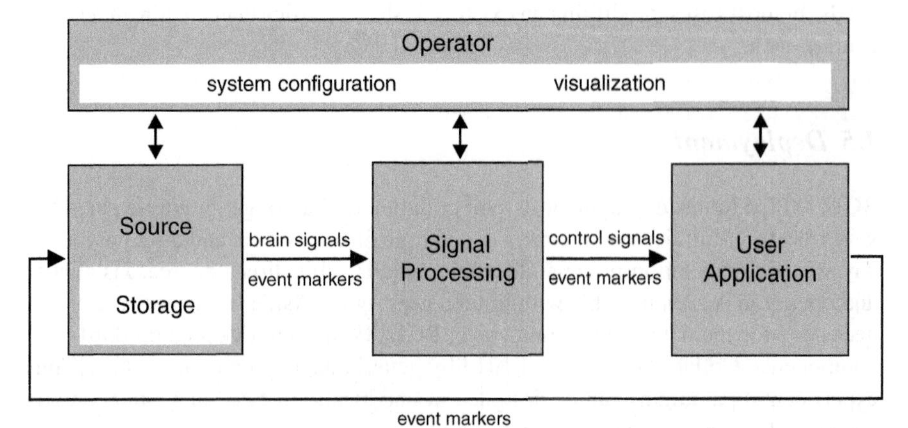

Fig. 1 BCI2000 System Model. Acquired data are processed sequentially by three modules (i.e., Source, Signal Processing, and User Application). Processing is controlled by an additional Operator module (from [27])

Once BCI2000 has been started, further configuration and control is performed through a dedicated *Operator Module*. That module provides the experimenter's user interface. While it may contain different parameters that are specific to a particular experiment, the Operator module is the same program for all possible configurations of BCI2000. In other words, the BCI2000 graphical interface to the investigator does not have to be rewritten for different configurations. The Operator module provides a user interface to start, pause, and resume system operation. In addition, it is able to display logging information, signal visualization, and a reduced copy of the user's screen during system operation (Fig. 2).

The functional details of individual modules are configured by parameters that are requested by the individual components that make up each module. For example, the data acquisition component of the BCI2000 Source module typically requests a parameter that sets the device's sampling rate. The parameters of all modules are displayed, modified, and organized in the Operator module's configuration dialog (Fig. 3).

To perform experiments using BCI2000, one needs a Source module that corresponds to the utilized hardware used; a Signal Processing module that extracts features relevant to the BCI paradigm of choice; and a User Application module that implements the particular BCI paradigm. For simple stimulation experiments that do not require real-time feedback, the Signal Processing module may be replaced with an empty "dummy" module.

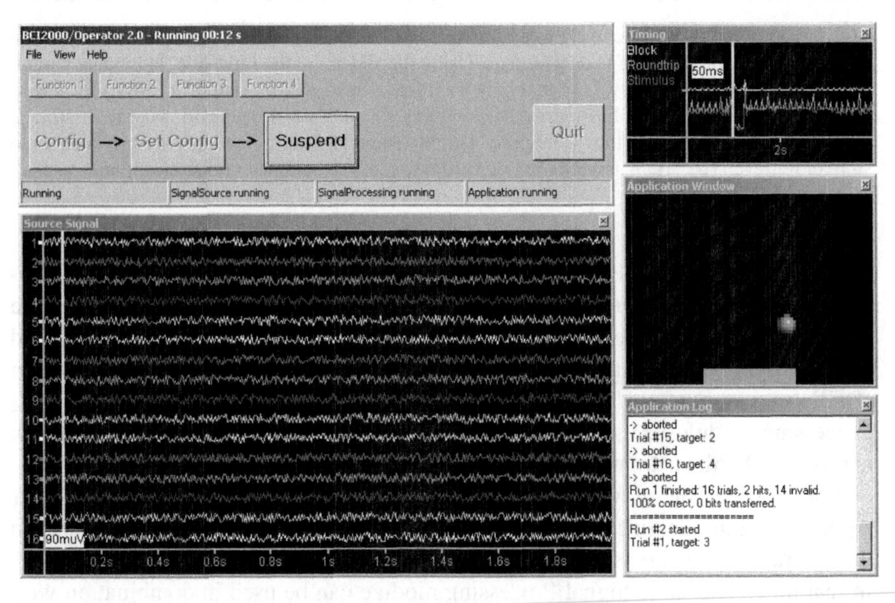

Fig. 2 BCI2000 Operator module, displaying source signal (*bottom left*), timing information (*top right*), a downsampled copy of the application window (*medium right*), and an application log (*bottom right*)

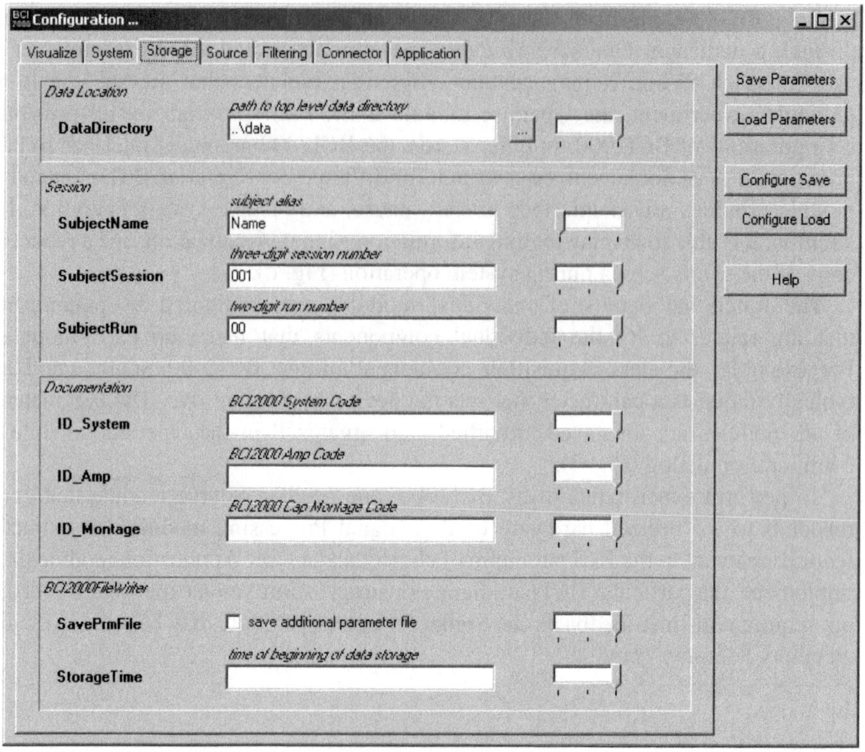

Fig. 3 The BCI2000 parameter configuration dialog. The "Storage" tab is selected and reveals the configuration details relating to data storage. Using the sliders on the right side, individual parameters can be assigned user levels so that only a subset of them will be visible at lower user levels; this is useful for clinical applications of BCI systems. "Save" and "Load" buttons provide storage and retrieval of parameters, the "Configure" buttons allow to determine a subset of parameters to store or retrieve. The "Help" button provides instant access to parameter documentation

In more detail, BCI experiments based on sensorimotor rhythms are supported by two signal processing modules that extract spectral features, i.e., the ARSignalProcessing module that uses an autoregressive spectral estimator, and the FFTSignalProcessing module that uses short-term Discrete Fourier Transform (DFT) to compute an amplitude spectrum. Typically, output of these Signal Processing modules is used to control a cursor on a feedback screen; such feedback is provided by the CursorTask User Application module (Fig. 4 left).

For presentation of visual or auditory stimuli, we provide a dedicated User Application module. In conjunction with a corresponding Signal Processing module that averages data across epochs, this module is used to classify evoked responses in real time. The same Signal Processing module can be used in combination with the P300 matrix speller application module (Fig. 4; right).

Fig. 4 Examples of application screens. *Left*: A cursor moving towards a target, controlled by sensorimotor rhythms (Cursor Task). *Right*: The P300 speller in copy spelling mode

2.2 Software Components

Inside core modules, software components act on chunks of brain signal data in sequence, forming a chain of "filters" (Fig. 5). Module operation may be adapted by writing new filters, modifying existing ones, or simply by rearranging them. These filters are written in C++, so their adaptation requires some programming skills. However, BCI2000 provides a programming framework that is designed to achieve slim and transparent coding inside filter components, and thus simplifies filter modification. Also, as described in more detail later, we provide a filter that employs user-provided Matlab^{TM} code for online data processing.

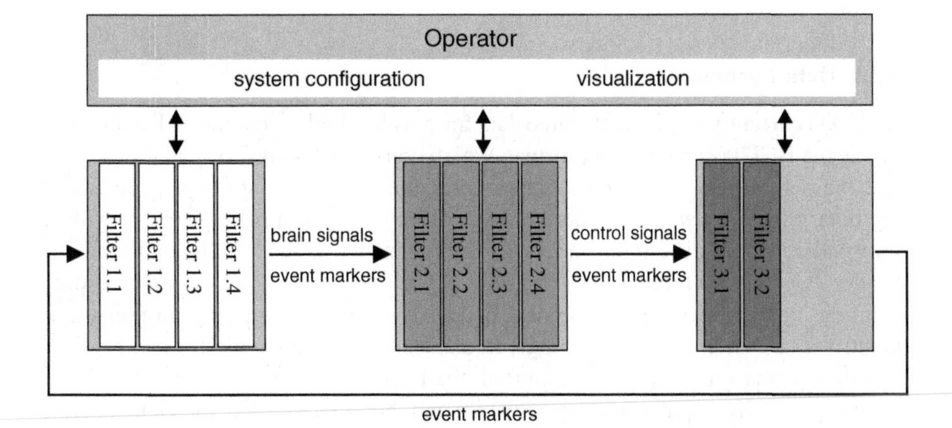

Fig. 5 BCI2000 Filter Chain. Inside core modules, individual filter components are indicated

Filters do not have their own user interface. Rather, they typically request configuration *parameters* from the programming framework, which will then be displayed and organized in the Operator module, with no specific programming effort required. Parameters may be of numeric or string type, and may be single values, lists, or matrices. These parameters are typically constant during one particular experimental recording session. In addition to parameters, BCI2000 supports event markers called *states*. These state variables encode various aspects of the state of the system during operation. They can, and typically do, change their values during a recording session. A typical application of states is to use them as event markers, encoding information about stimulation or user task for later data analysis. Similarly to parameters, states are requested by filters, and then propagated and recorded transparently by the framework. During operation, a filter may access parameters and states using a simple programming interface that allows setting and obtaining parameter and state values in a manner similar to assigning values to or from variables. The advantage of using this software infrastructure is that requested parameters automatically appear in particular tabs in the configuration dialog, that they are automatically stored in the header of the data file (so that all aspects of system configuration can be retrieved offline), and that state variables are automatically associated with data samples and stored in the data file. In other words, BCI2000 system operation, data storage, and real-time data interchange, can be adapted substantially to the needs of particular situations by only making local changes to individual BCI2000 filters within modules.

2.3 Interfacing Components

For greater flexibility, BCI2000 supports interoperability with other software in various ways, both at the time of online operation and offline data analysis. These interfaces fall into the following categories (Fig. 6):

2.3.1 Data Formats

BCI2000 currently implements three data formats available for storage. The first format is the BCI2000 native data format, which consists of a human-readable header, and the actual data in compact binary form. In addition to specifying details of the binary file format, the header contains all configuration parameters, providing information about all circumstances of the recording. Due to its human readable header, the BCI2000 data format is especially suited for integration into recording databases, and allows for the use of full-text searching techniques to organize and identify a large number of recordings. In addition to the native file format, two further data formats are currently supported: the European Data Format (EDF, [10]), a format especially popular in sleep research; and, building on EDF, the General Data Format for Biosignals (GDF, [29]), which is designed mainly with BCI applications in mind. Implementation of file formats is realized in particular components of the

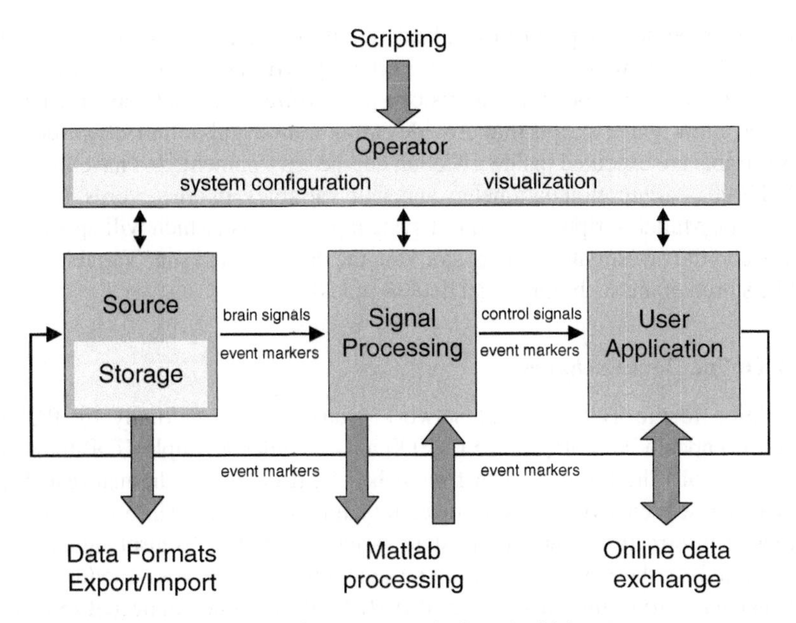

Fig. 6 BCI2000 interfaces to external programs are indicated with large arrows

Source module. Thus, further file formats may be added by creating new components; the effort of writing such a component is limited to implementing the output format itself, by providing code that writes incoming data into an output stream.

2.3.2 Data Exchange

BCI2000 provides a Matlab™ extension (i.e., a "mex file") that supports reading recorded BCI2000 data files directly into Matlab. Using this component, data from multiple files may be concatenated, and restricted to a range of samples within files. Parameters and states are mapped to Matlab variables, allowing to access parameter and state values by name, which results in well-understandable Matlab code. Building on that mex file, BCI2000 provides an EEGlab extension [8] that enables EEGlab to directly operate on BCI2000 data files. Other data exchange components include an export tool that translates BCI2000 data files into ASCII data, or into binary format suited for import into BrainProducts™ BrainVision Analyzer. In addition to the components provided by BCI2000 itself, the BIOSIG toolbox [30] also includes support for BCI2000 data files.

2.3.3 Matlab Filter Scripts

As an alternative to writing filter code in C++, online processing of brain signals may be done using Matlab code. BCI2000 comes with a "MatlabFilter" that exposes BCI2000's full signal processing filter interface to Matlab™. Thus, a new signal processing module may profit from all existing BCI2000 components, and additionally introduce a new filter component, purely by writing Matlab code rather than C++. Such a filter component consists of a *processing* script that acts on chunks of data, and a few helper scripts that provide configuration and initialization functions. These scripts are executed inside a Matlab engine environment, and have full access to BCI2000 configuration parameters and state variables. In other words, it is possible to write Matlab scripts that request system parameters (which will appear at the interface to the investigator and are stored in the data file) and state variables (which will be stored along with individual data samples).

2.3.4 Online Data Exchange

In addition to the TCP/IP-based network protocol used internally by BCI2000 (which is capable but complex), BCI2000 also includes a simple UDP-based network protocol. The protocol data format is ASCII-based and human-readable; it consists of a sequence of names followed by numerical values. This protocol is used to transfer information contained in state variables, and the "control signal" that is transferred from the Signal Processing module into the application module.

This data sharing interface is useful if BCI2000 is to be connected to external application programs such as word processors (so that the external application is controlled by the output of the Signal Processing module), or to hardware such as environmental control interfaces, wheelchairs, or limb prostheses. The use of that interface requires writing a small interfacing application in any programming language, or modifying an existing application to read data over a network connection.

2.3.5 Operator Module Scripting

As described above, the Operator module is the experimenter's graphical user interface (GUI) for configuration and control of the BCI2000 system. Some of the operations of the Operator module can be scripted to execute programmatically. Most importantly, scripts may be tied to events such as system startup, or the end of an experimental run. Using Operator scripting, it is possible to completely wrap up the experimenter side of BCI2000 into scripts, and to remote-control it from a separate application. Thus, for example, a complex series of stimulus presentation paradigms can be realized using combinations of BCI2000 and external applications.

2.4 Important Characteristics of BCI2000

In summary, BCI2000 provides a number of characteristics that make it a good choice for many research applications. First, it comes with full source code in C++ and can incorporate signal processing routines written in Matlab™. Second, BCI2000 does not rely on third-party components for its execution; modifications to the system only require the Borland C++ Builder environment. BCI2000 may be used at no cost for academic and educational purposes. Third, BCI2000 defines its own file format, and a generic concept of parameters and states, which encode configuration and run-time event information. The format's versatility, and human readability of its file header, are most advantageous in large research programs that involve multi-site cooperation, or managing large amounts of recordings.

2.5 Getting Started with BCI2000

Getting started with BCI2000 first requires appropriate hardware. This consists of a computer system, recording equipment, and the BCI2000 software. The computer system may be standard desktop or a laptop computer running Windows.[1] The computer should come with a dedicated 3D-capable video card. For most experiments, it is advisable to use separate video screens for the experimenter and the subject. When using a desktop computer, this implies a two-monitor configuration; when using a laptop computer, a separate monitor for the subject, i.e., the BCI user.

The recording equipment consists of an amplifier, a digitizer, and appropriate sensors. Today, many amplifiers come with an integrated digitizer. For example, this includes the g.tec™ amplifiers supported in the core distribution of BCI2000: the 16-channel g.USBamp that connects via USB, and the 8-channel g.MOBIlab+ that connects via Bluetooth. Many other EEG systems and digitizer boards are supported through additional Source modules that have been provided by BCI2000 users. Examples comprise EEG/ECoG systems by Tucker-Davis, Biosemi, Brainproducts (BrainAmp, V-Amp) and Neuroscan, as well as digitizer boards by National Instruments, Measurement Computing, and Data Translation. In most situations in humans, BCI systems will record from sensors placed on the scalp (electroencephalography (EEG)). EEG caps are usually offered in conjunction with amplifiers, or may be purchased separately.

3 Research Scenarios

3.1 BCI Classroom

Teaching relevant aspects of brain–computer interfacing is important not only in teaching environments, but also when building or maintaining a BCI research laboratory.

Much like students, research staff will need to be introduced into the field's practical and theoretical aspects. While many important publications on BCI research

[1] We have comprehensively tested BCI2000 on Windows NT, Windows 2000, and Windows XP. While BCI2000 functions well under Windows Vista, Vista's timing performance, in particular with regards to audio and video output, is reduced compared to Windows XP.

exist, and textbooks are now beginning to be published, introductory material is currently still often not available. Technical knowledge from fields as distant as electroencephalography, electrical engineering, computer science, neuroscience, and psychology must be available to successfully perform BCI experiments [35]. In the laboratories pioneering this research, this knowledge has been acquired over decades, often cumbersome, on basis of trial and error. Obviously, when starting in BCI research, re-enacting this trial-and-error process is not an efficient way to introduce research staff to BCI methods.

In this scenario, BCI2000 helps in providing an all-in-one BCI research software system for recording and analysis. It comes with practical, step-by-step tutorials covering the relevant aspects of the two most important brain signals used for human BCI experimentation, i.e., sensorimotor rhythms (SMR) and P300 evoked potentials. In the following paragraphs, we will discuss the resources required in a typical BCI teaching scenario, and how BCI2000 fits into such a scenario.

3.1.1 EEG Hardware

As discussed further below, BCI2000 connects to a number of standard EEG amplifiers, and facilitates connection of currently unsupported systems. For the present classroom scenario, we assume that no EEG hardware is present at the research site, and measurement stations need to be purchased. Such a measurement station, suited for teaching BCI experiments, will typically consist of an EEG amplifier with 16–64 channels, a standard laptop computer running Windows™, and a separate flatscreen monitor to be used as the display for the subject. While experiments may be run from desktop machines, and single-monitor installations as well, the advantage of the suggested laptop-monitor combination in terms of versatility is great enough, and its additional cost low enough, to suggest it for most situations. See Fig. 7 for an example hardware configuration.

3.1.2 Software

BCI2000 is provided free of charge for research and academic purposes. Its binary distribution or source code may be downloaded from the BCI2000 website. Installing BCI2000 consists of the extraction of a single directory hierarchy to the target machine, and does not even require administrative privileges. This way, it is easy to maintain synchronized BCI2000 installations by copying a reference installation from a central server. Thus, setup time for a BCI2000 installation is minimal.

3.1.3 Getting Acquainted

Before doing actual experiments, instructors will need to get acquainted with operating BCI2000 itself. BCI2000 supports this part of the preparation by providing an introductory 30–45 min “BCI2000 Tour” that takes a novice user through the most important aspects of the user interface.

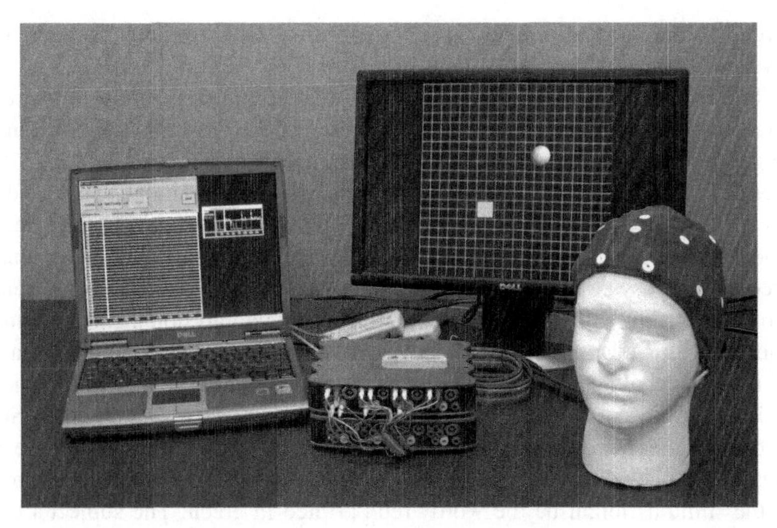

Fig. 7 BCI2000 running on a laptop-based 32-channel EEG system. The system is connected to two 16-channel g.tec^{TM} g.USBamps and an EEG electrode cap

3.1.4 Tutorial Experiments

BCI2000 provides tutorials for SMR and P300 BCI experiments. These tutorials are comprised of short theoretical introductions into the physiological and physical background of the respective brain signal, a practical introduction to EEG measurements, and step-by-step instructions on how to perform BCI sessions and doing data analysis using these paradigms. In a typical BCI teaching context, instructors will first work through this material, and do the experiments themselves, and then use it as teaching material to instruct their students or laboratory staff. Depending on the amount of prior knowledge in the fields of electrophysiology and EEG, working through the introductory material and doing experiments and analyses will require several hours each for the SMR and P300 paradigms.

3.2 Performing Psychophysiological Experiments

BCI2000 can also be valuable for laboratories specialized in the fields of electrophysiology and psychophysiology. Typical psychophysiological experiments involve visual, auditory, or other types of stimulation, and simultaneous recording of brain signal responses. In such a research environment, BCI2000 may be used as a stimulation and recording software.

Usually, psychophysiological experiments require the appropriate integration of three different elements. The first is recording software, which is typically provided by the vendor of the measurement equipment. The second is stimulation software. Such software is typically an independent product and must be connected to the recording device or software in order to obtain an account of stimulation in the recorded data. The final element is data analysis, which requires mastery of an analysis software environment.

While BCI2000 is not intended to replace a dedicated stimulation software system (such as Presentation or E-Prime), it provides a flexible stimulation environment with good timing behavior, and it integrates stimulation and data recording. It also comes with tools that provide easy-to-use, basic statistical data analyses, and provides a number of interfaces to commercial as well as non-commercial analysis toolboxes (Sect. 2.3). Finally, BCI2000 comes with a standardized timing evaluation procedure that may be run on the target hardware to ensure quality of measurement.

To illustrate how BCI2000 could be useful for psychophysiological experiments, we describe how to configure BCI2000 for a Stroop task. The Stroop task is a classical neuropsychological paradigm designed to measure the ability of a subject to override strongly automated behavior [31]. In a computerized version of the Stroop task, color names are displayed as textual stimuli on a computer screen. These color names are displayed in color; color names and actual colors do not always match, e.g., the stimulus might be the word "red" printed in green. The subject's task is to respond to the actual color, not the color name, by pressing a key. In data analysis, performance may be analyzed in terms of response time, and correctness of responses.

Implementation of a Stroop task experiment in BCI2000 requires three steps. First, graphics files for stimuli need to be created, which contain the names of colors and are printed in different color. These graphics files may be created using any graphics application such as the *Paint* program that comes with Windows™. For five color names, and five actual colors, up to 25 graphics files need to be created, depending on whether all possible combination are to be used.

The second step is to configure BCI2000 display the stimuli. Using the StimulusPresentation application module contained in the BCI2000 core distribution, this requires creating a table listing all stimuli, specifying parameters for a pseudo-random sequence in which stimuli will be presented, and deciding on stimulus duration, inter-stimulus intervals, and inter-stimulus interval randomization.

The final step is to configure BCI2000 to record key presses during the experiment. The respective BCI2000 component is provided within all source modules, and is activated by adding the command-line option "–Logkeyboard=1" when starting up a source module.

3.3 Patient Communication System

As a real-world application of BCI technology, a patient communication system uses a BCI as an input device similar to a computer keyboard or mouse. Once a particular BCI system has been designed and tested in laboratory experiments, BCI researchers may want to apply the resulting system to patients. Typically, this target population is constrained to their beds and depends on caregivers. A BCI may improve their quality of life by increasing their independence and amount of control over their environment.

For use in a patient communication system, it is necessary to make the system more robust (since it needs to be operated by caregivers who are typically not experts on BCI systems), and to make it simpler to set up and use. In addition, it may also be necessary to integrate the BCI system with existing augmentative communication technologies, such as predictive spellers. In such a context, the role of the BCI is reduced to an input device, and, as such, its user interface needs to be reduced to a minimum of options.

BCI2000 facilitates its application as part of a patient communication system mainly in two ways. First, by integration of its control interface, i.e., its Operator module, into a larger software system (Fig. 6); and second, by connecting its output to external devices or software. Integration of BCI2000's control interface is possible through a number of means. First, the parameter configuration dialog may be restricted to only show those parameters that are to be changed by operators. Second, system behavior can be controlled via command-line parameters. Using command-line parameters, it is possible to automatically load parameter files, start system operation, and quit BCI2000 at the end of a run. In addition, beginning with BCI2000 version 3, the system will be further modularized so that the graphical interface to the operator can be completely replaced with an application-specific (rather than generic) interface.

BCI2000 may be integrated into a larger software system by connecting it to external devices or applications. One way to accomplish this is to use BCI2000's external application interface, which provides a bi-directional link to exchange information with external processes running on the same or a different machine. Via the external application interface, read/write access to BCI2000 state information and to the control signal is possible. For example, in an SMR-based BCI, an external application may read the classification result, control the user's task, or get access to the control signal that is calculated by the signal processing module so as to control an external output device (such as a robotic arm or a web browser).

As an example of BCI2000's inter-operability capabilities, we will discuss the scenario of a patient control system that allows paralyzed patients to use brain activity to control a standard word processor. This scenario comprises the following specifications:

- In an initial training phase, the BCI needs to be configured and adapted to the patient. This usually requires expert supervision.
- In further sessions, the system should be operated by nursing staff, with a minimum of interactions.
- The system should be based on the P300 speller paradigm, and choosing individual matrix entries should correspond to entering letters into a standard word processor (Fig. 8).

In this scenario, the standard P300 speller configuration will serve as a starting point. First, implementing the required connectivity to external devices, one

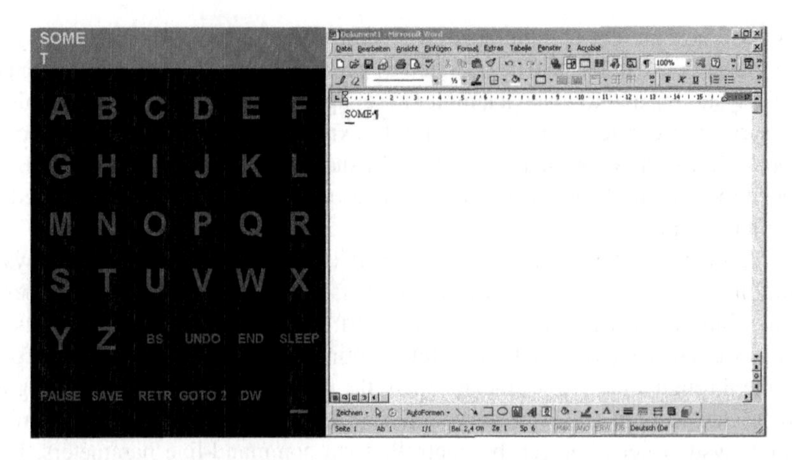

Fig. 8 A P300-Speller based patient communication system entering text into a word processor

would begin by extending the existing speller system to recognize additional selection commands, and act upon them by sending appropriate device, or application, commands. In the present example, no software modification is necessary, since the standard P300 speller module is capable of sending information about letter selection to the same or another computer via UDP, where it may be picked up by any external program. Thus, to connect speller selections to a standard word processor, it is only necessary to create a small interfacing program that reads letter selections from the speller's UDP port, and uses the operating system to send appropriate keypress events to the word processing application.

Once basic functionality is established, one would wrap up BCI2000, the interfacing program, and the word processor into a larger system. This system will use a simplified user interface to be operated by nursing staff. The simplified user interface may request a user name at startup, and a very limited number of additional parameters. Once acquired, these parameters will then be concatenated with a user-specific parameter file that contains the actual BCI configuration. Then, the system will start up the word processor, the interfacing program, and BCI2000. Using operator module command line parameters, it will automatically load the temporarily created parameter file, and begin BCI operation. Once the user selects a "quit" command, BCI2000 will quit automatically. By configuring BCI2000 appropriately, it is possible to avoid all references to the BCI2000 user interface on the operator's screen, and to only retain the application module's user screen (i.e., the speller matrix).

3.4 Multi-Site Research

Cooperation across multiple research sites requires agreement on standards regarding experimental design, documentation, data formats. Here, BCI2000 helps by providing such standards, and is versatile enough to accommodate a wide range of needs. Data files consist of a human-readable header that provides extensive documentation, and compact binary data in a single file. Unlike the numeric event coding present in other file formats (such as GDF), BCI2000 parameters and state variables (event markers) intentionally avoid a high level of formalization while enforcing the presence of all aspects of configuration and stimulus presentation in each single data file.

A second important aspect of multi-site cooperation is deployment. Maintaining software installation and configurations on a central data storage connected to the Internet, and keeping these synchronized in an automated fashion, may greatly facilitate synchronization of experiments across sites and quality control, and will avoid duplication of maintenance effort (Fig. 9).

Here, the advantage of BCI2000 is that its binary distribution consists of a single directory tree. Aside from hardware drivers for the data acquisition device, BCI2000 does not require any third-party components. Also, as there is no system-wide installation procedure required, installing or updating BCI2000 can be done without administrative privileges, simply by synchronizing with a reference installation.

Finally, multi-site cooperation requires that all data be available for analysis by researchers at one or multiple sites. Because BCI2000 organizes its data into session directories, and includes comprehensive documentation of recordings that is present inside data files themselves, it is rather simple to centrally maintain a database of recordings by synchronizing with data directories that exist on measurement stations.

In summary, the potential of BCI2000 in conjunction with a central data storage and synchronization allows for: deploying software and experimental setups to a

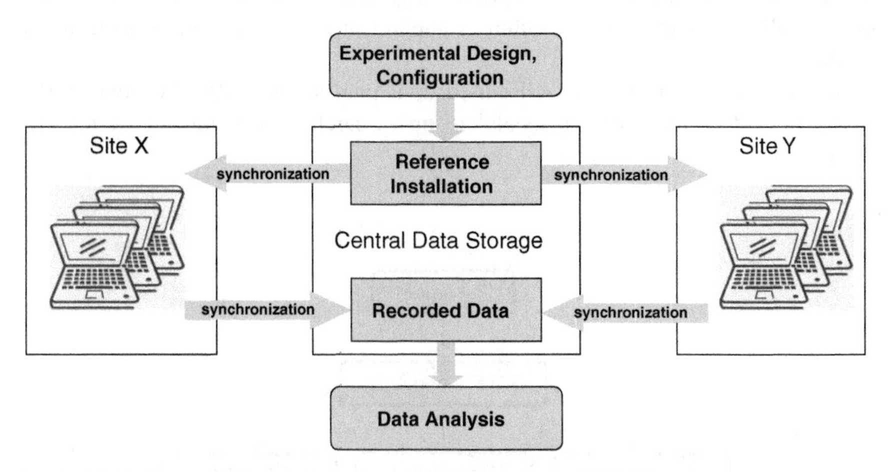

Fig. 9 Multi-site research data flow. On a central location, a BCI2000 reference installation is maintained, which contains experiment configuration as well. This installation is synchronized with measurement equipment located at a number of sites (equipment symbolized by laptop computers; exemplarily, two sites are depicted). Data recordings are synchronized back to the central storage, where all data are then available for analysis. Typically, synchronization will take place over Internet connections

number of measurement stations at different sites; maintaining an overview of a cooperation project's progress; managing aspects of quality control; and managing and analyzing of large amounts of data at virtually no extra cost when compared to single-site projects.

4 Research Trajectories

In conclusion, we would like to review how research activities discussed in this chapter may be combined to form paths of research, or research trajectories. In Fig. 10, smooth transitions between research activities as supported by BCI2000 are indicated by arrows. Potential BCI2000 users may become familiar with the system in the "classroom" scenario. From this, a number of research trajectories are possible when following the arrows. All of these scenarios may profit from BCI2000's existing components, and from its advantages in multiple-site settings as discussed in the previous section.

All suggested trajectories may begin with the "classroom" scenario, and then progress into one of three main directions of research. As a first direction, shifting research focus from BCI research to psychophysiological experiments in general may profit from available BCI2000 knowledge when BCI2000 is used to perform such experiments. As a second direction of research, development of real-world BCI applications may profit from using BCI2000 to first establish a reliable BCI, and then connect it to external software, or devices. Similarly, the third area of research, development of signal processing algorithms, may profit from a development circle in which first data is recorded in BCI experiments using a standard algorithm; then, an improved algorithm is tested off-line on existing data; finally the improved algorithm is applied in an on-line setting to prove its viability.

In summary, BCI2000 is well suited as a platform to enter the area of BCI research, and to then further specialize on research into a number of research directions.

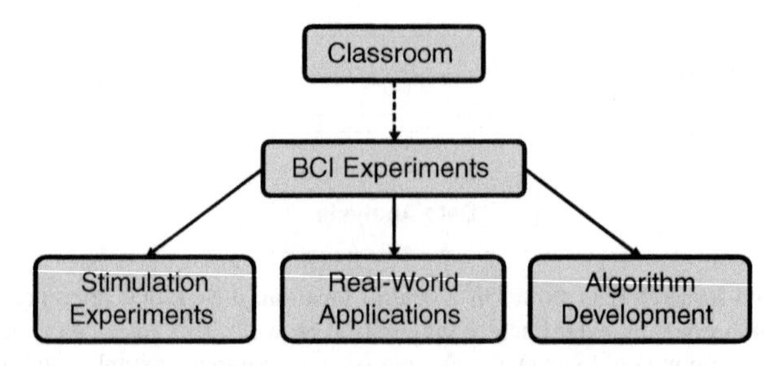

Fig. 10 Research scenarios and trajectories. Starting with the classroom scenario, smooth transitions into a number of BCI research areas are possible

5 Dissemination and Availability

To date, the BCI2000 project has organized a number of workshops on the theory and application of the system: Albany, New York, June 2005; Beijing, China, July 2007; Rome, Italy, December 2007; Utrecht, The Netherlands, July 2008; Bolton Landing, New York, October 2009; Asilomar, California, June 2010. During those workshops, talks provide introductions to practical aspects of BCI experiments, use and modification of BCI2000; and participants also gain BCI experience in hands-on tutorials.

The main web site for the BCI2000 project is http://www.bci2000.org. All source code is available on http://source.bci2000.org; and full documentation is available on http://doc.bci2000.org. Access to BCI2000 source code and executables is free of charge for research and educational purposes.

Since version 3.0, BCI2000 may be compiled using either the Visual C++ compiler, or gcc/MinGW. Compilation of source, signal processing, and application modules is also possible with a recent Borland/CodeGear compiler.

Acknowledgments BCI2000 has been in development since 2000. Since the project's inception, Gerwin Schalk has been responsible for the direction and implementation of the project. Dennis McFarland and Thilo Hinterberger contributed greatly to the initial system design and its implementation, and Drs. Wolpaw and Birbaumer provided support and useful advice in earlier stages of this project. Since 2002, Jürgen Mellinger has been responsible for software design and architecture. Since 2004, Adam Wilson and Peter Brunner have contributed system components and much needed testing. The following individuals or parties have also contributed to the development of BCI2000 (in alphabetical order):

Erik Aarnoutse, Brendan Allison, Maria Laura Blefari, Simona Bufalari, Bob Cardillo, Febo Cincotti, Joshua Fialkoff, Emanuele Fiorilla, Dario Gaetano, g.tec^{TM}, Sebastian Halder, Jeremy Hill, Jenny Hizver, Sam Inverso, Vaishali Kamat, Dean Krusienski, Marco Mattiocco, Griffin "The Geek" Milsap, Melody M. Moore-Jackson, Yvan Pearson-Lecours, Christian Puzicha, Thomas Schreiner, Chintan Shah, Mark Span, Chris Veigl, Janki Vora, Shi Dong Zheng.

Initial development of BCI2000 has been sponsored by an NIH Bioengineering Research Partnership grant (EB00856) to Jonathan Wolpaw. Current development is sponsored by a NIH R01 grant (EB006356) to Gerwin Schalk.

References

1. B.Z. Allison, D.J. McFarland, G. Schalk, S.D. Zheng, M.M. Jackson, and J.R. Wolpaw, Towards an independent brain-computer interface using steady state visual evoked potentials. Clin Neurophysiol, 119(2), 399–408, Feb (2008).
2. C.J. Bell, P. Shenoy, R. Chalodhorn, and R.P. Rao, Control of a humanoid robot by a noninvasive brain-computer interface in humans. J Neural Eng, 5(2), 214–220, Jun (2008).
3. P. Brunner, A.L. Ritaccio, T.M. Lynch, J.F. Emrich, J.A. Wilson, J.C. Williams, E.J. Aarnoutse, N.F. Ramsey, E.C. Leuthardt, H. Bischof, and G Schalk. A practical procedure for real-time functional mapping of eloquent cortex using electrocorticographic signals in humans. Epilepsy Behav, 15(3), 278–286, July (2009).
4. E. Buch, C. Weber, L.G. Cohen, C. Braun, M.A. Dimyan, T. Ard, J. Mellinger, A. Caria, S. Soekadar, A. Fourkas, and N. Birbaumer, Think to move: A neuromagnetic brain-computer interface (BCI) system for chronic stroke. Stroke, 39(3), 910–917, Mar (2008).

5. A.F. Cabrera and K. Dremstrup, Auditory and spatial navigation imagery in brain-computer interface using optimized wavelets. J Neurosci Methods, 174(1), 135–146, Sep (2008).
6. F. Cincotti, D. Mattia, F. Aloise, S. Bufalari, L. Astolfi, F. De Vico Fallani, A. Tocci, L. Bianchi, M.G. Marciani, S. Gao, J. Millan, and F. Babiloni, High-resolution EEG techniques for brain-computer interface applications. J Neurosci Methods, 167(1), 31–42, Jan (2008).
7. F. Cincotti, D. Mattia, F. Aloise, S. Bufalari, G. Schalk, G. Oriolo, A. Cherubini, M.G. Marciani, and F. Babiloni, Non-invasive brain-computer interface system: towards its application as assistive technology. Brain Res Bull, 75(6), 796–803, Apr. (2008).
8. A. Delorme and S. Makeig, EEGLAB: An open source toolbox for analysis of single-trial EEG dynamics including independent component analysis. J Neurosci Methods, 134(1), 9–21, (2004).
9. E.A. Felton, J.A. Wilson, J.C. Williams, and P.C. Garell, Electrocorticographically controlled brain-computer interfaces using motor and sensory imagery in patients with temporary subdural electrode implants. report of four cases. J Neurosurg, 106(3), 495–500, Mar (2007).
10. B. Kemp, A. Värri, A.C. Rosa, K.D. Nielsen, and J. Gade, A simple format for exchange of digitized polygraphic recordings. Electroenceph Clin Neurophysiol, 82, 391–393, (1992).
11. A. Kübler, F. Nijboer, J. Mellinger, T.M. Vaughan, H. Pawelzik, G. Schalk, D.J. McFarland, N. Birbaumer, and J. R. Wolpaw, Patients with ALS can use sensorimotor rhythms to operate a brain-computer interface. Neurology, 64(10), 1775–1777, May (2005).
12. E.C. Leuthardt, K. Miller, N.R. Anderson, G. Schalk, J. Dowling, J. Miller, D.W. Moran, and J.G. Ojemann, Electrocorticographic frequency alteration mapping: a clinical technique for mapping the motor cortex. Neurosurgery, 60(4 Suppl 2), 260–270, Apr (2007).
13. E.C. Leuthardt, K.J. Miller, G. Schalk, R.P. Rao, and J.G. Ojemann, Electrocorticography-based brain computer interface–the Seattle experience. IEEE Trans Neural Syst Rehabil Eng, 14(2), 194–198, Jun (2006).
14. E.C. Leuthardt, G. Schalk, J.R. Wolpaw, J.G. Ojemann, and D.W. Moran, A brain-computer interface using electrocorticographic signals in humans. J Neural Eng, 1(2), 63–71, Jun (2004).
15. S.G. Mason and G.E. Birch, A general framework for brain-computer interface design. IEEE Trans Neur Syst Rehabil Eng, 11(1), 70–85, (2003).
16. D.J. McFarland, D.J. Krusienski, W.A. Sarnacki, and J.R. Wolpaw, Emulation of computer mouse control with a noninvasive brain-computer interface. J Neural Eng, 5(2), 101–110, Mar (2008).
17. D.J. McFarland, W.A. Sarnacki, and J.R. Wolpaw, Electroencephalographic (EEG) control of three-dimensional movement. J Neural Eng, 7(3), 036007, June (2010).
18. J. Mellinger and G. Schalk, BCI2000: A general-purpose software platform for BCI. In: G. Dornhege, J.d.R. Millan, T. Hinterberger, D.J. McFarland, and K.R. Müller, (Eds), *Toward brain-computer interfacing*, MIT Press, Cambridge, MA, (2007).
19. J. Mellinger, G. Schalk, C. Braun, H. Preissl, W. Rosenstiel, N. Birbaumer, and A. Kübler. An MEG-based brain-computer interface (BCI). Neuroimage, 36(3), 581–593, Jul (2007).
20. K.J. Miller, M. denNijs, P. Shenoy, J.W. Miller, R.P. Rao, and J.G. Ojemann, Real-time functional brain mapping using electrocorticography. Neuroimage, 37(2), 504–507, Aug (2007).
21. K.J. Miller, E.C. Leuthardt, G. Schalk, R.P. Rao, N.R. Anderson, D.W. Moran, J.W. Miller, and J.G. Ojemann, Spectral changes in cortical surface potentials during motor movement. J Neurosci, 27(9), 2424–2432, Feb (2007).
22. F. Nijboer, E.W. Sellers, J. Mellinger, M.A. Jordan, T. Matuz, A. Furdea, S. Halder, U. Mochty, D.J. Krusienski, T.M. Vaughan, J.R. Wolpaw, N. Birbaumer, and A. Kübler, A P300-based brain-computer interface for people with amyotrophic lateral sclerosis. Clin Neurophysiol, 119(8), 1909–1916, Aug (2008).
23. A.S. Royer and B. He., Goal selection versus process control in a brain-computer interface based on sensorimotor rhythms. J Neural Eng, 6(1), 16005–16005, Feb (2009).

24. G. Schalk, P. Brunner, L.A. Gerhardt, H. Bischof, and J.R. Wolpaw, Brain-computer interfaces (bcis): detection instead of classification. J Neurosci Methods, 167(1), 51–62, Jan (2008).
25. G. Schalk, J. Kubánek, K.J. Miller, N.R. Anderson, E.C. Leuthardt, J.G. Ojemann, D. Limbrick, D. Moran, L.A. Gerhardt, and J.R. Wolpaw, Decoding two-dimensional movement trajectories using electrocorticographic signals in humans. J Neural Eng, 4(3), 264–275, Sep (2007).
26. G. Schalk, E.C. Leuthardt, P. Brunner, J.G. Ojemann, L.A. Gerhardt, and J.R. Wolpaw, Real-time detection of event-related brain activity. *Neuroimage*, 43(2), 245–249, Nov (2008).
27. G. Schalk, D.J. McFarland, T. Hinterberger, N. Birbaumer, and J.R. Wolpaw, BCI2000: A general-purpose brain-computer interface (BCI) system. IEEE Trans Biomed Eng, 51, 1034–1043, (2004).
28. G. Schalk, K.J. Miller, N.R. Anderson, J.A. Wilson, M.D. Smyth, J.G. Ojemann, D.W. Moran, J.R. Wolpaw, and E.C. Leuthardt, Two-dimensional movement control using electrocorticographic signals in humans. J Neural Eng, 5(1), 75–84, Mar 2008.
29. A. Schlögl, GDF – A general dataformat for biosignals. ArXiv Comput Sci arXiv:cs/0608052v6 [cs.DL], August (2006).
30. A. Schlögl, G.R. Müller, R. Scherer, and G. Pfurtscheller, BioSig – an open source software package for biomedical signal processing. 2nd OpenECG Workshop, Berlin, 1–3 Apr (2004).
31. J.R. Stroop, Studies of interference in serial verbal reactions. J Exp Psych, 18(6), 643–662, Dec 1935.
32. T.M. Vaughan, D.J. McFarland, G. Schalk, W.A. Sarnacki, D.J. Krusienski, E.W. Sellers, and J.R. Wolpaw, The Wadsworth BCI Research and Development Program: At home with BCI. IEEE Trans Neural Syst Rehabil Eng, 14(2), 229–233, Jun 2006.
33. J.A. Wilson, E.A. Felton, P.C. Garell, G. Schalk, and J.C. Williams, ECoG factors underlying multimodal control of a brain-computer interface. IEEE Trans Neural Syst Rehabil Eng, 14(2), 246–250, Jun 2006.
34. K.J. Wisneski, N. Anderson, G. Schalk, M. Smyth, D. Moran, and E.C. Leuthardt, Unique cortical physiology associated with ipsilateral hand movements and neuroprosthetic implications. Stroke, 39(12), 3351–3359, Dec 2008.
35. J.R. Wolpaw, N. Birbaumer, D.J. McFarland, G. Pfurtscheller, and T.M. Vaughan, Brain-computer interfaces for communication and control. Electroenceph Clin Neurophysiol, 113(6), 767–791, June 2002.
36. J.R. Wolpaw and D.J. McFarland., Control of a two-dimensional movement signal by a noninvasive brain-computer interface in humans. Proc Natl Acad Sci U S A, 101(51), 17849–17854, Dec 2004.
37. N. Yamawaki, C. Wilke, Z. Liu, and B. He, An enhanced time-frequency-spatial approach for motor imagery classification. IEEE Trans Neural Syst Rehabil Eng, 14(2), 250–254, Jun 2006.

The First Commercial Brain–Computer Interface Environment

Christoph Guger and Günter Edlinger

1 Introduction

The first commercial brain–computer interface environment has been developed so research centers could easily and quickly run BCI experiments to test algorithms and different strategies. A first BCI system was available on the market in 1999, and was continuously improved to the system available today, which is now used in more than 60 countries worldwide.

Many factors influence the design of a BCI system, as shown in Fig. 1. There are technical issues and issues concerning the individual subject that must be addressed. Different brain signals are used in BCIs, such as the Electroencephalogram (EEG) recorded non-invasively from the scalp, or the Electrocorticogram (ECoG), which requires invasive electrodes. Therefore, different safety issues, sampling frequencies and electrode types are required. The applied signal processing algorithms have to work in on-line mode to provide real-time capabilities and decision making in real life situations. However, the algorithms must also work in off-line mode to support in depth analysis of already acquired EEG data. The signal acquisition and processing unit has to ensure high data quality using e.g. over-sampling techniques providing a high signal to noise ratio (SNR). One type of BCI approach is based on motor imagery [1] (see also Chapters "Brain Signals for Brain–Computer Interfaces" and "Dynamics of Sensorimotor Oscillations in a Motor Task" in this book). This approach can be realized with only a few electrodes over the sensorimotor cortex. Some other approaches that use spatial filtering techniques [2] use about 16–128 electrodes. A BCI based on ECoG signals also requires many channels (64–128) because dense electrode grids overlaying parts of the cortex [3] are utilized.

C. Guger (✉)
Guger Technologies OG/g.tec medical engineering GmbH, Herbersteinstrasse 60, 8020 Graz, Austria
e-mail: guger@gtec.at

B. Graimann et al. (eds.), *Brain–Computer Interfaces*, The Frontiers Collection,
DOI 10.1007/978-3-642-02091-9_16, © Springer-Verlag Berlin Heidelberg 2010

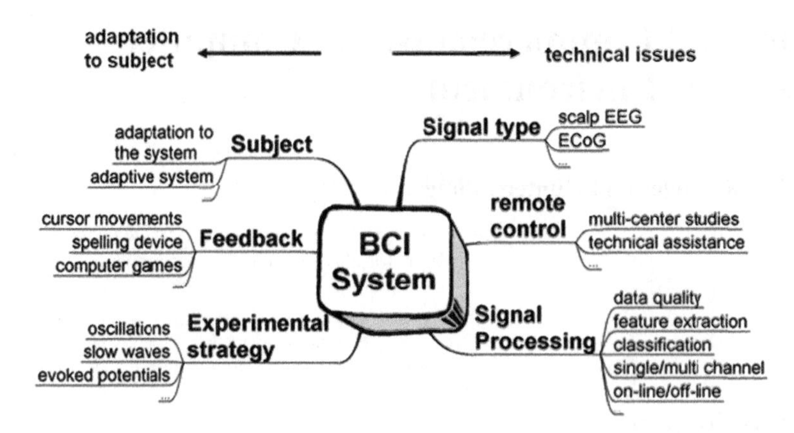

Fig. 1 Components that influence BCI development

For higher numbers of channels, the real-time capabilities of the BCI system must be also guaranteed for accurately timed generation of the desired control signals for feedback to e.g. external devices.

On the subject side, it is very important that the BCI system adapts itself to the learning process of the user. This can either be done in on-line mode, where the classification algorithm is retrained after each decision, or in a quasi on-line mode, where e.g. 40 decisions are made and the classifier is then retrained to adapt to the new control signals [4]. However, it is also important that the subject adapts to the BCI system and learns how to control the system. Hence, the provided feedback strategy is very important for a successful learning process. Immediate feedback of a cursor movement, fast control in computer games or correct responses of a spelling device are therefore vital requirements. Very important for a subject's success in BCI control is the choice of the optimal BCI approach which can be based on: (i) oscillations in the alpha and beta range, (ii) slow waves, (iii) P300 components or (iv) steady-state visual evoked potentials (SSVEP) [5–8]. For some subjects, oscillations in the alpha and beta band during motor imagery work better than the evoked potential approach and vice versa. Therefore, it is very important that the BCI system enables testing the subject with all these different approaches to find the optimal one.

A traditional development approach is shown in Fig. 2, left side. Here, feature extraction algorithms (e.g. recursive least square (RLS) estimation of adaptive autoregressive parameters – AAR) and classification algorithms (e.g. linear discriminant analysis – LDA) are typically implemented in a first step for the off-line analysis of already acquired EEG data. After adaptation and optimization of the algorithms, the hardware for the EEG acquisition and the real-time processing environment is developed, leading to a new implementation process of the off-line algorithms and to many iterations between the algorithm and hardware design. On the other side, the rapid prototyping process combines the algorithm design with the real-time tests on the hardware (see Fig. 2, right side). Typically, an already proven hardware and software environment is used for such an approach. The diagrams in

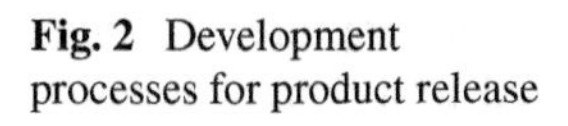

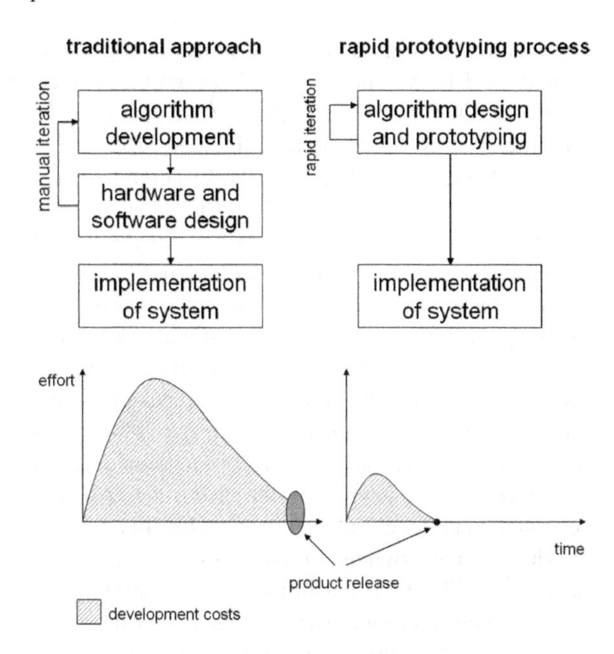

Fig. 2 Development processes for product release

Fig. 2 illustrate the effort and the development costs of the traditional approach, which are typically much higher than for the rapid prototyping process [4].

Based on the discussed above requirements, a rapid prototyping environment for BCI systems has been developed and upgraded since 1996 [9]. The most important features of this BCI environment will be discussed in detail in the next sections, and the realization of important BCI applications based on this environment will be shown.

2 Rapid Prototyping Environment

2.1 Biosignal Amplifier Concepts

One of the key components of a BCI system is the biosignal amplifier for EEG or ECoG data acquisition. Figure 3 illustrates 3 different devices with different specific

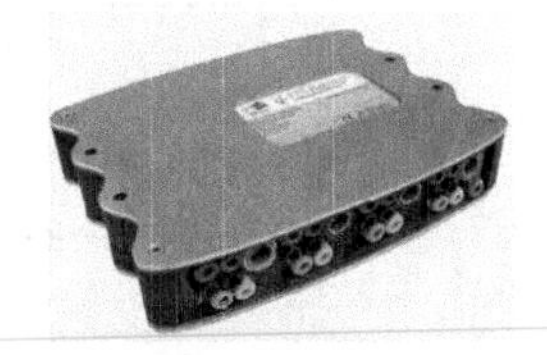

Fig. 3 *Left*: 16 channel stand-alone amplifier g.BSamp. Mid: 8 channel wireless amplifier g.MOBIlab+. *Right*: 16 channel amplifier g.USBamp with USB connection

key features. g.BSamp is a stand-alone analog amplifier which amplifies the input signals to ±10 V. The output of the amplifier is connected to a data acquisition board (DAQ) for analog to digital conversion (ADC). g.MOBIlab+ is a portable amplifier that transmits already digitized EEG data via Bluetooth to the processing unit. g.USBamp sends the digitized EEG via USB to the processing unit.

A block diagram of g.USBamp is given in Fig. 4. This device has 16 input channels, which are connected over software controllable switches to the internal amplifier stages and anti-aliasing filters before the signals are digitized with sixteen 24 Bit ADCs. The device is also equipped with digital to analog converters (DAC) enabling the generation of different signals like sinusoidal waves, which can be sent to the inputs of the amplifiers for system testing and calibration. Additionally, the impedance of each electrode can be checked by applying a small current via the individual electrodes and measuring the voltage drops. All these components are part of the so-called applied part of the device, as a subject or patient is in contact to this part of the device via the electrodes. All following parts of the device are separated via optical links from the subject/patient.

The digitized signals are passed to a digital signal processor (DSP) for further processing. The DSP performs an over-sampling of the EEG data, band pass filtering, Notch filtering to suppress the power line interference and calculates bipolar derivations. These processing stages eliminate unwanted noise from the EEG signal, which helps to ensure accurate and reliable classification. Then the pre-processed

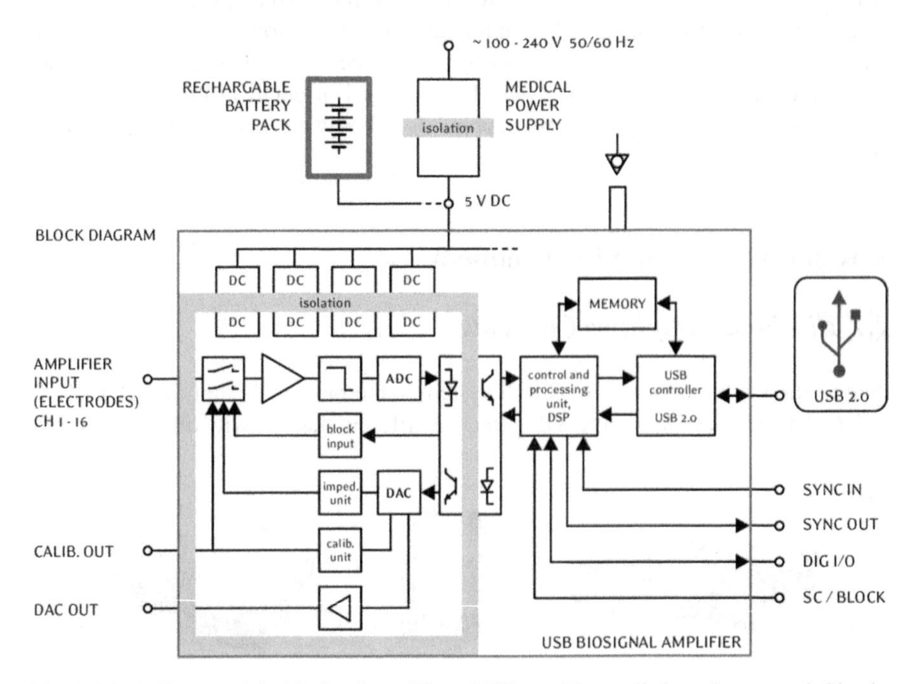

Fig. 4 Block diagram of the biosignal amplifier g.USBamp. The applied part is surrounded by the dark gray frame

EEG data is sent to a controller which transmits the data via USB 2.0 to the PC. One important feature of the amplifier is the over-sampling capability. Each ADC is sampling the EEG at 2.4 MHz. Then the samples are averaged to the desired sampling frequency of e.g. 128 Hz. Here a total of 19.200 samples are averaged, which improves the signal to noise ratio by the square root of 19.200 = 138.6 times.

For EEG or ECoG recordings with many channels, multiple devices can be daisy chained. One common synchronization signal is utilized for all ADCs, yielding a

Table 1 Technical key properties of biosignal amplifiers for BCI operation

	g.BSamp	g.MOBIlab+	g.USBamp
Signal type	EEG/EP/ECoG	EEG/EP	EEG/EP/ECoG
Channels number N	8/16	8	16
Stackable	32–80	–	32–256
Sampling rate [Hz]	250 kHz/N	256	64–38.4 k
Simultaneous sample and hold	No	No	Yes
ADC inside amplifier	No	Yes	Yes
#ADCs	1	1	16
ADC resolution [Bit]	16	16	24
Over sampling	–	–	19.400 at 128 Hz
Conversion time [μs]	4 μs	43 μs	26 μs
Time delay between 1st and last channel	Conversion time * N	Conversion time * 8	Conversion time
Interface	PCI/PCMCIA	Bluetooth	USB 2.0
Range [m]	2	30	3
Power supply	12 V medical power supply or battery	4 AA batteries	5 V medical power supply or battery
Operation time on battery [h]	8	36	8
Input Sensitivity	±5 mV/±500 μV	±500 μV	±250 mV
Minimum high pass [Hz]	0.01	0.01	0
Maximum low pass [Hz]	5 k	100	6.8 k
Band pass filter	Analog	Analog	Digital (DSP)
Notch filter	Analog	–	Digital (DSP)
Derivation	Bipolar	Monopolar	Monopolar/bipolar (DSP)
# ground potentials	2	1	4
Connectors	1.5 mm safety	1.5 mm safety/system connectors	1.5 mm safety/system connectors
Impedance test	External	External	Internal
Applied part	CF	BF	CF
Isolation of applied part	Isolation amplifier	Opto-coupler	Opto-coupler
Digital I/Os	8	8	8
Scanned with inputs	No	Yes	Yes

perfect non delayed acquisition of all connected amplifiers. This is especially important for evoked potential recordings or recordings with many EEG channels. If only one ADC with a specific conversion time is used for many channels, then a time lag between the first channel and the last channel could be the result (e.g. 100 channels * 10 μs = 1 ms). Important is also that biosignal acquisition systems provide trigger inputs and outputs to log external events in synchrony to the EEG data or to send trigger information to other external devices such as a visual flash lamp. Digital outputs can also be used to control external devices such as a prosthetic hand or a wheelchair. An advantage here is to scan the digital inputs together with the biosignals to avoid time-shifts between events and physiological data. A medical power supply that works with 220 and 110 V is required for BCI systems that are used mainly in the lab. For mobile applications like the controlling a wheelchair, amplifiers which run on battery power are also useful.

Table 1 compares key technical properties of the 3 amplifiers shown in Fig. 3 (g.BSamp, g.MOBIlab+ and g.USBamp). The most important factor in selecting an appropriate amplifier is whether EEG or ECoG data should be processed. For ECoG, only devices with an applied part of type CF are allowed. For EEG measurements, both BF and CF type devices can be used. The difference here is the maximum allowed leakage current. Leakage current refers to electric current that is lost from the hardware, and could be dangerous for people or equipment. For both systems, the character F indicates that the applied part is isolated from the other parts of the amplifier. This isolation is typically done based on opto-couplers or isolation amplifiers. For a BF device, the ground leakage current and the patient leakage current must be =100 μA according to the medical device requirements, such as IEC 60601 or EN 60601. These refer to widely recognized standards that specify details of how much leakage current is allowed, among other details. For a CF device, the rules are more stringent. The ground leakage current can also be =100 μA, but the patient leakage current must be =10 μA only. The classification and type of the device can be seen from the labeling outside. Details for BF and CF devices can be seen in Fig. 6.

The next important feature is the number of electrodes used. For slow wave approaches or oscillations in the alpha and beta range and P300 systems, a total of 1–8 EEG channels are sufficient [10–12]. BCIs that use spatial filtering, such as common spatial pattern (CSP), require more channels (16–128) [13]. For ECoG recordings, 64–128 channel montages are typically used [3]. Therefore, stack-able systems might be advantageous because they can extend the functionality with future applications. A stack-able e.g. 64 channel system can also be split into four 16 channels systems if required for some experiments. USB 2.0 provides a much higher bandwidth than Bluetooth and therefore allows higher sampling frequencies and more channels. Two clear advantages of Bluetooth devices are portability and mobility. Subjects can move freely within a radius of about 30 m. USB based wired devices have cable lengths of about 3 m, and the distance between a stand-alone amplifier and a DAQ board should be as short as possible (<2 m). Another advantage is that moving the ADC as close as possible to the amplification unit yields a higher signal to noise ratio.

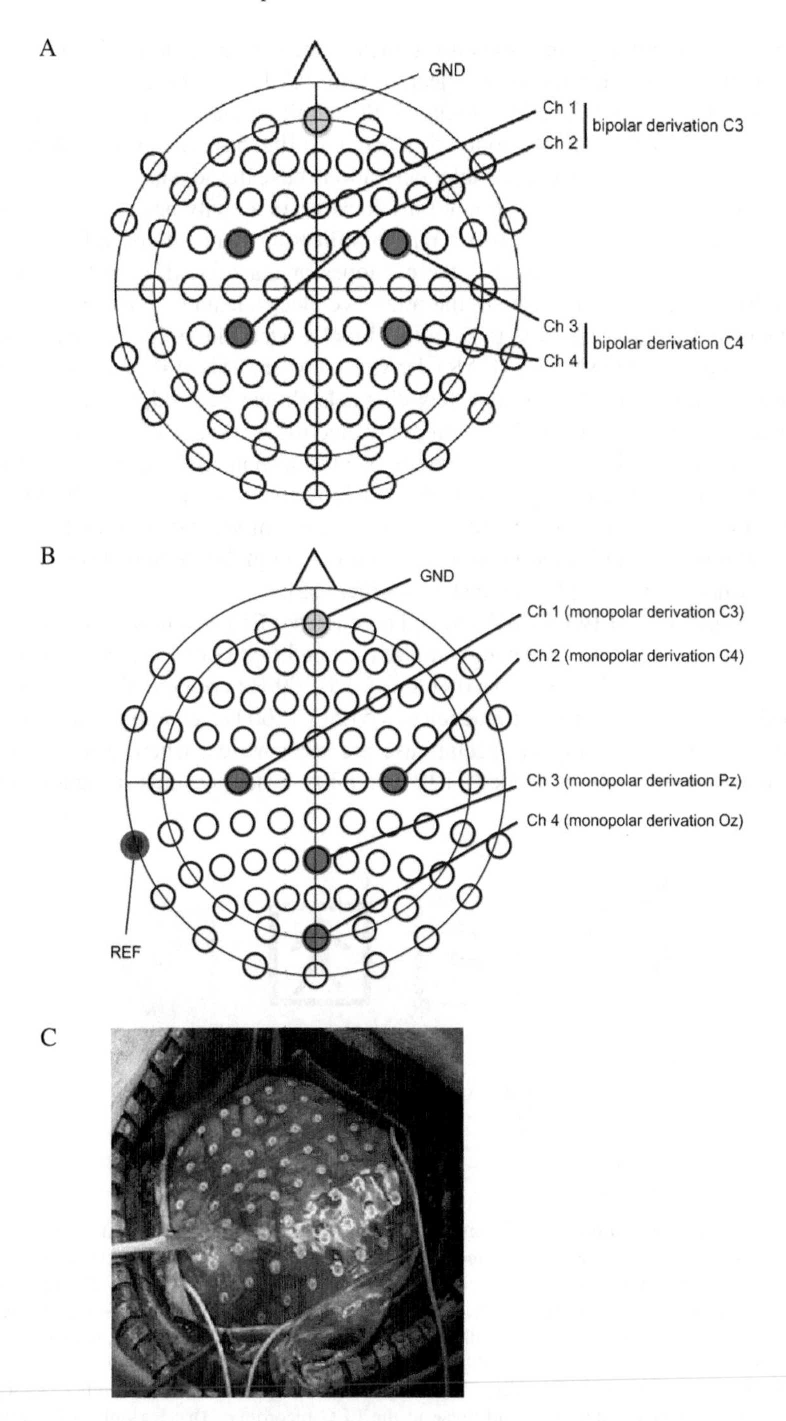

Fig. 5 Scalp electrode montage with bipolar (**a**) and monopolar recordings (**b**). (**c**): ECoG electrode with 64 electrodes (picture from Gerwin Schalk Wadsworth Center, USA, Kai Miller and Jeff Ojemann University of Washington, USA)

Amplifiers with bipolar inputs use typically instrumentation amplifiers as input units with a high common mode rejection ratio (CMRR). The electrodes are connected to the plus and the minus inputs of the amplifier and electrodes are mounted on the scalp in a distance of about 2.5–10 cm (see Fig. 5a). The ground electrode is mounted e.g. on the forehead. Bipolar derivations have the advantage of suppressing noise and artifacts very well, so that only local brain activity near the electrodes is picked up. In contrast, monopolar input amplifiers have a common reference electrode that is typically mounted at the ear lobes or mastoids (Fig. 5b). Monopolar recordings refer measurements to the reference electrode and are more sensitive to artifacts, but make it possible to calculate bipolar, small/large Laplacian, Hjorth's, or common average reference (CAR) derivations afterwards [14]. Typically, bipolar derivations are preferred if only a few channels should be used, while monopolar channels are used for recordings with many electrodes such as ECoG or when spatial filters are applied (Fig. 5C). Groups of ground potential separated amplifiers allow the simultaneous acquisition of other biosignals like ECG and EMG along with EEG without any interference. Another benefit of separated ground potentials is the ability to record multiple subjects with one amplifier, which allows e.g. BCI games where users can play against each other [15].

The signal type (EEG, ECoG, evoked potentials – EP) also influences the necessary sampling frequency and bandwidth of the amplifier. For EEG signals, sampling frequencies of 256 Hz with a bandwidth of 0.5–100 Hz are typically used [4]. For ECoG recordings, sampling frequencies of 512 or 1200 Hz are applied with a bandwidth of 0.5–500 Hz [3]. A special case are slow waves, where a lower cut-off frequency of 0.01 Hz is needed [10]. For P300 based systems, a bandwidth of

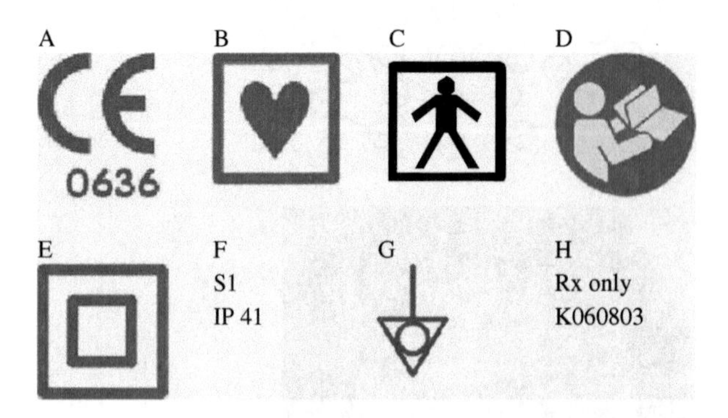

Fig. 6 Labeling on the outside of biosignal amplifiers. (**a**): In the European Union, only amplifiers with the CE mark are allowed. The number indicates the notified body that tested the product according to EN 60601. (**b**): Applied part of type CF. (**c**): Applied part of type BF. (**d**): Follow instructions for use. (**e**): Safety class II. (**f**): S1 – Permanent operation., IP 41: Protected against particulate matter of degree 4 (wire with diameter of 1 mm). Protection against ingress of water of degree 1 (protection against vertical falling water drops). (**g**): Potential equalization connector – should be connected to the shielding of the experimental room, to the PC and other metal parts surrounding the BCI system to avoid noise in the EEG recordings. (**h**): Rx only – Prescription device in the USA and example K-number for FDA approved device

0.1–30 Hz is typically used [16]. Notch filters are used to suppress the 50 Hz or 60 Hz power line interference. A notch filter is typically a narrow band-stop filter having a very high order. Digital filtering has the advantage that every filter type (Butterworth, Bessel, etc), filter order, and cut-off frequency can be realized. Analog filters inside the amplifier are predefined and can therefore not be changed. The high input range of g.USBamp of $\pm$250 mV combined with a 24 Bit converter (resolution of 29 nV) allows measuring all types of biosignals (EMG, ECG, EOG, EPs, EEG, ECoG) without changing the amplification factor of the device. For 16 Bit AD converters, the input range must be lower in order to have a high enough ADC resolution.

Figure 6 shows usual labels at the outside of biosignal amplifiers.

2.2 Electrode Caps

EEG electrodes are normally distributed on the scalp according to the international 10–20 electrode system [17]. Therefore, the distance from the Inion to the Nasion is first measured. Then, electrode Cz on the vertex of the cap is shifted exactly to 50% of this distance, as indicated in Fig. 7a. Figure 7b shows a cap with 64 positions. The cap uses screwable single electrodes to adjust the depth and optimize electrode impedance. Each electrode has a 1.5 mm safety connector which can be directly connected to the biosignal amplifiers. Active electrodes have system connectors to supply the electronic components with power. There are two main advantages of a single electrode system: (i) if one electrode breaks down it can be removed immediately and (ii) every electrode montage can be realized easily. The disadvantage is that all electrodes must be connected separately each time. Hence, caps are also available with integrated electrodes. All the electrodes are combined in one ribbon cable that can be directly connected to system connectors of the amplifiers. The main disadvantage is the inflexibility of the montage, and the whole cap must be removed if one electrode breaks down.

2.3 Programming Environment

BCI systems are constructed under different operating systems (OS) and programming environments. Windows is currently the most widely distributed platform, but there are also implementations under Window Mobile, Linux and Mac OS. C++, LabVIEW (National Instruments Corp., Austin, TX, USA) and MATLAB (The MathWorks Inc., Natick, USA) are mostly used as programming languages. C++ implementations have the advantages that no underlying software package is needed when the software should be distributed, and allow a very flexible system design. Therefore, a C++ Application Program Interface (API) was developed that allows the integration of the amplifiers with all features into programs running under Windows or Windows Mobile. The main disadvantage is the longer development time as shown in Table 2. BCI2000 was developed with the C++ API as platform

Table 2 Features of programming environments

	Flexibility	Development speed	Required SW
C API	+++	+	none
MATLAB API	++	++	Matlab
LabVIEW High-Speed	+	++	LabVIEW
Simulink High-speed	+	+++	Matlab/Simulink

to eliminate some of the overhead costs of BCI system development (see [14] and Chapter "Using BCI2000 in BCI Research").

Under the MATLAB environment, several specialized toolboxes such as signal processing, statistics, wavelets, and neural networks are available, which are highly useful components for a BCI system. Signal processing algorithms are needed for feature extraction, classification methods are needed to separate EEG patterns into distinct classes, and statistical functions are needed e.g. for performing group studies. Therefore, a MATLAB API was also developed, which is seamlessly integrated into the Data Acquisition Toolbox (for the different concepts see Fig. 8). This allows direct control of the amplification unit from the MATLAB command window to capture the biosignal data in real-time and to write user specific m-files for the data processing. Furthermore, standard MATLAB toolboxes can be used for processing, as well as self-written programs. The MATLAB processing engine is based upon highly optimized matrix operations, allowing very high processing speed. Such a processing speed is very difficult to realize with self-written C code.

The signal flow oriented Simulink which runs under MATLAB allows people to create block diagrams for signal processing. Therefore, a hardware-interrupt driven device driver was implemented which sends the data in real-time to the Simulink model (High-speed On-line Processing for Simulink toolbox). Therefore,

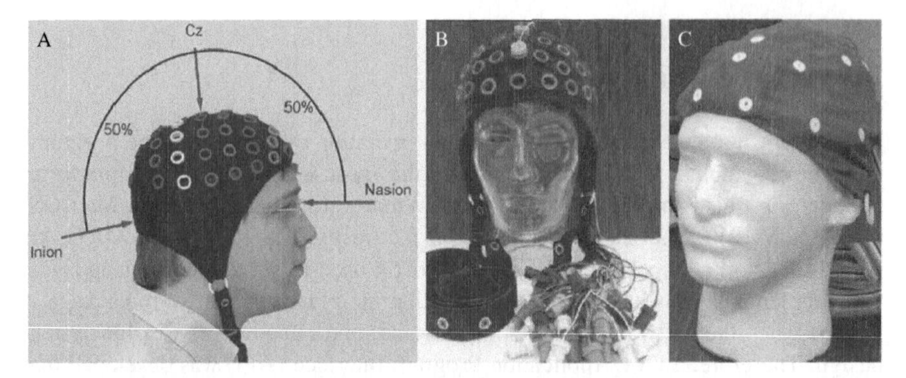

Fig. 7 (**a**): Electrode positioning according to the 10/20 electrode system. (**b**): Electrode cap with screwable single passive or active electrodes. (**c**): Electrode cap with build-in electrodes with a specific montage

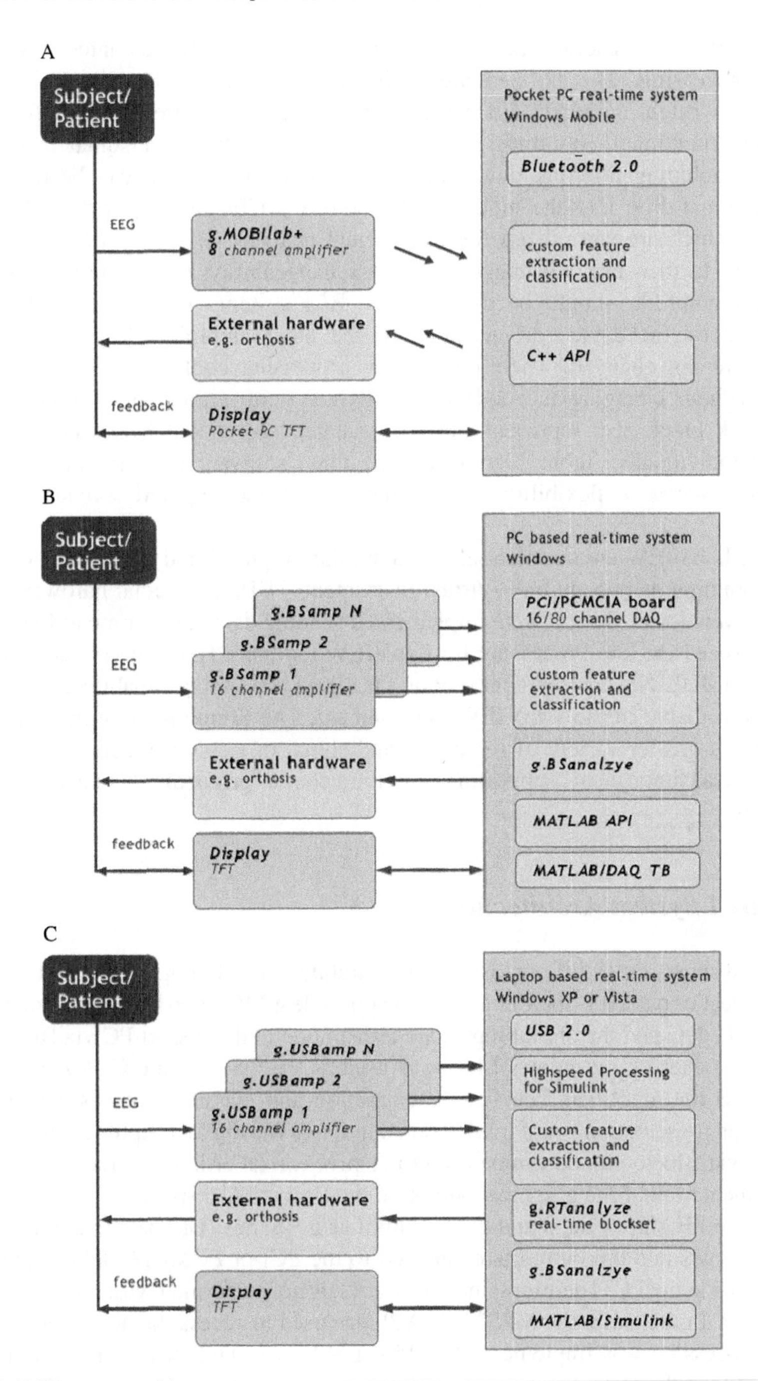

Fig. 8 BCI system architectures

the amplifier is pacing the whole block model, which guarantees the real-time processing. The utilized algorithms represented via the Simulink blocks can be written in MATLAB code or in C code and are called s-functions (system functions). Simulink provides also a large library of signal processing blocks including filtering, down-sampling, online FFT, which can be integrated by drag-and-drop into the models. The toolbox g.RTanalyze provides additional blocks which are specialized for EEG signal analysis (Hjorth, Barlow, RLS,...) [18]. A further big advantage is the exchangeability of the blocks. If e.g. another amplifier should be used for the BCI system, then only this block is replaced by the device driver block of the other amplifier. All other components are not changed. Therefore, a new processing configuration can be realized without changing the code of the BCI system within only minutes. Each Simulink block also represents an encapsulated object which can be used and tested individually. Table 2 compares the three software programming environments in terms of flexibility, development speed and required software (except the OS).

The LabVIEW environment allows a similar graphical and signal flow oriented programming as in Simulink. Virtual instruments (VI), i.e. special hardware interrupt driven device drivers were implemented sending data in real-time to LabVIEW (High-speed On-line Processing for LabVIEW toolbox). Hence users can similarly to the SIMULINK environment rely on the signal processing capabilities and algorithms available in the LabVIEW environment. The Simulink environment offers perhaps a greater variety of toolboxes and signal processing functions; LabView has the advantage of supporting communication protocols between different devices.

2.4 BCI System Architectures

Figure 8 shows three different BCI system architectures. In Fig. 8a, the EEG data of the subject or patient is measured with the portable g.MOBIlab+ biosignal amplifier. The EEG data and the digital inputs are transmitted to the Pocket PC via Bluetooth. On the Pocket PC, Windows Mobile is used as the OS, and the C++ API is used to access the amplifier. The feature extraction and classification, as well as the paradigm presentation, were implemented in C++. The digital outputs are also transmitted via Bluetooth and can be used to control e.g. an orthosis. Important is a fast implementation of the graphical output on the Pocket PC display.

Figure 8B shows the stand-alone amplifier g.BSamp. The device amplifies the input signals and the output is connected to the PCI or PCMCIA data acquisition board inside the PC. Therefore, the digitization is not performed within the biosignal amplifier. In this case, the MATLAB API was used to access the amplifier, and the BCI processing was implemented in MATLAB code. Digital outputs can also be accessed via the Data Acquisition toolbox.

Figure 8c uses g.USBamps, which are connected with a USB hub to the laptop. The device sends the data via USB 2.0. In this case, the g.USBamp High-speed

On-line Processing for Simulink toolbox was used for the signal processing and paradigm presentation in real-time.

3 BCI Training

3.1 Training for a Motor Imagery BCI Approach

For BCI training, EEG data must first be acquired in an off-line session. This means that an experimental paradigm is presented to the subject but no feedback is provided. Such a training session for motor imagery is illustrated in Fig. 9.

A subject sits in front of the Pocket PC display and waits until an arrow pointing either to the right or left side of the screen occurs. The direction of the arrow instructs the subject to imagine a right or left hand movement for 3 s. Then, after some delay, the next arrow appears. The direction of the arrows is randomly chosen, and about 30–200 trials are typically used for further processing. The EEG data, together with the time points of the appearance of the arrows on the screen, are loaded for off-line analysis. This is done with a package called g.BSanalyze. Then the EEG data are divided into epochs of 2 s before and 6 s after the arrow, yielding to a trial length of 8 s. Then class labels (right or left) have to be assigned to each of the trials. Afterwards, reactive frequency components of the EEG data during the motor imagery should be investigated [1]. Therefore, the power spectrum of the EEG data for the resting period (e.g. 1–2 s before the arrow) and during the imagination (e.g. 2–3 s after the arrow) are calculated. This allows finding spectral differences between the reference and active period. Also ERD/ERS maps

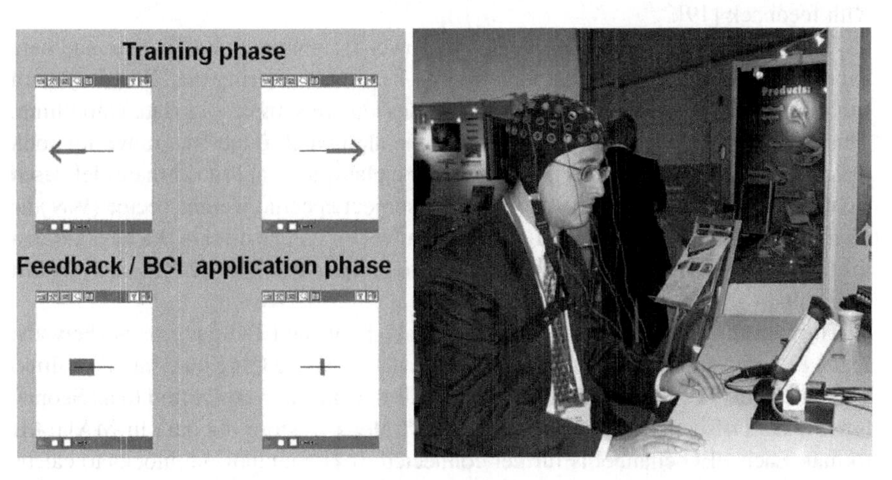

Fig. 9 *Left, top row*: BCI training with arrows pointing to the left and right side; *Left, bottom row*: Feedback with cursor movement to the left and right side. *Right*: BCI training with a Pocket PC implementation of the BCI

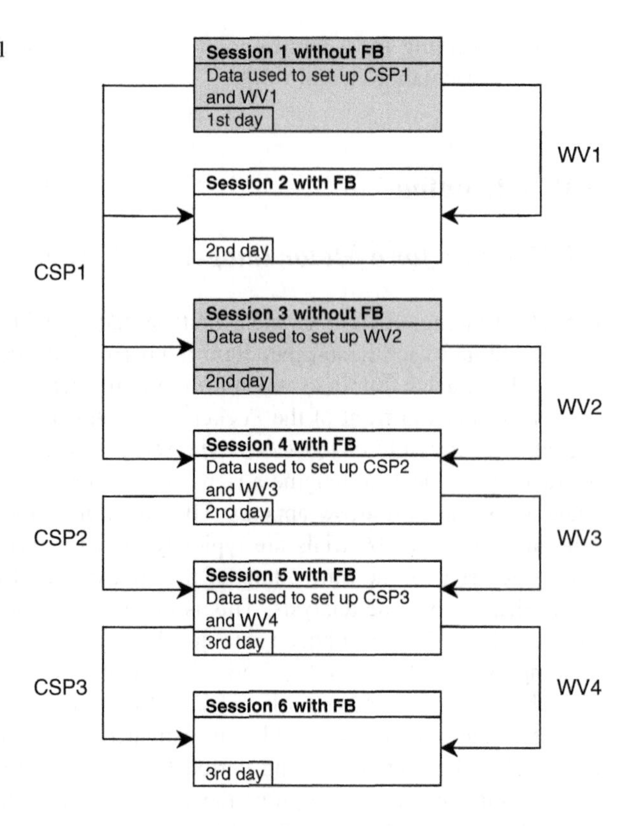

Fig. 10 Typical experimental workflow for EEG data acquisition using spatial pattern calculations (CSP) without feedback (FB) and weight vector (WV) calculation for the sessions with feedback

(event-related desynchronization and synchronization) can be calculated to identify ERD/ERS components during the imagination for further real-time experiments with feedback [19].

Based on this knowledge e.g. the band power is computed in the alpha and beta bands of the EEG data. This is done by first band pass filtering the EEG data, then squaring each sample and averaging over consecutive samples for data smoothing. This results in a band power estimation in the alpha and in the beta range for each channel. These signal features are then sent to a classifier that discriminates left from right movement imagination. As a result, a subject specific weight vector (WV) as illustrated in Fig. 10 is computed. This weight vector can be used in the next session to classify in real-time the EEG patterns and to give feedback to the subject as shown in Fig. 9.

The Simulink model for the real-time analysis of the EEG patterns is shown in Fig. 11. Here "g.USBamp" represents the device driver reading data into Simulink. Then the data is converted to "double" precision format and connected to a "Scope" for raw data visualization and to a "To File" block to store the data in MATLAB format. Each EEG channel is further connected to 2 "Bandpower" blocks to calculate the power in the alpha and beta frequency range (both ranges were identified with the ERD/ERS and spectral analysis). The outputs of the band power calculation are connected to the "BCI System", i.e. the real-time LDA implementation which

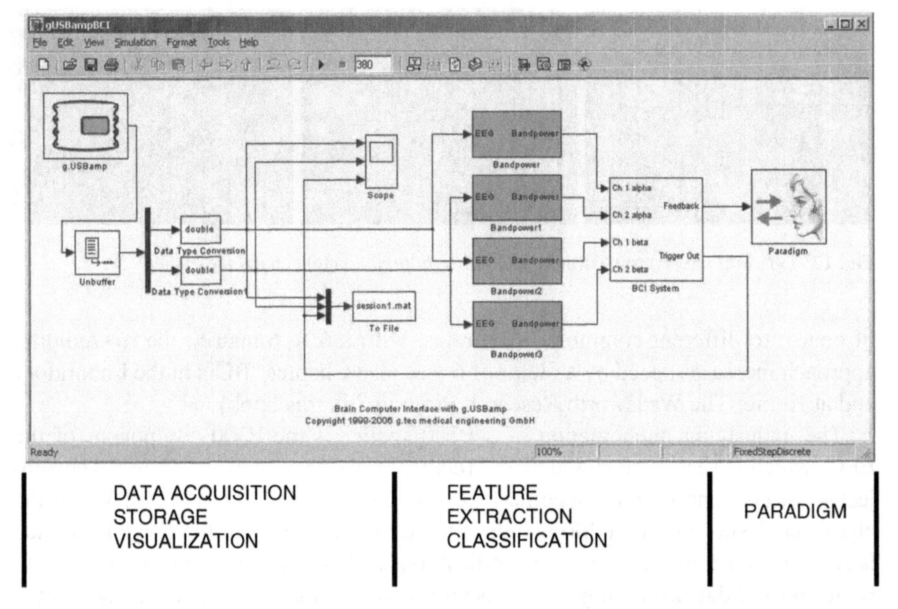

Fig. 11 Simulink model for the real-time feature extraction, classification and paradigm presentation

multiplies the features with the WV. The "Paradigm" block is responsible for the presentation of the experimental paradigm in this case the control of the arrows on the screen and the feedback.

In addition to the parameter estimation and classification algorithms, spatial filters such as common spatial patterns (CSP), independent component analysis (ICA), or Laplacian derivation can also be applied [2]. In this case, spatial patterns are also calculated from the first session to filter the EEG data before feature extraction and classification. This also means the subject specific WV is trained on this specific spatial filter. During the real-time experiments with feedback, the EEG data is influenced and changed because of the BCI system feedback and the subject has the chance to learn and adapt to the BCI system. However, it is necessary to retrain the BCI system based on the new EEG data. Important is that both the spatial filter and the classifier are calculated based on feedback data if they are used for feedback sessions. As illustrated in Fig. 10, several iterations are necessary to allow both systems to adapt. The subject that performed the experiment described in Fig. 10 was the first subject ever who reached 100% accuracy in 160 trials [2] of BCI control.

3.2 Training with a P300 Spelling Device

A P300 spelling device can be based on a 6 × 6 matrix of different characters displayed on a computer screen. The row/column speller flashes a whole row or a whole column of characters at once in a random order as shown in Fig. 12. The single character speller flashes only one single character at an instant in time. This yields

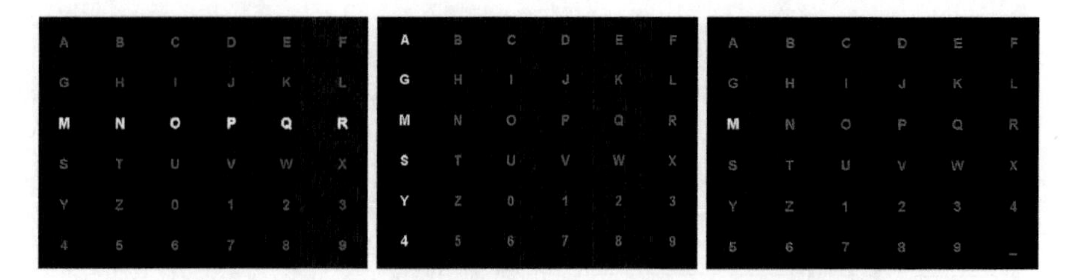

Fig. 12 *Left, mid panels*: row/column speller. *Right panel*: single character speller

of course to different communication rates; with a 6 × 6 matrix, the row/column approach increases speed by a factor of 6 (see also Chapter "BCIs in the Laboratory and at Home: The Wadsworth Research Program" in this book).

The underlying phenomenon of a P300 speller is the P300 component of the EEG, which is seen if an attended and relatively uncommon event occurs. The subject must concentrate on a specific letter he wants to write [7, 11, 20]. When the character flashes on, the P300 is induced and the maximum in the EEG amplitude is reached typically 300 ms after the flash onset. Several repetitions are needed to perform EEG data averaging to increase the signal to noise ratio and accuracy of the system. The P300 signal response is more pronounced in the single character speller than in the row/column speller and therefore easier to detect [7, 21].

For training, EEG data are acquired from the subject while the subject focuses on the appearance of specific letters in the copy spelling mode. In this mode, an arbitrary word like LUCAS is presented on the monitor. First, the subject counts whenever the L flashes. Each row, column, or character flashes for e.g.100 ms per flash. Then the subject counts the U until it flashes 15 times, and so on. These data, together with the timing information of each flashing event, are then loaded into g.BSanalyze. Then, the EEG data elicited by each flashing event are extracted within a specific interval length and divided into sub-segments. The EEG data of each segment are averaged and sent to a step-wise linear discriminant analysis (LDA). The LDA is trained to separate the target characters, i.e. the characters the subject was concentrating on (15 flashes × 5 characters), from all other events (15 × 36–15 × 5). This yields again a subject specific WV for the real-time experiments. It is very interesting for this approach that the LDA is trained only on 5 characters representing 5 classes and not on all 36 classes. This is in contrast to the motor imagery approach where each class must also be used as a training class. The P300 approach allows minimizing the time necessary for EEG recording for the setup of the LDA. However, the accuracy of the spelling system increases also with the number of training characters [21].

After the setup of the WV (same principle as in Fig. 10) real-time experiments can be conducted with the Simulink model shown in Fig. 13.

The device driver "g.USBamp" reads again the EEG data from the amplifier and converts the data to double precision. Then the data is band pass filtered ("Filter") to remove drifts and artifacts and down sampled to 64 Hz (Downsample 4:1'). The "Control Flash" block generates the flashing sequence and the trigger signals for

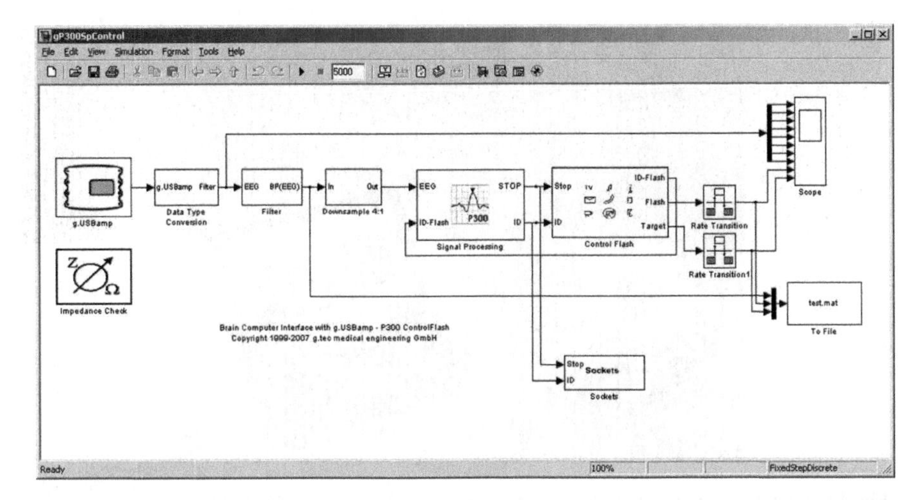

Fig. 13 Simulink model for P300 analysis and smart home control

each flashing event and sends the "ID" to the "Signal Processing" block. The "Signal Processing" block creates a buffer for each character. After all the characters flashed, the EEG data is used as input for the LDA and the system decides which letter was most likely investigated by the subject. Then this character is displayed on the computer screen. Nowadays, the P300 concept allows very reliable results with high information transfer rates of up to 1 character per second [7, 21, 22].

4 BCI Applications

4.1 IntendiX

intendiX® is designed to be installed and operated by caregivers or the patient's family at home. The system consists of active EEG electrodes to avoid abrasion of the skin, a portable biosignal amplifier and a laptop or netbook running the software under Windows (see Fig. 14). The electrodes are integrated into the cap to allow a fast and easy montage of the intendiX equipment.

The intendiX software allows viewing the raw EEG to inspect data quality, but indicates automatically to the unexperienced user if the data quality on a specific channel is good or bad. If the system is started up for the first time, a user training has to be performed. Therefore usually 5–10 training characters are entered and the user has to copy spell the characters. The EEG data is used to calculate the user specific weight vector which is stored for later usage. Then the software switches automatically into the spelling mode and the user can spell as many characters as wanted.

The user can perform different actions: (i) copy the spelled text into an Editor, (ii) copy the text into an email, (iii) send the text via text-to-speech facilities to the

Fig. 14 Intendix running on the laptop and user wearing the active electrodes

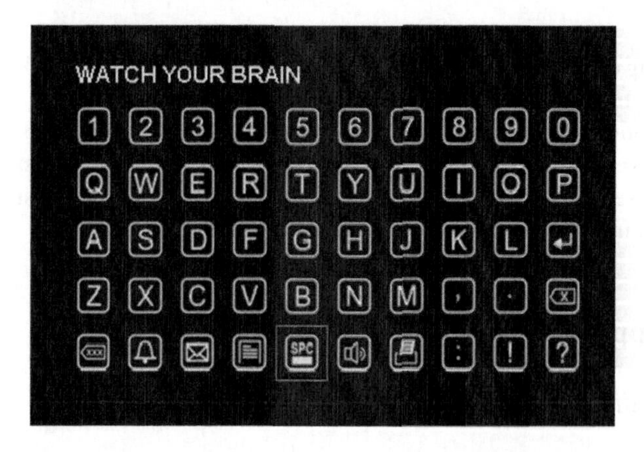

Fig. 15 User interface with 50 characters and computer keyboard like layout

loud speakers, (vi) print the text or (v) send the text via UDP to another computer. For all these services a specific icon exists as shown in Fig. 15.

The number of flashes for each classification can be selected by the user or the user can also use a statistical approach that detects automatically if the user is working with the BCI system. The latter one has the advantage that no characters are selected if the user is not looking at the matrix or does not want to use the speller.

4.2 *Virtual Reality Smart Home Control with the BCI*

Recently, BCI systems were also combined with Virtual Reality (VR) systems. VR systems use either head mounted displays (HMDs) or highly immersive

Fig. 16 *Left*: CAVE system at University College London. *Right*: Smart home VR realization from Chris Groenegress and Mel Slater from Universitat Politècnica de Catalunya, Barcelona

back-projection systems (CAVE like systems) as shown in Fig. 16. Such a CAVE has 3 back-projectors for the walls and one projector on top of the CAVE for the floor. The system projects two images which are separated by shutter glasses to achieve 3D effects.

There are several issues that must be solved to use of a BCI system in such an environment: (i) the biosignal amplifiers must be able to work in such a noisy environment, (ii) the recordings should ideally be done without wires to avoid collisions and irritations within the environment, (iii) the BCI system must be coupled with the VR system to exchange information fast enough for real-time experiments (iv) a special BCI communication interface must be developed to have enough degrees of freedom available to control the VR system.

Figure 17 illustrates the components in detail. A 3D projector is located behind a projection wall for back projections. The subject is located in front the projection wall to avoid shadows and wears a position tracker to capture movements, shutter glasses for 3D effects and the biosignal amplifier including electrodes for EEG recordings. The XVR PC is controlling the projector, the position tracker controller and the shutter glass controller. The biosignal amplifier is transmitting the EEG data wirelessly to the BCI system that is connected to the XVR PC to exchange control commands.

In order to show that such a combination is possible, a virtual version of a smart home was implemented in XVR (VRmedia, Italy). The smart home consists of a living room, a kitchen, a sleeping room, a bathroom, a floor and a patio as shown in Figs. 16 and 18. Each room has several devices that can be controlled: TV, MP3 player, telephone, lights, doors, etc. Therefore, all the different commands were summarized in 7 control masks: a light mask, a music mask, a phone mask, a temperature mask, a TV mask, a move mask and a go to mask. Figure 18 shows the TV mask and as an example the corresponding XVR image of the living room. The subject can e.g. switch on the TV by looking first at the TV symbol. Then, the station and the volume can be regulated. The bottom row of Fig. 18 shows the go to mask with an underlying plan of the smart home. Inside the mask, there are letters indicating the different accessible spots in the smart home which flash during

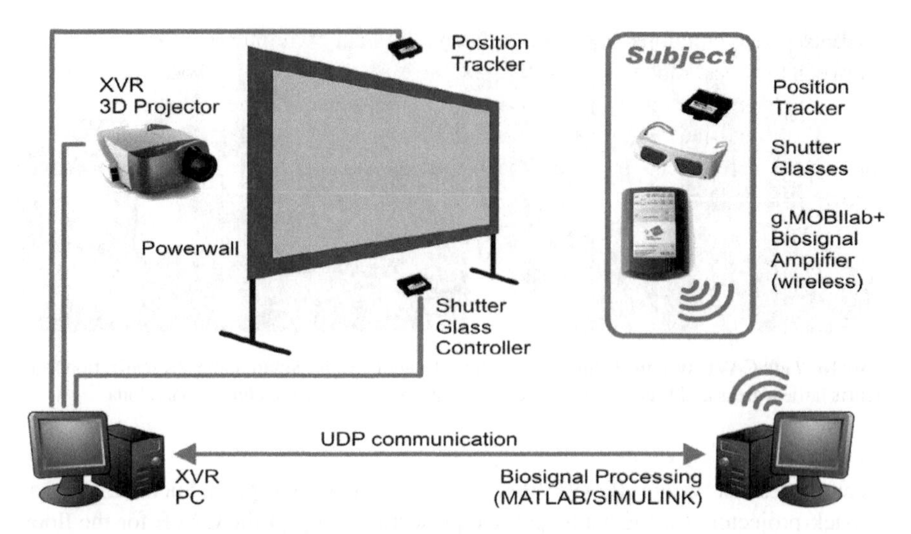

Fig. 17 Components of a virtual reality system linked to a BCI system

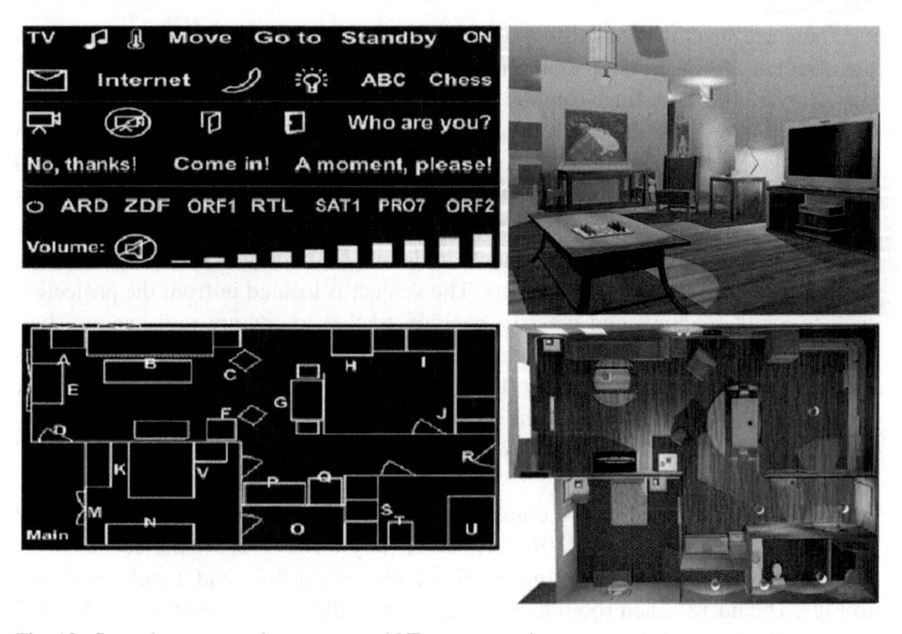

Fig. 18 Smart home control sequence and VR representation

the experiment. Therefore, the subject has to focus on the spot where he wants to go. After the decision of the BCI system, the VR program moves to a bird's eye view of the apartment and zooms to the spot that was selected by the user. This is a goal oriented BCI control approach, in contrast to navigation task, where each small navigational step is controlled [23].

Each mask contained 13–50 commands. In such an application, precise timing between the appearance of the symbol on the screen and the signal processing unit is very important. Therefore, the flashing sequence was implemented under Simulink where the real-time BCI processing was also running. A UDP interface was also programmed that sent the data to the XVR system to control the smart home, as shown in Fig. 13.

4.3 Avatar Control

Another interesting application inside VR is the control of avatars. In this application, a human subject was rendered in 3D. To create more realistic motions of the avatar, a motion capture system was used. Hence, special motion patterns such as walking, kick-boxing, dancing, or jumping are available for the VR animations. A control mask was then implemented that allowed 15 commands, as shown in Figure 19. In this case, the commands were not goal-oriented like in the smart home

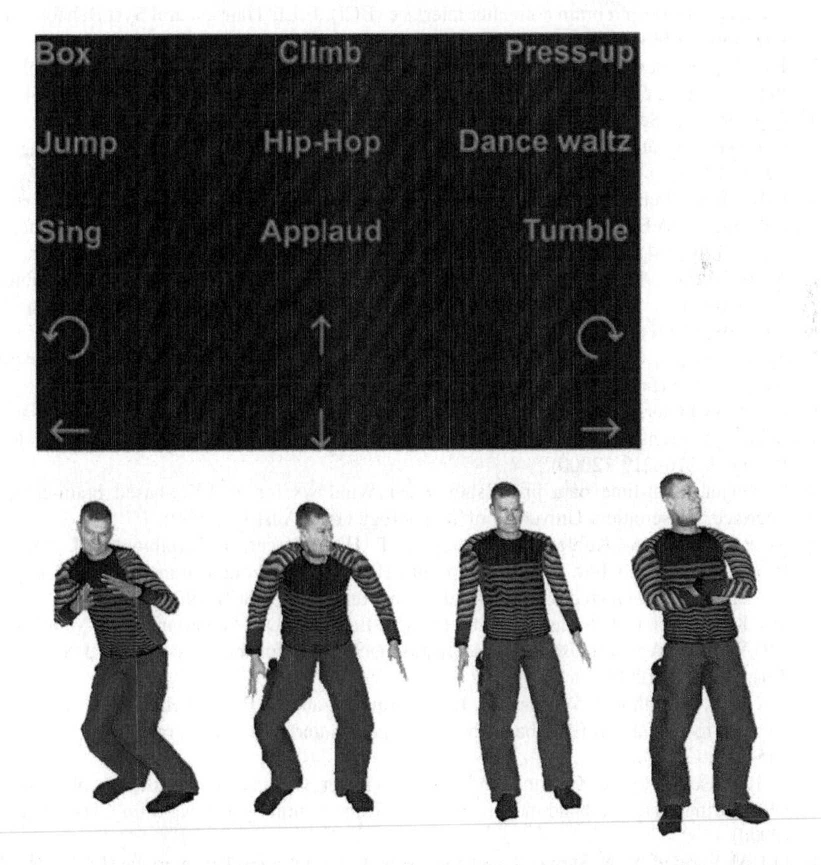

Fig. 19 *Top*: Avatar control mask. *Bottom*: Typical behaviors of the avatar include standing, being bored, dancing. Realized by Chris Christus and Mel Salter from University College London

realization, but task-oriented. If the subject selected e.g. the forward arrow then the avatar moved forward until another command was initiated. Therefore, not only the accuracy of the BCI system, but also the decision-making speed was important.

The smart home and avatar control are just two examples that show applications of BCI systems in Virtual Reality. Launching new applications in the near future, including bringing the systems in the patient's home, requires a flexible development environment.

Acknowledgments This work was supported by the EC projects Presenccia, SM4all, Brainable, Decoder, Better and Vere.

References

1. G. Pfurtscheller, C. Neuper, D. Flotzinger, and M. Pregenzer, EEG-based discrimination between imagination of right and left hand movement. Electroenceph clin Neurophysiol, 103, 642–651, (1997).
2. C. Guger, H. Ramoser, and G. Pfurtscheller, Real-time EEG analysis with subject-specific spatial patterns for a brain computer interface (BCI). IEEE Trans Neural Syst Rehabil Eng, 8, 447–456, (2000).
3. E.C. Leuthardt, G. Schalk, J.R. Wolpaw, J.G. Ojemann, and D.W. Moran, A brain-computer interface using electrocorticographic signals in humans. J Neural Eng, 1, 63–71, (2004).
4. C. Guger, A. Schlögl, C. Neuper, D. Walterspacher, T. Strein, and G. Pfurtscheller, Rapid prototyping of an EEG-based brain-computer interface (BCI). IEEE Trans Rehab Engng, 9(1), 49–58, (2001).
5. G.R. Muller-Putz, R. Scherer, C. Brauneis, and G. Pfurtscheller, Steady-state visual evoked potential (SSVEP)-based communication: impact of harmonic frequency components. J Neural Eng, 2(4), 123–130, (2005).
6. N. Birbaumer, N. Ghanayim, T. Hinterberger, I. Iversen, B. Kotchoubey, A. Kubler, J. Perelmouter, E. Taub, and H. Flor, A spelling device for the paralysed. Nature, 398(6725), 297–298, (1999).
7. M. Thulasidas, G. Cuntai, and W. Jiankang, Robust classification of EEG signal for brain-computer interface. IEEE Trans Neural Syst Rehabil Eng, 14(1), 24–29, 2006.
8. G. Pfurtscheller, C. Neuper, C. Guger, B. Obermaier, M. Pregenzer, H. Ramoser, and A. Schlögl, Current trends in Graz brain-computer interface (BCI) research. IEEE Trans Rehab Engng, 8, 216–219, (2000).
9. C. Guger, Real-time data processing under Windows for an EEG-based brain-computer interface. Dissertation, University of Technology Graz, Austria, (1999).
10. N. Birbaumer, A. Kubler, N. Ghanayim, T. Hinterberger, J. Perelmouter, J. Kaiser, I. Iversen, B. Kotchoubey, N. Neumann, and H. Flor, The thought translation device (TTD) for completely paralyzed patients. IEEE Trans Rehabil Eng, 8(2), 190–193, 2000.
11. D.J. Krusienski, E.W. Sellers, F. Cabestaing, S. Bayoudh, D.J. McFarland, T.M. Vaughan, and J.R. Wolpaw, A comparison of classification techniques for the P300 Speller. J Neural Eng, 3(4), 299–305, (2006).
12. C. Guger, G. Edlinger, W. Harkam, I. Niedermayer, and G. Pfurtscheller, How many people are able to operate an EEG-based brain computer interface? IEEE Trans Rehab Engng, 11, 145–147, (2003).
13. H. Ramoser, J. Muller-Gerking, and G. Pfurtscheller, Optimal spatial filtering of single trial EEG during imagined hand movement. IEEE Trans Neural Syst Rehabil Eng, 8(4), 441–446, (2000).
14. D.J. McFarland, W.A. Sarnacki, and J.R. Wolpaw, Brain-computer interface (BCI) operation: optimizing information transfer rates. Biol Psychol, 63(3), 237–251, (2003).

15. G. Edlinger, and C. Guger: Laboratory PC and mobile pocket PC brain-computer interface architectures. Conf Proc IEEE Eng Med Biol Soc, 5, 5347–5350, (2005).
16. E.W. Sellers, D.J. Krusienski, D.J. McFarland, T.M. Vaughan, and J.R. Wolpaw, A P300 event-related potential brain-computer interface (BCI): the effects of matrix size and inter stimulus interval on performance. Biol Psychol, 73(3), 242–252, (2006).
17. G. Klem, H. Lüders, H. Jasper, and C. Elger, The ten-twenty electrode system of the International Federation. The International Federation of Clinical Neurophysiology. Cleveland Clinic Foundation, 52, 3–6, (1999).
18. B. Obermaier, C. Guger, C. Neuper, and G. Pfurtscheller, Hidden Markov Models for online classification of single trial EEG data. Pattern Recogn Lett, 22, 1299–1309, (2001).
19. C.Neuper, G. Pfurtscheller, C. Guger, B. Obermaier, M. Pregenzer, H. Ramoser, and A. Schlögl, Current trends in Graz brain-computer interface (BCI) research. IEEE Trans Rehab Engng, 8, 216–219, (2000).
20. C. Guan, M. Thulasida, and W. Jiankang, High performance P300 speller for brain-computer interface. IEEE Int Workshop Biomed. Circuits Syst, S3, 13–16, (2004).
21. M. Waldhauser, Offline and online processing of evoked potentials. Master thesis, FH Linz, (2006).
22. E.W. Sellers and E. Donchin, A P300-based brain-computer interface: initial tests by ALS patients. Clin Neurophysiol, . 117(3), 538–548, (2006).
23. C. Guger, C. Groenegress, C. Holzner, G. Edlinger, and M. Slater, Brain-computer interface for controlling virtual environments. 2nd international conference on applied human factors and ergonomics. . Las Vegas, NV, USA, (2008).

Digital Signal Processing and Machine Learning

Yuanqing Li, Kai Keng Ang, and Cuntai Guan

Any brain–computer interface (BCI) system must translate signals from the users brain into messages or commands (see Fig. 1). Many signal processing and machine learning techniques have been developed for this signal translation, and this chapter reviews the most common ones. Although these techniques are often illustrated using electroencephalography (EEG) signals in this chapter, they are also suitable for other brain signals.

This chapter first introduces the architecture of BCI systems, followed by signal processing and machine learning algorithms used in BCIs. The signal processing sections address data acquisition, followed by preprocessing such as spatial and temporal filtering. The machine learning text primarily discusses feature selection and translation. This chapter also includes many references for the interested reader on further details of these algorithms and explains step-by-step how the signal processing and machine learning algorithms work in an example of a P300 BCI.

The text in this chapter is intended for those with some basic background in signal processing, linear algebra and statistics. It is assumed that the reader understands the concept of filtering, general eigenvalue decomposition, statistical mean and variance.

1 Architecture of BCI systems

A brain–computer interface (BCI) allows a person to communicate or to control a device such as a computer or prosthesis without using peripheral nerves and muscles. The general architecture of a BCI system is shown in Fig. 1, which includes 4 stages of brain signal processing:

Data acquisition: Brain signals are first acquired using sensors. For example, the electroencephalogram (EEG) measures signals acquired from electrodes placed on the scalp, as shown in Fig. 1. These signals are then amplified and digitalized by

Y. Li (✉)
School of Automation Science and Engineering, South China University of Technology, Guangzhou, China, 510640
e-mail: auyqli@scut.edu.cn

B. Graimann et al. (eds.), *Brain–Computer Interfaces*, The Frontiers Collection, DOI 10.1007/978-3-642-02091-9_17, © Springer-Verlag Berlin Heidelberg 2010

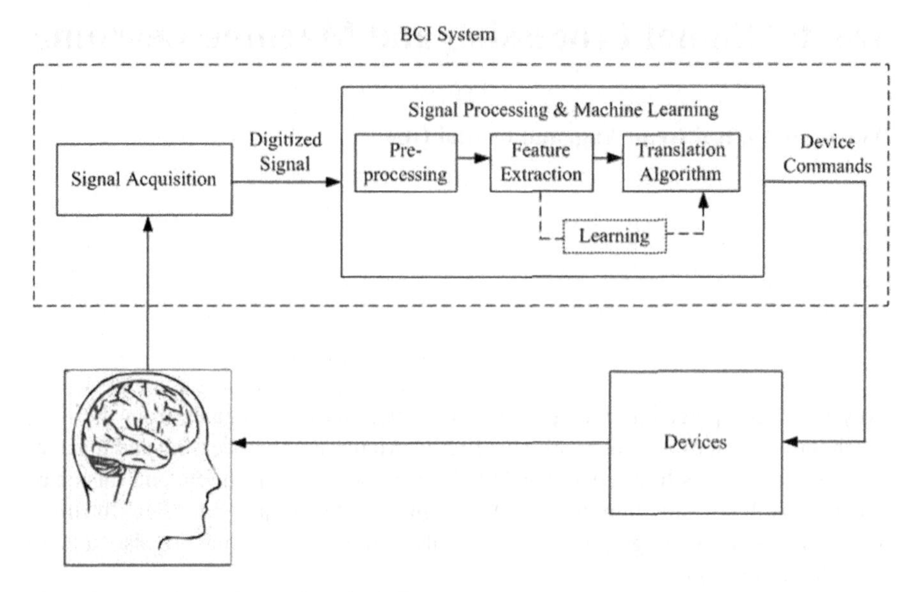

Fig. 1 Basic design and operation of a BCI system

data acquisition hardware before being processed on a computer. An accurate setup of the signal acquisition system is important. In the case of an EEG based BCI, this includes skin preparation and the application of electrode gel. This is necessary because large and unequal electrode impedances can compromise signal quality. However, good signal quality is important to simplify the already difficult task of extracting information from brain signals.

Preprocessing: "Noise" and "artifacts" refer to information that reduces signal quality. In BCI signal processing, many signal components are noisy. This may include brain signals that are not associated with brain patterns generated by the user's intent, or signals added by the hardware used to acquire the brain signals. Examples of artifacts are signals generated by eye or muscle activity, which are recorded simultaneously with the brain signals. Preprocessing methods such as spatial filtering and temporal filtering can improve signal quality by greatly reducing noise and artifacts.

Feature extraction: Useful signal features reflecting the user's intent are then extracted from the preprocessed brain signals. For non-invasive BCIs, the feature extraction is mainly based on brain patterns such as slow cortical potentials (SCP) [1, 2], event related potentials (ERPs), event-related desynchronization (ERD) and event-related synchronization (ERS) reflected by mu and beta rhythms. Examples of ERPs include: P300 [3, 4], and steady-state visual evoked potential (SSVEP) [5–6]. Mu and beta rhythms include 8–13 Hz and 18–23 Hz EEG signals from the motor cortex. Details on the neurophysiological and neuropsychological principles behind these features are provided in Chapter "Brain signals for Brain–Computer Interfaces" and "Dynamics of Sensorimotor Oscillations in a Motor Task" in this book. Depending on the underlying brain patterns invoked or induced by the mental strategy of the BCI, one can differentiate between features extracted in the time or frequency domain. Features reflecting the P300, for instance, are usually extracted in the time domain. ERD/ERS features are often extracted from the frequency domain.

Translation algorithms: Finally, the extracted features are translated into discrete or continuous control signals that operate devices such as a computer, wheelchair, or prosthesis.

2 Preprocessing

The preprocessing of brain signals plays an important role in BCIs. Components of brain signals that reflect the user's intentions are useful, whereas the remaining components may be considered noise. For example, in P300-based BCI speller, the useful component is the P300 potential, and noise components include high frequency signals, alpha waves, EMG, EOG, and interference from power sources and other devices. Preprocessing is necessary for most BCI systems because an effective preprocessing method improves the signal-to-noise ratio and spatial resolution, which subsequently improves the performance of BCIs. Spatial and temporal filtering are common preprocessing methods used in most BCIs [7–9]. These methods are described in more detail in the following:

2.1 Spatial Filtering

This section first introduces linear transformations and then describes five spatial filtering methods, namely, common average reference (CAR), Laplacian reference [7], common spatial patterns (CSP), principle component analysis (PCA) [6] and independent component analysis (ICA) [10, 11]. All spatial filtering methods described here are linear transformations.

2.1.1 Linear Transformations

A linear transformation is an operation that transforms the original signals by linearly combining the samples of all channels or a certain subset of channels and for each individual sample time. The procedure to calculate the transformed value of the current sample value of a specific channel is as follows: First, all sample values of the channels to be considered are multiplied by individual weight values derived from the method applied (e.g. the CAR or CSP method, see below). Second, the resulting values of these multiplications are added together. These two steps are called linear combination and the result of the summation is the transformed value. Mathematically, linear transformation is modeled by

$$\mathbf{y} = \mathbf{W}\mathbf{x}, \tag{1}$$

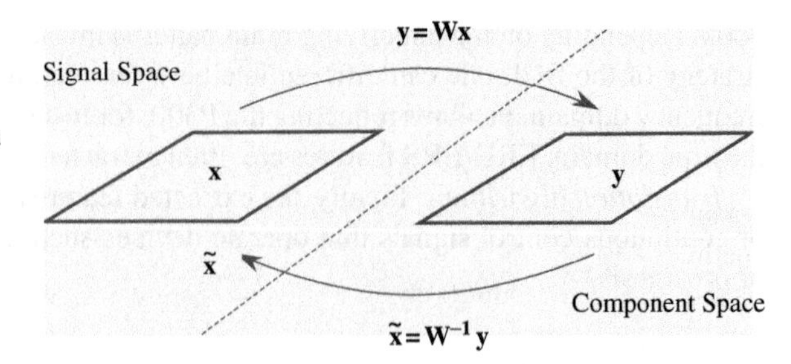

Fig. 2 A linear transformation from a signal space to a component space and its inverse transformation from the component space to the signal space

where $\mathbf{x}$ and $\mathbf{y}$ represent an original signal vector and a transformed signal vector respectively, $\mathbf{W}$ is a transformation matrix.

A linear transformation (1) is depicted in Fig. 2 using the concept of spaces. In this case the signal is linearly transformed from the signal space to the component space. It is hoped that the transformed channels, which are called components, are improved somehow. Improvement in this context can mean various signal qualities such as signal-to-noise ratio, spatial resolution, or separability. Often the improvement is not restricted to only one quality. For instance, a transformation which transforms the recorded channels into components that contain artifacts and components that contain brain signals improves the signal-to-noise ratio and most likely the separability as well. The subsequent signal processing, such as temporal filtering, trial averaging, or frequency power analysis is done in the component or in the signal space. In the latter case, the signal is transformed back by an inverse transformation after a specific manipulation in the transformed space. In both cases, the goal of the transformation of the spatial filter is to simplify subsequent signal processing.

Figure 2 shows that the transformation matrix $\mathbf{W}$ is the spatial filter applied on the multichannel signal $\mathbf{x}$. The transformed or spatially filtered signal $\mathbf{y}$ is a result of linear combinations of the filter coefficients contained in the rows of $\mathbf{W}$ and the signal $\mathbf{x}$. Due to these linear combinations, the direct association of the components to the channels is lost. However, the values of the columns of the inverse of the transformation matrix $\mathbf{W}$ determine the contributions of each individual component from the channels in the signal space. Thus these values and the topographic information of the channels can be used to generate the so called scalp or topographic maps. Examples of such maps are provided in Fig. 4.

The transformation matrix $\mathbf{W}$ in (1) is selected based on constraints or desired attributes of the time series $\mathbf{y}(t)$ [12]. These various constraints result in different spatial filters including CAR, Laplacian reference, CSP, PCA and ICA described in the following.

2.1.2 Common Average Reference (CAR)

The common average reference (CAR) method adjusts the signal at each electrode by subtracting the average of all electrodes

$$y_i = x_i - \frac{1}{N}(x_1 + \cdots + x_N), \tag{2}$$

where $i = 1, \cdots, N$.

CAR is a linear transformation method in which all entries of the ith row of the transformation matrix $\mathbf{W}$ in (1) are $-\frac{1}{N-1}$ except that the ith entry is 1. The CAR method is effective in reducing the noise that are common to all electrodes, such as the interference of 50 or 60 Hz power sources. Since the useful brain signals are generally localized in a few electrodes, this method enhances the useful brain signals from the averaged signals. However, the CAR method is not effective in reducing noise that are not common to all electrodes, such as electrooculogram (EOG) from eye blinks and electromyogram (EMG) from musclclclcle contractions. The EOG signal is stronger at the frontal cortex, and the EMG is stronger near the relevant muscles. Therefore, other methods such as regression methods or ICA have to be used to reduce these artifacts.

2.1.3 Laplacian Reference

Laplacian reference adjusts the signal at each electrode by subtracting the average of the neighboring electrodes. The Laplacian method is effective in reducing noise that should be more focused at a specific region. Here we describe two types of Laplacian references: small and large. For example, Fig. 3 presents a 64 channel electrode cap with small and large Laplacian references.

The small Laplacian reference subtracts the averaged EEG signals of the nearest four electrodes from the signal being preprocessed as shown in the left subplot of Fig. 3. In this example, the 9th preprocessed EEG signal (y_9) computed using the small Laplacian reference is given by

$$y_9 = x_9 - \frac{1}{4}[x_2 + x_8 + x_{10} + x_{16}]. \tag{3}$$

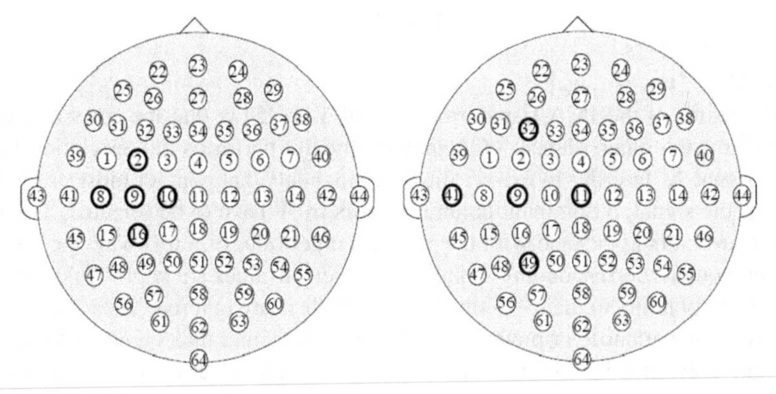

Fig. 3 Laplacian reference preprocessing for the 9th channel of EEG signal. *Left*: small Laplacian reference. *Right*: large Laplacian reference

The large Laplacian reference subtracts the averaged EEG signals of the next-nearest four electrodes from the signal being preprocessed as shown in the right subplot of Fig. 3. In this example, the 9th preprocessed EEG signal (y_9) computed using the large Laplacian reference is given by

$$y_9 = x_9 - \frac{1}{4}[x_{11} + x_{32} + x_{41} + x_{49}]. \tag{4}$$

2.1.4 Principal Component Analysis (PCA)

Principal component analysis (PCA) and independent component analysis (ICA) are preprocessing methods that separate the brain signals into components. Some of these components components are of interest, such as brain patterns associated with motor imagery are of interest, but the remaining components are of no interest, such as noise. Hence, PCA and ICA methods are often used to separate the useful components from the noise signals. This section explains the PCA method and the following section explains the ICA method.

Mathematically, PCA performs an orthogonal linear transformation to a new coordinate system such that the original signals are decomposed in uncorrelated components which are ordered according to decreasing variance. That is, the first component contains most of the variance of the original signals, the second component contains second most of the variance and so forth. If $\mathbf{X} \in R^{n\times m}$ is a brain signal matrix whereby each row corresponds to the signal acquired from a specific electrode with zero mean, and each column corresponds to a signal at a specific sampling time. Then, the PCA transformation matrix $\mathbf{W} = [\mathbf{w}_1, \cdots, \mathbf{w}_n]$ can be obtained by performing a general eigenvalue decomposition of the covariance matrix $\mathbf{R} = \mathbf{X}\mathbf{X}^T$, where $\mathbf{w}_1, \cdots, \mathbf{w}_n$ are n normalized orthogonal eigenvectors of $\mathbf{X}\mathbf{X}^T$ that correspond to n different eigenvalues $\lambda_1, \cdots, \lambda_n$ in descending order. The PCA transformation of $\mathbf{X}$ is then given by:

$$\mathbf{Y} = \mathbf{W}^T\mathbf{X}, \tag{5}$$

where $\mathbf{W} = [\mathbf{w}_1, \cdots, \mathbf{w}_n]$.

The result of the PCA transformed signal $\mathbf{Y}$ in (5) is that the rows are uncorrelated to each other. Hence, PCA mathematically performs de-correlation on the input signal $\mathbf{X}$. In order to extract those components that contain most of the variance of the signal, p dominant column vectors in $\mathbf{W}$ have to be selected. These are the eigenvectors associated with the p largest eigenvalues. In this way, the p principal components corresponding to the p largest eigenvalues are kept, while the $n-p$ ones corresponding to the $n-p$ smallest eigenvalues are ignored. These eigenvalues represent the variances or power of the brain signals, and thus the largest eigenvalues may correspond to the useful components, while the lowest eigenvalues may correspond to the noise components.

2.1.5 Independent Component Analysis (ICA)

ICA can also be used to extract or separate useful components from noise in brain signals. ICA is a common approach for solving the blind source separation problem that can be explained with the classical "cocktail party effect", whereby many people talk simultaneously in a room but a person has to pay attention to only one of the discussions. Humans can easily separate these mixed audio signals, but this task is very challenging for machines. However, under some quite strict limitations, this problem can be solved by ICA. Brain signals are similar to the "cocktail party effect", because the signal measured at a particular electrode originates from many neurons. The signals from these neurons are mixed and then aggregated at the particular electrode. Hence the actual brain sources and the mixing procedure is unknown.

Mathematically, assuming there are n mutually independent but unknown sources $s_1(t), \cdots, s_n(t)$ in the brain signals denoted as $\mathbf{s}(t) = [s_1(t), \cdots, s_n(t)]^T$ with zero mean, and assuming there are n electrodes such that the sources are instantaneously linearly mixed to produce the n observable mixtures $\mathbf{x}(t) = [x_1(t), \cdots, x_n(t)]^T$, $x_1(t), \cdots, x_n(t)$ [13, 14], then

$$\mathbf{x}(t) = \mathbf{A}\mathbf{s}(t), \tag{6}$$

where $\mathbf{A}$ is an $n \times n$ time-invariant matrix whose elements need to be estimated from observed data. $\mathbf{A}$ is called the mixing matrix, which is often assumed to be full rank with n linearly independent columns.

The ICA method also assumes that the components s_i are statistically independent, which means that the source signals s_i generated by different neurons are independent to each other. The ICA method then computes a demixing matrix $\mathbf{W}$ using the observed signal $\mathbf{x}$ to obtain n independent components as follows,

$$\mathbf{y}(t) = \mathbf{W}\mathbf{x}(t), \tag{7}$$

where the estimated independent components are $y_1, \cdots, y_n$, denoted as $\mathbf{y}(t) = [y_1(t), \cdots, y_n(t)]^T$.

After decomposing the brain signal using ICA, the relevant components can then be selected or equivalently irrelevant components can be removed and then projected back into the signal space using

$$\tilde{\mathbf{x}}(t) = \mathbf{W}^{-1}\mathbf{y}(t). \tag{8}$$

The reconstructed signal $\tilde{\mathbf{x}}$ represents a cleaner signal than $\mathbf{x}$ [15].

There are many ICA algorithms such as the information maximization approach [13] and the second order or high order statistics based approaches [16]. Several of them are freely available on the net. There is another source separation technique called sparse component analysis (SCA) which assumes that the number of source components is larger than the number of observed mixtures and the components can be dependent to each other. Interested readers can refer to [17, 18] for more details on SCA.

2.1.6 Common Spatial Patterns (CSP)

In contrast to PCA, which maximizes the variance of the first component in the transformed space, the common spatial patterns (CSP) maximizes the variance-ratio of two conditions or classes. In other words, CSP finds a transformation that maximizes the variance of the samples of one condition and simultaneously minimizes the variance of the samples of the other condition. This property makes CSP one of the most effective spatial filters for BCI signal processing provided the user's intent is encoded in the variance or power of the associated brain signal. BCIs based on motor imagery are typical examples of such systems [19–21]. The term conditions or classes refer to different mental tasks; for instance, left hand and right hand motor imagery. The CSP algorithm requires not only the training samples but also the information to which condition the samples belong to compute the linear transformation matrix. In contrast, PCA and ICA do not require this additional information. Therefore, PCA and ICA are unsupervised methods, whereas CSP is a supervised method, which requires the condition or class label information for each individual training sample.

For a more detailed explanation of the CSP algorithm, we assume that $\mathbf{W} \in R^{n \times n}$ is a CSP transformation matrix. Then the transformed signals are $\mathbf{WX}$, where $\mathbf{X}$ is the data matrix of which each row represents an electrode channel. The first CSP component, i.e. the first row of $\mathbf{WX}$, contains most of the variance of class 1 (and least of class 2), while the last component i.e. the last row of $\mathbf{WX}$ contains most of the variance of class 2 (and least of class 1), where classes 1 and 2 represent two different mental tasks.

The columns of $\mathbf{W}^{-1}$ are the common spatial patterns [22]. As explained earlier, the values of the columns represent the contribution of the CSP components to the channels, and thus can be used to visualize the topographic distribution of the CSP components. Figure 4 shows two common spatial patterns of an EEG data analysis example on left and right motor-imagery tasks, which correspond to the first and the last columns of $\mathbf{W}^{-1}$ respectively. The topographic distributions of these components correspond to the expected contralateral activity of the sensorimotor rhythms induced by the motor imagery tasks. That is, left hand motor imagery induces sensorimotor activity patterns (ERD/ERS) over the right sensorimotor areas, while right hand motor imagery results in activity patterns over the left sensorimotor areas.

CSP and PCA are both based on the diagonalization of covariance matrices. However, PCA diagonalizes one covariance matrix, whereas CSP simultaneously diagonalizes two covariance matrices $\mathbf{R}_1$ and $\mathbf{R}_2$, which correspond to the two different classes. Solving the eigenvalue problem is sufficient for PCA. For CSP, the generalized eigenvalue problem with $\mathbf{R}_1^{-1}\mathbf{R}_1$ has to be solved [12] to give the transformation matrix $\mathbf{W}$ that simultaneously diagonalizes the two covariance matrices:

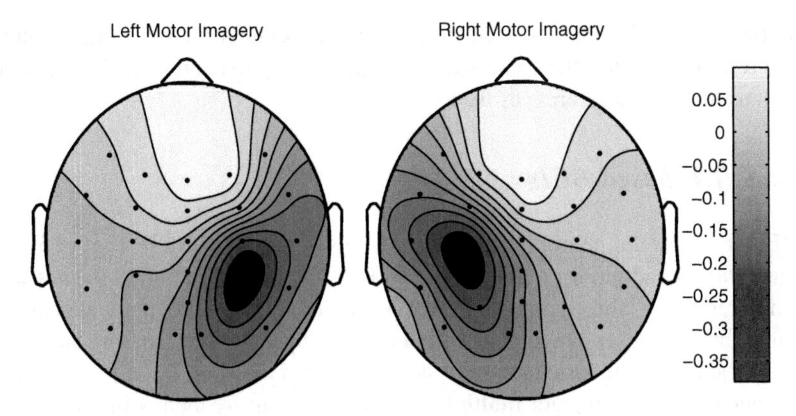

Fig. 4 Two common spatial patterns of an example corresponding to *left* and *right* motor-imagery tasks

$$\mathbf{W}\mathbf{R}_1\mathbf{W}^T = \mathbf{D} \text{ and } \mathbf{W}\mathbf{R}_2\mathbf{W}^T = \mathbf{I} - \mathbf{D}. \quad (9)$$

The basic CSP algorithm can only deal with two conditions. Extensions to more than two conditions exist. For details about the CSP algorithm and possible extensions, please see [23].

2.2 Temporal Filtering

Temporal filtering, also known as frequency or spectral filtering, plays an important role in improving the signal-to-noise ratio in BCI systems [7]. For example, temporal filtering can be used to remove the 50 Hz (60 Hz) noise from power sources. Temporal filtering can also be used to extract the components representing motor imagery from the brain signals in a specific frequency band such as the mu or beta band. In mu/beta-based BCIs, brain signals such as EEG, ECoG and MEG are generally temporally filtered using (for instance) band-pass filters of 8–12 Hz for the mu rhythm, and 18–26 Hz for the beta rhythm [7]. Although P300-based BCIs do not rely primarily on specific frequency bands, brain signals are often band-pass filtered using 0.1–20 Hz. The lower frequency signals are generally related to eye blinks [24] and other artifacts such as amplifier drift or changes in skin resistance due to sweat. Hence, the lower frequency brain signals are generally filtered away to remove the noise.

3 Feature Extraction

Although the signal-to-noise ratio of preprocessed brain signals is improved, useful features still have to be extracted for the translation algorithms to work effectively. This section describes commonly used feature extraction methods in SSVEP-based,

P300-based, and ERD/ERS-based BCIs. For more details on these brain patterns, please refer to Chapter "Brain signals for Brain–Computer Interfaces" of this book as well as the related references listed below.

3.1 SSVEP-based BCIs

SSVEP is one of the neurophysiological signals suitable for BCIs. Figure 5 shows a stimulation paradigm for a SSVEP-based BCI in which each button flashes with a specific frequency labeled in the figure. The labeled frequencies do not appear in the real interface. The user can select a number by gazing at a corresponding button. The SSVEP elicited by the visual stimulus is composed of a series of components whose frequencies are exact integer multiples of the flickering frequency of the stimulus. Several SSVEP-based BCI systems have been developed [6, 25, 26]. The amplitude and phase of SSVEP are highly sensitive to stimulus parameters such as the flashing rate and contrast of stimulus modules [5]. Gaze shifting may improve performance with SSVEP-based BCIs, but is not required, at least for some users. The importance of gaze shifting depends heavily on the display and task. For example, if there are many targets, or if they are located outside the fovea, gaze shifting may be essential [25, 26]. One possible feature extraction method used in SSVEP-based BCIs is as follows [6]:

First, two EEG channels are recorded from electrodes placed over the visual cortex. Next, the EEG signal is filtered using a band-pass filter for removing noise. The EEG signal is subsequently segmented using a Hamming window for estimating the dominant frequency that corresponds to the flashing rate of the button of interest. The spectra of each segment is estimated using a fast Fourier transformation (FFT). Finally, a feature vector is then constructed from the spectra for classification.

3.2 The P300-based BCI

P300 is a positive displacement of EEG amplitude occurring around 300 ms after a stimulus appears. The P300, like SSVEP, is based on selective attention. This mental

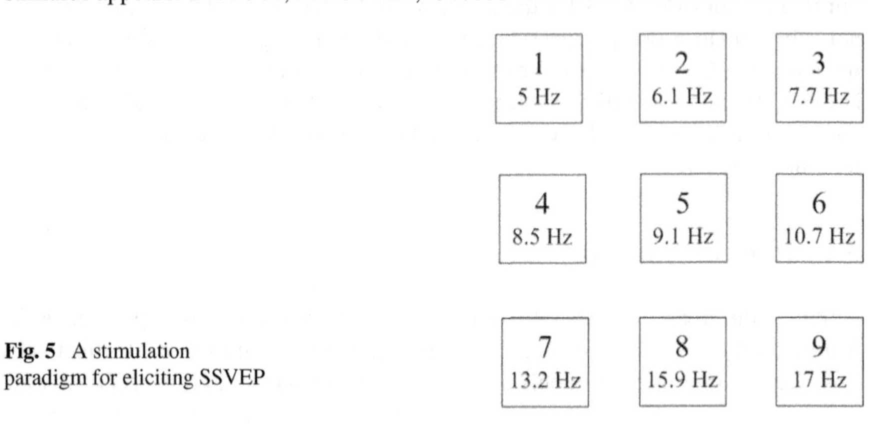

Fig. 5 A stimulation paradigm for eliciting SSVEP

strategy does usually not require much of a training, in contrast to other mental strategies like motor imagery [3, 4]. As the amplitude of P300 is only a few micro volts compared to EEG, which is on the order of a few tens of microvolts, many trials must often be averaged to distinguish the P300 from other activity. Hence. a major challenge in P300-based BCIs is to classify the P300 response from the background EEG with a minimum number of averaging. One possible feature extraction method used in P300-based BCI is described as follows:

Firstly, the EEG signals are filtered using a band pass-filter with the range 0.1–20 Hz. Segments of the filtered EEG signal are then extracted from the start of the stimulus. The feature vector is formed by concatenating the segments from all the electrodes. A smaller feature vector can be extracted by using a lowpass filter and downsampling. The reader is referred to the illustrative example in Sect. 7 for more details.

3.3 ERD/ERS-based BCI

In ERD/ERS-based BCIs, the power or the amplitude of filtered brain signals is used as features that reflect ERD/ERS. For example, the amplitude of mu/beta rhythms is used in a cursor control BCI system in [27]. By modulating the amplitude of mu/beta rhythms, the user can move the cursor to the target. The following describes two feature extraction methods for ERD/ERS-based BCIs, namely, band power feature extraction and autoregressive model coefficients. Note that features describing the power of the signals are often used in combination with CSP. This is because CSP maximizes the power (variance) ratio between two conditions, and thus power features are ideally suited [22, 28, 29].

3.3.1 Power Feature Extraction Based on Band-Pass Filter

Since ERD/ERS is defined as the the change in power of a frequency band (mu/beta band) in brain signals, the band pass power feature is a simple method of extracting ERD/ERS features. This method generally comprises the following steps [30]:

1. Apply a spatial filter such as Common average reference, Laplacian, or CSP on subject-specific EEG channels.
2. Band-pass filter on subject-specific frequency ranges.
3. Square the EEG samples to obtain power samples.
4. Average over time samples to smooth the data and reduce variability.
5. Log transform to make the distribution more Gaussian.

The subject-specific EEG channels and frequency ranges may be determined from training data, and there may be multiple band power features for each subject [30].

3.3.2 Autoregressive Model Coefficients

An autoregressive model is the model of the current value of the time series from previous values. The principle behind autoregressive modeling is to extract certain statistics such as the variance of the data for modeling the time series. Hence, the coefficients of an AR model can be used as features of brain signals. Given k channels of brain signal $s_1(t), \cdots, s_k(t)$, the brain signals can be modeled with p the model order as follows

$$s_i(t) = a_{i1}s_i(t-1) + \cdots + a_{ip}s_i(t-p) + \epsilon_i(t), \tag{10}$$

where $i = 1, \cdots, k$. $a_{i1}, \cdots, a_{ip}$ are the coefficients of the AR model, and ϵ_i is a zero-mean white noise process. The reader is referred to [31] for the process of computing the coefficients of the AR model.

Given N trials of data, the feature vector of the nth trial can be constructed as $[a_{11}^{(n)}, \cdots, a_{1p}^{(n)}, \cdots, a_{k1}^{(n)}, \cdots, a_{kp}^{(n)}, \sigma_{n1}^2 \cdots, \sigma_{nk}^2]^T$, where n represents the Nth trial, $n = 1, \cdots, N$. This is a simple method for feature extraction based on the AR model. However, there exist other more advanced methods. For example, a feature extraction method based on multivariate AR model is presented in [31], where a sum-square error method is used to determine the order P of the AR model. An adaptive AR model is used for feature extraction, of which the coefficients are time-varying in [32], and also Chapter "Adaptive Methods in BCI Research–An Introductory Tutorial" in this volume.

4 Feature Selection

The number of channels recorded by a BCI system may be large. The number of features extracted from each individual channel can be large too. In the instance of 128 channels and features derived from an AR model with order 6, the number of features in total, i.e. the dimension of the feature space, would be 6 times 128. This is a fairly large amount of features. Although it can be assumed that data recorded from a large number of channels contain more information, it is likely that some of it is redundant or even irrelevant with respect to the classification process. Further, there are at least two reasons why the number of features should not be too large. First, the computational complexity may become too large to fulfill the real-time requirements of a BCI. Second, an increase of the dimension of the feature space may cause a decrease of the performance of the BCI system. This is because the pattern recognition methods used in BCI systems are set up (trained) using training data, and the BCI system will be affected much by those redundant or even irrelevant dimensions of the feature space. Therefore, BCIs often select a subset of features for further processing. This strategy is called feature or variable selection. Using feature selection, the dimension of the feature space can be effectively reduced. Feature selection may also be useful in BCI research. When using new mental tasks or new BCI paradigms, it may not be always clear which channel locations and which features are the most suitable. In such cases, feature selection can be helpful.

Initially, many features are extracted from various methods and from channels covering all potentially useful brain areas. Then, feature selection is employed to find the most suitable channel locations and feature extraction methods. This section describes two aspects of feature selection for BCIs: channel selection and frequency band selection.

4.1 Channel Selection

As mentioned before, not all electrodes distributed over the whole scalp are useful in BCI systems in general. Hence, channel selection is performed to select the useful channels according to the features used in a BCI system. For SSVEP-based BCIs, the EEG signals from the visual cortex are selected. For mu/beta-based BCIs, the channels in the sensorimotor area are selected. For P300-based BCIs, the channels exhibiting an obvious P300 are selected. For P300 or SSVEP-based BCIs, channel selection is often manually performed before other processing steps. Apart from physiological considerations, channel selection can be performed by applying a feature selection method to a training dataset. Examples of such a feature selection method include optimizing statistical measures like Student's t-statistics, Fisher criterion [33] and bi-serial correlation coefficient [34] and genetic algorithms [35].

In the following, channel selection based on the bi-serial correlation coefficient is described [34]. Let $(x_1, y_1), \cdots, (x_K, y_K)$ be a sequence of one-dimensional observations (i.e. a single feature) with labels $y_k \in \{1, -1\}$. Define X^+ as the set of observations x_k with label of 1, and X^- the set of observations x_k with label of -1. Then bi-serial correlation coefficient r is calculated as

$$r_X = \frac{\sqrt{N^+N^-}}{N^+ + N^-} \frac{\text{mean}(X^-) - \text{mean}(X^+)}{\text{std}(X^+ \bigcup X^-)}, \tag{11}$$

where N^+ and N^- are the numbers of observations of X^+ and X^- respectively. r^2-coefficient r_X^2, which reflects how much of the variance in the distribution of all samples is explained by the class affiliation, is defined as a score for this feature.

For each channel of a BCI system, a score is calculated as above using the observations from this channel. If the objective is to choose the N most informative channels, one would choose the channels with the top N scores.

4.2 Frequency Band Selection

As mentioned in Sect. 2.2, temporal filtering plays an important role in improving the signal-to-noise ratio in BCI systems [7], and temporal filtering is also important to extract band power features. Before temporal filtering, one or several frequency bands need to be determined that affect the performance of BCIs. The optimal frequency bands are generally subject-specific in BCIs. Frequency band selection is often performed using a feature selection method. As an example, we use the bi-serial correlation coefficient presented in Sect. 4.1 to carry out frequency band selection as follows, a wide frequency band is first divided into several sub-bands which may overlap. For each sub-band, a r^2-coefficient is calculated as a score using the observations in this sub-band. Based on this score, one or several sub-bands can be selected.

5 Translation Methods

After the features that reflect the intentions of the user are extracted, the next step is to translate these discriminative features into commands that operate a device.

The translation methods employed in BCIs convert the features extracted into device control commands [7]. These commands can be of discrete-value such as letter selection, or of continuous-value such as vertical and horizontal cursor movement. The translation methods in BCIs often employ machine learning approaches in order to train a model from the training data collected. Translation methods can be broadly classified into classification and regression methods. Sect. 5.1 describes the former, which translates features into discrete-value commands, while Sect. 5.2 describes the latter, which translates features into continuous-value commands.

5.1 Classification Methods

A classification algorithm is defined as one that involves building a model from the training data so that the model can be used to classify new data that are not included in the training data [36].

The "No Free Lunch" theorem states that there is no general superiority of any approach over the others in pattern classification; furthermore, if one approach seems to outperform another in a particular situation, it is a consequence of its fit to the particular pattern recognition problem [33]. Hence, a wide variety of classification methods are used in BCIs. Prevailingly, the Fisher linear discriminant (FLD) [37–41], support vector machine (SVM) [37, 39, 40, 41], Bayesian [42], and hidden Markov model (HMM) [43] are commonly used as classification methods in BCIs.

FLD and SVM classification algorithms are described in the following subsections. The classification algorithms use the training data that comprises n samples denoted $\mathbf{x}_j$ with class label y_j where $j = 1, \cdots, n$, to form the training model. The classifier then calculates an estimate of the class label y of an unseen sample using the training model. These algorithms (as well as many others) are implemented for Matlab in the Statistical Pattern Recognition ToolBox (STPRTool) from http://cmp.felk.cvut.cz/cmp/software/stprtool/.

5.1.1 Fisher Linear Discriminant

Linear discrimination is a method that projects the high dimensional feature data onto a lower dimensional space. The projected data is then easier to separate into two classes. In Fisher linear discriminant (FLD), the separability of the data is measured by two quantities: the distance between the projected class means (which should be large), and the size of the variance of the data in this direction (should be small) [34]. This can be achieved by maximizing the ratio of between-class scatter to within-class scatter given by [33]

$$J(\mathbf{w}) = \frac{\mathbf{w}^T \mathbf{S}_B \mathbf{w}}{\mathbf{w}^T \mathbf{S}_W \mathbf{w}}, \tag{12}$$

$$\mathbf{S}_B = (\mathbf{m}_1 - \mathbf{m}_2)(\mathbf{m}_1 - \mathbf{m}_2)^T. \tag{13}$$

$$\mathbf{S}_W = \mathbf{S}_1 + \mathbf{S}_2, \tag{14}$$

where $\mathbf{S}_B$ is the between class scatter matrix for two classes is given in Eq. (13), $\mathbf{S}_W$ is the within class scatter matrix for two classes given in Eq. (14), $\mathbf{w}$ is an adjustable weight vector or projection vector.

In Eq. (14) and (13), the scatter matrix $\mathbf{S}_\omega$ and the sample mean vector $\mathbf{m}_\omega$ of class ω is given in Eq. (15) and (16) respectively, $\omega = 1, 2$.

$$\mathbf{S}_\omega = \sum_{j \in I_\omega} (\mathbf{x}_j - \mathbf{m}_\omega)(\mathbf{x}_j - \mathbf{m}_\omega)^T, \tag{15}$$

$$\mathbf{m}_\omega = \frac{1}{n_\omega} \sum_{j \in I_\omega} \mathbf{x}_j, \tag{16}$$

where n_ω is the number of data samples belonging to class ω, I_ω is the set of indices of the data samples belonging to class ω.

The weight vector $\mathbf{w}$ in (12), which is one of the generalized eigenvectors which jointly diagonalize $\mathbf{S}_B$ and $\mathbf{S}_W$, can be computed by

$$\mathbf{w} = \mathbf{S}_W^{-1} (\mathbf{m}_1 - \mathbf{m}_2). \tag{17}$$

The classification rule of the FLD is given by

$$y = \begin{cases} 1 & \text{if } \mathbf{w}^T \mathbf{x} \geq b, \\ -1 & \text{if } \mathbf{w}^T \mathbf{x} < b, \end{cases} \tag{18}$$

where $\mathbf{w}$ is given by Eq. (17), and b is given by

$$b = \frac{1}{2} \mathbf{w}^T (\mathbf{m}_1 + \mathbf{m}_2). \tag{19}$$

For a c-class problem, there are several ways of extending binary classifiers to multi-class [33, 44]. In the one-versus-rest approach, c binary classifiers are constructed whereby the kth classifier is trained to discriminate class y_k from the rest. In the pairwise approach, $c(c-1)$ binary classifiers are constructed whereby each classifier is trained to discriminate two classes. An unseen sample x is then classified by each classifier in turn, and a majority vote is taken.

5.1.2 Support Vector Machine

The support vector machine (SVM) [45] is a linear discriminant tool that maximizes the margin of separation between two classes based on the assumption that it improves the classifier's generalization capability. In contrast, Fisher linear discriminant maximizes the average margin, i.e., the margin between the class means [34]. Figure 6 illustrates a typical linear hyperplane learned by a SVM.

From articles such as [34, 45], an optimal classifier for unseen data is that with the largest margin $\frac{1}{||\mathbf{w}||^2}$, i.e. of minimal Euclidean norm $||\mathbf{w}||^2$. For a linear SVM, the large margin (i.e. the optimal hyperplane $\mathbf{w}$) is realized by minimizing the cost function on the training data

$$J(\mathbf{w}, \xi) = \frac{1}{2}\|\mathbf{w}\|^2 + C\sum_{i=1}^{n} \xi_i, \tag{20}$$

under the constraints

$$y_i \left(\mathbf{w}^T \mathbf{x}_i + b\right) \geq 1 - \xi_i, \ \ \xi_i \geq 0, \ \ \forall i = 1, \cdots, n, \tag{21}$$

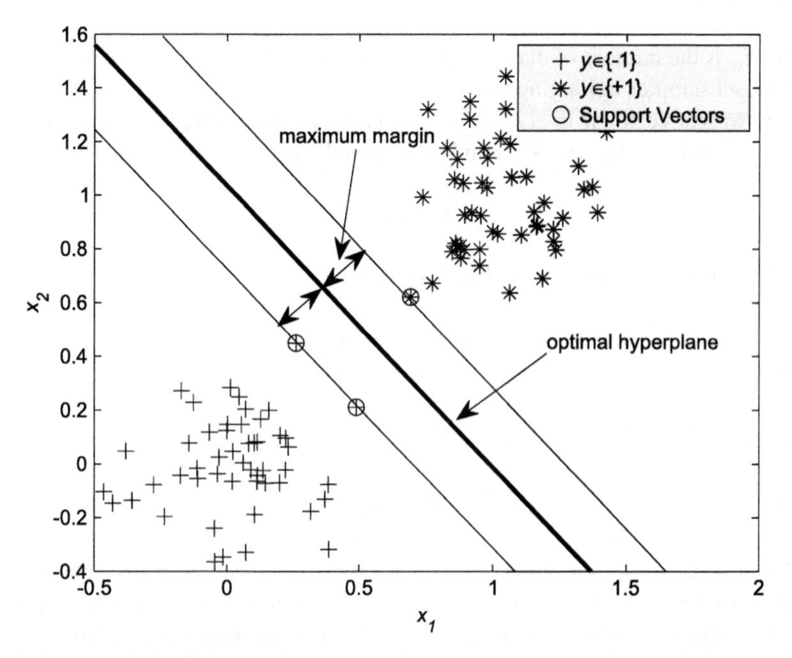

Fig. 6 An example that illustrates the training of a support vector machine that finds the optimal hyperplane with maximum margin of separation from the nearest training patterns. The nearest training patterns are called the support vectors

where $\mathbf{x}_1, \mathbf{x}_2, \cdots, \mathbf{x}_n$ are the training data, $y_1, y_2, \cdots, y_n \in \{-1, +1\}$ are the training labels, $\xi_1, \xi_2, \cdots, \xi_n$ are the slack variables, c is a regularization parameter that controls the tradeoff between the complexity and the number of non-separable points, and b is a bias. The slack variables measure the deviation of data points from the ideal condition of pattern separability. The parameter C can be user-specified or determined via cross-validation. The SVM employs the linear discriminant classification rule given in Eq. (18).

Similar to FLD or any other binary classificator, the one-versus-rest and pairwise approaches can be used to extend the binary SVM classifier to multi-class [33, 44]. Other generalization of SVM to c-class problems includes the generalized notion of the margin as a constrained optimization problem with a quadratic objective function [46].

5.2 Regression Method

A regression algorithm is defined as one that builds a functional model from the training data in order to predict function values for new inputs [33]. The training data comprises n samples $\{\mathbf{x}_1, \mathbf{x}_2, ..., \mathbf{x}_n\}$ that are similar to the classification method, but with continuous-valued outputs $\{y_1, y_2, \cdots, y_n\}$ instead. Given a data sample $\mathbf{x} = [x_1, x_2, ..., x_d]^T$ with d features, the regression method predicts its output $\hat{y}$ using the functional model.

Linear regression involves predicting an output $\hat{y}$ of $\mathbf{x}$ using the following linear model

$$\hat{y} = \mathbf{w}^T \mathbf{x} + b. \tag{22}$$

where $\mathbf{w} \in R^d$ and $b \in R$ are to be determined by an algorithm e.g. the least mean square (LMS) algorithm [33].

6 Parameter Setting and Performance Evaluation for a BCI System

Several parameters of a BCI system must be set in advance, such as the frequency band for filtering, the length of the time segment of brain signal to be extracted, and the various parameters of spatial filters and classifiers, etc. These parameters can be determined by applying machine learning approaches to the training data collected. Parameter setting also plays an important role for improving the performance of a BCI system.

In training a classifier, over-fitting can occur. Using a training data set, a classifier f is trained which satisfies $f(\mathbf{x}) = y$ where $\mathbf{x}$ is a feature vector in the training data set and y is its label. However, this classifier may not generalize well to unseen examples. The over-fitting phenomenon may happen if the number of training data is small and the number of parameters of the classifier is relatively large. The cross-validation method described next is a useful method of alleviating this over-fitting problem.

A large body of training data is often needed for determining the parameters of BCIs. However, a tedious and lengthy data collection process can be demanding for the subject. More often, there is not the time for that. Thus, tools that work with small data sets are also important for BCIs [47]. Two semi-supervised learning algorithms were proposed to deal with this problem in [48, 29].

6.1 K–folds Cross-Validation

Cross-validation is often used to determine the parameters of a BCI. K–fold cross-validation is performed by dividing the training set D into K equal parts denoted as $D_j, j = 1, \cdots, K$. In the jth fold of the cross-validation process, D_j is used as a test set, while the remaining $K-1$ parts are used as a training set. Based on the predicted results and the true labels in D_j, a prediction accuracy rate r_j can be computed. Figure 7 shows the procedure of the jth fold. After all K folds of cross-validation are performed, an average prediction accuracy rate r over the K folds is obtained. The objective of performing cross-validation is to determine a set of parameters that maximizes the average prediction accuracy rate r.

An illustrative example is used to describe the cross-validation process of parameter selection using the training data set D. Assuming that a support vector machine (SVM) is used as the classifier, SVM includes a regularization parameter c to be determined. Furthermore, if n is large, and the feature vector $\mathbf{x} = [x_1, \cdots, x_n]^T$ contains redundant entries, then dimension reduction or feature selection is necessary to be performed. Assuming that the significance of the entries of x_k decreases with respect to k, it suffices to choose a number L such that $\forall \mathbf{x} = [x_1, \cdots, x_n]^T$, $\bar{\mathbf{x}} = [x_1, \cdots, x_L]^T$ is the final feature vector used in classification.

K-fold cross-validation is then used to select the parameters L and C simultaneously that maximize the average prediction accuracy rate r. L and C are searched on a two-dimensional grid in which $L \in \{1, \cdots, n\}$ and $C \in \{C_1, \cdots, C_q\}$. For each

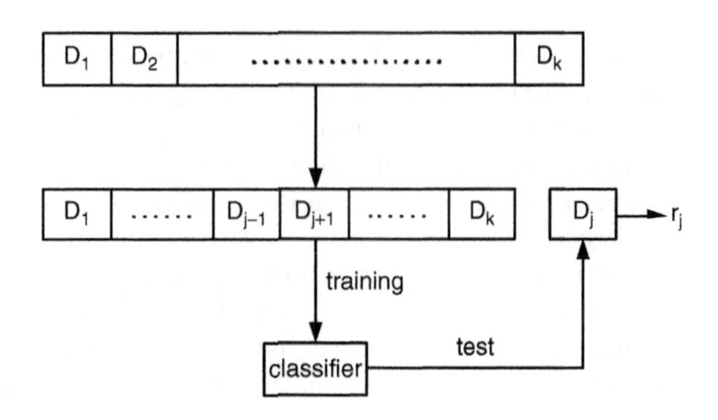

Fig. 7 The procedure of the jth fold of cross-validation

combination (L_k, C_l), K-fold cross-validation is performed and the average prediction accuracy rate denoted as $r(L_k, C_l)$ is computed. The combination (L_{k_0}, C_{l_0}) is chosen such that $r(L_{k_0}, C_{l_0})$ is the largest.

6.2 Performance Evaluation of a BCI System

An effective BCI system relies on several factors that include short training time, a small electrode montage, high accuracy, high speed and adaptability. The following describes some measures for evaluating the performance of a BCI System, see also Chapter Toward Ubiquitous BCIs.

6.2.1 Speed and Accuracy

For many BCI systems, speed and accuracy are used as the performance measures. For example, in a P300-based BCI speller, accuracy is the ratio of the number of correct character inputs over the entire number of character inputs. Speed is measured as the average time time required to send one signal. In cursor control systems with targets, accuracy reflects the probability that the cursor can hit a correct target. Speed is reflected by the average time required for the cursor to hit the target [49].

6.2.2 Information Transfer Rate

Information transfer rate (ITR) is defined by Shannon [50] as the amount of information communicated per unit of time. This measure encompasses speed, classification accuracy and the number of classes (e.g. the number of mental tasks in a BCI paradigm) in a single value and is used for comparing different BCI approaches and for measuring system improvements.

The ITR in bits per trial is given by [47]

$$B = log_2 N + P log_2 P + (1 - P) log_2 \frac{1 - P}{N - 1}, \tag{23}$$

where N is the number of different types of mental tasks and P the accuracy of classification.

If a trial lasts t seconds, then the ITR is $\frac{B}{t}$ bits per second. Considering the above definition of ITR is related to trials, thus it is not suitable for the performance evaluation of self-paced BCIs, which do not run trial by trial. In addition, it is assumed in the above definition of ITR that the accuracy of classification P is constant for all trials. Hence, the statistical error distribution with trials is not considered.

6.2.3 ROC Curve

ROC curves are suitable for evaluating asynchronous BCIs and two-class problems [51, 52]. For a convenient description of the receiver operating characteristic

(ROC) analysis approach, we first introduce several concepts. Let us consider a two-class prediction problem (binary classification), in which the outcomes are labeled as either a positive (p) or a negative (n) class. There are four possible outcomes from a binary classifier. If the outcome from a prediction is p and the actual value is also p, then it is called a true positive (TP); however, if the actual value is n then it is said to be a false positive (FP). Similarly, a true negative (TN) implies that both the prediction outcome and the actual value are n, while the false negative (FN) implies the prediction outcome is n however the actual value is p. A ROC curve is a graphical plot of sensitivity versus (1-specificity), or equivalently, a plot of the true positive rate versus the false positive rate for a binary classifier system as its discrimination threshold is varied (http://en.wikipedia.org/wiki/Receiver_operating_characteristic).

ROC analysis [53] has become a standard approach to evaluate the sensitivity and specificity of detection procedures. For instance, if a threshold θ is assumed to be a detection criterion, then a smaller θ indicates a higher sensitivity to events but a lower specificity to the events of interest. This is because those events that are of no interest are also easily detected. ROC analysis estimates a curve that describes the inherent tradeoff between sensitivity and specificity of a detection test. Each point on the ROC curve is associated with a specific detection criterion. The area under the ROC curve has become a particularly important metric for evaluating detection procedures because it is the average sensitivity over all possible specificities.

An example is presented here to illustrate the use of ROC curve analysis in evaluating asynchronous BCIs [52]. The two axes of the ROC curves are the true positive rate (TPR) and the false positive rate (FPR). These quantities are captured by the following equations:

$$TPR(\theta) = \frac{TP(\theta)}{TP(\theta) + FN(\theta)}, \quad FPR(\theta) = \frac{FP(\theta)}{TN(\theta) + FP(\theta)}, \tag{24}$$

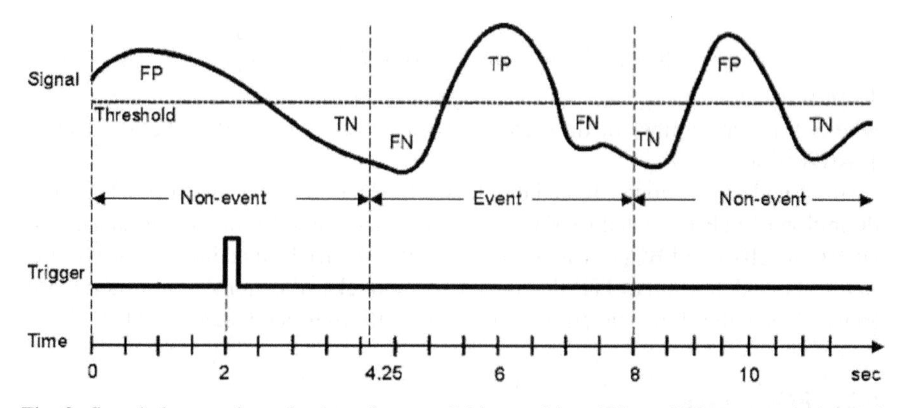

Fig. 8 Sample-by-sample evaluation of true and false positives (TPs and FPs, respectively) and true and false negatives (TNs and FNs, respectively). Events are defined as the period of time from 2.25 to 6.0 s after the rising edge of the trigger, which occurs 2.0 s after the start of each trial [52]

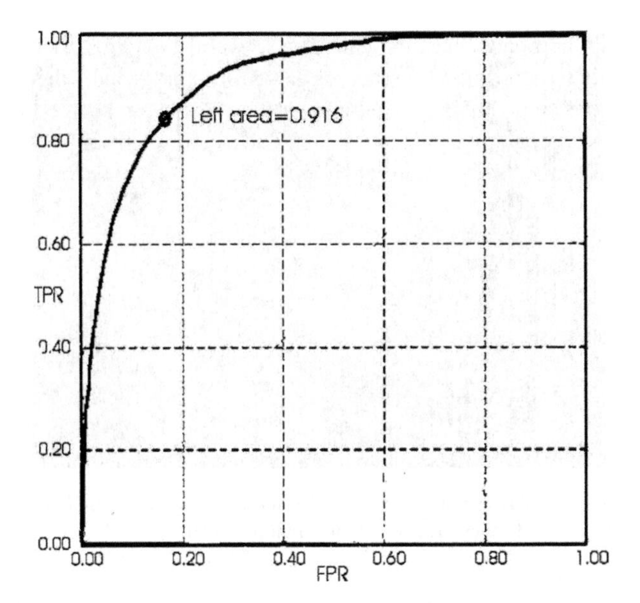

Fig. 9 A ROC curve extracted from [52]. The circled point on the curve marks the point of equal balance of sensitivity and selectivity

where parameter θ represents the threshold, $TP(\theta)$, $FN(\theta)$, $TN(\theta)$, and $FP(\theta)$ are the numbers of true positives, false negatives, true negatives, and false positives respectively, which are determined by θ. All these values are counted in samples.

Figure 8 illustrates the above parameters. The corresponding ROC curve is presented in Fig. 9.

7 An Example of BCI Applications: A P300 BCI Speller

The previous sections described some important signal processing and machine learning techniques used in BCIs. This section presents an example with a P300 BCI speller. The data were collected from 10 subjects using a P300-based speller paradigm as described in [54]. In the experiment, a 6-by-6 matrix that includes characters such as letters and numbers is presented to the user on a computer screen (Fig. 10). The user then focuses on a character while each row or column of the matrix flashes in a random order. Twelve flashes make up one run that covers all the rows and columns of the matrix. Two of these twelve flashes in each run contain the desired symbol, which produces a P300. The task of the speller is to identify the user's desired character based on the EEG data collected in the these flashes.

The dataset contains training data and testing data collected from 10 subjects. The system is trained separately for each user using a common phrase with 41 characters "THE QUICK BROWN FOX JUMPS OVER LAZY DOG 246138 579". The same phrase is also used in testing data collection with random word order.

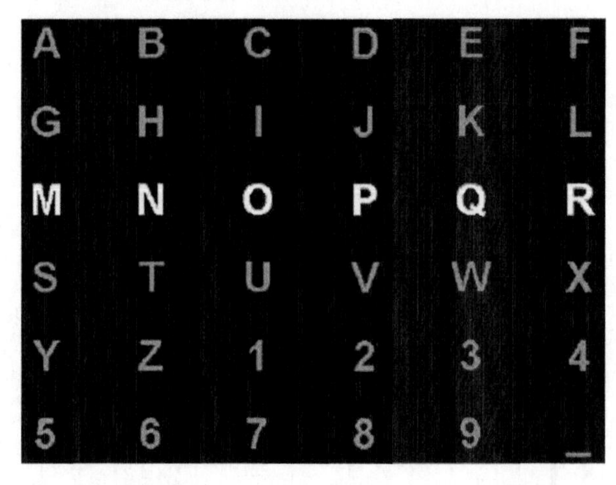

Fig. 10 The stimulus matrix shown on a computer monitor for a P300 BCI speller. One row or one column of the matrix flashes successively in a random order

A feature vector for each flash is first constructed according to the following steps:

Step 1 L channel EEG signals are filtered in the range of 0.1–20 Hz ($L = 24$ in this experiment).

Step 2 Cut the segment (e.g. from 150 to 500 ms after a flash) of EEG data from each channel for every flash, i.e. each segment is 350 ms and contains 87 data points (the sampling rate is 250 Hz).
Next, these segments are downsampled by a rate of 5. 17 data points are obtained for each segment. A feature data vector $Fe_{k,q}$ is then constructed for each flash to comprise all the 17 data points from all the L channels.

Step 3 Repeat step 2 to construct feature vectors $Fe_{k,q}$ for all the flashes on the screen for $k = 1, \cdots, 12$ and $q = 1, \cdots, 10$ where K represents a row or column in Fig. 10 ($k = 1, \cdots, 6$ refer to rows, $k = 7, \cdots, 12$ refer to columns), and q represents each round of flashing.

According to the experimental settings, $10 \times 12 = 120$ flashes occur for each character. Therefore, a character corresponds to 120 feature vectors. These feature vectors are categorized into two classes. The first class contains 20 feature vectors with their corresponding 20 flashes occur on the row or column containing the desired character. Physiologically, a P300 can be detected in each of these 20 feature vectors. The other 100 feature vectors which do not contain a P300 belong to the second class. A SVM classifier is then trained according to the following steps to detect P300 for a new trial:

Step 1 Train a SVM classifier using the training data set which contains two classes of features.

Step 2 For the qth round in a new trial, feature vectors $Fe_{k,q}$ ($k = 1, \cdots, 12$, $q = 1, \cdots, 10$) are extracted. Applying the SVM to these feature vectors, 120 scores denoted as $s_{k,q}$ are obtained. These scores are the values of the objective function of the SVM.

Step 3 P300 detection. Sum the scores for all the rows or columns $k = 1, \cdots, 12$ across all the q rounds using $ss_k = \sum_{q=1}^{10} s_{k,q}$. The row ($k = 1, \cdots, 6$) and column ($k = 7, \cdots, 12$) with the respective maximum summed scores are then identified. The decision is then made on the character at the intersection of the identified row and column. This character forms the output of the BCI speller.

Using this BCI speller, the 10 users were able to write the test phrase of 42 characters with an average accuracy of 99% [54].

8 Summary

This chapter discusses signal processing and machine learning methods that are commonly used in BCIs. However, there are many methods that are not discussed. Furthermore, signal processing often involves several stages and there is no best method for each stage. Nevertheless, the most suitable method and parameters depend on the brain patterns used in BCIs (such as P300, SSVEP, or ERD/ERS), the quantity and quality of the training data acquired, the training time, and various user factors.

Acknowledgments The authors are grateful to G. Townsend, B. Graimann, and G. Pfurtscheller for their permission to use Figures 7 and 8 in this chapter. The authors are also grateful to anonymous reviewers and the editors B. Graimann, G. Pfurtscheller, B. Allison of this book for their contributions to this chapter. Yuanqing Li's work was partially supported by National Natural Science Foundation of China under Grant 60825306, Natural Science Foundation of Guangdong Province, China under Grant 9251064101000012. Kai Keng Ang and Cuntai Guan's work were supported by the Science and Engineering Research Council of A*STAR (Agency for Science, Technology and Research), Singapore.

References

1. B. Rockstroh, T. Elbert, A. Canavan, W. Lutzenberger, and N. Birbaumer, Eds., *Slow cortical potentials and behavior*, 2nd ed. Urban and Schwarzenberg, Baltimore, MD, (1989).
2. N. Birbaumer, Slow cortical potentials: their origin, meaning and clinical use. In G.J.M. van Boxtel and K.B.E. Böcker, (Eds.) *Brain and behavior:past, present, and future*, Tilburg University Press, Tilburg, pp. 25–39, (1997).
3. E. Donchin, 'Surprise!...surprise?' Psychophysiology, 18(5) 493–513, (1981).
4. L.A. Farwell and E. Donchin, Talking off the top of your head: Toward a mental prosthesis utilizing event-related brain potentials. Electroencephalogr Clin Neurophysiol, 70(6) 510–523, (1998).
5. T.F. Collura, Real-time filtering for the estimation of steady-state visual evoked brain potentials. IEEE Trans Biomed Eng, 37(6), 650–652, (1990).

6. M. Cheng, X. Gao, S. Gao, and D. Xu, Design and implementation of a Brain–computer interface with high transfer rates. IEEE Trans Biomed Eng, 49(10), 1181–1186, (2002).
7. J.R. Wolpaw, N. Birbaumer, D.J. McFarland, G. Pfurtscheller, and T.M. Vaughan, Brain–computer interfaces for communication and control. Clin Neurophysiol, 113(6), 767–791, (2002).
8. A. Kostov and M. Polak, Parallel man-machine training in development of EEG-based cursor control. IEEE Trans Rehabil Eng, 8(2), 203–205, (2000).
9. D.J. McFarland, L.M. McCane, S.V. David, and J.R. Wolpaw, Spatial filter selection for EEG-based communication. Electroencephalogr Clin Neurophysiol, 103(3), 386–394, (1997).
10. T.-P. Jung, C. Humphries, T.-W. Lee, S. Makeig, M.J. McKeown, V. Iragui, and T.J. Sejnowski, 'Extended ICA removes artifacts from electroencephalographic recordings.' In D. Touretzky, M. Mozer, and M. Hasselmo, (Eds.) *Advances in neural information processing systems*, MIT Press, Cambridge, MA, vol. 10, pp. 894–900, (1998).
11. T.-P. Jung, S. Makeig, C. Humphries, T.-W. Lee, M. J. McKeown, V. Iragui, and T. Sejnowski, 'Removing electroencephalographic artifacts by blind source separation.' *Psychophysiology*, 37(2), 163–178, (2000).
12. L. Parra, C. Spence, A. Gerson, and P. Sajda, Recipes for the linear analysis of EEG. NeuroImage, 28(2), 326–341, (2005).
13. A.J. Bell and T.J. Sejnowski, An information-maximization approach to blind separation and blind deconvolution. Neural Comput, 7 (6), 1129–1159, (1995).
14. A. Hyvarinen and E. Oja, Independent component analysis: Algorithms and applications. Neural Netw, 13(4–5), 411–430, (2000).
15. N. Xu, X. Gao, B. Hong, X. Miao, S. Gao, and F. Yang, BCI competition 2003-data set IIb: enhancing P300 wave detection using ICA-based subspace projections for BCI applications. IEEE Trans Biomed Eng, 51(6), 1067–1072, (2004).
16. A. Cichocki and S.-I. Amari, *Adaptive blind signal and image processing: learning algorithms and applications*. John Wiley, New York, (2002).
17. A. Li, Y. Cichocki and S.-I. Amari, Analysis of sparse representation and blind source separation. Neural Comput, 16(6), 1193–1234, (2004).
18. Y. Li, A. Cichocki, and S.-I. Amari, Blind estimation of channel parameters and source components for EEG signals: A sparse factorization approach. IEEE Trans Neural Netw, 17(2), 419–431, (2006).
19. H. Ramoser, J. Muller-Gerking, and G. Pfurtscheller, Optimal spatial filtering of single trial EEG during imagined hand movement. IEEE Trans Rehabil Eng, 8(4), 441–446, (2000).
20. C. Guger, H. Ramoser, and G. Pfurtscheller, Real-time EEG analysis with subject-specific spatial patterns for a brain–computer interface (BCI). IEEE Trans Rehabili Eng, 8(4), 447–456, (2000).
21. B. Blankertz, G. Dornhege, M. Krauledat, K.-R. Müller, V. Kunzmann, F. Losch, and G. Curio, The Berlin brain–computer interface: EEG-based communication without subject training. IEEE Trans Neural Syst Rehabili Eng, 14(2), 147–152, (2006).
22. B. Blankertz, G. Dornhege, M. Krauledat, K.-R. Muller, and G. Curio, The non-invasive Berlin brain–computer interface: Fast acquisition of effective performance in untrained subjects. NeuroImage, 37, 2, 539–550, (2007).
23. M. Grosse-Wentrup and M. Buss, Multiclass common spatial patterns and information theoretic feature extraction. IEEE Trans Biomed Engi. 55(8), 1991–2000, (2008).
24. G. Gratton, M. G. H. Coles, and E. Donchin, A new method for off-line removal of ocular artifact. Electroencephalogr Clin Neurophysiol, 55(4), 468–484, (1983).
25. S. Kelly, E. Lalor, R. Reilly, and J. Foxe, "Visual spatial attention tracking using high-density SSVEP data for independent brain–computer communication," *IEEE Transactions on Neural Systems and Rehabilitation Engineering*, vol. 13, no. 2, pp. 172–178, (2005).

26. B. Allison, D. McFarland, G. Schalk, S. Zheng, M. Jackson, and J. Wolpaw, Towards an independent brain–computer interface using steady state visual evoked potentials. Clin Neurophysiol, 119(2), 399–408, (2008).
27. J. R. Wolpaw and D. J. McFarland, Control of a two-dimensional movement signal by a non-invasive brain–computer interface in humans, Proc Nat Acad Sci, 101(51), 17849–17854, (2004).
28. G. Blanchard and B. Blankertz, BCI competition 2003-data set IIa: spatial patterns of self-controlled brain rhythm modulations. IEEE Trans Biomed Eng, 51(6), 1062–1066, (2004).
29. Y. Li and C. Guan, Joint feature re-extraction and classification-using aniterative-semi-supervised support vector machine algorithm. Mach Learn, vol. online, 71, 33–53, (2008).
30. G. Pfurtscheller, C. Neuper, D. Flotzinger, and M. Pregenzer, EEG-based discrimination between imagination of right and left hand movement. Electroencephalogr Clin Neurophysiol, 103(6), 642–651, (1997).
31. C.W. Anderson, E.A. Stolz, and S. Shamsunder, Multivariate autoregressive models for classification of spontaneous electroencephalographic signals during mental tasks. IEEE Trans Biomed Eng, 45(3), 277–286, (1998).
32. G. Pfurtscheller, C. Neuper, A. Schlogl, and K. Lugger, Separability of EEG signals recorded during right and left motor imagery using adaptive autoregressive parameters. IEEE Trans Rehabil Eng, 6(3), 316–325, (1998).
33. R.O. Duda, P.E. Hart, and D.G. Stork. Patt Classifi, 2nd ed. John Wiley, New York, (2001).
34. K. Muller, M. Krauledat, G. Dornhege, G. Curio, and B. Blankertz, Machine learning techniques for brain–computer interfaces. Biomed Tech, 49(1), 11–22, (2004).
35. D. Goldberg, *Genetic algorithms in search, optimization and machine learning*. Addison-Wesley Longman, Boston, MA, (1989).
36. J. Han and M. Kamber, *Data mining: Concepts and techniques*. Morgan Kaufmann, San Francisco, CA, (2001).
37. B. Blankertz, G. Curio, and K.-R. Müller, Classifying single trial EEG: Towards brain computer interfacing. In T. G. Diettrich, S. Becker, and Z. Ghahramani, (Eds.) *Advances in neural information processing systems (NIPS 01)*, MIT Press, Cambridge, MA, vol. 14, pp. 157–164, (2002).
38. B. Blankertz, G. Dornhege, S. Lemm, M. Krauledat, G. Curio, and K. Müller, "The Berlin brain–computer interface: Machine learning based detection of user specific brain states. J Uni Comput Sci, 12(6), 581–607, (2006).
39. R. Boostani, B. Graimann, M. Moradi, and G. Pfurtscheller, A comparison approach toward finding the best feature and classifier in cue-based BCI. Med Biolo Engi Comput, 45(4), 403–412, (2007).
40. R. Krepki, G. Curio, B. Blankertz, and K.-R. Müller, Berlin brain–computer interface-The HCI communication channel for discovery. Int J. Human-Comput Stud, 65(5), 460–477, (2007).
41. C.-I. Hung, P.-L. Lee, Y.-T. Wu, L.-F. Chen, T.-C. Yeh, and J.-C. Hsieh, Recognition of motor imagery electroencephalography using independent component analysis and machine classifiers. Ann Biomed Eng, 33(8), 1053–1070, (2005).
42. S. Lemm, S. Lemm, C. Schafer, and G. Curio, BCI competition 2003-data set III: probabilistic modeling of sensorimotor /spl mu/ rhythms for classification of imaginary hand movements. IEEE Trans Biomed Eng, 51(6), 1077–1080, (2004).
43. R. Sitaram, H. Zhang, C. Guan, M. Thulasidas, Y. Hoshi, A. Ishikawa, K. Shimizu, and N. Birbaumer, Temporal classification of multichannel near-infrared spectroscopy signals of motor imagery for developing a brain–computer interface. NeuroImage, 34(4), 1416–1427, (2007).
44. A. R. Webb, *Statistical Pattern Recognition*, 2nd ed. Wiley, West Sussex, England, (2002).
45. V. N. Vapnik, *Statistical learning theory*, ser. Adaptive and learning systems for signal processing, communications, and control. Wiley, New York, (1998).

46. K. Crammer and Y. Singer, On the algorithmic implementation of multiclass kernel-based vector machines. J Mach Learn Res, 2, 265–292, (2002), 944813.
47. G. Dornhege, B. Blankertz, G. Curio, and K.-R. Müller, Boosting bit rates in noninvasive EEG single-trial classifications by feature combination and multiclass paradigms. IEEE Trans Biomed Eng, 51(6), 993–1002, (2004).
48. Y. Li and C. Guan, An extended EM algorithm for joint feature extraction and classification in brain–computer interfaces. Neural Comput, 18(11), 2730–2761, (2006).
49. D. J. McFarland and J. R. Wolpaw, EEG-based communication and control: Speed-accuracy relationships. Appl Psychophysiol Biofeedback, 28(3), 217–231, (2003).
50. C. E. Shannon and W. Weaver, *The mathematical theory of communication*. University of Illinois Press, Urbana, (1971).
51. S. G. Mason and G. E. Birch, A brain-controlled switch for asynchronous control applications. IEEE Trans Biomed Eng, 47(10), 1297–1307, (2000).
52. G. Townsend, B. Graimann, and G. Pfurtscheller, Continuous EEG classification during motor imagery-simulation of an asynchronous BCI. IEEE Trans Neural Syst Rehabili Eng, 12(2), 258–265, (2004).
53. J.A. Swets, ROC analysis applied to the evaluation of medical imaging techniques," Investigative Radiol, 14(2), 109–121, (1979).
54. M. Thulasidas, C. Guan, and J. Wu, Robust classification of eeg signal for brain–computer interface. IEEE Trans Neural Syst Rehabil Eng, 14(1), 24–29, (2006).

Adaptive Methods in BCI Research - An Introductory Tutorial

Alois Schlögl, Carmen Vidaurre, and Klaus-Robert Müller

This chapter tackles a difficult challenge: presenting signal processing material to non-experts. This chapter is meant to be comprehensible to people who have some math background, including a course in linear algebra and basic statistics, but do not specialize in mathematics, engineering, or related fields. Some formulas assume the reader is familiar with matrices and basic matrix operations, but not more advanced material. Furthermore, we tried to make the chapter readable even if you skip the formulas. Nevertheless, we include some simple methods to demonstrate the basics of adaptive data processing, then we proceed with some advanced methods that are fundamental in adaptive signal processing, and are likely to be useful in a variety of applications. The advanced algorithms are also online available [30]. In the second part, these techniques are applied to some real-world BCI data.

1 Introduction

1.1 Why We Need Adaptive Methods

All successful BCI systems rely on efficient real-time feedback. Hence, BCI data processing methods must be also suitable for online and real-time processing. This requires algorithms that can only use sample values from the past and present but not the future. Such algorithms are sometimes also called causal algorithms. Adaptive methods typically fulfill this requirement, while minimizing the time delay. The data processing in BCIs consists typically of two main steps, (i) signal processing and feature extraction, and (ii) classification or feature translation (see also Fig. 1). This work aims to introduce adaptive methods for both steps; these are also closely related to two types of non-stationarities - namely short-term changes related to different mental activities (e.g. hand movement, mental arithmetic, etc.), and less

---

A. Schlögl (✉)
Institute of Science and Technology Austria (IST Austria), Am Campus 1, A-3400 Klosterneuburg, Austria
e-mail: alois.schloegl@gmail.com

B. Graimann et al. (eds.), *Brain–Computer Interfaces*, The Frontiers Collection,
DOI 10.1007/978-3-642-02091-9_18,

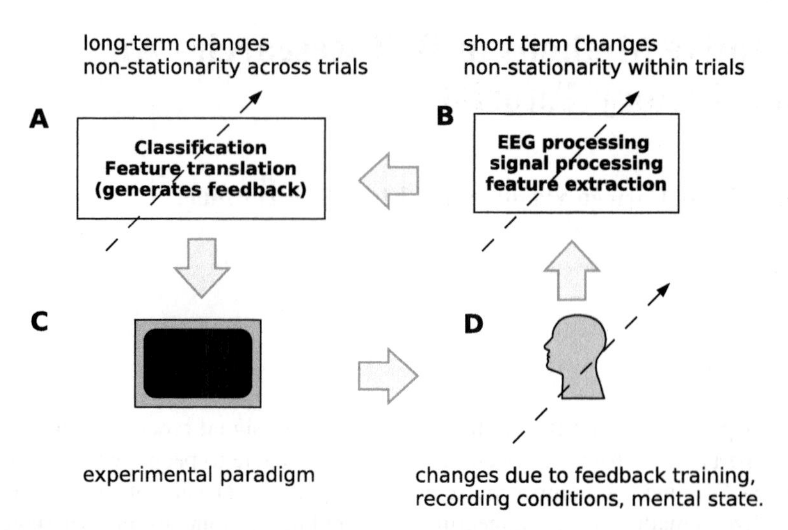

Fig. 1 Scheme of a Brain–Computer Interface. The brain signals are recorded from the subject (**d**) and processed for feature extraction (**b**). The features are classified and translated into a control signal (**a**), and feedback is provided to the subject. The arrows indicate a possible variation over time (see also the explanation in the text)

specific long term changes related to fatigue, changes in the recording conditions, or effects of feedback training.

The first type of changes (i.e. short-term changes) is addressed in the feature extraction step (B in Fig. 1). Typically, these are changes within each trial that are mainly due to the different mental activities for different tasks. One could also think of short-term changes unrelated to the task, which are typically the cause for imperfect classification and are often difficult to distinguish from the background noise, so these are not specifically addressed here.

The second type of non-stationarities are long-term changes caused by e.g. a feedback training effect. More recently, adverse long-term changes (e.g. due to fatigue, changed recording conditions) have been discussed. These non-stationarities are addressed in the classification and feature translation step (part a in Fig. 1).

Accordingly, we do see class-related short-term changes (due to the different mental tasks) (e.g. Fig. 4), class-related long-term changes (due to feedback training), and unspecific long-term changes (e.g. due to fatigue). The source of the different non-stationarities are the probands and its brain signals as well as the recording conditions (part D in Fig. 1) [24, 50, 51]. Specifically, feedback training can modify the subjects' EEG patterns, and this might require an adaptation of the classifier which might change again the feedback. The possible difficulties of such a circular relation have been also discussed as the "man–machine learning dilemma" [5, 25]. Theoretically, a similar problem could also occur for short-term changes. These issues will be briefly discussed at the end of this chapter.

Segmentation-type approaches are often used to address non-stationarities. For example, features were extracted from short data segments (e.g. FFT-based

Bandpower [23, 25, 27], AR-based spectra in [18], slow cortical potentials by [2], or CSP combined with Bandpower [4, 6, 7, 16]). Also classifiers were obtained and retrained from specific sessions (e.g. [4, 25]) or runs. A good overview on various methods is provided in chapter "Digital signal Processing and Machine Learning" in this volume [17].

Segmentation methods may cause sudden changes from one segment to the next one. Adaptive methods avoid such sudden changes, but are continuously updated to the new situation. Therefore, they have the potential to react faster, and have a smaller deviation from the true system state. A sliding window approach (segmentation combined with overlapping segments) can also provide a similar advantage, however, we will demonstrate that this comes with increased computational costs.

In the following pages, some basic adaptive techniques are first presented and discussed, then some more advanced techniques are introduced. Typically, the stationary method is provided first, and then the adaptive estimator is introduced. Later, a few techniques are applied to adaptive feature extraction and adaptive classification methods in BCI research, providing a comparison between a few adaptive feature extraction and classification methods.

A short note about the notation: first, all the variables that are a function of time will be denoted as $f(t)$ until Sect. 1.3. Then, the subindex k will be used to denote sample-based adaptation and n to trial-based adaptation.

1.2 Basic Adaptive Estimators

1.2.1 Mean Estimation

Let us assume the data as a stochastic process $x(t)$, that is series of stochastic variables x ordered in time t; at each instant t in time, the sample value $x(t)$ is observed, and the whole observed process consists of N observations. Then, the (overall) mean value μ_x of $x(t)$ is

$$mean(x) = \mu_x = \frac{1}{N}\sum_{t=1}^{N} x(t) = E\langle x(t)\rangle \tag{1}$$

In case of a time-varying estimation, the mean can be estimated with a sliding window approach using

$$\mu_x(t) = \frac{1}{\sum_{i=0}^{n-1} w_i}\sum_{i=0}^{n-1} w_i \cdot x(t-i) \tag{2}$$

where n is the width of the window and w_i are the weighting factors. A simple solution is using a rectangular window i.e. $w_i = 1$ resulting in

$$\mu_x(t) = \frac{1}{n}\sum_{i=0}^{n-1} x(t-i) \tag{3}$$

For the rectangular window approach ($w_i = const$), the computational effort can be reduced by using this recursive formula

$$\mu_x(t) = \mu_x(t-1) + \frac{1}{n} \cdot (x(t) - x(t-n)) \quad (4)$$

Still, one needs to keep the n past sample values in memory. The following adaptive approach needs no memory for its past sample values

$$\mu_x(t) = (1 - UC) \cdot \mu_x(t-1) + UC \cdot x(t) \quad (5)$$

whereby UC is the update coefficient, describing an exponential weighting window

$$w_i = UC \cdot (1 - UC)^i \quad (6)$$

with a time constant of $\tau = 1/(UC \cdot F_s)$ if the sampling rate is F_s. This means that an update coefficient UC close to zero emphasizes the past values while the most recent values have very little influence on the estimated value; a larger update coefficient will emphasize the most recent sample values, and forget faster the earliers samples. Accordingly, a larger update coefficient UC enables a faster adaptation. If the update coefficient UC becomes too large, the estimated values is based only on a few samples values. Accordingly, the update coefficient UC can be used to determine the tradeoff between adaptation speed and estimation accuracy.

All mean estimators are basically low pass filters whose bandwidths (or edge frequency of the low pass filter) are determined by the window length n or the update coefficient UC. The relationship between a rectangular window of length n and an exponential window with $UC = \frac{1}{n}$ is discussed in [36] (Sect. 3.1). Thus, if the window length and the update coefficient are properly chosen, a similar characteristic can be obtained.

Table 1 shows the computational effort for the different estimators. The stationary estimator is clearly not suitable for a real-time estimation; the sliding window approaches require memory that is proportional to the window size and are often computationally more expensive than adaptive methods. Thus the adaptive method has a clear advantage in terms of computational costs.

Table 1 Computational effort of mean estimators. The computational and the memory effort per time step are shown by using the O-notation, with respect to the number of samples N and the window size n.[1]

Method	Memory effort	Computational effort
stationary	$O(N)$	$O(N)$
weighted sliding window	$O(n)$	$O(N \cdot n)$
rectangular sliding window	$O(n)$	$O(N \cdot n)$
recursive (only for rectangular)	$O(n)$	$O(N)$
adaptive (exponential window)	$O(N)$	$O(N)$

[1]The O-notation is frequently used in computer science to show the growth rate of an algorithm's usage of computational resources with respect to its input.

1.2.2 Variance Estimation

The overall variance σ_x^2 of x_t can be estimated with

$$var(x) = \sigma_x^2 = \frac{1}{N}\sum_{t=1}^{N}(x(t) - \mu)^2 = E\langle(x(t) - \mu)^2\rangle \tag{7}$$

$$= \frac{1}{N}\sum_{t=1}^{N}(x(t)^2 - 2\mu x(t) + \mu^2) = \tag{8}$$

$$= \frac{1}{N}\sum_{t=1}^{N}x(t)^2 - \frac{1}{N}\sum_{t=1}^{N}2\mu x(t) + \frac{1}{N}\sum_{t=1}^{N}\mu^2 = \tag{9}$$

$$= \frac{1}{N}\sum_{t=1}^{N}x(t)^2 - 2\mu\frac{1}{N}\sum_{t=1}^{N}x(t) + \frac{1}{N}N\mu^2 = \tag{10}$$

$$= \sigma_x^2 = \frac{1}{N}\sum_{t=1}^{N}x(t)^2 - \mu_x^2 \tag{11}$$

Note: this variance estimator is biased. To obtain an unbiased estimator, one must multiply the result by $N/(N-1)$.

An adaptive estimator for the variance is this one

$$\sigma_x(t)^2 = (1 - UC) \cdot \sigma_x(t-1)^2 + UC \cdot (x(t) - \mu_x(t))^2 \tag{12}$$

Alternatively, one can also compute the adaptive mean square

$$MSQ_x(t) = (1 - UC) \cdot MSQ_x(t-1) + UC \cdot x(t)^2 \tag{13}$$

and obtain the variance by

$$\sigma_x(t)^2 = MSQ_x(t) - \mu_x(t)^2 \tag{14}$$

When adaptive algorithms are used, we also need initial values and a suitable update coefficient. For the moment, it is sufficient to assume that initial values and the update coefficient are known. Various approaches to identify suitable values will be discussed later (see Sect. 1.5).

1.2.3 Variance-Covariance Estimation

In case of multivariate processes, also the covariances between the various dimensions are of interest. The (stationary) variance-covariance matrix (short covariance matrix) is defined as

$$cov(x) = \boldsymbol{\Sigma}_x = \frac{1}{N}\sum_{t=1}^{N}(\boldsymbol{x}(t) - \boldsymbol{\mu}_x)^T \cdot (\boldsymbol{x}(t) - \boldsymbol{\mu}_x) \tag{15}$$

whereby ^{T} indicates the transpose operator. The variances are the diagonal elements of the variance-covariance matrix, and the off-diagonal elements indicate the covariance $\sigma_{i,j} = \frac{1}{N}\sum_{t=1}^{N}\left((x_i(t)-\mu_i)\cdot(x_j(t)-\mu_j)\right)$ between the i-th and j-th element. We define also the so-called *extended covariance matrix* (ECM) $\boldsymbol{E}$ as

$$ECM(x) = \boldsymbol{E}_x = \sum_{t=1}^{N_x}[1,\boldsymbol{x}(t)]^T\cdot[1,\boldsymbol{x}(t)] = \left[\begin{array}{c|c} a & \boldsymbol{b} \\ \hline \boldsymbol{c} & \boldsymbol{D}\end{array}\right] = N_x\cdot\left[\begin{array}{c|c} 1 & \boldsymbol{\mu}_x \\ \hline \boldsymbol{\mu}_x{}^T & \boldsymbol{\Sigma}_x + \boldsymbol{\mu}_x^T\boldsymbol{\mu}_x\end{array}\right] \quad (16)$$

One can obtain from the ECM $\boldsymbol{E}$ the number of samples $N = a$, the mean $\boldsymbol{\mu} = \boldsymbol{b}/a$ as well as the variance-covariance matrix $\boldsymbol{\Sigma} = \boldsymbol{D}/a - (\boldsymbol{c}/a)\cdot(\boldsymbol{b}/a)$. This decomposition will be used later.

The adaptive version of the ECM estimator is

$$\boldsymbol{E}_x(t) = (1-UC)\cdot\boldsymbol{E}_x(t-1) + UC\cdot[1,\boldsymbol{x}(t)]^T\cdot[1,\boldsymbol{x}(t)] \quad (17)$$

where t is the sample time, UC is the update coefficient. The decomposition of the ECM $\boldsymbol{E}$, mean μ, variance σ^2 and covariance matrix $\boldsymbol{\Sigma}$ is the same as for the stationary case; typically is $N = a = 1$.

1.2.4 Adaptive Inverse Covariance Matrix Estimation

Some classifiers like LDA or QDA rely on the inverse $\boldsymbol{\Sigma}^{-1}$ of the covariance matrix $\boldsymbol{\Sigma}$; therefore, adaptive classifiers require an adaptive estimation of the inverse covariance matrix. The inverse covariance matrix $\boldsymbol{\Sigma}$ can be obtained from Eq. (16) with

$$\boldsymbol{\Sigma}^{-1} = a\cdot\left(\boldsymbol{D} - \boldsymbol{c}\cdot a^{-1}\cdot\boldsymbol{b}\right)^{-1}. \quad (18)$$

This requires an explicit matrix inversion. The following formula shows how the inverse convariance matrix $\boldsymbol{\Sigma}^{-1}$ can be obtained without an explicit matrix inversion. For this purpose, the block matrix decomposition [1] and the matrix inversion lemma (20) is used. Let us also define $iECM = \boldsymbol{E}^{-1} = \left[\begin{array}{c|c} A & B \\ \hline C & D\end{array}\right]^{-1}$ with $S = D - CA^{-1}B$. According to the block matrix decomposition [1]

$$\begin{aligned}\boldsymbol{E}_x^{-1} = \left[\begin{array}{c|c} A & B \\ \hline C & D\end{array}\right]^{-1} &= \left[\begin{array}{c|c} A^{-1}+A^{-1}BS^{-1}CA^{-1} & -A^{-1}BS^{-1} \\ \hline -S^{-1}CA^{-1} & S^{-1}\end{array}\right] \\ &= \left[\begin{array}{c|c} 1+\boldsymbol{\mu}_x\boldsymbol{\Sigma}_x^{-1}\boldsymbol{\mu}_x & -\boldsymbol{\mu}_x^T\boldsymbol{\Sigma}_x^{-T} \\ \hline -\boldsymbol{\Sigma}_x^{-1}\boldsymbol{\mu}_x^T & \boldsymbol{\Sigma}_x^{-1}\end{array}\right]\end{aligned} \quad (19)$$

The inverse extended covariance matrix $iECM = \boldsymbol{E}^{-1}$ can be obtained adaptively by applying the matrix inversion lemma (20) to Eq. (17). The matrix inversion lemma (also know as Woodbury matrix identity) states that the inverse $\boldsymbol{A}^{-1}$ of a given matrix $\boldsymbol{A} = (\mathbf{B}+\mathbf{UDV})$ can be determined by

$$\begin{aligned} A^{-1} &= (\mathbf{B}+\mathbf{UDV})^{-1} = 20 \\ &= B^{-1} + B^{-1}U\left(D^{-1} + VB^{-1}U\right)^{-1} VB^{-1} \end{aligned} \tag{20}$$

To adaptively estimate the inverse of the *extended covariance matrix*, we identify the matrices in (20) as follows:

$$A = E(t) \tag{21}$$

$$B^{-1} = (1 - UC) \cdot E(t-1) \tag{22}$$

$$U^T = V = x(t) \tag{23}$$

$$D = UC \tag{24}$$

where UC is the update coefficient and $x(t)$ is the current sample vector. Accordingly, the inverse of the covariance matrix is:

$$E(t)^{-1} = \frac{1}{(1-UC)} \cdot \left(E(t-1)^{-1} - \frac{1}{\frac{1-UC}{UC} + x(t) \cdot v} \cdot v \cdot v^T\right) \tag{25}$$

with $v = E(t-1)^{-1} \cdot x(t)^T$. Since the term $x(t) \cdot v$ is a scalar, no explicit matrix inversion is needed.

In practice, this adaptive estimator can become numerically unstable (due to numerical inaccuracies, the iECM can become asymmetric and singular). This numerical problem can be avoided if the symmetry is enforced, e.g. in the following way:

$$E(t)^{-1} = \left(E(t)^{-1} + E(t)^{-1,T}\right)/2 \tag{26}$$

Now, the inverse covariance matrix Σ^{-1} can be obtained by estimating the extended covariance matrix with Eq. (25) and decomposing it according to Eq. (19).

Kalman Filtering and the State Space Model

The aim of a BCI is to identify the state of the brain from the measured signals. The measurement itself, e.g. some specific potential difference at some electrode, is not the “brain state” but the result of some underlying mechanism generating different patterns depending on the state (e.g. alpha rhythm EEG). Methods that try to identify the underlying mechanism are called system identification or model identification methods. There are a large number of different systems and different methods in this area. In the following, we’ll introduce an approach to identify a state-space model (Fig. 2). A state-space model is a general approach and can be used to describe a large number of different models. In this chapter, an autogregressive model and a linear discriminant model will be used, but a state-state space model can be also used to describe more complex models. Another useful advantage, besides the general

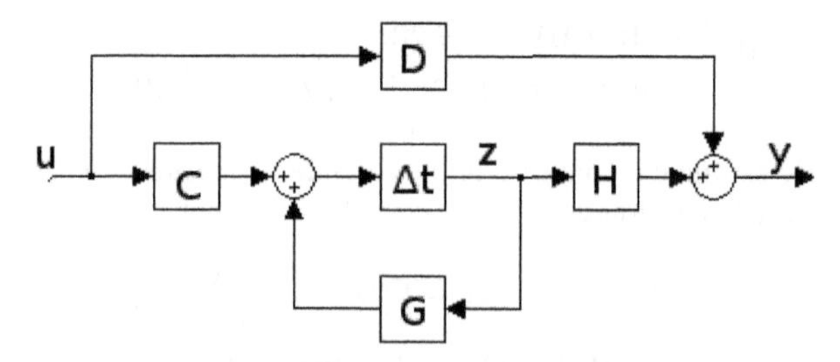

Fig. 2 State Space Model. G is the state transition matrix, H is the measurement matrix, Δt denotes a one-step time delay, C and D describe the influence of some external input u to the state vector and the output y, respectively. The system noise w and observation noise v are not shown

nature of the state-space model, is the fact that efficient adaptive algorithms are available for model identification. This algorithm is called the Kalman filter.

Kalman [12] and Bucy [13] presented the original idea of Kalman filtering (KFR). Meinhold et al. [20] provided a Bayesian formulation of the method. Kalman filtering is an algorithm for estimating the state (vector) of a state space model with the system Eq. (27) and the measurement (or observation) Eq. (28).

$$z(t) = \boldsymbol{G}(t, t-1) \cdot z(t-1) + \boldsymbol{C}(t) \cdot \boldsymbol{u}(t) + \boldsymbol{w}(t) \tag{27}$$

$$\boldsymbol{y}(t) = \boldsymbol{H}(t) \cdot z(t) + \boldsymbol{D}(t) \cdot \boldsymbol{u}(t) + \boldsymbol{v}(t) \tag{28}$$

$\boldsymbol{u}(t)$ is an external input. When identifying the brain state, we usually ignore the external input. Accordingly $\boldsymbol{C}(t)$ and $\boldsymbol{D}(t)$ are zero, while $z(t)$ is the state vector and depends only on the past values of $\boldsymbol{w}(t)$ and some initial state z_0. The observed output signal $\boldsymbol{y}(t)$ is a combination of the state vector and the measurement noise $\boldsymbol{v}(t)$ with zero mean and variance $\boldsymbol{V}(t) = \boldsymbol{E}\langle \boldsymbol{v}(t) \cdot \boldsymbol{v}(t)^T \rangle$. The process noise $\boldsymbol{w}(t)$ has zero mean and covariance matrix $\boldsymbol{W}(t) = \boldsymbol{E}\langle \boldsymbol{w}(t) \cdot \boldsymbol{w}(t)^T \rangle$. The state transition matrix $\boldsymbol{G}(t, t-1)$ and the measurement matrix $\boldsymbol{H}(t)$ are known and may or may not change with time.

Kalman filtering is a method that estimates the state $z(t)$ of the system from measuring the output signal $y(t)$ with the prerequisite that $\boldsymbol{G}(t, t-1)$, $\boldsymbol{H}(t)$, $\boldsymbol{V}(t)$ and $\boldsymbol{W}(t)$ for $t > 0$ and z_0 are known. The inverse of the state transition matrix $\boldsymbol{G}(t, t-t)$ exists and $\boldsymbol{G}(t, t-1) \cdot \boldsymbol{G}(t-1, t) = \boldsymbol{I}$ is the unity matrix $\boldsymbol{I}$. Furthermore, $\boldsymbol{K}(t, t-1)$, the a-priori state-error correlation matrix, and $\boldsymbol{Z}(t)$, the a posteriori state-error correlation matrix are used; $\boldsymbol{K}_{1,0}$ is known. The Kalman filter equations can be summarized in this algorithm

$$\begin{aligned}
\boldsymbol{e}(t) &= \boldsymbol{y}(t) - \boldsymbol{H}(t) \cdot \hat{z}(t) \\
\hat{z}(t+1) &= \boldsymbol{G}(t, t-1) \cdot \hat{z}(t) + \boldsymbol{k}(t-1) \cdot \boldsymbol{e}(t) \\
\boldsymbol{Q}(t) &= \boldsymbol{H}(t) \cdot \boldsymbol{K}(t, t-1) \cdot \boldsymbol{H}^T(t) + \boldsymbol{V}(t) \\
\boldsymbol{k}(t) &= \boldsymbol{G}(t, t-1) \cdot \boldsymbol{K}(t, t-1) \cdot \boldsymbol{H}^T(t) / \boldsymbol{Q}(t) \\
\boldsymbol{Z}(t) &= \boldsymbol{K}(t, t-1) - \boldsymbol{G}(t-1, t) \cdot \boldsymbol{k}(t) \cdot \boldsymbol{H}(t) \cdot \boldsymbol{K}(t, t-1) \\
\boldsymbol{K}(t+1, t) &= \boldsymbol{G}(t, t-1) \cdot \boldsymbol{Z}(t) \cdot \boldsymbol{G}(t, t-1)^T + \boldsymbol{W}(t)
\end{aligned} \tag{29}$$

Using the next observation value $\boldsymbol{y}(t)$, the one-step prediction error $\boldsymbol{e}(t)$ can be calculated using the current estimate $\hat{z}(t)$ of the state $z(t)$, and the state vector $z(t+1)$ is updated (29). Then, the estimated prediction variance $\boldsymbol{Q}(t)$ that can be calculated which consists of the measurement noise $\boldsymbol{V}(t)$ and the error variance due to the estimated state uncertainty $\boldsymbol{H}(t) \cdot \boldsymbol{K}(t, t-1) \cdot \boldsymbol{H}(t)^T$. Next, the Kalman gain $\boldsymbol{k}(t)$ is determined. Finally, the a posteriori state-error correlation matrix $\boldsymbol{Z}(t)$ and the a-priori state-error correlation matrix $\boldsymbol{K}(t+1, t)$ for the next iteration are obtained.

Kalman filtering was developed to estimate the trajectories of spacecrafts and satellites. Nowadays, Kalman filtering is used in a variaty of applications, including autopilots, economic time series prediction, radar tracking, satellite navigation, weather forecasts, etc.

1.3 Feature Extraction

Many different features can be extracted from EEG time series, like temporal, spatial, spatio-temporal, linear and nonlinear parameters [8, 19]. The actual features extracted use first order statistical properties (i.e. time-varying mean like the slow cortical potential [2]), or more frequently the second order statistical properties of the EEG are used by extracting the frequency spectrum, or the autoregressive parameters [18, 31, 36]. Adaptive estimation of the mean has been discussed in Sect. 1.2. Other time-domain parameters are activity, mobility and complexity [10], amplitude, frequency, spectral purity index [11], Sigma et al. [49] and brainrate [28]. Adaptive estimators of these parameters have been implemented in the open source software library Biosig [30].

Spatial correlation methods are PCA, ICA and CSP [3, 4, 16, 29]; typically these methods provide a spatial decomposition of the data and do not take into account a temporal correlation. Recently, extensions of CSP have been proposed that can construct spatio-temporal filters [4, 7, 16]. To address non-stationarities, covariate shift compensation approaches [38, 40, 41] have been suggested and adaptive CSP approaches have been proposed [42, 43]. In order to avoid the computational expensive eigendecomposition after each iteration, an adaptive eigenanalysis approaches as suggested in [21, 39] might be useful.

Here, the estimation of the adaptive autoregressive (AAR) parameters is discussed in greater depth. AAR parameters can capture the time-varying second order statistical moments. Almost no a priori knowledge is required, the model order p is not very critical and, since it is a single coefficient, it can be easily optimized. Also, no expensive feature selection algorithm is needed. AAR parameters provide a simple and robust approach, and hence provide a good starting point for adaptive feature extraction.

1.3.1 Adaptive Autoregressive Modeling

A univariate and stationary autoregressive (AR) model is described by any of the following equations

$$\begin{aligned} y_k &= a_1 \cdot y_{k-1} + \cdots + a_p \cdot y_{k-p} + x_k = \\ &= \sum_{i=1}^{p} a_i \cdot y_{k-i} + x_k = \\ &= [y_{k-1}, \ldots, y_{k-p}] \cdot [a_1, ..., a_p]^T + x_k = \\ &= \boldsymbol{Y}_{k-1}^T \cdot \boldsymbol{a} + x_k \end{aligned} \tag{30}$$

with innovation process $x_k = N(\mu_x = 0, \sigma_x^2)$ having zero mean and variance σ_x^2. For a sampling rate f_0, the spectral density function $P_y(f)$ of the AR process y_k is

$$P_y(f) = \frac{\sigma_x^2/(2\pi f_0)}{|1 - \sum_{k=1}^{p}(a_i \cdot exp^{-jk2\pi f/f_0})|^2} \tag{31}$$

There are several estimators (Levinson-Durbin, Burg, Least Squares, geometric lattice) for the stationary AR model. Estimation of adaptive autoregressive (AAR) parameters can be obtained with Least Mean Squares (LMS), Recursive Least Squares (RLS) and Kalman filters (KFR) [36]. LMS is a very simple algorithm, but typically performs worse (in terms of adaptation speed and estimation accuracy) than RLS or KFR [9, 31, 36]. The RLS method is a special case of the more general Kalman filter approach. To perform AAR estimation with the KFR approach, the AAR model needs to be adapted – in a suitable way – to the state space model.

The aim is to estimate the time-varying autoregressive parameters; therefore, the AR parameters become state vectors $z_k = \boldsymbol{a}_k = [a_{1,k}, \cdots, a_{p,k}]^T$. Assuming that the AR parameters follow a mutltivariate random walk, the state transition matrix becomes the identity matrix $\boldsymbol{G}_{k,k-1} = \boldsymbol{I}_{p\times p}$ and the system noise w_k allows for small alterations. The observation matrix $\boldsymbol{H}_k$ consists of the past p sampling values $y_{k-1}, \ldots, y_{k-p}$. The innovation process $v_k = x_k$ with $\sigma_x^2(k) = V_k$. The AR model (30) is translated into the state space model formalism (27-28) as follows:

$$\begin{aligned} &\textit{State Space Model} \Leftrightarrow \textit{Autoregressive Model} \\ z_k &= \boldsymbol{a}_k = [a_{1,k}, ...a_{p,k}]^T \\ \boldsymbol{H}_k &= \boldsymbol{Y}_{k-1} = [y_{k-1}, ..., y_{k-p}]^T \\ \boldsymbol{G}_{k,k-1} &= \boldsymbol{I}_{p\times p} \\ \boldsymbol{V}_k &= \sigma_x^2(k) \\ \boldsymbol{Z}_k &= E\langle(\boldsymbol{a}_k - \hat{\boldsymbol{a}}_k)^T \cdot (\boldsymbol{a}_k - \hat{\boldsymbol{a}}_k)\rangle \\ \boldsymbol{W}_k &= \boldsymbol{A}_k = E\langle(\boldsymbol{a}_k - \boldsymbol{a}_{k-1})^T \cdot (\boldsymbol{a}_k - \boldsymbol{a}_{k-1})\rangle \\ \boldsymbol{v}_k &= x_k \end{aligned} \tag{32}$$

Accordingly, the Kalman filter algorithm for the AAR estimates becomes

$$\begin{aligned} e_k &= y_k - \boldsymbol{Y}_{k-1} \cdot \hat{\boldsymbol{a}}_{k-1} \\ \hat{\boldsymbol{a}}(k) &= \hat{\boldsymbol{a}}_{k-1} + \boldsymbol{k}_{k-1} \cdot e_k \\ \boldsymbol{Q}_k &= \boldsymbol{Y}_k \cdot \boldsymbol{A}_{k-1} \cdot \boldsymbol{Y}_k^T + V_k \\ \boldsymbol{k}_k &= A_{k-1} \cdot \boldsymbol{Y}_{k-1}/\boldsymbol{Q}_k \\ \boldsymbol{Z}_k &= \boldsymbol{A}_{k-1} - \boldsymbol{k}_k \cdot \boldsymbol{Y}_k^T \cdot A_{k-1} \\ \boldsymbol{A}_k &= \boldsymbol{Z}_k + \boldsymbol{W}_k \end{aligned} \tag{33}$$

$\boldsymbol{W}_k$ and $\boldsymbol{V}_k$ are not determined by the Kalman equations, but must be known. In practice, some assumptions must be made which result in different algorithms

[36]. For the general case of KFR, the equation with explicit W_k is used $A_k = Z_k + W_k$ with $W_k = q_k \cdot I$. In the results with KFR-AAR below, we used $q_k = UC \cdot trace(A_{k-1})/p$. The RLS algorithm is characterized by the fact that $W_k = UC \cdot A_{k-1}$. Numerical inaccuracies can cause instabilities in the RLS method [36]; these can be avoided by enforcing a symmetric state error correlation matrix A_k. For example, Eq. (33) can be choosen as $A_k = (1 + UC) \cdot \left((Z_k + Z_k^T\right)/2$. The AAR parameters calculated using this algorithm are referred as RLS-AAR. In the past, KFR was usually used for stability reasons. With this new approach, RLS-AAR performs best among the various AAR estimation methods, as shown by results below.

Typically, the variance of the prediction error $V_k = (1 - UC) \cdot V_{k-1} + UC \cdot e_k^2$ is adaptively estimated from the prediction error (33) according to Eq. (12).

Kalman filters require initial values, namely the initial state estimate $z_0 = a_0$, the initial state error correlation matrix A_0 and some guess for the variance of innovation process V_0. Typically, a rough guess might work, but can also yield a long lasting initial transition effect. To avoid such a transition effect, a more sensible approach is recommended. A two pass approach was used in [33]. The first pass was based on some standard initial values, these estimates were used to obtain the initial values for the second pass $a_0 = \mu_a$, $A_0 = cov(a_k)$, $V_0 = var(e_k)$. Moreover, $W_k = W = cov(\alpha_k)$ with $\alpha_k = a_k - a_{k-1}$ can be used for the KFR approach.

For an adaptive spectrum estimation (31), not only the AAR parameters, but also the variance of the innovation process $\sigma_x^2(k) = V_k$ is needed. This suggests that the variance can provide additional information. The distribution of the variance is χ^2-distribution. In case of using linear classifiers, this feature should be "linearized" (typically with a logarithmic transformation). Later, we will show some experimental results comparing AAR features estimates with KFR and RLS. We will further explore whether including variance improves the classification.

1.4 Adaptive Classifiers

In BCI research, discriminant based classifiers are very popular because of their simplicity and the low number of parameters needed for their computation. For these reasons they are also attractive candidates for on-line adaptation. In the following, linear (LDA) and quadratic (QDA) discriminant analysis are discussed in detail.

1.4.1 Adaptive QDA Estimator

The classification output $D(x)$ of a QDA classifier in a binary problem is obtained as the difference between the square root of the Mahalanobis distance to the two classes i and j as follows:

$$D(x) = d_{\{j\}}(x) - d_{\{i\}}(x) \tag{34}$$

where the Mahalanobis distance is defined as:

$$d_{\{i\}}(\boldsymbol{x}) = ((\boldsymbol{x} - \boldsymbol{\mu}_{\{i\}})^T \cdot \boldsymbol{\Sigma}_{\{i\}}^{-1} \cdot (\boldsymbol{x} - \boldsymbol{\mu}_{\{i\}}))^{1/2} \tag{35}$$

where $\boldsymbol{\mu}_{\{i\}}$ and $\boldsymbol{\Sigma}_{\{i\}}$ are the mean and the covariance, respectively, of the class samples from class i. If $D(\boldsymbol{x})$ is greater than 0, the observation is classified as class i and otherwise as class j. One can think of a minimum distance classifier, for which the resulting class is obtained by the smallest Mahalanobis distance $argmin_i(d_{\{i\}}(\boldsymbol{x}))$. As seen in Eq. (35), the inverse covariance matrix (16) is required. Writing the mathematical operations in Eq. (35) in matrix form yields:

$$d_{\{i\}}(\boldsymbol{x}) = ([1;\ \boldsymbol{x}] \cdot \boldsymbol{F}_{\{i\}} \cdot [1;\ \boldsymbol{x}]^T)^{1/2} \tag{36}$$

with

$$\boldsymbol{F}_{\{i\}} = \left[\frac{-\boldsymbol{\mu}_{\{i\}}^T}{\boldsymbol{I}}\right] \cdot \boldsymbol{\Sigma}_{\{i\}}^{-1} \cdot \left[-\boldsymbol{\mu}_i | \boldsymbol{I}\right] = \left[\begin{array}{c|c} \boldsymbol{\mu}_{\{i\}}^T \boldsymbol{\Sigma}_{\{i\}}^{-1} \boldsymbol{\mu}_{\{i\}} & -\boldsymbol{\mu}_{\{i\}}^T \boldsymbol{\Sigma}_{\{i\}}^{-T} \\ \hline -\boldsymbol{\Sigma}_{\{i\}}^{-1} \boldsymbol{\mu}_{\{i\}} & \boldsymbol{\Sigma}_{\{i\}}^{-1} \end{array}\right] \tag{37}$$

Comparing Eq. (19) with (37), we can see that the difference between $\boldsymbol{F}_{\{i\}}$ and $\boldsymbol{E}_{\{i\}}^{-1}$ is just a 1 in the first element of the matrix, all other elements are equal. Accordingly, the time-varying Mahalanobis distance of a sample $\boldsymbol{x}(t)$ to class i is

$$d_{\{i\}}(\boldsymbol{x}_k) = \left\{[1, \boldsymbol{x}_k] \cdot \left(\boldsymbol{E}_{\{i\},k}^{-1} - \begin{bmatrix} 1 & 0_{1\times M} \\ 0_{M\times 1} & 0_{M\times M} \end{bmatrix}\right) \cdot [1, \boldsymbol{x}_k]^T\right\}^{1/2} \tag{38}$$

where $\boldsymbol{E}_{\{i\}}^{-1}$ can be obtained by Eq. (25) for each class i.

1.4.2 Adaptive LDA Estimator

Linear discriminant analysis (LDA) has linear decision boundaries (see Fig. 3). This is the case when the covariance matrices of all classes are equal; that is, $\boldsymbol{\Sigma}_{\{i\}} = \boldsymbol{\Sigma}$ for all classes i. Then, all observations are distributed in hyperellipsoids of equal shape and orientation, and the observations of each class are centered around their corresponding mean $\boldsymbol{\mu}_{\{i\}}$. The following equation is used in the classification of a two-class problem:

$$D(\boldsymbol{x}) = \boldsymbol{w} \cdot (\boldsymbol{x} - \boldsymbol{\mu}_x)^T = [b, \boldsymbol{w}] \cdot [1, \boldsymbol{x}]^T \tag{39}$$

$$\boldsymbol{w} = \Delta\boldsymbol{\mu} \cdot \boldsymbol{\Sigma}^{-1} = (\boldsymbol{\mu}_{\{i\}} - \boldsymbol{\mu}_{\{j\}}) \cdot \boldsymbol{\Sigma}^{-1} \tag{40}$$

$$b = -\boldsymbol{\mu}_x \cdot \boldsymbol{w}^T = -\frac{1}{2} \cdot (\boldsymbol{\mu}_{\{i\}} + \boldsymbol{\mu}_{\{j\}}) \cdot \boldsymbol{w}^T \tag{41}$$

where $D(\boldsymbol{x})$ is the difference in the distance of the feature vector $\boldsymbol{x}$ to the separating hyperplane described by its normal vector $\boldsymbol{w}$ and the bias b. If $D(\boldsymbol{x})$ is greater than 0, the observation x is classified as class i and otherwise as class j.

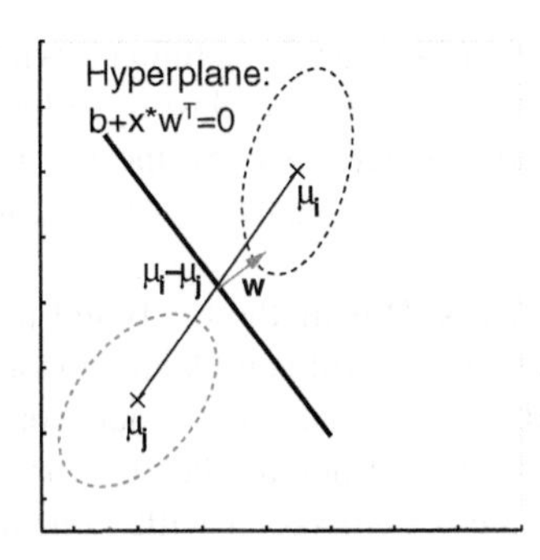

Fig. 3 Concept of classification with LDA. The two classes are reprensented by two ellipsoids (the covariance matrices) and the respective class mean values. The hyperplane is the boundary of decision, with $D(\boldsymbol{x}) = b + \boldsymbol{x} \cdot \boldsymbol{W}^T = 0$. A new observation $\boldsymbol{x}$ is classified as follows: if $D(\boldsymbol{x})$ is greater than 0, the observation x is classified as class i and otherwise as class j. The normal vector to the hyperplane, $\boldsymbol{w}$, is in general not in the direction of the difference between the two class means

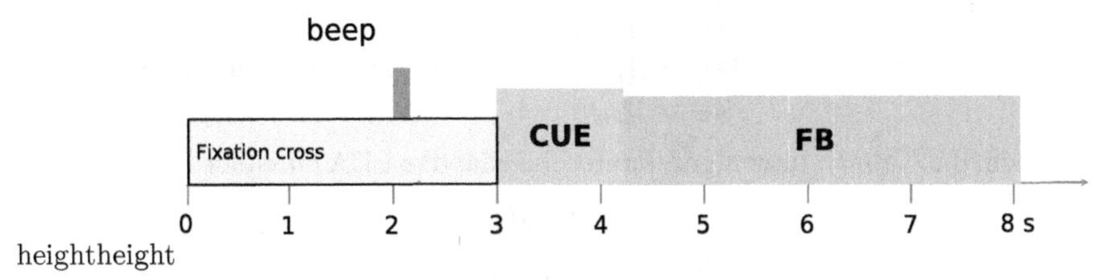

heightheight

Fig. 4 Paradigm of cue-based BCI experiment. Each trial lasted 8.25 s. A cue was presented at $t = 3s$, feedback was provided from **t**=4.25 to 8.25 s

The methods to adapt LDA can be divided in two different groups. First, using the estimation of the covariance matrices of the data, for which the speed of adaption is fixed and determined by the update coefficient. The second group is based on Kalman Filtering and has the advantage of having a variable adaption speed depending on the properties of the data.

Fixed Rate Adaptive LDA Using (19), it can be shown that the distance function (Eq. 39) is

$$D(\boldsymbol{x}_k) = [b_k, \boldsymbol{w}_k] \cdot [1, \boldsymbol{x}_k]^T \tag{42}$$

$$= b_k + \boldsymbol{w}_k \cdot \boldsymbol{x}_k^T \tag{43}$$

$$= -\Delta\boldsymbol{\mu}_k \cdot \boldsymbol{\Sigma}_k^{-1} \cdot \boldsymbol{\mu}_k^T + \Delta\boldsymbol{\mu}_k \cdot \boldsymbol{\Sigma}_k^{-1} \cdot \boldsymbol{x}_k^T \tag{44}$$

$$= [0, \boldsymbol{\mu}_{\{i\},k} - \boldsymbol{\mu}_{\{j\},k}] \cdot \boldsymbol{E}_k^{-1} \cdot [1, \boldsymbol{x}_k] \tag{45}$$

with $\Delta\boldsymbol{\mu}_k = \boldsymbol{\mu}_{\{i\},k} - \boldsymbol{\mu}_{\{j\},k}$, $b = -\Delta\boldsymbol{\mu}(t) \cdot \boldsymbol{\Sigma}(t)^{-1} \cdot \boldsymbol{\mu}(t)^T$ and $\boldsymbol{w} = \Delta\boldsymbol{\mu}(t) \cdot \boldsymbol{\Sigma}^{-1}$.

Accordingly, the adaptive LDA can be estimated with Eq. (45) using (25) for estimating E_k^{-1} and (5) for estimating the class-specific adaptive mean $\mu_{\{i\},k}$ and $\mu_{\{j\},k}$. The adaptation speed is determined by the update coefficient UC used in the Eq. (5) and (25). For a constant update coefficient, the adaptation rate is also constant.

Variable Rate Adaptive LDA This method is based in Kalman Filtering and its speed of adaptation depends on the Kalman Gain, shown in Eq. (29), which varies with the properties of the data. The state space model for the classifier case is summarized in (46), where c_k is the current class label, z_k are the classifier weights, the measurement matrix $\boldsymbol{H}_k$ is the feature vector with a one added in the front $[1;\ \boldsymbol{x}_k]$, and $D_k(\boldsymbol{x})$ is the classification output.

$$\begin{aligned}
&\textit{State Space Model} \Leftrightarrow \textit{LinearCombiner} \\
y_k &= c_k \\
z_k &= [b_k, \boldsymbol{w}_k]^T \\
\boldsymbol{H}_k &= [1; \boldsymbol{x}_k]^T \\
\boldsymbol{G}_{k,k-1} &= \boldsymbol{I}_{M\times M} \\
\boldsymbol{Z}_k &= E\langle(\boldsymbol{w}_k - \hat{w}_k)^T \cdot (\boldsymbol{w}_k - \hat{w}_k)\rangle \\
\boldsymbol{W}_k &= \boldsymbol{A}_k = E\langle(\boldsymbol{w}_k - \boldsymbol{w}_{k-1})^T \cdot (\boldsymbol{w}_k - w_{k-1})\rangle \\
\boldsymbol{e}_k &= D_k(\boldsymbol{x}) - c_k
\end{aligned} \tag{46}$$

Then, the Kalman filter algorithm for the adaptive LDA classifier is

$$\begin{aligned}
e_k &= y_k - \boldsymbol{H}_k \cdot z_{k-1}^T 47 \\
z_k &= z_{k-1} + \boldsymbol{k}_k \cdot e_k \\
Q_k &= \boldsymbol{H}_k \cdot \boldsymbol{A}_{k-1} \cdot \boldsymbol{H}_k^T + V_k \\
\boldsymbol{k}_k &= A_{k-1} \cdot \boldsymbol{H}_k^T / Q_k \\
\boldsymbol{Z}_k &= \boldsymbol{A}_{k-1} - \boldsymbol{k}_k \cdot \boldsymbol{H}_k^T \cdot A_{k-1} \\
\boldsymbol{A}_k &= \boldsymbol{Z}_k + \boldsymbol{W}_k 47
\end{aligned} \tag{47}$$

The variance of the prediction error $\boldsymbol{V}_k$ was estimated adaptively from the prediction error (47) according to Eq. (12). The RLS algorithm was used to estimate $\boldsymbol{A}_k$.

As the class labels are bounded between 1 and -1, it would be convenient to also bound the product $\boldsymbol{H}_k \cdot z_{k-1}^T$ between these limits. Hence, a transformation in the estimation error can be applied, but then the algorithm is not a linear filter anymore:

$$e_k = y_k + 1 - \frac{2}{(1 + \exp(-\boldsymbol{H}_k \cdot z_{k-1}^T))} \tag{48}$$

1.5 Selection of Initial Values, Update Coefficient and Model Order

All adaptive algorithms need some initial values and must select some update coefficients. Some algorithms like adaptive AAR need also a model order p. Different approaches are available to select these parameters. The initial values can be often obtained by some a priori knowledge. Either it is known that the data has zero mean (e.g. because it is low pass filtered), or a reasonable estimate can be obtained from previous recordings, or a brief segment in the beginning of the record is used to estimate the initial values. If nothing is known, it is also possible to use some random initial values (e.g. zero) and wait until the adaptive algorithm eventually converges to the proper range. For a state space model [9], we recommend starting with a diagonal matrix weighted by the variance of previous data and multiplied by a factor δ, which can be very small or very large $\boldsymbol{\Sigma}_0 = \delta\sigma^2\boldsymbol{I}$.

Of course one can also apply more sophisticated methods. For example, to apply the adaptive classifier to new (untrained) subjects, a general classifier was estimated from data of seven previous records from different subjects. This had the advantage that no laborious training sessions (i.e. without feedback) were needed, but the new subjects could work immediately with BCI feedback. Eventually, the adaptive classifier adapted to the subject specific pattern [44–48].

A different approach was used in an offline study using AAR parameters [33]. Based on some preliminary experiments, it became obvious that setting the initial values of the AAR parameters to zero can have some detrimental influence on the result. The initial transient took several trials, while the AAR parameters were very different than the subsequent trials. To avoid this problem, we applied the AAR estimation algorithm two times. The first run was initialized by $\vec{a_0} = [0, ..., 0], \boldsymbol{A}_0 = I_{pp}$, $V_k = 1 - UC$, $\boldsymbol{W}_k = \boldsymbol{I} \cdot UC \cdot trace(\boldsymbol{A}_{k-1})/p$. The resulting AAR estimates were used to estimate more reasonable initial values $\vec{a_0} = mean(\vec{a_t})$, $\boldsymbol{A}_0 = cov\vec{a_t}$, $\boldsymbol{V}_k = var e_t$, $\boldsymbol{W}_k = cov\Delta\vec{a_t}$ with $\Delta\vec{a_t} = \vec{a_t} - \vec{a_{t-1}}$.

The selection of the update coefficient is a trade-off between adaptation speed and estimation accuracy. In case of AAR estimation in BCI data, a number of results [31, 35, 36] suggest, that there is always a global optimum to select the optimum update coefficient, which makes it rather easy to identify a reasonable update coefficient based on some representative data sets. In case of adaptive classifiers, it is more difficult to identify a proper update coefficient from the data; therefore we determined the update coefficient based on the corresponding time constant. If the classifier should be able to adapt to a new pattern within 100 trials, the update coefficient was chosen such that the corresponding time constant was about 100 trials.

The order p of the AAR model is another free parameter that needs to be determined. Traditional model order selection criteria like the Akaike Information Criterion (AIC) and similar ones are based on stationary signal data, which is not the case for AAR parameters. Therefore, we have developed a different approach to select the model order which is based on the one-step prediction error [35, 36] of the AAR model. These works were mostly motivitated by the principle of uncertainty between time and frequency domain suggesting model orders in the range from 9 to 30. Unfortunately, the model order obtained with this approach was not necessarily the best for single trial EEG classification like in BCI data, often much smaller orders gave much better results. We have mostly used model orders of 6 [27, 31, 34] and 3 [33, 44, 45, 47]. These smaller orders are prefered by the classifiers, when the number of trials used for classification is rather small. A simple approach is the use the rule of thumb that the nunber of features for the classifier should not exceed a 1/10 of the number of trials. So far the number of studies investigating the most suitable strategy for selecting model order, update coefficient and initial values are rather limited, future studies will be needed to address this open issues.

1.6 Experiments with Adaptive QDA and LDA

Traditional BCI experiments use a block-based design for training the classifiers. This means that some data must be recorded first before a classifier can be estimated; and the classifier can be only modified after a "run" (which is typically about 20 or 40 trials) is completed. Typically, this procedure also involve a manual decision whether the previous classifier should be replaced by the new one or not. Adaptive classifiers overcome this limitation, because the classifier is updated with every trial. Accordingly, an adaptive classifier can react much faster to a change in recording conditions, or when the subject modifies its brain patterns. The aim of the study was to investigate whether such adaptive classifiers can be applied in practical BCI experiments.

Experiments were carried out with 21 able-bodied subjects without previous BCI experience. They performed experiments using the "basket paradigm" [15]. At the bottom of a computer screen, a so-called basket was presented either on the left side or the right side of the screen. A ball moved from the top of the screen to the bottom at a constant velocity. During this time (typically 4 s), the subject controls the horizontal (left-right) position with the BCI system (Fig. 4 shows the timing of each trial). The task was to control the horizontal position of the ball to move the ball into the displayed basket. Each subject conducted three different sessions, with 9 runs per session and 40 trials per run. 1080 trials were available for each of them (540 trials for each class). Two bipolar channels, C3 and C4, were recorded.

The system was a two-class cue-based and EEG-based BCI, and the subjects had to perform motor imagery of the left or right hand depending on the cue. More specifically, they were not instructed to imagine any specific movement, but they were free to find their own strategy. Some of them reported that the imagination of movements that involve several parts of the arm were more successful. In any case, they were asked to maintain their strategy for at least one run.

In the past, KFR-AAR was the best choice because it was a robust and stable method; other methods were not stable and required periodic reinitialization. With the enforcing of a symetric system matrix (Eq. 1.3), RLS could be stabilized. Moreover, based on the compostion of AR spectra, it seems reasonable to also include the variance of the innovation process as a feature. To compare these methods, Kalman based AAR parameters (KFR-AAR) (model order $p = 6$), RLS-AAR ($p = 6$) parameters, RLS-AAR ($p = 5$) combined with the logarithm of the variance (RLS-AAR+V) and the combination of RLS-AAR($p = 4$) and band power estimates (RLS-AAR+BP) are compared. The model order p was varied to maintain 6 features per channels. The classifier was LDA without adaptation, and

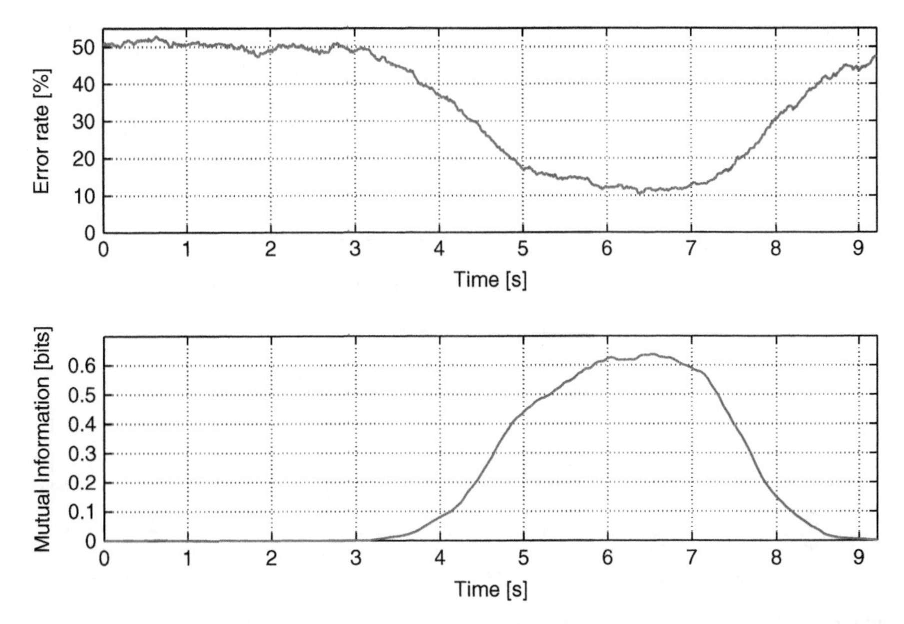

Fig. 5 Time course of the performance measurements. These changes are caused by the short-term nonstationarities of the features; the classifier was constant within each trial

a leave-one-out cross-validation procedure was used selecting the order of the AR model.

The performance results were obtained by single trial analysis of the online classification output of each experimental session. For one data set, the time courses of the error rate and the mutual information (MI are shown in Fig. 5, cf. timing with Fig. 4). The mutual information *MI* is a measure the transfered information and is defined $MI = 0.5 \cdot log_2(1 + SNR)$ with the signal-to-noise ratio $SNR = \frac{\sigma^2_{\text{signal}}}{\sigma^2_{\text{noise}}}$. The signal part of the BCI output is the variance from the class-related differences, and the noise part is the variance of the background activity described by the variability within one class. Accordingly, the mutual information can be determined from the total variability (variance of the BCI output among all classes), and the average within-class variability by [32, 34, 36]

$$MI = 0.5 \cdot log_2 \frac{\sigma^2_{\text{total}}}{\sigma^2_{\text{within}}} \tag{49}$$

For further comparison, the minimum error rate and the maximum mutual information are used. Figure 6 depicts the performance scatter plots of different AAR-based features against KFR-AAR. The first row shows ERR and the second MI values. For ERR (MI), all values under the diagonal show the superiority of the

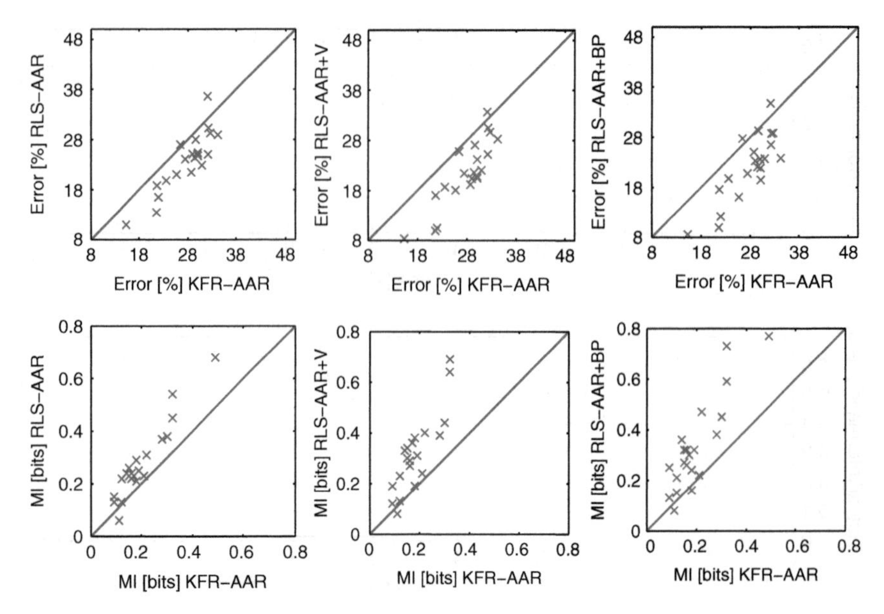

Fig. 6 The first row shows the scatter plots of error rates of different adaptive feature types using an LDA classifier and cross-validation based on leave-one (trial)-out procedure. The second row are scatter plots for mutual information results. The used methods are (i) bandpower values for the bands 8–14 and 16–28 Hz (BP) with a 1-second rectangular sliding window, (ii) AAR estimates using Kalman filtering (KFR), (iii) AAR estimates using the RLS algorithm, RLS-based AAR estimates combined with the logarithm of the variance V (AAR+V), and RLS-based AAR estimates combined with bandpower (AAR+BP). In the first row, values below the diagonal show the superiority of the method displayed in the y-axis. In the second row (MI values), the opposite is true. This figure shows that all methods outperform AAR-KFR

method displayed in the y-axis. Looking at these scatter plots, one can see that KFR-AAR is clearly inferior to all other methods. For completition of the results, and to compare the performance of each feature type, the mean value and standard error of each investigated feature were computed and presented in Table 2.

Table 2 Summary results from 21 subjects. The mean and standard error of the mean (SEM) of minimum ERR and maximum MI are shown. The AAR-based results are taken from the results shown in Fig. 6. Additionally, results from standard bandpower (10–12 and 16–24 Hz) and the bandbower of selected bands (8–14 and 16–28 Hz) are included

Feature	ERR[%]	MI[bits]
BP-standard	26.16±1.90	0.258±0.041
BP	25.76±1.86	0.263±0.041
KFR-AAR	27.85±1.04	0.196±0.021
RLS-AAR	23.73±1.27	0.277±0.031
RLS-AAR+V	21.54±1.45	0.340±0.041
RLS-AAR+BP	22.04±1.44	0.330±0.040

The results displayed in Table 2, with a threshold p-value of 1.7%, show similar performance in ERR and MI for RLS-AAR+V and RLS-AAR+BP; both were found significantly better than KFR-AAR and RLS-AAR. Also RLS-AAR was significantly better than KFR-AAR. The bandpower values are better then the Kalman filter AAR estimates, but are worse compared to RLS-AAR and RLS-AAR+V.

Using the features RLS-AAR+V, several adaptive classifiers were tested. To simulate a causal system, the time point when the performance of the system was measured was previously fixed, and the ERR and MI calculated at these previously defined time-points. The set of parameters for the classifiers in the first session were common to all subjects and computed from previously recorded data from 7 subjects during various feedback sessions [26]. The set of parameters in the second session were found by subject specific optimization of the data of the first session. The same procedure was used for the parameters selected for the third session.

Table 3 shows that all adaptive classifiers outperform the no adaptation setting. The best performing classifier was aLDA, which outperformed the Adaptive QDA and Kalman LDA. Kalman LDA also was found statistically better than Adaptive QDA.

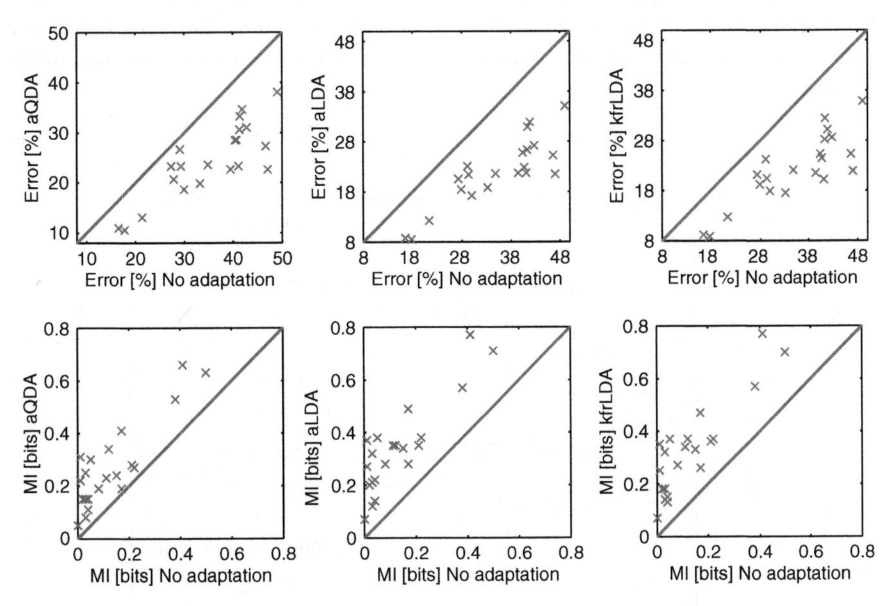

Fig. 7 Scatter plots of performance (error rates in first row and mutual information in second row) of different adaptive classifiers using RLS-AAR+V feature. "No adaptation" uses the initial classifier computed from previously recorded data from 7 different subjects. "aQDA" is the adaptive QDA approach, aLDA is the adaptive LDA approach with fixed update rate, and kfrLDA is the (variable rate) adaptive LDA using Kalman filtering. In the first row, values below the diagonal show the superiority of the method displayed in the y-axis. In the second row (MI values), the contrary is true. These results suggest the superiority of adaptive classification versus no adaptation

Table 3 Average and SEM of ERR and MI at predefined time points. Error rate values were taken from results shown also in Fig. 7

Classifier	ERR[%]	MI[bits]
No adapt	35.17±2.08	0.132±0.031
Adaptive QDA (aQDA)	24.30±1.60	0.273±0.036
Adaptive LDA (aLDA)	21.92±1.48	0.340±0.038
Kalman LDA (kfrLDA)	22.22±1.51	0.331±0.039

Figure 8 depicts how the weights of the adaptive classifier change in time, and we can see a clear long-term change in their average value. This change can be largely explained by the improved separability due to feedback training. To present the changes in the feature space, the features were projected into a two-dimensional subspace defined by the optimal separating hyperplanes similar to [14, 37]. Figure 9 shows how the distributions (means and covariances of the features) change from session 1 to 2 and from session 2 to 3. In this example, the optimal projection changes and some common shift of both classes can be observed. The change of the optimal projection can be explained by the effect of feedback training. However, the common shift of both classes indicates also other long-term changes (e.g. fatigue, new electrode montage, or some other change in recording conditions).

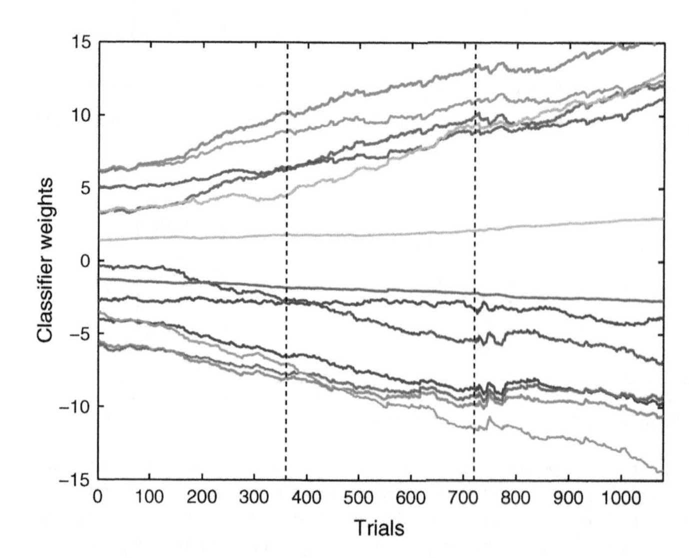

Fig. 8 Classifier weights changing in time of subject S11. These changes indicate consistent long-term changes caused by an improved separability due to feedback training. The data is from three consecutive sessions, each session had 360 trials. The changes after trial 720 probably indicate some change in the recording conditions (e.g. due to the new electrode montage)

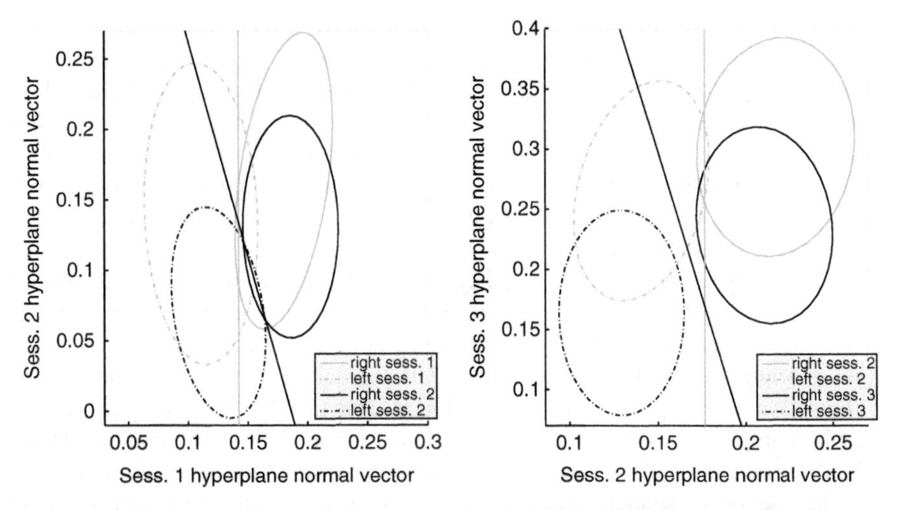

Fig. 9 Changes in feature distributions from session 1 to session 2 (*left*) and session 2 to session 3 (*right*). The separating hyperplanes of two sessions were used to find an orthonormal pair of vectors, and the features were projected in this subspace. The averages and covariances of each class and each session are projected into a common 2-dimensional subspace defined by the two separating hyperplanes

1.7 Discussion

Compensating for non-stationarities in complex dynamical systems is an important topic in data analysis and pattern recognition in EEG and many other analysis. While we have emphasized and discussed the use of adaptive algorithms for BCI, there are further alternatives to be considered when dealing with non-stationarities: (a) segmentation into stationary parts where each stationary system is modeled separately (e.g. [22]), (b) modeling invariance information, i.e. effectively using an invariant feature subspace that is stationary for solving the task, (c) modeling the non-stationarity of densities, which can so far be remedied only in the covariate shift setting, where the conditional $p(y|x)$ stays stable and the densities $p(x)$ exhibit variation [40, 41].

An important aspect when encountering non-stationarity is to measure and quantify the degree of non-stationary behavior, e.g. as done in [37]. Is non-stationarity behavior caused by noise fluctuations, or is it a systematic change of the underlying system? Depending on the answer, different mathematical tools are suitable [33, 36, 40, 47].

Several adaptive methods have been introduced and discussed. The differences between rectangular and exponential windows are exemplified in the adaptive mean estimation. The advantage of an exponential window is shown in terms of computational costs, the memory requirements and the computational efforts are independent of the window size and the adaptation time constant. This advantage holds not only for the mean but also for all other estimators.

The extended covariance matrix was introduced, which makes the software implementation more elegant. An adaptive estimator for the inverse covariance matrix was introduced; the use of the matrix inversion lemma enables avoiding an explicit (and computational costly) matrix inversion. The resulting algorithm was suitable for adaptive LDA and adaptive QDA classifiers. The Kalman filtering method for the general state-space model was explained and applied to two specific models, namely (i) the autoregressive model and (ii) the linear combiner (adaptive LDA) in the translation step.

All techniques are causal (that is, they use samples only from the past and present but not from the future) and are therefore suitable for online and real-time application. This means that no additional time delay is introduced, but the total response time is determined by the window size (update coefficient) only. The presented algorithms have been implemented and tested in M-code (available in Biosig for Octave and Matlab [30]), as well as in the real-time workshop for Matlab/Simulink. These algorithms were used in several BCI studies with real-time feedback [46–48].

The use of subsequent adaptive steps can lead, at least theoretically, to an unstable system. To avoid these pitfalls, several measures were taken in the works described here. First, the feature extraction step and the classification step used very different time scales. Specifically, the feature extraction step takes into account only changes within each trial, and the classification step takes into account only the long-term changes. A more important issue might be the simultaneous adaptation of the subject and the classifier. The results of [46, 47, 48] also demonstrate that the used methods provide a robust BCI system, since the system did not become unstable. This was also supported by choosing conservative (i.e. small) update coefficients. Nevertheless, there is no guarantee that the BCI system will remain stable under all conditions. Theoretical analyses are limited by the fact that the behavior of the subject must be considered. But since the BCI control is based on deliberate actions of the subject, the subject's behavior can not be easily described. Therefore, it will be very difficult to analyse the stability of such a system from a theoretical point of view.

The present work did not aim to provide a complete reference for all possible adaptive methods, but it provides a sound introduction and several non-trivial techniques in adaptive data processing. These methods are useful for future BCI research. A number of methods are also available from BioSig - the free and open source software library for biomedical signal processing [30].

Acknowledgments This work was supported by the EU grants "BrainCom" (FP6-2004-Mobility-5 Grant No 024259) and "Multi-adaptive BCI" (MEIF-CT-2006 Grant No 040666). Furthermore, we thank Matthias Krauledat for fruitful discussions and tools for generating Fig. 5.

References

1. Block matrix descompositions (2008–2010). http://ccrma.stanford.edu/~jos/lattice/Block_matrix_decompositions.html. Accessed on Sept 2010.
2. N. Birbaumer et al., The thought-translation device (TTD): Neurobehavioral mechanisms and clinical outcome. IEEE Trans Neural Syst Rehabil Eng, 11(2), 120–123, (2003).

3. B. Blankertz et al., The BCI competition. III: Validating alternative approaches to actual BCI problems. IEEE Trans Neural Syst Rehabil Eng, 14(2), 153–159, (2006).
4. B. Blankertz et al., Optimizing spatial filters for robust EEG single-trial analysis. IEEE Signal Proc Mag, 25(1), 41–56, (2008).
5. del R. Millán and Mouriño J.del, R. Millán and J. Mouriño, Asynchronous BCI and local neural classifiers: An overview of the adaptive brain interface project. IEEE Trans Neural Syst Rehabil Eng 11(2), 159–161, (2003).
6. G. Dornhege et al., Combining features for BCI. In S. Becker, S. Thrun, and K. Obermayer (Eds.), *Advances in neural information processing systems,* MIT Press, Cambridge, MA, pp. 1115–1122, (2003).
7. G. Dornhege et al., Combined optimization of spatial and temporal filters for improving brain-computer interfacing. IEEE Trans Biomed Eng, 53(11), 2274–2281, (2006).
8. G. Dornhege et al., *Toward brain-computer interfacing*, MIT Press, Cambridge, MA, (2007).
9. S. Haykin, *Adaptive filter theory*, Prentice Hall International, New Jersey, (1996).
10. B. Hjorth, "EEG analysis based on time domain properties." Electroencephalogr Clin Neurophysiol, 29(3), 306–310, (1970).
11. I.I. Goncharova and J.S. Barlow, Changes in EEG mean frequency and spectral purity during spontaneous alpha blocking. Electroencephalogr Clin Neurophysiol, 76(3), 197–204; Adaptive methods in BCI research – an introductory tutorial 25, (1990).
12. R. Kalman and E. Kalman, A new approach to Linear Filtering and Prediction Theory. J Basic Eng Trans ASME, 82, 34–45, (1960).
13. B.R.E. Kalman, and R.S. Bucy, New results on linear filtering and prediction theory. J Basic Eng, 83, 95–108, (1961).
14. K. M. Krauledat, Analysis of nonstationarities in EEG signals for improving Brain-Computer Interface performance, Ph. D. diss. Technische Universität Berlin, Fakultät IV – Elektrotechnik und Informatik, (2008).
15. G. Krausz et al., Critical decision-speed and information transfer in the 'graz brain-computer-interface'. Appl Psychophysiol Biofeedback, 28, 233–240, (2003).
16. S. Lemm et al. Spatio-spectral filters for robust classification of single-trial EEG. IEEE Trans Biomed Eng, 52(9), 1541–48, (2005).
17. A. Li, G. Yuanqing, K. Li, K. Ang, and C. Guan, Digital signal processing and machine learning. In B. Graimann, B. Allision, and G. Pfurtscheller (Eds.), *Advances in neural information processing systems*, Springer, New York, (2009).
18. J. McFarland, A.T. Lefkowicz, and J.R. Wolpaw, Design and operation of an EEG-based brain-computer interface with digital signal processing technology. Behav Res Methods, Instruments Comput, 29, 337–345, (1997).
19. D.J. McFarland et al., BCI meeting 2005-workshop on BCI signal processing: feature extraction and translation." IEEE Trans Neural Syst Rehabil Eng 14(2), 135–138, (2006).
20. R.J. Meinhold and N.D. Singpurwalla, Understanding Kalman filtering. Am Statistician, 37, 123–127, (1983).
21. E. Oja and J. Karhunen, On stochastic approximation of the eigenvectors and eigenvalues of the expectation of a random matrix. J Math Anal Appl, 106, 69–84, (1985).
22. K. Pawelzik, J. Kohlmorgen and K.-R. Muller, Annealed competition of experts for a segmentation and classification of switching dynamics. Neural Comput, 8, 340–356, (1996).
23. G. Pfurtscheller et al., On-line EEG classification during externally-paced hand movements using a neural network-based classifier. Electroencephalogr Clin Neurophysiol, 99(5), 416–25, (1996).

24. N.G. Pfurtscheller and C. Neuper, Motor imagery and direct brain-computer communications, Proceedings IEEE 89, 1123–1134, (2001).
25. G. Pfurtscheller et al., EEG-based discrimination between imagination of right and left hand movement. Electroencephalogr Clin Neurophysiol, 103, 642–651, (1997).
26. G. Pfurtscheller et al., Current trends in Graz brain-computer interface (BCI) research. IEEE Trans Rehabil Eng, 8, 216–219, (2000).
27. G. Pfurtscheller et al., "Separability of EEG signals recorded during right and left motor imagery using adaptive autoregressive parameters." IEEE Trans Rehabil Eng, 6(3), 316–25, (1998).
28. N. Pop-Jordanova and J. Pop-Jordanov, Spectrum-weighted EEG frequency (Brainrate) as a quantitative indicator of arousal Contributions. Sec Biol Med Sci XXVI(2), 35–42, (2005).
29. H. Ramoser, J. Müller-Gerking, G. Pfurtscheller, "Optimal spatial filtering of single trial EEG during imagined hand movement. IEEE Trans Rehabil Eng. 8(4), 441–446, (2000).
30. A. Schlogl, BioSig – an open source software library for biomedical signal processing, http://biosig.sf.net, 2003–2010.
31. A. Schlogl, D. Flotzinger and G. Pfurtscheller, Adaptive autoregressive modeling used for single-trial EEG classification. Biomedizinische Technik, 42, 162–167, (1997).
32. A. Schlögl et al. "Information transfer of an EEG-based brancomputer interface." First international IEEE EMBS conference on neural engineering, 2003, 641–644, (2003).
33. A. Schlögl et al. "Characterization of four-class motor imagery eeg data for the bci-competition 2005. J Neural Eng, 2(4), L14–L22, (1997).
34. C. Neuper and G. Pfurtscheller, Estimating the mutual information of an EEG-based brain-computer-interface. Biomedizinische Technik, 47(1–2), 3–8, (2002).
35. S.J. Roberts and G. Pfurtscheller, A criterion for adaptive autoregressive models. Proceedings Ann Int Conf IEEE Eng Med Biol, 2, 1581–1582, (2000).
36. A. Schlögl, *The electroencephalogram and the adaptive autoregressive model: theory and applications*, Shaker Verlag, Aachen, Germany, (2000).
37. M.P. Shenoy, R. P. N. Rao, M. Krauledat, B. Blankertz and K.-R. Müller, Towards adaptive classification for BCI. J Neural Eng, 3(1), R13–R23, (2006).
38. H. Shimodaira, Improving predictive inference under covariate shift by weighting the log likelihood function. J Stat Plan Inference, 90, 227–244, (2000).
39. V. Solo and X. Kong, Performance analysis of adaptive eigenanalysis algorithms, IEEE Trans Signal Process, 46(3), 636–46, (1998).
40. M. Sugiyama, M. Krauledat and K.-R. Müller, Covariate shift adaptation by importance weighted cross validation. J Mach Learning Res 8, 1027–1061, (2007).
41. M.M. Sugiyama and K.-R. Müller, Input-dependent estimation of generalization error under covariate shift. Stat Decis, 23(4), 249–279, (2005).
42. S. Sun and C. Zhang, Adaptive feature extraction for EEG signal classification. Med Bio Eng Comput, 44(2), 931–935, (2006).
43. R. Tomioka et al., Adapting spatial filtering methods for nonstationary BCIs. Proceedings of 2006 Workshop on Information-Based Induction Sciences (IBIS2006), 2006, 65–70, (2006).
44. C. Vidaurre et al., About adaptive classifiers for brain computer interfaces. Biomedizinische Technik, 49(Special Issue 1), 85–86, (2004).
45. C. Vidaurre et al., A fully on-line adaptive brain computer interface. Biomedizinische Technik, 49(Special Issue 2), 760–761, (2004).
46. C. Vidaurre et al., Adaptive on-line classification for EEG-based brain computer interfaces with AAR parameters and band power estimates. Biomedizinische Technik, 50, 350–354. Adaptive methods in BCI research – an introductory tutorial, 27, (2005).
47. C. Vidaurre et al., A fully on-line adaptive BCI, IEEE Trans Biomed Eng, 53, 1214–1219, (2006).
48. C. Vidaurre et al., Study of on-line adaptive discriminant analysis for EEG-based brain computer interfaces. IEEE Trans Biomed Eng, 54, 550–556, (2007).
49. J. Wackermann, Towards a quantitative characterization of functional states of the brain: from the non-linear methodology to the global linear descriptor. Int J Psychophysiol, 34, 65–80, (1999).

50. J.R. Wolpaw et al., Brain-computer interface technology: A review of the first international meeting. IEEE Trans Neural Syst Rehabil Eng, 8(2), 164–173, (2000).
51. J.R. Wolpaw et al., Brain-computer interfaces for communication and control." Clin Neurophysiol, 113, 767–791, (2002).

Toward Ubiquitous BCIs

Brendan Z. Allison

1 Introduction

The preceding chapters in this book described modern BCI systems. This concluding chapter instead discusses future directions. While there are some specific predictions, I mainly analyze key factors and trends relating to practical mainstream BCI development. While I note some disruptive technologies that could dramatically change BCIs, this chapter focuses mainly on realistic, incremental progress and how progress could affect user groups and ethical issues.

There are no other BCI articles like this chapter. Some articles address BCIs further in the future [68], or mention future directions en passant [11, 59, 60, 74], but this chapter focuses largely on the mechanics of current and foreseeable change. BCIs are in the early stages of a slow but major revolution that will change BCIs from mostly impractical and inaccessible exotica for very specific user groups into increasingly ubiquitous tools.

This view is not conventional. People typically extrapolate a dim future for BCIs, since their views are based on current systems and user groups. BCIs exist primarily to enable communication for people who cannot communicate otherwise due to severe motor disability. Thus, BCIs are not useful to people who can communicate through other means.

I encounter this conventional perspective all the time, whether talking to other research scientists, media or business experts, or laypeople. It may be influenced by science fiction, which typically presents BCIs as ubiquitous and useful, but only because of advances far beyond conceivable technology. This perspective assumes that:

1. BCIs only provide communication and control
2. BCIs provide the same information otherwise available through other interfaces
3. BCIs are exclusive interfaces

B.Z. Allison (✉)
Institute for Knowledge Discovery, Laboratory of Brain–Computer Interfaces, Graz University of Technology, Krenngasse 37, 8010 Graz, Austria
e-mail: allison@tugraz.at

B. Graimann et al. (eds.), *Brain–Computer Interfaces*, The Frontiers Collection,
DOI 10.1007/978-3-642-02091-9_19, © Springer-Verlag Berlin Heidelberg 2010

It follows that BCIs are of no practical value to people who can otherwise communicate, and will not become more widely adopted unless they can become much faster.

This perspective is mistaken. Instead, BCIs may provide useful communication to many new user groups, including persons with less severe movement disabilities and healthy people. The very definition of a BCI as a tool for communication and control is beginning to change. BCIs might also provide other functions, such as rehabilitation of different disorders [59, 60] (see also chapters “The Graz Brain–Computer Interface” and “Brain–Computer Interface in Neurorehabilitation” in this book), which are either unattainable or problematic for other interfaces.

Furthermore, the critical progress that is needed is not in speed but other key factors like improved sensors, greater accessibility to nonexperts, and smoother integration with existing devices and software. These and some other key factors are progressing at a rapid and potentially revolutionary pace. BCIs will therefore become practical tools for much wider user groups and goals than most people think. BCIs will not become ubiquitous in the next 10 years, but they will grow well beyond communication for severely disabled users.

2 Key Factors in BCI Adoption

Schalk [68] discusses BCI adoption much like adoption of other consumer electronics, such as radio, television, or DVDs. This approach, while unconventional within the BCI research community, is becoming increasingly popular with other EEG-based applications such as neuromarketing, lie detection, and sleep or alertness monitoring. These applications are not BCIs because they do not provide realtime feedback and/or do not allow intentional communication (see also chapter “Brain–Computer Interfaces: A Gentle Introduction”). However, progress with these and related systems could affect BCI development.

Schalk [68] depicts the relationship between BCI price and performance, by which Schalk means bits/second. While a BCI’s information transfer rate (ITR) is important, other factors may be at least as important to potential users [11, 36, 83]. The right table in Fig. 1 below summarizes these key factors. Each of these factors is influenced by both hardware and software considerations. The two tables on the left side of Fig. 1 present three catalysts unique to BCIs and several related catalysts. The arrows indicate that these two types of catalysts influence each other and interact bidirectionally with the key factors. The arrows represent the strength of this influence. While progress in BCI catalysts and their key performance indices will influence related disciplines, this influence will be weaker than its reverse. There will be a lot more research in related catalysts than BCI-specific catalysts. While new sensors developed for BCIs might inspire improved ExG sensors,[1] electronics,

[1]This term is a contraction of electro-(anything)-gram. ExG sensors refer to any devices used to record people’s electrophysiological activity. The word “sensor” is used in its colloquial meaning throughout this chapter, not the stricter definition within electrical engineering.

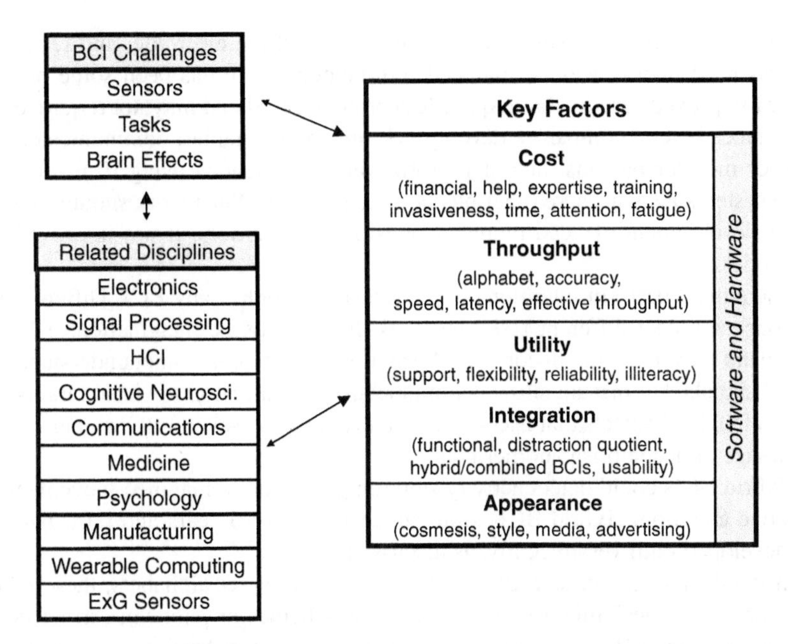

Fig. 1 Key factors in BCI adoption and catalysts

manufacturing, or wearable computing, improvements in these or other related disciplines will have a greater effect on BCI sensors.

2.1 BCI Catalysts

Sensors refer to devices used to monitor brain activity. Most BCIs rely on noninvasive electrical measures of brain activity [44] – that is, brainwaves recorded from the surface of the scalp. In ten years, most BCIs will probably still use scalp recorded EEGs, but this view is far from universal. There has been great attention in the BCI literature to BCIs based on other brain monitoring technologies such as fMRI, MEG, or fNIR [84]. These approaches are overrated for BCI applications. None of these approaches has allowed an ITR comparable to EEG-based BCIs. fMRI and MEG require equipment that is not remotely portable and costs millions of dollars. fNIR is a bit more promising, since it is less expensive and more portable. Similarly, there has been more work describing invasive BCI systems recently [16, 31, 34, 38, 71] and corresponding enthusiasm for providing much greater benefits to patients (see Chapters "Intracortical BCIs: A Brief History of Neural Timing", BCIs Based on Signals from Between the Brain and Skull", and "A simple, Spectral–Change Based, Electrocorticographic Brain–Computer Interface" in this book).

How can BCI sensors be improved? Modern laboratory EEG sensors have several problems. An expert must rub the scalp to clean the skin and remove dead skin cells, precisely position the electrodes, squirt electrode gel to get a good contact, check the impedance between the skin and the electrode, and repeat this process until impedances are reduced. Fortunately, there has been some progress toward improved sensors that require less or no gel [65] and may not require expert assistance. Several companies develop systems to play games, communicate, diagnose or monitor patients, and/or perform various functions using EEG. Some of these systems also measure other physiological signals. Such work should facilitate wider development of head-mounted physiological sensors, if not improved BCI electrodes.

Users must perform intentional mental tasks to control any BCI. Different BCI approaches reflect different task alphabets. In some BCIs, imagination of left vs. right hand movement might convey information. Other BCIs cannot understand such neural signals, but instead allow communication via attention to visual targets. The term "alphabet" reflects that users must usually combine several BCI signals to send a message or accomplish another goal.

Hybrid BCIs could detect activity stemming from both imagined movement and selective attention. Hybrid BCIs were suggested over 10 years ago [10], but were not developed until very recently, as discussed below.

In the distant future, as BCIs learn to recognize larger alphabets, these mental tasks or "cognemes" might be compared to graphemes or phonemes in written or spoken language [4]. Of course, most written and spoken languages have vocabularies with many thousands of different elements. Modern BCIs recognize very few different signals. It is unlikely that BCIs will have alphabets comparable to natural languages for a very long time.

However, there is some progress toward task alphabets that are larger and more consistent with natural languages. Some recent work describes a BCI that can identify some vowels and consonants that a user is imagining [31, 69]. This feat requires an invasive BCI. Efforts to reliably identify individual words or sentences through the EEG have failed [78]. While noninvasive BCIs will develop larger alphabets, a noninvasive system will never allow a vocabulary as large as a natural language without a dramatic revolution in neuroimaging technology. Similarly, a "literal" BCI that can directly translate all of your thoughts into messages or commands is very far from reality.

Recent work has discussed BCIs based on new or better defined tasks, such as quasi-movements, first person movement imagery, or imagined music [23, 52, 56]. Brain activity relating to perceived error, image recognition, or other experiences might be used in other systems based on the EEG, perhaps combined with other signals [13, 18, 25, 29, 54]. Ideally, the mental tasks used in BCIs should be easy to generate, difficult to produce accidentally, and easily mapped on to desired commands. BCIs might also use existing tasks for novel goals.

The last BCI Catalyst, Brain Effects, reflects changes that may occur in the brain or body because of BCI use. There could be negative side effects, which may be similar to the negative side effects of watching too much TV or playing too many computer games. However, some newer BCI systems are based on a revolutionary principle: using a BCI to produce positive effects. Although the user does send command and control signals, the system's main goal is not to enable communication, but to facilitate changes in the brain that can alleviate disorders such as stroke, autism, addiction, epilepsy, and emotional or mental problems [58, 60, 63, 77] (see also Chapters "Neurofeedback Training for BCI Control" and "The Graz Brain–Computer Interface").

For example, a user might use an ERD BCI to control an image of a car on a monitor. The system encourages the brain areas that produce ERD to adapt in ways that reduce the symptoms of autism. In another example, a stroke patient might learn to use an ERD BCI to control hand orthosis (see Chapter "Non Invasive BCIs for Neuroprostheses Control of the Paralysed Hand"). Over months of training, the user could develop new ERD patterns while opening and closing the hand many times. Because this process encourages neural plasticity in structures related to motor activity, the patient might recover some movement. Recovery could well involve changes to the spinal cord and other nerves, not just the brain.

A BCI or other system that can change the nervous system in new, significant, desirable, and otherwise unattainable ways could represent a major breakthrough and unanticipated disruptive technology. Many obstacles to wider adoption may seem insignificant amidst new tools that can produce otherwise unattainable effects on the brain or body. If a new system could really help people with certain disorders and/or desires, then consumer demand would encourage innovative new ways to get such systems to users or treatment centers. Imagine a device that can reliably provide rehabilitation, health, nirvana, transcendence, forgiveness, forgetfulness, obedience, faith, epiphany, joy, fulfilment, or peace. These devices are unlikely in the next 10 years, but it's easy to see how they could encourage ubiquitous BCIs even if other obstacles to wider adoption have not been overcome. For example, such devices could become widely adopted with existing sensor technologies, even invasive or nonportable ones.

This issue begs a key question: what is, and isn't a BCI? There is no standardized definition of a BCI, and different groups use different definitions. The most widely cited definition today is from the most heavily cited BCI review: "A BCI is a communication system in which messages or commands that an individual sends to the external world do not pass through the brain's normal output pathways of peripheral nerves and muscles [83]."

TU Graz and other groups have clarified that a BCI must use realtime feedback and must involve intentional, goal-directed control [57]. Devices that passively monitor changes in brain activity that users do not intend to produce are not BCIs. Otherwise, a BCI might include a device that sounds an alarm when a patient undergoing surgery awakens too early. Such a device does present realtime feedback based on direct measures of brain activity, but few experts would consider it a BCI.

This chapter addresses BCIs in the emerging future, and the definition of BCIs may change as related systems become more practical and widely adopted. While it is important to have clear definitions of key terms, these efforts will be lost on most users. Most people would change the channel if it aired an airy academic trumpeting terminological trivialities. Those of us who enjoy a science article with morning coffee recognize we're atypical. Details of BCI research may not percolate to the general public.

The key factors for BCI adoption are outlined below. Each factor is briefly defined, followed by the current status and emerging changes. All factors depend on both hardware and software.

2.2 Cost

Cost reflects not just the costs of purchasing hardware and software, but making and keeping them operational. In addition to financial cost, BCI use can cost time, attention, and energy. These costs may apply to caregivers as well as the BCI user. There may be ongoing costs if another person is needed to help with daily preparation and cleanup. Cost may increase substantially if this caregiver must be an expert. Invasive BCIs incur special costs due to the risk, scarring, and personal intrusion associated with implanting electrodes on or in the brain.

The financial cost of BCI hardware depends on the type of BCI used. A common hardware configuration for a BCI research laboratory is a USB amplifier with a 16 channel cap from a reputable manufacturer. Such a system currently costs about €15000. Cost increases with more electrodes or if another brain imaging approach is used; MEG or fMRI cost millions. It is possible to get an operational laboratory-grade BCI for under €1000, however, by using fewer electrodes. This reduces the cost of both the amp and cap. [64] collected data through a three-electrode headband that cost about $10 and a customized mini-amplifier that cost about $800. Subjects used this device to control a computer game with a BCI. Manufacturers have said that the main reason for the high cost is that they only sell a few hundred amplifiers per year.

Some companies currently advertise BCIs for home use, and the information in this paragraph is based on their websites as of Sep 2010. NeuroSky sells a one-channel "MindSet" system for about $199. OCZ advertises a multichannel "Neural Impulse Actuator" that seems available through numerous resellers for about $150. Emotiv's website advertises its multichannel EPOC Headset for $299. Uncle Milton and Mattel sell a Force Trainer and Mindflex System for less than $100. Consumer BCI systems from other manufacturers, such as Brain Actuated Technologies, Cyberlink, Advanced Brain Monitoring or SmartBrainGames, are more expensive, but still cost less than a typical laboratory BCI. The Swedish group Smart Studio sells fairly inexpensive games and toys based on BCIs, including a relaxation game called BrainBall and a game that lets people mix drinks with a BCI! Some people have asked me whether using this system affects performance. I never used it, but I'm sure I could round up several BCI experts to study this vital issue if provided with a free system (including the drinks).

Noteworthily, all of these systems evidently rely on activity from the brain, eyes, and facial muscles. This will probably remain prevalent with consumer BCIs. EEG information is much easier to acquire from the forehead, since electrodes placed there do not have to contend with hair, and hence it's no surprise that most of these consumer BCIs seem to rely heavily on forehead electrodes. Any forehead electrode that can detect brain activity could also read signals from the eyes and facial muscles, which might carry useful information such as when a user blinks, moves her eyes, frowns, or smiles.

The European Commission recently introduced the term "BNCI" to include devices that detect physiological signals after they have passed through peripheral nerves. The "N" in this acronym means "Neuronal." Many people in the BCI research community are used to thinking of such signals as cheating, since we typically develop BCIs as systems for patients who cannot move and thus could not use peripheral signals. However, this view will change as BCIs gain attention among larger audiences. Users such as gamers will want to use whatever signal gives them an advantage. Blinking, frowning, smiling, teeth clenching, and even touching an electrode can convey different signals that might be mapped on to spells, attacks, defenses, movement, or other game commands.

Unlike hardware, there is presently not much competition for BCI software. The dominant software platform for BCIs is called BCI2000 [46] (see also chapter "Using BCI2000 in BCI Research"). This software contains contributions from many different groups and continues to grow through academic funding. OpenVibe is another BCI software platform that (unlike others) is open source. Both BCI2000 and OpenVibe are free for academic use. Telemonitoring tools developed by TU Graz [47] have reduced the need for researchers to travel to patients' homes, which can reduce the need for experts' time. Jose Millán, Febo Cincotti, and their colleagues within the TOBI grant are currently working on a common platform for BCI systems. As BCIs blossom, more people will want to pay for BCI software, which will further improve its quality, flexibility, and accessibility.

The financial cost of an invasive BCI is much higher than a non-invasive system because neurosurgery requires very expensive experts. There are no easy ways to eliminate this requirement for expertise! Checkups to prevent infection increase the ongoing cost in time and money. However, invasive BCIs do not require preparation and cleanup with each session, which reduces some ongoing costs. Invasive BCI software is somewhat different from other BCI software because invasive BCIs produce different signals (see chapters "Intracortical BCIs: A Brief History of Neural Timing" and "BCIs Based on Signals form Between the Brain and Skull").

Like any interface, using any BCI incurs costs in time and attention. Noninvasive BCIs require users or caretakers to put the cap on the head, get a good signal, and perhaps clean the electrodes and hair later. If the BCI needs to be updated according to changes in brain activity or the environment, that takes more time. Some BCIs require no training, while others require days or weeks of training. Using a BCI takes time and requires mental resources that could be directed elsewhere. Nobody has really studied the extent to which BCI use precludes other activity. This inattention to attention is a major oversight in BCI research and is addressed further in the discussion about integration.

BCIs do not require much electricity. However, if BCI use produces fatigue [2, 37], then this fatigue reflects a cost in terms of personal energy. I used a P300 and later SSVEP BCI for about eight hours per day while reporting only mild fatigue [4, 7], but a few subjects, especially older subjects, found even short term SSVEP BCI use annoying [2]. Other reports have shown that patients can use P300 or other

BCI systems in their home environments for hours per day [14, 80]; please also see chapter "Brain–Computer Interfaces for Communication and Control in Locked-in Patients". Further, the patient described in [80] is capable of using an eye tracking system, but prefers to communicate via his P300 BCI. Devices such as eye trackers, EMG switches, sip and puff switches, or tongue systems may be tiring for individuals with severe disabilities. Some people may prefer BCIs over other assistive communication technologies that rely on motor control because BCIs seem less tiring or easier to use to them.

2.3 Information Transfer Rate (ITR)

A BCI's ITR depends on three factors:

Alphabet: The number of possible selections (N),
Accuracy: The probability of identifying the target selection (P), and
Speed: The number of selections per minute (S).

Many researchers have described BCIs with an ITR (also called throughput) of at least 25 bits per minute [27, 80, 85]. Such reports typically utilize healthy subjects who are not representative of the general population. The subjects in [27, 80, 85], for example, were the "best of the best." That is, the four subjects presented in that paper were among the top trainees in that laboratory.

It is difficult to identify the "true" average ITR for a given BCI in a field setting. As a rough estimate, it varies between 0–30 bits per minute, and the distribution is skewed toward the slower end (e.g., [83]). In the next 5–10 years, ITR will increase considerably, especially in invasive BCIs. However, the ITR of BCIs will remain well below speech or typing for considerably longer than that.

As noted above, ITR may be less important than other factors [11, 14, 36, 49, 83]. Further, some users may not prefer a BCI with the highest possible ITR, perhaps because they prefer excellent accuracy even if it makes the system slower. The formula for ITR does not account for latency – the delay between the user's decision to send a command and its execution. Latency differs from speed because only the latter incorporates the time between selections. In some situations that require relatively rapid communication, such as gaming, emergencies, online chatting, or some robotic applications, a BCI with an excellent ITR but poor latency would be impractical.

Another problem is that ITR is simply a number that does not account for clever interface features that could allow the user to convey more information with the same ITR. Realtime error detection (whether through the EEG, spellcheckers, or other sources), letter or word completion, error corrective coding, predictive interfaces such as DASHER, menu-based BCIs [3, 80], goal oriented protocols that do not burden the user with unnecessary task details [55, 82], and other features could allow a high effective throughput even with a relatively low ITR.

The best opportunity for increasing effective throughput is through these clever interface features [3, 80]. This is relatively virgin territory within BCI research. While improving raw ITR is still important, this task has consumed the bulk of BCI research. Many articles about BCIs present novel signal processing tools or combinations that are applied offline to existing data, often from a different lab, and produce a marginal performance improvement in some subjects. ITR could be improved more by increasing the number of mental tasks used either simultaneously or sequentially, such as with a hybrid BCI, but the main focus should be on interfaces that can do more with a specific ITR.

Hardware innovations could also improve the effective throughput. For example, improved amplifiers or sensors that provide a cleaner picture of brain activity could result in fewer errors. Hardware improvements are worthwhile, but the best overall opportunity for increasing effective throughput remains software with clever interface features that can allow people to convey more useful information with fewer bits per minute.

2.4 Utility

What can you do with a BCI? Right now, most BCIs control simple monitor-based applications, such as a speller or cursor with a basic interface and graphics. More advanced monitor and VR applications are emerging, such as BCIs to control Smart Homes, immersive games, or virtual environments [20, 41, 55, 64, 72, 74] (see also chapters "The Graz Brain–Computer Interface" and "The First Commercial Brain–Computer Interface Environment" in this book). So too are BCIs to control devices such as orthoses, prostheses, robotic arms, mobile robots, or advanced rehabilitation robotic systems such as a combined wheelchair and robotic arm [26, 30, 60]. Hence, both software and hardware advances are being made, but BCIs still control a lot less than mainstream interfaces. BCI utility must and will increase well before BCIs become ubiquitous.

BCI utility reflects how useful a user considers a BCI. Can the BCI meet a potential user's needs or desires? If not, then s/he will probably not use it, even if it is extremely inexpensive, easy to use, portable, etc. BCI utility is subjective on an individual level, but can be measured more objectively over larger groups.

A BCI will not seem useful to a user who cannot get it to work. Support availability is a major issue today. Right now, an expert is typically required to identify the right components, put them together, customize different parameters, and fix problems (see chapter "Brain–Computer Interfaces for Communication and Control in Locked-in Patients"). Efforts to reduce dependence on outside help are essential. A BCI tech support infrastructure must develop to support new and sometimes frustrated users.

Modern BCIs are relatively inflexible; one BCI can spell or move a cursor or browse the internet or control a robot arm, but not switch between these applications. Flexibility could be improved with a Universal Application Interface (UAI) that could let users easily switch between BCI applications using any of the major BCI approaches (P300, SSVEP, ERD,[2] see chapters "Brain Signals for Brain–Computer Interfaces" and "Dynamics of Sensorimotor Oscillations in a Motor Task" for details). Users might spell, use communications and entertainment tools, operate wheelchairs, and control Smart Home devices through the same interface. The UAI should adapt to each user's needs, abilities, desires, environment, and brain activity [3]. Ongoing BCI2000 and OpenVibe developments are increasing the number of available input and output devices. When BCIs attain the same flexibility as keyboards or mice, users will be able to think anything you could type or click. This milestone will result partly from better integration with existing software.

Panel B of Fig. 2 below shows an example of how a BCI could be modified to increase its flexibility. The display is identical to the clever Hex-O-Spell system used by the Berlin BCI group (chapter "Detecting Mental States by Machine Learning Techniques: The Berlin Brain–Computer Interface"), except that two arrows have been added to the top of the display. These arrows might rotate the Hex-O-Spell display forward or back to reveal additional displays with added functionality, such as Hex-O-Play, Hex-O-Home, or Hex-O-Move. The idea of grouping commands within hexes to allow two levels of selection might be extended within these new applications. For example, in a Hex-O-Home system, the hexes might each represent a different Smart Home device, such as lights, windows and blinds, cameras, or heaters, and the commands that could be issued to each device would be within each hex. A Hex-O-Select system in which hexes represent different applications such as Smart Home, Internet, Music, or Game might serve as a type of UAI.[3]

As BCI research increases, and more diverse people try out BCIs, reliability is becoming an increasingly obvious problem. Ideally, a BCI system should work for any subject, any time, in any environment. No such BCI has been developed.

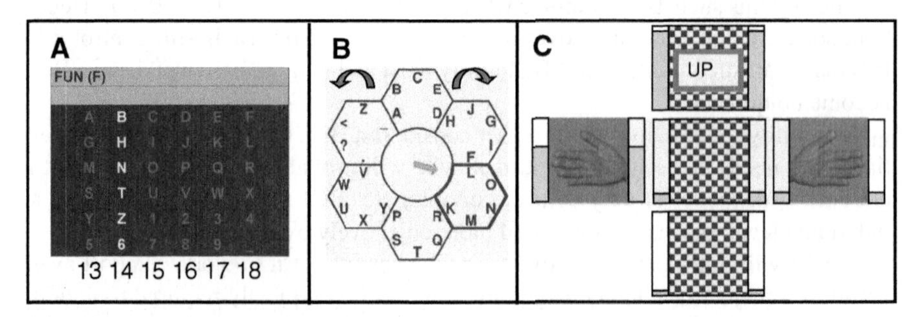

Fig. 2 Three possible hybrid BCI systems. Panel (**a**): A P300/SSVEP BCI based on the Donchin matrix. Panel (**b**): A hybrid ERD/P300 or SSVEP BCI based on the Berlin Hex-O-Spell. Panel (**c**): An original system that could use all three BCI approaches

[2]Some groups use the term mu or SMR instead of ERD

[3]Our team with the BrainAble research project at www.brainable.org is working toward implementing Hex-O-Select.

Across any of the major BCI non-invasive BCI approaches (P300, SSVEP, ERD), about 10–30% of people cannot attain effective control [5, 6, 11, 14, 53, 55, 58, 60, 66, 75, 80]. This problem of reliability across subjects has been called BCI illiteracy by some groups [35, 55, 66]. This term has various problems, including the implication that BCI illiteracy reflects a lack of learning by the user [5, 8].

Work with a large group of subjects that were not pre-screened stated that most untrained people can use an ERD BCI, even if their ITR is only a few bits per minute [33]. However, that article assumed that subjects who attained greater than 60% accuracy in a 2 choice task could use a BCI effectively. If the more common threshold of 70% accuracy were used instead, more subjects would have been illiterate. Similar work validated P300 and SSVEP BCIs across many users [2, 32].

It is true that, in some cases, people who cannot initially use a BCI can attain control after training (see chapters "Neurofeedback Training for BCI Control" and "BCIs in the Laboratory and at Home: The Wadsworth Research Program" in this book). Improved signal processing or different instructions can also help. However, some subjects are unable to produce the brain activity patterns needed to control a specific BCI approach (such as ERD) no matter what. We recently showed that switching to a different BCI approach could help such users [5]. This paper, and other recent work [2, 8, 32, 54] also suggested that illiteracy may be worse in ERD BCIs than SSVEP or P300 BCIs.

A recent major study addressed reliability across subjects in a noisy environment. Our team ran 106 subjects in 6 days on an SSVEP BCI. All subjects had no prior BCI experience and were recruited freely from visitors to our booth at the CeBIT exhibition in Hannover in March 2008. We did not reject any potential subjects; everyone who wanted to be a subject got to be a subject. CeBIT was a huge public exhibition with extreme electrical noise and many distractions.

There were no problems with preparing, running, or cleaning any subject. Questionnaires showed that the system was not annoying nor fatiguing to most users. Most subjects could spell with this system. Some subjects who initially could not attain effective control performed better after lighting, display, or signal processing parameters were changed. Performance tended to be worse among older and male subjects [2].

BCIs should also work reliably any time a user wants to communicate. Relevant EEG patterns can change within and across days. BCIs need to adapt to such changes.

Another facet of reliability is robustness to environmental noise. Many of us in the BCI research community had a field BCI recording session go awry because of an air conditioner motor, generator, refrigerator motor, or medical device (see chapter "Brain–Computer Interfaces for Communication and Control in Locked-in Patients"). This is becoming less of a concern with better amplifiers, sensors, noise reduction software, and wireless systems that avoid the electrical noise that can be produced along cables. As noted above, many studies and companies have described BCIs that can operate in very noisy settings.

While BCI reliability needs to improve, it does not need to be perfect before BCIs gain wider adoption. Mainstream interfaces also do not work for many millions of users who have physical or other disabilities. They do not work in many environments, and even veteran typists sometimes make mistakes.

2.5 Integration

The most fascinating and underappreciated revolution involves BCI integration. Software integration refers to creating tools that let the user perform desired tasks with conventional software tools as smoothly and easily as possible. Hardware integration reflects the challenge of seamlessly combining the gear needed for a BCI with other headmounted devices and any hardware needed to implement tasks. BCIs could also be better integrated with other things people typically put on their heads, such as glasses, headbands, caps, or hats.

Early BCI software was unique to that BCI and was typically written by someone in the lab or by a colleague. Newer BCIs have been used to navigate virtual environments or Google maps, play console games, and otherwise interact with existing software [41, 64, 72, 74]. The increasing sophistication and distribution of the dominant BCI software platform, BCI2000, has facilitated interaction with other software platforms [46]. SmartBrainGames sells a BCI system that can be plugged in to a Playstation console so that users can convey the equivalent of pressing the "A" button with EEGs. Such integration is likely to continue and make it much easier to control conventional software with a BCI. For example, although a BCI can provide functionality comparable to a mouse with 2D movement, click, and grasp [21, 45], there is no software to control a Windows cursor with a BCI. Similarly, most modern computer games allow users to change mappings such that any keyboard, mouse, or joystick command can control any function.

Eventually, manufacturers will normatively add BCIs to that list of inputs. Further, drivers will be developed such that BCI input can be treated the same as other input to any software. While this is possible today, the low speed, precision, and accuracy of modern BCIs would mean that the resulting mouse movements would be slow, choppy, and/or prone to errors. A modern BCI mouse would be best suited as an assistive communication technology for disabled users or applications that do not require high bandwidth or specificity.

Software integration involves integration of functionality – developing a combination of tasks, goals, and interface features to create an easy, effective, enjoyable user experience. This integration is rarely discussed because it is typically assumed that BCIs are "exclusive" interfaces. In other words, using the BCI means you are not using any other interface, and doing little or nothing else. Recent work proposed BCIs as supplementary rather than exclusive interfaces [55], but this remains an underdeveloped idea.

It is possible to use a BCI while distracted, at least while distractions do not require active concentration. In 2004 and 2005, I used a BCI at a public exposition called NextFest while fielding questions from onlookers. I was unable to answer questions in realtime without distracting myself from the task of BCI use. Instead, I remembered questions and answered them during breaks, unless a labmate answered the question. Thus, for one subject without extensive training, BCI use required enough attention to prevent speech production, but was not distracting enough to prevent language processing or working memory [4, 7]. Other work also showed that people could use a BCI in a distracting environment such as a public exhibition [2, 33, 48]. [81] showed that subjects could control a robotic arm with a BCI even across different levels of distraction. Further research is necessary to explore the "distraction quotient" of different BCI approaches and parameters with different users, simultaneous tasks, training, and environments [5, 17].

This distraction quotient will gain more attention in the near future. Some results of this research direction can be predicted from classic research on skill learning [76], since learning to use BCIs is like learning other skills [11, 14, 36, 51, 83]. Practice should increase automaticity, meaning that people will become more capable of multitasking with increasing BCI experience. The distraction quotient will decrease considerably in the first few hours of training, with decreasing marginal benefits from ongoing practice. Distraction will be worse if people have to perform tasks that require the same sensory modality or mental processes. Hence, designers should avoid obvious oversights, such as requiring people to use an SSVEP BCI while doing other demanding visual tasks.

An especially fascinating possibility is integration of BCIs with other interfaces that measure physiological signals, including other BCIs [5, 10, 11, 49, 61, 83]. Initial work suggests that "hybrid" BCIs are viable. Subjects could turn a BCI on through voluntary heart rate changes [73]. More recent work showed that subjects could produce both SSVEP and ERD activity simultaneously, in a paradigm that used tasks similar to online BCIs [5].

A hybrid BCI might combine different imaging approaches, such as an invasive and noninvasive approach [83]. For example, an invasive BCI might provide very fine details of finger movement [38], while a noninvasive system provides less precise information about other areas of the body.

BCIs could also be combined with conventional interfaces. I recently visited the headquarters of NeuroSky and Emotiv and played their games that include BCI signals. Both games allowed control of a character with a conventional handheld gaming controller while moving virtual objects through brain activity. The immersive and engaging game design created a reward for simultaneous manual and BCI control. Players could use the handheld control to run while simultaneously using the BCI to levitate objects and launch them at my opponent. The integration was initially difficult, but became easier over about 10 min of practice.

Similarly, the BCI system described in [64] used an immersive first person shooter game environment that allowed subjects to either turn or fire with ERD activity. In unpublished work, some of Prof. Pineda's research assistants simultaneously used both an ERD BCI and conventional interfaces (a keyboard and mouse). We did not have trouble using both interfaces simultaneously, although it was a simple game. More recent work has shown that immersive BCI environments reduce frustration, errors, and training time [41]. However, nobody has yet demonstrated a mixed system with a BCI and conventional interface that was more effective than the conventional interface alone.

The commercial ramifications of overcoming this problem are considerable. If such a system could offer improved performance, it could appeal to anyone who uses conventional interfaces. If a gamer using an integrated BCI/hand controller can defeat an otherwise equally skilled gamer using hand controllers alone on a popular game – even periodically – the resulting shockwave through the massive and growing gaming community will hearken the end of gaming as we know it.

BCIs could be used in combination with other assistive communication technologies. The second (complementary) interface could influence one or more of the three components of ITR. Accuracy could be improved, and/or selection time reduced, if the BCI can help identify the desired message. For example, an SSVEP system might facilitate an eye tracking system's decision about where a user is looking/attending. If a user can send one of four signals by moving his tongue in one of four directions, and five seconds are needed per selection due to unreliable motor control, a system might simultaneously present the four options via a P300 BCI. Or, perhaps an ERD BCI could provide additional information about a user's motor intent. Alternately, the add-on interface could increase the number of selections. If a 4-choice BCI allows a user to select one of four wheelchair directions, an EMG switch might be used to indicate "discrete" or "continuous" motion. (That is, instruct the wheelchair to either go forward 100 cm, or continue going forward.) This would effectively double the N, the number of signals available. The add-on signal might also be used in simultaneous or interleaved fashion, meaning that users might use both interfaces at once, as in the above examples, or might use them separately.

For example, if an eye tracking system is fatiguing, a user might use the eye tracker to move a cursor around the screen, and then use a BCI to choose what to do at a target location. This would allow the user frequent rest periods.

Very recent work showed that people can use two different BCIs at the same time [5, 57]. "Hybrid" BCIs like these are different from earlier work that described efforts to combine different SSVEP approaches with multiple targets [6, 50], or ERD BCIs in which users imagine multiple movements [28, 45, 62]. However, the idea of combining different BCIs, which was first presented in a citable format in [10], has received little attention until recently, and is revolutionary in the Kuhnian sense. This means that the idea is not simply new, but a major change from the dominant paradigm that is not currently envisioned or accepted by mainstream expert researchers[4].

In a recent study, 14 subjects participated in three conditions. In the first condition, subjects would imagine left hand movement after a left cue, and right hand movement after a right cue. In the second condition, subjects would focus on a left LED after a left cue, and a right LED after a right cue. In the third condition, subjects performed both the imagined movement and visual attention tasks simultaneously in response to left or right cues. Results from this third condition showed that subjects could simultaneously produce SSVEP and ERD activity that should be adequate for

[4]This chapter was originally written in 2008, and this sentence was true through 2009. However, numerous groups began reporing hybrid BCI research in 2010.

controlling a BCI. Furthermore, this "hybrid" task was not considered especially difficult, according to subjects' questionnaires [5]. We later validated this approach in online work with 12 subjects, and then extended this work with a two dimensional hybrid BCI. Subjects could move a cursor horizontally with SSVEP activity, and simultaneously control vertical position with ERD activity [1, 12]. Related work from TU Graz showed that subjects can use ERD activity to turn an SSVEP BCI on or off [61]. [55] presents these and other examples of hybrid BCIs-all of which were developed within the last year.

Figure 2 shows three examples of possible hybrid BCI systems. Panel A is based on the Donchin matrix, the canonical approach to a P300 BCI [9, 75]. Subjects focus on a particular letter or other character, called the target, and count each time a row or column flash illuminates the target. For example, if the subject wanted to spell the letter F, she would not count the column flashed in Panel A, but would count each time the top row or rightmost column flashed. This system could be hybridized if the rows and/or columns oscillated in addition to flashing. The added information from SSVEP activity could improve accuracy or reduce selection time. Panel B shows a variant on the Hex-O-Spell system that might include two arrows at the top that flash or oscillate, allowing people to choose within an application with ERD and use P300 or SSVEP to switch applications. These added signals might also be used to confirm or jump to specific selections. Panel C shows a new BCI system that could train users to combine P300, ERD, and SSVEP activity. Each box is bracketed by two feedback bars that reflect the strength of the user's brain activity associated with that box. Three boxes contain oscillating checkerboxes, and thus the feedback bars represent the SNR of the relevant EEG features. The other two boxes depend on left or right hand motor imagery, and the feedback bars thus reflect ERD over relevant areas and frequencies. With this system, users could send one of five signals, such as moving a cursor in two dimensions plus select, with SSVEP and ERD activity. Each box might also flash a word containing the position of that box. If the user silently counts the flash, then this produces a P300 that could confirm attention to that box. Like other hybrid BCIs, the added signal could allow the user to choose from more possible targets, thus increasing N, or send information more accurately. Some approaches might yield stronger brain activity. Furthermore, if the BCI included software that could learn how to best combine the different signals, then hybrid BCIs might work in people who lack one type of brain signal. This could substantially improve BCI reliability.

Hardware integration has typically foundered on the same assumption that hampered software and functional integration: that BCIs are standalone systems. The necessary hardware, such as sensors, an amplifier, and computer, are typically devoted only to BCI operation. This has begun to change as EEG sensors have been mounted in glasses, headphones, gamer microphone systems, and other headwear [65, 79]. As head-mounted devices become more ubiquitous, integrated EEG and other sensors will become more common. This integration of BCI hardware, software, and functionality is essential for wider BCI adoption. Integration is likely to continue, yielding increasingly transparent and ubiquitous BCIs.

The integration of the mouse followed a similar pattern. Mice were initially greeted with scepticism. They were not well integrated with other hardware nor software. There was little consideration of how the functionality offered by mice could be integrated with other computing functions.

BCI developers today face a challenge similar to that faced by Doug Engelbart, the developer of the mouse, in the 60s: how do you get a new interface adopted when nobody has the needed hardware, no software supports it, and nobody thinks about how to incorporate the added functionality? This creates the fun challenge of considering how BCIs could facilitate modern computing environments. Just as the dropdown menu is a tool developed around the functionality afforded by mice, new tools could be developed according to the new options presented by BCIs and similar systems. Mice cannot directly detect which specific events or regions on a monitor are grabbing attention, or whether the user believes s(he) just made an error.

Many futurists envision a world in which people control computers with a variety of inputs, including their fingers, voices, and direct measures from the brain, eyes, or elsewhere. Such control could allow the benefits of one input modality to supplement the weaknesses of another one, but intelligently combining different input modalities into a usable, effective system entails many underappreciated challenges.

Consider this hypothetical scenario describing hardware, software, and functional integration of activity from BCIs and similar systems:

A project manager is typing a project summary. While doing so, she mistypes a word. The word processor identifies this unfamiliar word and underlines it. Normally, she would have to disengage from the keyboard and use the mouse to right click on the word to call a list of alternate words. However, the headset system she uses identifies a combination of EEG signals, including the ERN, P300, and changes in specific rhythmic activity, reflecting her belief that she just made an error. Through effective integration with the word processing system, the software identifies what she was doing when she believed she made the error and automatically presents a list of alternate words, eliminating the need to disengage from the keyboard. Five minutes later, she correctly types an acronym that the software does not recognize. This time, the system detects no indication that she thought she made an error. Thus, rather than requiring her to inform the word processor that no error was made, the software automatically so concludes and ignores the error.

That afternoon, she is navigating web pages to evaluate equipment from different manufacturers. While typing a URL, she realizes that she missed some information from the last page, and thinks "back" rather than using the mouse. Because a BCI can also determine which events or regions are of interest to her, she can select links by thinking instead of clicking. This is sometimes annoying as unscrupulous advertisers try various tricks to draw attention to their pages, but her software can usually overcome this.

After work, she travels home, where her son is playing computer games. He sends information through his handheld game console and the microphone and sensors on his head. He uses EEG activity to provide more information than his console alone could provide. He initially found this cumbersome, but with some practice and a recently downloaded patch, he can usually beat his friends who use only handheld controls. He hears rumors that the online BCI game champion cheats because he has an invasive BCI that provides cleaner signals, or by hacking the data stream to introduce signals that did not come from his BCI.

A scenario like this may be feasible after 10 years or so, and might involve other physiological signals too. For example, BNCI information from eye activity, heart rate, breathing, and facial muscles could help detect errors, since errors could cause people to blink, frown, sigh, or otherwise react.

Effective integration requires greater attention to usability. In the above example, the BCIs do not demand too much attention or valuable real estate on the users' faces or heads. BCIs are ready to draw more heavily from ample research in human–computer interaction (HCI) and effective software design [3]. Clunky interfaces that are hard to learn or understand will be harder to sell. Designers should avoid the "Midas touch problem," in which a user accidentally sends a message or command [49]. This risk depends on the BCI approach. With ERD BCIs, people might accidentally send a command if they are thinking about sports; with P300 BCIs, a wrong message may be sent if the user is distracted or surprised, which can produce a P300. Hardware usability includes developing headwear that is comfortable, lightweight, easy to don and remove, and does not interfere with vision or other headworn devices [20].

While the recent discussion has focused on BCIs and modern computing environments, future BCIs will need to interact with other emerging technologies. How can BCIs best be integrated with next generation operating systems, input mechanisms, headsets and other worn devices, and pervasive computing environments? How can a BCI contribute to the expected rise in virtual and augmented reality environments? Consider an augmented reality environment in which users simply look at specific devices, locations, or people, and then think specific thoughts to manipulate those devices, go to those locations, or message those people. Other thoughts might allow other actions, such as learning more about a target object, storing an image of the target, or making notes to oneself. The merits of BCI input depend on which emerging technologies become widespread.

2.6 Appearance

Both hardware and software are becoming more cosmetically appealing, another necessary trend that will likely continue. Early BCIs, like the systems used today in most labs and patients' homes, require electrode caps over the entire head, studded with electrodes oozing gooey electrode gel that look ridiculous. Newer systems with better looking sensors that are smoothly integrated into unobtrusive head mounted devices are more likely to be adopted. In fact, there seem to be at least three diverging styles: the "futuristic" look by game manufacturers and others going after enthusiastic technophiles; the "conservative" look by medical device manufacturers and others who instead want to seem safe and established; and the "minimalist" approach of some dry electrode groups and others of trying to embed electrodes in glasses, microphones, and other established headwear.

These design trends also apply to BCI software. BCI games and virtual applications are the beginning of much more immersive, futuristic BCI environments. Virtual Reality (VR) might also be used by medical companies going for a conservative look, such as making virtual limbs seem as realistic as possible for immersive stroke rehabilitation [41]. The most advanced development will be minimalist BCI software that provides smooth and solid functionality without requiring significant monitor real estate.

People's perception of BCIs and their users also depends heavily on prevailing views, which stem partly from news and media. Unfortunately, most articles in the popular media are very sloppy. Over 20 articles said that a BCI was the first BCI. In the most notorious fracas, the senior author of a 2003 paper claimed his work was "way beyond" anyone else's, and several reputable media entities published this view without a second opinion (for a well-written reply, see [85]). BCI research could suffer a backlash if a charlatan triggers a misguided media frenzy akin to Fleishman and Pons' claims of cold fusion or the Raelians' claims of human cloning. Journalists do not have to write a PhD thesis before producing an article about BCIs, but should check their facts with a quick online search and consultation with a few objective experts in the field.

Unlike journalists, authors of BCI-related science fiction (which might be dubbed bci fi) are not required to be factually correct. Nonetheless, these authors are urged to learn about the real world of BCIs when writing and developing bci fi. There are many true stories from BCI research that could inspire good fiction or nonfiction. Most bci fi presents BCIs and their developers as unrealistically powerful, invasive, and evil, but the "Star Trek" reference to BCIs in the opening chapter shows that BCIs can be portrayed as helpful tools as well. Furthermore, claims that sci fi is grounded in reality are common by people trying to develop hype for a film. For example, the director of "Surrogates" claimed that he researched BCIs when developing his movie, although this claim is probably hype, and this movie does not reflect any insights from contact with experts.

Due to poor media and other factors, BCIs certainly have an image problem today. A guy who walks into a bar wearing an electrode cap and talking about BCIs will not meet many women (though I keep trying anyway). A cool hat and slick interface probably wouldn't help much. The bigger problem is the common perception that BCIs are nerdy, strange, and useless.

This perception will change as BCIs become increasingly widespread and more people perceive them as being useful. There was a time when a man walking down the street talking to a small box in his hand would be labelled a freak. As cellphones became more popular, the same person with the same phone might seem hip, successful, common, and finally boring. BCIs may follow a similar image cycle.

Similarly, a gamer today wearing BCI headgear might not gain much respect among peers. The same gamer might earn more clout when his BCI enables him to blow away his pesky nemeses (virtually, of course). Or consider a child who has to attend an outpatient clinic for treatment of an attention disorder with an MEG BCI. Today, the child might be mocked for spending two hours a week with his head inside a huge white beehive that seems designed for contorting secretaries' hair in the 60s. However, the mockery would be reduced if such treatment were commonplace and produced real improvement.

Commercial forces will also encourage more positive views of BCIs. As companies try to make systems that not only look and act cool, but are perceived as cool, they will spend more money on advertising efforts to bolster BCIs.

Accurate information, especially BCI success stories, also help encourage a more positive view of BCIs. One of our goals with this book was to make BCIs accessible to the general public. Some colleagues and I recently established a website at www.future-bnci.org with a similar goal. Efforts such as these could counteract media gaffes, encourage new users and researchers, and help dispel the myth that you need a PhD to understand or enjoy BCI research.

3 Other Incipient BCI Revolutions

Many of the key factors in BCI adoption will change in the near to medium future, some only slightly, some dramatically. These changes intertwine like tangled neurons with revolutions in funding and user groups.

3.1 Funding

Most BCI research and development today is funded by government entities devoted to supporting scientific research and development. This fact reflects the general perception – which was true until very recently – that BCIs and related systems are only useful for very limited user groups. The US government has focused more heavily on invasive BCIs and noninvasive EEG-based systems for "Augmented Cognition" such as monitoring alertness or error, while the EU has focused primarily on noninvasive BCIs. The EU funded three major international research efforts beginning in 2008 called TOBI, BRAIN, and TREMOR, and funded seven more major BCI-related projects beginning in 2010 called BrainAble, Better, MindWalker, Mundus, Decoder, Asterics, and Future BNCI. All of these ten projects focus on noninvasive BCIs, sometimes including BNCI signals that do pass through peripheral nerves and muscles. Some national projects within the EU, such as the massive Dutch BrainGain project, also focus on noninvasive BCIs. Since invasive BCIs are unlikely to become ubiquitous, the funding emphasis by the European Commission seems likely to increase European dominance at this critical juncture in BCI research.

In many cases, funds from other sources are used to support similar research mechanisms and goals. For example, in 2005, a nongovernmental organization called the ALTRAN Foundation gave $1 Million to one of the most active and respected BCI research groups, Dr. Wolpaw's Wadsworth Research Lab in New York, to support their BCI research efforts.

Similarly, most BCI hardware, such as amplifiers and electrode caps, is purchased by research laboratories to conduct research. A slowly growing number of

BCI systems are purchased by patients who need them. Since many patients cannot afford a BCI system, there have been some instances when a research lab provides them with a system for free.

The funding sources for both BCI research and equipment purchases are beginning to change dramatically. Commercial entities such as Mattel, Uncle Milton, NeuroSky, Emotiv, g.tec, Starlab, and Philips develop and sell BCI systems. These and other companies will likely focus on different types of BCIs, and will develop hardware and software that will be sold to individual end users rather than given freely to research groups. As total sales increase, equipment costs will plummet, thereby making BCIs more practical for different buyers. The increasing involvement of industry should also greatly accelerate development of more practical, wearable, usable, cosmetically appealing hardware systems.

Birbaumer and Sauseng (see chapter "Brain–Computer Interface in Neurorehabilitation") argue that insurance companies should be forced to pay for BCI systems if they are necessary for patients. This is an excellent idea, but will probably require legislation. In countries with more socialist economies, governments are urged to consider providing funds for BCIs if needed.

3.2 User Groups Today

Who uses BCIs, and why? Today, most BCI users fall in to one of three categories (see Table 1):

BCI research scientists and developers: Our main goals are to develop, test, improve, describe, and disseminate our BCIs and views.

Their friends, family, or colleagues: These users may have similar goals, but may also be motivated by friendly cajoling or begging.

Research subjects: In addition to some of the reasons above, subjects may be motivated by cash payment or academic credit.

Most of these people are only BCI users for less than one hour, then never use any BCI again. Although a few research subjects participate in dozens of training sessions [15, 45], many BCI subjects participate in only one recording session. Furthermore, most research subjects do not typically plan on using the BCI after the research study is complete. As BCIs gain popularity outside of laboratory settings, some people will use BCIs much more than ever before, creating new opportunities to study changes in brain activity and develop new BCI paradigms. Similarly, if the number of BCI users grows to thousands or more, and their raw data is available to researchers, we could learn a lot more about how different brains learn and adapt.

What about patients? Since most BCI research focuses on helping patients, shouldn't they be in this table?

Table 1 Dominant BCI user groups today

BCI user groups	Type of disability	Failing(s) of other interfaces or substitutes
BCI researchers	Varies	Goal is to test or use BCIs
BCI researchers' friends and family	None	Goal is to appease curiosity or nagging friends/relatives
University research subjects	None	Can't otherwise get research credit for Psychology class

They should, and the fact that BCIs are not helping many patients is widely regarded as a major problem in our field. One of the most common future directions listed in BCI articles is extending laboratory work or demonstrations to help patients [11, 36, 58, 60, 83]. The very theme of the 2005 International BCI conference was "Moving Beyond Demonstrations."

Unfortunately, getting systems to patients is much harder than it sounds. The chapter by Nijboer and Broermann (chapter "Brain–Computer Interfaces for Communication and Control in Locked-in Patients") provides a great overview of the challenges of getting a BCI to work with an individual patient in her home. This chapter shows why it is often necessary to have an expert in the home for a myriad of different tasks. Dr. Nijboer gracefully avoided any complaint about the very hard work that she and her colleagues also had to do back in the laboratory to set up and optimize the BCI. But, this is also a time consuming part of our job.

Furthermore, just like healthy users, the overwhelming majority of patients who use a BCI never get to do so again. Once the study ends, the patient is again left in silence.

There are a few exceptions. One patient used a BCI to communicate for over 10 years. He was first trained to use a BCI in 1996 by scientists from the Birbaumer lab [15], and successfully used a BCI until he passed away in 2007. A few other patients have successfully used a BCI to communicate for over one year [80, 14].

The CBS program "60 Min" aired a story in November 2008 about a patient with ALS who relies on a P300 BCI developed by the Wolpaw lab. This patient chose to waive confidentiality in this story; his name is Dr. Scott Mackler. Dr. Mackler is himself a professional neuroscientist who uses his BCI to run his lab. Many BCI researchers (including Dr. Wolpaw and others in his lab) said they felt that CBS did a great job with this story. You can find it by searching online, or by clicking on the link from Dr. Wolpaw's foundation: http://www.braincommunication.org/. The main goal of this foundation is to help patients who need BCIs to communicate, and readers who wish to help are encouraged to donate.

3.3 User Groups Tomorrow

Who will use BCIs, and why? A world of ubiquitous BCIs implies a lot more end users. People like Drs. Broermann and Mackler, who could not effectively communicate without a BCI, will grow in number but not proportion. That is, while

Table 2 Examples of future BCI user groups

BCI user groups	Type of disability	Failing(s) of other interfaces or substitutes
ALS, injury, stroke	Severe motor	All require the user to move
Injury, stroke, CP, MS, some diseases	Less severe motor	Cost, bandwidth, usability, flexibility, fatigue
Persons seeking rehabilitation	Stroke, ADD, autism, Psychiatric other?	Cannot affect behaviour or repair damaged brain as well
People who may not have any disability	Situational, or none	Hands/voice unavailable, busy, too difficult, undesirably overt, too mundane, or provide inadequate bandwidth

the total number of severely disabled patients will grow, the proportion of them to other end users will eventually greatly diminish because there will be so many other end users.

Future BCI users will also differ critically from modern BCI users in another key way: a BCI expert will not be needed to identify or purchase the necessary hardware or software, get the system working, configure it to individual users, etc. This group is currently the least numerous BCI user group, and will become the dominant group.

Table 2 summarizes potential future BCI user groups. Some disabilities are repeated in the table because the severity or nature of the disorder, and the user's goals, could substantially affect the type of BCI they use and why.

Severe motor disability for communication: These people use BCIs because they cannot otherwise communicate.

Less severe motor disability for communication: These people might be able to communicate with other tools, but prefer a BCI. This group will become more prevalent as BCIs become faster, easier, or otherwise better than other assistive technologies.

Rehabilitation: Persons with motor or other disabilities might rely on BCIs to help alleviate their symptoms, either temporarily or permanently. A related group might include healthy people who use BCIs for other Brain Effects that might induce personal wellness, relaxation, or other improvements to their cognitive or emotive states.

Healthy users: Please see our article titled "Why use a BCI if you're healthy?" for more details [55]. Briefly, a major reason is "situational disability." Healthy people are often in situations that prevent them from using their hands or voices to communicate. A driver, soldier, mechanic, astronaut, or surgeon may need to use his hands for his work, and might be unable to speak effectively because of background noise or because he must remain silent. Similarly, these or other users might want a BCI to supplement existing communication mechanisms. A driver or surgeon might not want to interrupt a conversation to send a simple command, such as changing a radio station or video display. A gamer or cell phone user might also want the additional bandwidth that a BCI could provide. A "supplemental" BCI such as this would of course only be useful if it does not overly distract the user from other tasks, which requires effective functional integration.

Lazy people might effectively perceive themselves as situationally disabled. Who would buy a television without a remote control? It seems like the remote control is a tool to help people who are unable to use the controls on the front of a TV. Yet this is rarely the case; nearly all remote control users simply don't want to get up.

BCIs are also the stealthiest form of communication possible, since they require no movement that a competitor or enemy could intercept. BCIs and similar systems might provide information or possibilities that are otherwise unavailable, such as automatic error detection or rehabilitation. In the distant future, BCIs might be faster or easier to use than other interfaces. In contrast, while BCIs may not be novel in the distant future, they may seem very exciting and new today. That's a major selling point today and in the near future.

As noted, there are already some examples of patients and healthy people who bought BCIs rather than used them through a research lab. Many of these modern end users are tinkers, futurists, bored rich people, new age enthusiasts, and/or gamers. Several companies sell home EEG systems intended to facilitate alertness, relaxation, and personal wellness through neurofeedback. Neuromarketing companies such as NeuroFocus and EMSense use EEG and other brain imaging tools to study how people feel about products, speeches, advertisements, or other stimuli. Other companies such as NoLieMRI and Brain Fingerprinting Laboratories provide lie detection via fMRI or EEG measures, respectively. Military, academic, and commercial research efforts have developed brain monitoring systems to assess fatigue, alertness, or attention that can provide some record or warning when people are unfit to work, drive, study, or perform other tasks.

It's hard to say which BCI or related system will dominate. This is not because there is no opportunity in the near or medium future; many different possibilities are promising. Many of the applications presented so far require some progress that may not occur, and disruptive technologies with BCI research or related fields could dramatically change the landscape. The "killer application" remains elusive.

4 BCI Ethics Today and Tomorrow

BCI research involves many ethical issues, especially when working with patients. Common ethical considerations in BCI articles include choosing the right BCI for a given patient (especially invasive vs. non-invasive BCIs), proper handling of research subjects (such as handling of confidential data and informed consent), and proper discussion and dissemination of results [14, 22, 24, 36, 83]. There has been less attention to larger ethical issues that may arise as BCIs become mainstream, such as privacy, liability [68], and laziness. We should consider whether and how a changing BCI landscape will, or should, affect ethical questions.

Choosing the right BCI is a relatively minor consideration today, since it is so hard to get any BCI in the first place. However, this is changing as BCIs become more common and flexible. While key factors such as cost or utility will heavily influence these decisions, the ethical concerns raised by invasive BCIs cannot be ignored.

The decision to use an invasive BCI should stem from consideration of the costs and risks as well as the non-invasive alternatives. Today, non-invasive BCIs can do anything that invasive BCIs can do. Despite repeated efforts to suggest that noninvasive approaches are inadequate for some tasks [19, 34], repeated rebuttals in the literature [45, 85] have shown that noninvasive BCIs offer performance equal or better than invasive BCIs.

This will probably change within ten years. Many new advances could foreshadow invasive BCIs that offer performance that non-invasive counterparts cannot. A recent article in Nature described an invasive system in monkeys that could detect many details of desired movements and thereby allow up to 6.5 bits per second [67]. Taylor and Stetner (Chapter "Intracortical BCIs: A Brief History of Neural Timing") describe how invasive BCIs could provide even more precise information than actual movements (see also [70]). Tasks such as controlling a BCI by imagining specific finger movements may soon be feasible with invasive BCIs only. Invasive BCIs might someday allow people to precisely and independently control each finger on an artificial hand.

Thus, the ethical quandaries raised when deciding whether to recommend an invasive BCI to a patient will, and should, change if invasive BCIs offer new options. This may extend beyond BCIs for communication if, for example, invasive systems turn out to be the best way to treat stroke or other disorders.

However, invasive BCIs still require surgery, frequent checkups, and uncertain long-term damage, limiting potential users to people with a real medical need. Dr. Wolpaw opened the 2003 Society for Neuroscience symposium with: "Your brain or mine?" Invasive BCIs will offer better and better options to the people who need them, and should be researched and used. But, as BCIs become more ubiquitous, the practical problems inherent in invasive or nonportable BCIs will seriously limit adoption among user groups without a medical need for BCIs.

Will this change? Within the BCI community, there is no serious consideration of implanting BCIs in persons without a medical need. BCI ethics can draw on conventional medical ethics, which typically oppose surgery (especially brain surgery) unless there is a medical need. These ethical standards might change if there is some dramatic advancement that obviates surgery, such as Star Trek style "hyposprays" that can somehow inject devices without surgery. In the distant future, if invasive BCIs have little history of medical complications and can provide substantial benefits over other options, surgery might be considered less dramatic. People who wear glasses often turn to eye surgery to improve their vision, even though it was considered too risky a generation ago, and is still not medically necessary. Similarly, people might someday consider some types of neurosurgery less dramatic – or, they may downplay the risks if they care enough about the potential benefits. The drive to win a high-profile game competition could crack a gamer's soul and skull like some athletes turn to steroids or doping.

Privacy is a major concern that will keep growing as BCIs become both more common and more capable. BCI use may leave a trail of EEG data that could be mined to reveal more information than the user intended. As BCIs become more common, people may grow more casual about sending or storing data. BCIs currently lack the capability to reveal anything too insidious, but this is beginning to change. Companies like BrainwaveScience and NoLieMRI are working on lie detection tools based on EEG or fMRI. Systems that analyze mood, fatigue, medication or illegal drugs, or other characteristics could have therapeutic applications, but could also be misused.

Liability issues will arise in a BCI-enabled society. If future BCIs could interpret thoughts more quickly than we consciously realize them, or send commands that have not been thoroughly considered, who is responsible? What about the "Midas touch" problem [49], in which a user might send an unintended command? What about systems that write to the brain – who is liable for resulting actions?

The privacy concern is more serious than liability right now, and both concerns can be addressed through comparisons to other recent technologies. Regarding privacy, regulations and standards must be established, with an informed and effective enforcement entity, to ensure that people's brain activity will be kept confidential and will only be used as the user intended. This is not so different from rules regarding cell phone transmissions, data sent online, research data, or medical files. These protections are not perfect, and some burden remains on the end user to avoid sending potentially damaging information with inadequate protection or through unscrupulous software or businesses. Many people today send credit card information through public internet terminals, loudly announce it in public places, or leave written copies. And just as many people have software to prevent popups, viruses, cookies, and sniffing, future software might somehow limit or screen any brain imagery used in a BCI to maintain privacy, prevent overly alluring "think here" links, or warn of inadvisable downloads or other software.

Liability is a more distant concern. People cannot really use BCIs for critical decisions now, and BCIs remain so slow that any messages sent must be deliberate. Computer-to-brain interfaces (CBIs) that write to the brain are currently used for helping with vision, hearing, or movement restoration, and not so much changing what you think. Liability issues should also be addressed via current perspectives. In general, the user is responsible for any messages that are sent. Typing errors are the user's fault, and the best way to avoid significant problems is to make sure that any major decision must be verified.

A related issue is liability for side effects. Many BCIs train the user's brain. Systems designed to treat disorders rely on such training to produce positive changes in the brain. Dobkin [24] suggests possible negative side effects such as a facial tic, unwanted thoughts, or perceptual distortions. Negative side effects have never been reported, even in people who have used BCIs for years.

Liability concerns will also require effective regulation, clinical trials, and an enforcement infrastructure. Commercial entities need to take strong measures to avoid prematurely introducing BCIs. Scandals with medications that turned out to have harmful side effects are expensive as well as unethical.

The final ethical concern is with societal side effects. What are the negative implications of a BCI-enabled society? Could people become too dependent on BCIs, and unable to address realworld problems? Greater BCI ubiquity implies decreased connection to the physical world and all its perceived burdens – work, pain, mud, commuting, getting stuck in the rain. Who would fardels bear, to grunt and sweat under a weary life, when he himself might his quietus make with a bare BCI? Would the convenience of ubiquitous BCIs ultimately make us not just physically but creatively lazy? If Shakespeare could write with a BCI, would he produce anything wittier than in-game taunts after splattering Marlowe with a plasma gun?

This everpresent ethical elephant in BCI research has been addressed much better by science fiction than the professional research community. I am aware of no journal publication, conference poster, or talk from any BCI researcher that addresses this – indeed, alarmists in the media even deserve more credit for addressing this issue, albeit with unresearched Chicken Little handwaving. Work has shown that reduced physical activity produces depression, and that mental and social challenges are important for alertness and problemsolving [39, 42]. Does this relate to BCIs?

This ethical concern depends on a related question: how much can BCIs reduce our dependence on the external world, compared to other technologies? In the near future, BCIs could not really foster sloth much more than other technologies and applications that already do so, such as an addictive TV series or computer game. This stems from the fundamental problem of BCIs today: they cannot really allow you to do something you could not do with another means of communication or control. Many people today choose to master keyboards, mice, and telephones to buy clothes or food, rather than learning to sew, hunt, or farm. Since a modern BCI requires more trouble than a conventional interface, a user who tried to order a new sweater or pizza through a BCI would end up doing more work than someone who uses conventional tools to accomplish the same goal.

Further in the future, BCIs could contribute heavily to unhealthy mental and physical sloth. Ultimately, BCIs may reduce or eliminate dependence not only on physical activity, but some cognitive activity. Marr [43] argues that problem solving involves three components: computational (identifying the problem); algorithmic (how to address the problem); and implementational (how to enact the solution). Modern BCIs, like television remote controls, simplify the last two stages. Very distant BCIs might eliminate at least these two stages: the user need only ponder a need, and it is (or seems) sated. Such an advanced system would indeed change human thinking in ways that external tools such as remote controls cannot. The current BCI research community could be compared to quantum physicists in the early part of the 20th century. Both groups had a core of motivated, interconnected people, genuinely fascinated by colleagues' progress and convinced they're contributing to the birth of something huge. We also have mindrending ethical dilemmas with the implications of our work. The problem is even greater with us. The worst quantum physicists could manage – even today – is destroying the earth. Nightmare scenarios of BCIs gone awry would leave the world intact while rotting humanity.

Fortunately, the nightmare scenario seen in some bci fi – a phlegmatic Orwellian dystopia of pale, unfulfilled, blindly dependent masses – can be proactively prevented. Many mechanisms for prevention stem from mechanisms already in place with other new technologies. BCI researchers are aware of ethical issues, discuss them openly, and are committed to helping people. We can have at least some confidence that parents, doctors, psychiatrists, and end users will generally behave ethically when choosing or recommending BCIs. People may revolutionize BCIs – many times – but BCIs will remain a minor part of humanity. Humanity may be changing technology, but that doesn't necessarily make us less human.

Acknowledgments This paper was supported in part by three grants: a Marie Curie European Transfer of Knowledge grant Brainrobot, MTKD-CT-2004-014211, within the 6th European Community Framework Program; the Information and Communication Technologies Coordination and Support action "FutureBNCI", Project number ICT-2010-248320; and the Information and Communication Technologies Collaborative Project action "BrainAble", Project number ICT-2010-247447. Thanks to Sara Carro-Martinez and Drs. Alida Allison, Clemens Brunner, Gary Garcia, Gaye Lightbody, Paul McCullagh, Femke Nijboer, Kai Miller, Jaime Pineda, Gerwin Schalk, and Jonathan Wolpaw for comments on this manuscript or on some ideas herein. The sentence about "fardels bear" is paraphrased from Hamlet.

References

1. B.Z. Allison, C. Brunner, S. Grissmann, and C. Neuper (2010). Toward a multidimensional "hybrid" BCI based on simultaneous SSVEP and ERD activity. Program No. 227.4. 2010 Neuroscience Meeting Planner. San Diego, CA: Society for Neuroscience, 2010. Online. Presentation accepted and scheduled for Nov 2010.
2. B.Z. Allison, D. Valbuena, T. Lueth, A. Teymourian, I. Volosyak, and A. Gräser, BCI demographics: How many (and what kinds of) people can use an SSVEP BCI? IEEE Trans Neural Syst Rehabil Eng, 18(2), 107–116, (2010).
3. B.Z. Allison, *Human-computer interaction: Novel interaction methods and techniques*, chapter The I of BCIs: next generation interfaces for brain – computer interface systems that adapt to individual users. Springer, New York, (2009).
4. B.Z. Allison, J.B. Boccanfuso, C. Agocs, L.A. McCampbell, D.S. Leland, C. Gosch, and M. Moore Jackson, Sustained use of an SSVEP BCI under adverse conditions. Cogn Neurosci Soc, 129, (2006).
5. B.Z. Allison, C. Brunner, V. Kaiser, G. Müller-Putz, C. Neuper, and G. Pfurtscheller. A hybrid brain-computer interface based on imagined movement and visual attention. J Neural Eng, 7(2), 26007, (2010).
6. B.Z. Allison, D.J. McFarland, G. Schalk, S.D. Zheng, M.M. Jackson, and J.R. Wolpaw, Towards an independent brain–computer interface using steady state visual evoked potentials. Clin Neurophysiol, 119, 399–408, (2008).
7. B.Z. Allison, and M.M. Moore (2004). Field validation of a P3 BCI under adverse conditions. Society for Neuroscience Conference. Program No. 263.9. San Diego, CA.
8. B.Z. Allison and C. Neuper, Could anyone use a BCI? In D.S. Tan and A. Nijholt, (Eds.), *(B+H)CI: The human in brain–computer interfaces and the brain in human-computer interaction*, volume in press. Springer, New York, (2010).
9. B.Z. Allison and J.A. Pineda, Effects of SOA and flash pattern manipulations on ERPs, performance, and preference: Implications for a BCI system. Int J Psychophysiol, 59, 127–140, (2006).

10. B.Z. Allison, A. Vankov, and J.A. Pineda., EEGs and ERPs associated with real and imagined movement of single limbs and combinations of limbs and applications to brain computer interface (BCI) systems. Soc Neurosci Abs, 25, 1139, (1999).
11. B.Z. Allison, E.W. Wolpaw, and J.R. Wolpaw, Brain–computer interface systems: progress and prospects. Expert Rev Med Devices, 4, 463–474, (2007).
12. C. Brunner, B.Z. Allison, C. Altstätter, and C. Neuper (2010). A hybrid brain-computer interface based on motor imagery and steady-state visual evoked potentials. Program No. 227.3. 2010 Neuroscience Meeting Planner. San Diego, CA: Society for Neuroscience, 2010. Online. Presentation accepted and scheduled for Nov 2010.
13. N. Bigdely-Shamlo, A. Vankov, R.R. Ramirez, and S. Makeig, Brain activity-based image classification from rapid serial visual presentation. IEEE Trans Neural Syst Rehabil Eng, 16(5), 432–441, Oct (2008).
14. N. Birbaumer and L. G. Cohen, brain–computer interfaces: communication and restoration of movement in paralysis. J Physiol, 579, 621–636, (2007).
15. N. Birbaumer, N. Ghanayim, T. Hinterberger, I. Iversen, B. Kotchoubey, A. Kübler, J. Perelmouter, E. Taub, and H. Flor, A spelling device for the paralysed. Nature, 398, 297–298, (1999).
16. T. Blakely, K.J. Miller, S.P. Zanos, R.P.N. Rao, and J.G. Ojemann, Robust, long-term control of an electrocorticographic brain–computer interface with fixed parameters. Neurosurg Focus, 27(1), E13, Jul (2009).
17. C. Brunner, B.Z. Allison, D.J. Krusienski, V. Kaiser, G.R. Müller-Putz, C. Neuper, and G. Pfurtscheller Improved signal processing approaches for a hybrid brain-computer interface simulation. J Neurosci Method, 188(1),165–73.
18. A. Buttfield, P. W. Ferrez, and J. del R. Millán, Towards a robust BCI: Error potentials and online learning. IEEE Trans Neural Syst Rehabil Eng, 14, 164–168, 2006.
19. J.M. Carmena, M.A. Lebedev, R.E. Crist, J.E. O'Doherty, D.M. Santucci, D.F. Dimitrov, P.G. Patil, C.S. Henriquez, and M.A.L. Nicolelis, Learning to control a brain-machine interface for reaching and grasping by primates. PLoS Biol, 1, E42, (2003).
20. F. Cincotti, D. Mattia, F. Aloise, S. Bufalari, G. Schalk, G. Oriolo, A. Cherubini, M.G. Marciani, and F. Babiloni, Non-invasive brain–computer interface system: towards its application as assistive technology. Brain Res Bull, 75(6), 796–803, Apr (2008).
21. L. Citi, R. Poli, C. Cinel, and F. Sepulveda, P300-based BCI mouse with genetically-optimized analogue control. IEEE Trans Neural Syst Rehabil Eng, 16, 51–61, (2008).
22. J. J. Daly and J. R. Wolpaw, brain–computer interfaces in neurological rehabilitation. Lancet Neurol, 7, 1032–1043, (2008).
23. B.J. de Kruif, R. Schaefer, and P. Desain, Classification of imagined beats for use in a brain computer interface. Conf Proc IEEE Eng Med Biol Soc, 2007, 678–681, (2007).
24. B. H. Dobkin, brain–computer interface technology as a tool to augment plasticity and outcomes for neurological rehabilitation. J Physiol, 579, 637–642, (2007).
25. P. W. Ferrez and J. del R. Millán, Error-related EEG potentials generated during simulated brain–computer interaction. IEEE Trans Biomed Engi, 55, 923–929, 2008.
26. F. Galán, M. Nuttin, E. Lew, P. W. Ferrez, G. Vanacker, J. Philips, and J. del R. Millán, A brain-actuated wheelchair: asynchronous and non-invasive brain?computer interfaces for continuous control of robots. Clin Neurophysiol, 119, 2159–2169, 2008.
27. X. Gao, D. Xu, M. Cheng, and S. Gao, A BCI-based environmental controller for the motion-disabled. IEEE Trans Neural Syst Rehabil Eng, 11, 137–140, 2003.
28. T. Geng, J. Q. Gan, M. Dyson, C. S. Tsui, and F. Sepulveda, A novel design of 4-class BCI using two binary classifiers and parallel mental tasks. Comput Intell Neurosci, 2008, 437306, (2008).
29. A. D. Gerson, L. C. Parra, and P. Sajda, Cortically coupled computer vision for rapid image search. IEEE Trans Neural Syst Rehabil Eng, 14(2), 174–179, Jun (2006).
30. Vora, J.Y., Allison, B.Z., & Moore, M.M. (2004). A P3 brain computer interface for robot arm control. Society for Neuroscience Abstract, 30, Program No. 421.19.

31. F.H. Guenther, J.S. Brumberg, E.J. Wright, A. Nieto-Castanon, J.A. Tourville, M. Panko, R. Law, S.A. Siebert, J.L. Bartels, D.S. Andreasen, P. Ehirim, H. Mao, and P.R. Kennedy., A wireless brain-machine interface for real-time speech synthesis. PLoS One, 4(12), e8218, (2009).
32. C. Guger, S. Daban, E. Sellers, C. Holzner, G. Krausz, R. Carabalona, F. Gramatica, and G. Edlinger. How many people are able to control a P300-based brain–computer interface (BCI)? Neurosci Lett, 462, 94–98, (2009).
33. C. Guger, G. Edlinger, W. Harkam, I. Niedermayer, and G. Pfurtscheller, How many people are able to operate an EEG-based brain–computer interface (BCI)? IEEE Trans Neural Syst Rehabil Eng, 11, 145–147, (2003).
34. L.R. Hochberg, M.D. Serruya, G.M. Friehs, J.A. Mukand, M. Saleh, A.H. Caplan, A. Branner, D. Chen, R.D. Penn, and J.P. Donoghue, Neuronal ensemble control of prosthetic devices by a human with tetraplegia. Nature, 442, 164–171, (2006).
35. A. Kübler and K.-R. Müller, An introduction to brain–computer interfacing. In G. Dornhege, J. del R. Millán, T. Hinterberger, D.J. McFarland, and K.-R. Müller, (Eds.), *Toward brain–computer interfacing*, MIT Press, Cambridge, MA, pp. 1–25, (2007).
36. A. Kübler, N. Neumann, J. Kaiser, B. Kotchoubey, T. Hinterberger, and N. Birbaumer, brain–computer communication: self-regulation of slow cortical potentials for verbal communication. Arch Phys Med Rehabil, 82, 1533–1539, (2001).
37. P.R. Kennedy, R.A. Bakay, M.M. Moore, K. Adams, and J. Goldwaithe, Direct control of a computer from the human central nervous system. IEEE Trans Rehabil Eng, 8(2), 198–202, Jun (2000).
38. J. Kubánek, K.J. Miller, J.G. Ojemann, J.R. Wolpaw, and G. Schalk., Decoding flexion of individual fingers using electrocorticographic signals in humans. J Neural Eng, 6(6), 066001, Dec (2009).
39. K. G. Lambert. Rising rates of depression in today's society: consideration of the roles of effort-based rewards and enhanced resilience in day-to-day functioning. Neurosci Behav Revi, 30, 497–510, (2006).
40. E.C. Lee; J.C. Woo, J.H. Kim et al. A brain-computer interface method combined with eye tracking for 3D interaction J Neurosci Method, 190(2), 289–298, (2010)
41. R. Leeb, R. Scherer, D. Friedman, F.Y. Lee, C. Keinrath, H. Bischof, M. Slater, and G. Pfurtscheller, Combining BCI and virtual reality: scouting virtual worlds. In G. Dornhege, J. del R. Millán, T. Hinterberger, D.J. McFarland, and K.-R. Müller, (Eds), *Toward brain–computer interfacing*, chapter 23, MIT Press, Cambridge, MA, pp. 393–408, (2007).
42. A. W. Mahncke, A. Bronstone, and M. M. Merzenich., Brain plasticity and functional losses in the aged: scientific basis for a novel intervention. *Progress in brain research: Reprogramming the brain*, Elsevier, New York, (2006).
43. D. Marr, *Vision*, W. H. Freeman, (1982). http://www.nature.com/nature/journal/v317/n6035/abs/317314a0.html
44. S.G. Mason, A. Bashashati, M. Fatourechi, K.F. Navarro, and G.E. Birch, A comprehensive survey of brain interface technology designs. Ann Biomed Eng, 35, 137–169, (2007).
45. D.J. McFarland, D. J. Krusienski, W. A. Sarnacki, and J. R. Wolpaw, Emulation of computer mouse control with a noninvasive brain–computer interface. J Neural Eng, 5, 101–110, (2008).
46. J. Mellinger, G. Schalk, C. Braun, H. Preissl, W. Rosenstiel, N. Birbaumer, and A. Kübler, An MEG-based brain–computer interface (BCI). NeuroImage, 36, 581–593, (2007).
47. G.R. Müller, C. Neuper, and G. Pfurtscheller, Implementation of a telemonitoring system for the control of an EEG-based brain–computer interface. IEEE Trans Neural Syst Rehabil Eng, 11, 54–59, (2003).
48. K.-R. Müller, M. Tangermann, G. Dornhege, M. Krauledat, G. Curio, and B. Blankertz, Machine learning for real-time single-trial EEG analysis: From brain–computer interfacing to mental state monitoring. J Neurosci Methods, 167, 82–90, (2008).
49. M. M. Moore, Real-world applications for brain–computer interface technology. IEEE Trans Neural Syst Rehabil Eng, 11, 162–165, (2003).

50. T.M. Srihari Mukesh, V. Jaganathan, and M. Ramasubba Reddy, A novel multiple frequency stimulation method for steady state VEP based brain computer interfaces. Physiol Measure, 27, 61–71, 2006.
51. N. Neumann, T. Hinterberger, J. Kaiser, U. Leins, N. Birbaumer, and A. Kübler, Automatic processing of self-regulation of slow cortical potentials: evidence from brain–computer communication in paralysed patients. Clin Neurophysiol, 115, 628–635, (2004).
52. C. Neuper, R. Scherer, M. Reiner, and G. Pfurtscheller, Imagery of motor actions: differential effects of kinaesthetic versus visual-motor mode of imagery on single-trial EEG. Brain Res Cogn Brain Res, 25, 668–677, 2005.
53. F. Nijboer, A. Furdea, I. Gunst, J. Mellinger, D.J. McFarland, N. Birbaumer, and A. Kübler. An auditory brain–computer interface (BCI). Neurosci Lett, 167, 43–50, (2008).
54. F. Nijboer, N. Birbaumer, and A. Kübler. The influence of psychological state and motivation on brain-computer interface performance in patients with amyotrophic lateral sclerosis–a longitudinal study. Front Neurosci, 4, (2010)
55. A. Nijholt, D. Tan, G. Pfurtscheller, C. Brunner, J. del R. Millán, B.Z. Allison, B. Graimann, F. Popescu, B. Blankertz, and K.-R. Müller, brain–computer interfacing for intelligent systems. IEEE Intell Syst, 23, 72–79, (2008).
56. V.V. Nikulin, F.U. Hohlefeld, A.M. Jacobs, and G. Curio. Quasi-movements: a novel motor-cognitive phenomenon. *Neuropsychologia*, 46:727–742, (2008).
57. G. Pfurtscheller, B.Z. Allison, C. Brunner, G. Bauernfeind, T. Solis Escalante, R. Scherer, T.O. Zander, G. Müller-Putz, C. Neuper, and N. Birbaumer,. The hybrid BCI. Front Neurosci, 4, 42. doi:10.3389/fnpro.2010.00003, (2010).
58. G. Pfurtscheller, R. Leeb, D. Friedman, and M. Slater, Centrally controlled heart rate changes during mental practice in immersive virtual environment: a case study with a tetraplegic. Int J Psychophysiol, 68, 1–5, (2008).
59. G. Pfurtscheller, G. Müller-Putz, R. Scherer, and C. Neuper, Rehabilitation with brain–computer interface systems. IEEE Comput Mag, 41, 58–65, (2008).
60. G. Pfurtscheller, G.R. Müller-Putz, A. Schlögl, B. Graimann, R. Scherer, R. Leeb, C. Brunner, C. Keinrath, F. Lee, G. Townsend, C. Vidaurre, and C. Neuper, 15 years of BCI research at Graz University of Technology: current projects. IEEE Trans Neural Syst Rehabil Eng, 14, 205–210, (2006).
61. G. Pfurtscheller, T. Solis-Escalante, R. Ortner, P. Linortner, and G.R. Müller-Putz, Self-paced operation of an SSVEP-based orthosis with and without an imagery-based "brain switch": A feasibility study towards a hybrid BCI. IEEE Trans Neural Syst Rehabil Eng, 18(4), 409–414, Aug (2010).
62. J.A. Pineda, B.Z. Allison, and A. Vankov, The effects of self-movement, observation, and imagination on mu rhythms and readiness potentials (RP's): toward a brain–computer interface (BCI). IEEE Trans Rehabil Eng, 8(2), 219–222, Jun (2000).
63. J.A. Pineda, D. Brang, E. Hecht, L. Edwards, S. Carey, M. Bacon, C. Futagaki, D. Suk, J. Tom, C. Birnbaum, and A. Rork, Positive behavioral and electrophysiological changes following neurofeedback training in children with autism. Res Autism Spect Disorders, 2(3), 557–581, (2008).
64. J.A. Pineda, D.S. Silverman, A. Vankov, and J. Hestenes, Learning to control brain rhythms: making a brain–computer interface possible. IEEE Trans Neural Syst Rehabil Eng, 11, 181–184, (2003).
65. F. Popescu, S. Fazli, Y. Badower, B. Blankertz, and K.-R. Müller, Single trial classification of motor imagination using 6 dry EEG electrodes. PLoS ONE, 2, e637, (2007).
66. C. Sannelli, M. Braun, M. Tangermann, and K.-R. Müller, Estimating noise and dimensionality in BCI data sets: Towards illiteracy comprehension. In G.R. Müller-Putz, C. Brunner, R. Leeb, G. Pfurtscheller, and C. Neuper, (Eds.), Proceedings of the 4th International brain–computer Interface Workshop and Training Course 2008, pp. 26–31, Verlag der Technischen Universität Graz, (2008).

67. G. Santhanam, S.I. Ryu, B.M. Yu, A. Afshar, and K.V. Shenoy, A high-performance brain–computer interface. Nature, 442, 195–198, (2006).
68. G. Schalk. brain–computer symbiosis. J Neural Eng, 5, P1–P15, (2008).
69. G. Schalk, X. Pei, N. Anderson, K. Wisneski, M.D. Smyth, W. Kim, D.L. Barbour, J.R. Wolpaw, and E.C. Leuthardt, *Decoding spoken and imagined phonemes using electrocorticographic (ECoG) signals in humans: Initial data. Program No. 778.6. 2008 Abstract Viewer/Itinerary Planner*. Society for Neuroscience, Washington, DC, (2008), Online.
70. G. Schalk, P. Brunner, L. A. Gerhardt, H. Bischof, and J. R. Wolpaw, brain–computer interfaces (BCIs): detection instead of classification. J Neurosci Methods, 167, 51–62, (2008).
71. G. Schalk, K.J. Miller, N.R. Anderson, J.A. Wilson, M.D. Smyth, J.G. Ojemann, D.W. Moran, J.R. Wolpaw, and E.C. Leuthardt, Two-dimensional movement control using electrocorticographic signals in humans. J Neural Eng, 5, 75–84, (2008).
72. R. Scherer, F. Lee, A. Schlögl, R. Leeb, H. Bischof, and G. Pfurtscheller, Toward self-paced brain–computer communication: navigation through virtual worlds. IEEE Trans Biomed Eng, 55, 675–682, (2008).
73. R. Scherer, G.R. Müller-Putz, and G. Pfurtscheller, Self-initiation of EEG-based brain–computer communication using the heart rate response. J Neural Eng, 4, L23–L29, (2007).
74. R. Scherer, G.R. Müller-Putz, and G. Pfurtscheller, Flexibility and practicality: Graz brain–computer interface approach. Int Rev Neurobiol, 86, 119–131, (2009).
75. E. W. Sellers, A. Kübler, and E. Donchin, brain–computer interface research at the University of South Florida Cognitive Psychophysiology Laboratory: The P300 speller. IEEE Trans Neural Syst Rehabil Eng, 14, 221–224, (2006).
76. R.M. Shiffrin and W. Schneider, Automatic and controlled processing revisited. Psychol Rev, 91, 269–276, 1984.
77. R. Sitaram, A. Caria, and N. Birbaumer, Hemodynamic brain–computer interfaces for communication and rehabilitation. Neural Netw, 22, 1320–1328, (2009).
78. P. Suppes, Z. L. Lu, and B. Han, Brain wave recognition of words. Proc Nat Acad Sci U S A, 94(26), 14965–14969, Dec (1997).
79. L.J. Trejo, N.J. McDonald, R. Matthews, and B.Z. Allison (2007) Experimental design and testing of a multimodal cognitive overload classifier. Automated Cognition International Conference, Baltimore, Maryland. Winner, best paper award. The conference was from 22–27 July: http://www.augmentedcognition.org/events.htm.
80. T.M. Vaughan, D.J. McFarland, G. Schalk, W.A. Sarnacki, D.J. Krusienski, E.W. Sellers, and J.R. Wolpaw, The Wadsworth BCI Research and Development Program: at home with BCI. IEEE Trans Neural Syst Rehabil Eng, 14(2), 229–233, Jun (2006).
81. Vora, J.Y., Allison, B.Z., & Moore, M.M. (2004). A P3 brain computer interface for robot arm control. Society for Neuroscience Abstract, 30, Program No. 421.19.
82. J.R. Wolpaw, brain–computer interfaces as new brain output pathways. J Physiol, 579, 613–619, (2007).
83. J.R. Wolpaw, N. Birbaumer, D.J. McFarland, G. Pfurtscheller, and T.M. Vaughan, brain–computer interfaces for communication and control. Clin Neurophysiol, 113, 767–791, 2002.
84. J.R. Wolpaw, G.E. Loeb, B.Z. Allison, E. Donchin, O. Feix do Nascimento, W.J. Heetderks, F. Nijboer, W.G. Shain, and J.N. Turner, BCI Meeting 2005–workshop on signals and recording methods. IEEE Trans Neural Syst Rehabil Eng, 14(2), 138–141, Jun (2006).
85. J.R. Wolpaw and D.J. McFarland, Control of a two-dimensional movement signal by a non-invasive brain–computer interface in humans. Proc Nat Acad Sci USA, 101, 17849–17854, (2004).

Index

A
AAR paramter, adaptive autoregressive parameter, 282
Action potential (AP), 31, 37–38, 65, 157, 172–173, 205–208, 213, 216–217
Adaptive autoregressive parameter (AAR), 282, 339–341, 344–349
Alpha rhythm, alpha band rhythm, alpha oscillation, alpha band, 47–49, 52–53, 67, 70, 138, 146, 156, 337
Amyotrophic lateral sclerosis (ALS), 5, 16, 29, 35, 40, 158, 166–167, 186–195, 197–198, 230, 259, 377–378
AR parameters, 86, 339–341, 345–346
Artefact, 163, 174, 197
Asynchronous BCI, self-paced BCI, 14, 80, 85, 91, 146, 323–324
Attention deficit hyperactivity disorder (ADHD), 22, 66
Autoregressive model (AR), 86, 250, 315–316, 333, 339–341, 346–347, 352

B
Band power, 50, 68, 70, 83, 86–87, 90, 116–118, 123–124, 130, 177–180, 246, 294, 315, 317, 346
BCI2000, 107, 233–234, 243, 253, 255, 259–278, 289–290, 363, 366, 368
BCI training, 50–51, 65–66, 70, 75, 88, 158–159, 161, 163, 177–179, 186, 190, 194–197, 293–295
Beta rebound, 54–55, 59, 84–85
Beta rhythm, beta band rhythm, beta oscillation, beta band, 21, 31–32, 34, 36, 47–49, 54–55, 57, 59, 66–67, 70, 72, 81, 83–84, 90–91, 98, 100–101, 137–139, 261, 282, 294, 306, 313, 315
Bipolar recording, 70, 90, 177
Blind source separation (BSS), 311
Blood oxygenation level dependent (BOLD), 7, 53, 84, 92, 119, 164–166
Brain oscillations, 11, 81

C
Center-out task, 37, 211, 215, 218
Central nervous system (CNS), 3–4, 30–31, 166, 172
Channel selection, 317
Classification, classifier, 13–15, 36, 48, 56–57, 65, 69–72, 74–75, 79–83, 85–88, 91–92, 98, 100–108, 114, 116–118, 127, 130, 143, 146, 149, 155, 166, 177, 179, 181, 223–224, 262, 273, 282, 284, 290, 292, 295, 298, 314, 316, 318–319, 321–324, 331–333, 341–345, 347, 349, 352
Cochlear implant, 4
Common average reference (CAR), 288, 307–309, 315
Common mode rejection ratio (CMRR), 288
Common spatial patterns (CSP), 56, 86–87, 116–117, 147, 286, 295, 307, 312–313
Completely locked-in state (CLIS), 158, 167, 188–191
Cross-validation, 130, 149, 321–323, 347–348

D
Data acquisition, 80, 259, 261–263, 275, 283–284, 290, 292, 294–295, 305–306
2D control, 100, 254, 256
3D control, 102, 108, 214
Desynchronization, 11, 36, 47–50, 52–54, 57–58, 70, 79, 81, 98, 116, 141, 177, 231, 245–246, 249, 294, 306

E
Electrocardiogram, electrocardiography (ECG), 288–289
Electrocorticogram, electrocorticography (ECoG), 8–9, 22, 31, 34, 37, 39–40, 48,

B. Graimann et al. (eds.), *Brain–Computer Interfaces,* The Frontiers Collection,
DOI 10.1007/978-3-642-02091-9, © Springer-Verlag Berlin Heidelberg 2010

65, 157, 163, 217, 221–237, 241–242, 244, 250, 253–254, 256, 259–260, 269, 281, 285–289, 313
Electroencephalogram, electroencephalography (EEG), 3, 6–9, 11–12, 14, 17, 21–22, 30–37, 39–40, 47–49, 51–57, 65–70, 72–74, 79, 81–83, 85–87, 89–92, 97–102, 106–108, 113–118, 125–126, 128–131, 137–152, 155–159, 163–164, 167, 174, 177, 180, 182, 194, 196, 217, 222–224, 226, 230, 233–237, 241, 244, 249, 259, 267, 269–271, 281–286, 288–290, 292–297, 299, 305–306, 309–310, 312–315, 317, 325–326, 332, 337, 339, 345–346, 351, 358–360, 362, 364, 367–368, 371–372, 375, 379, 381
Electromyogram, electromyography (EMG), 6, 37, 85, 115, 138, 158, 174, 223, 225, 231, 247, 288–289, 307, 309, 364, 370
Electrooculogram (EOG), 6, 85, 105, 138, 223, 225, 289, 307, 309
Event-related desynchronization (ERD), 11, 36, 47, 79, 81, 116, 141, 177, 231, 246, 249, 294, 306
Event-related potential (ERP), 32–35, 48, 97, 103, 115, 138, 156, 189, 229
Event-related synchronization (ERS), 11, 36, 47, 81, 116, 177, 231, 306
Evoked activity, 32–33, 38, 48, 138, 230
Evoked potential (EP), 10, 32–33, 38, 48, 106, 115, 137–138, 155, 160, 230, 259–261, 270, 282, 285–286, 288, 306

F

False positive, 85, 89, 92, 126, 146, 179, 232–233, 324–325
Fatigue, 129, 172, 176, 190, 192, 197, 227, 332, 350, 359, 363, 378–379, 381
Feature extraction, 13–14, 48, 85–87, 106, 114, 145, 149, 282, 290, 292, 295, 306, 313–317, 331–333, 339, 352
Feature selection, 36, 74, 87, 241, 247–250, 305, 316–317, 322, 339
Firing rate, 14, 34, 39, 208, 210–211, 216–218
Fourier transform, 141, 250, 264, 314
Frequency domain, 32, 37–38, 85, 108, 139, 229, 307, 345
Functional electrical stimulation (FES), 17, 181
Functional magnetic resonance imaging (fMRI), 7, 30, 53, 82, 163, 166
Functional near infrared (fNIR), 30–31, 359

G

Gamma activity, 11, 37

H

Human-computer interface (HCI), 5, 359, 373
Hybrid BCI, 21, 92, 360, 365–366, 369–371

I

Independent component analysis (ICA), 86, 116, 295, 307–312, 339
Induced activity, 55, 82, 84
Information transfer rate (ITR), 15–16, 71, 120–122, 125–126, 131, 142–143, 146, 150, 189, 212, 297, 323, 358–359, 364–365, 367, 370
International 10–20 system, 7–8
Intracortical electrodes, 8, 31, 38, 101, 216, 218
Intracortical recording, 8–9, 14, 22, 37, 207, 213

K

Kalman filter, 337–341, 343–344, 348–349, 352

L

Laplacian derivation, 85–87, 295
Linear discriminant analysis (LDA), 74, 86–88, 91, 103–104, 106–108, 117–119, 130, 149, 174–175, 177–180, 282, 294, 296–297, 336, 341–344, 346, 348–350, 352
Local field potential (LFP), 31, 37–39, 157, 217–218, 235, 237, 241
Locked-in state (LIS), 22, 79, 158–159, 167, 187–191, 198
Long-term potentiation (LTP), 68

M

Machine learning, 13–14, 56, 65, 68, 80, 97–98, 113–132, 141, 245, 249, 305–327, 332–333, 366
Magnetic resonance imaging (MRI), 7, 30, 53, 82–83, 163, 166, 245
Magnetoencephalogram, magentoencephalography (MEG), 7, 54–55, 58, 92, 148, 160–163, 259, 313, 359, 362, 374
Mahalanobis distance, 341–342
Microelectrode array, 37, 205, 216, 218
Monopolar recording, 287–288
Motor activity, 67, 160, 253, 312, 361
Motor cortex, 38–39, 50, 52, 54–55, 58–59, 67, 84, 123, 141, 148, 150, 157, 160, 163, 210, 216, 230, 236, 254, 306

Motor imagery (MI), 10–13, 15, 21, 36–37, 49–51, 54–59, 68–70, 72–73, 79–85, 87–88, 90–92, 98–99, 117, 123–125, 137, 139, 141–142, 145, 147–152, 156, 166, 174–175, 178–180, 234, 247, 256, 281–282, 293–296, 310, 312–313, 315, 346–350, 371
Motor neuron disease (MND), 186
Motor pathway, 3, 5
Movement-evoked potential (MEP), 38
Mu rhythm, 34, 36, 47–49, 51–53, 57–58, 67, 72, 98, 100–101, 138–139, 141–142, 148–149, 313

N
Near-infrared spectroscopy (NIRS), 7, 22, 92, 157, 166
Neurofeedback training, biofeedback training, neurofeedback, biofeedback, feedback, 3, 13, 22, 24, 39, 54, 59, 65–75, 80–81, 83, 91, 113–114, 120–126, 129, 139, 152, 155–157, 159, 161–165, 179, 182, 210, 214, 229, 231, 233–234, 236, 241, 246–247, 249–256, 261–264, 282, 293–295, 331–332, 343–345, 349–350, 352, 358, 361, 367, 371, 379
Neuromodulation, 68
Neuronal network, 48, 54, 82
Neuroprosthesis, neuroprostheses, 4, 17, 21, 59, 65, 69, 126, 171–182, 361
Neurorehabilitation, 5, 22, 69, 151, 155–167, 190, 235, 358, 376
Noise, 14, 65, 85, 91, 116, 138, 145, 148, 159, 197, 208, 213, 223–224, 226, 228, 233, 244, 281, 284–286, 288, 296, 306–310, 313–314, 316–317, 332, 338–340, 347, 351, 367, 378

O
OpenVibe, 363, 366
Operant conditioning, 13, 22, 35, 67–68, 73–74, 97, 114, 123, 155, 157, 159, 163–164, 189–190
Orthosis, orthotic device, 21, 59, 73, 126, 160–163, 182, 292, 361
Oscillatory activity, oscillations, 11, 31–32, 34–36, 38, 47–59, 67, 69, 72, 73, 81–84, 90, 97, 116, 159, 217, 281–282, 286, 306, 366

P
P300, 10–11, 16–17, 33–35, 40, 71, 97–98, 103–108, 114, 123, 137, 155–156, 189, 194, 196, 259, 261, 264–265, 270–271, 273–274, 282, 286, 288, 295–297, 305–307, 313–315, 317, 323, 325–327, 363–364, 366–367, 370–373, 377
Paradigm, 33, 53, 66, 68–75, 83, 91, 103, 105, 108, 114, 120–125, 129, 137, 149, 178–179, 224, 229, 235, 243, 247, 253, 260–261, 263, 268, 271–273, 292–293, 295, 314, 316, 323, 325, 343, 346, 369–370, 376
Pattern recognition, 3, 65, 113, 117, 316, 318, 351
Phase-locked, 48, 138, 144
Phase-locking value, 86, 148
Positron emission tomography (PET), 30, 163
Power spectral density (PSD), 141, 243, 247–248
Preprocessing, 13–14, 80, 85, 118, 121, 144, 149, 305–307, 309–310, 313
Principal component analysis (PCA), 86, 307–308, 310, 312, 339
Prosthesis, prosthetic device, prosthetic limb, 19–21, 125–126, 155, 157, 213–214, 305, 307

Q
Quadratic discriminant analysis (QDA), 336, 341–342, 346–352

R
Readiness potential, 115, 126–127
Real-time analysis, real-time signal processing, 13, 81, 294
Receiver operating characteristic (ROC), 323–325
Recursive least square (RLS), 282, 292, 340–341, 344, 346, 348–349
Regression, 100–102, 108, 118, 180, 309, 318, 321
Retinal implant, 4

S
Selective attention, 10–11, 13, 15, 53, 139, 314, 360
Sensorimotor area, 11, 47–48, 50, 52, 54, 69–70, 116, 177, 312, 317
Sensorimotor rhythm (SMR), 11, 14, 35, 57–59, 66–67, 72, 81, 97–102, 116–117, 123, 160, 189, 234, 259, 261–262, 264–265, 270, 312
Signal-to-noise ratio (SNR), 14, 138, 145–146, 148, 159, 233, 281, 307–309, 313, 317, 347, 371
Signal processing, 2–3, 10, 13–14, 22, 34, 56, 65, 80, 86, 117, 131, 138–141, 150, 177,

208, 236, 243, 245, 260–264, 267–268, 273, 276–277, 281, 290, 292–293, 297, 301, 305–327, 331, 333, 352, 359, 365, 367
Slow cortical potential (SCP), 32, 35, 48, 66, 68, 71, 97, 137, 155–156, 158, 189, 306, 333, 339
Smart home, 10, 297–302, 365–366
Sparse component analysis (SCA), 311–312
Spatial filtering, 56, 86–88, 116–117, 147, 245, 262, 281, 286, 288, 295, 306–308, 312, 315, 321
Spatial resolution, 8, 157, 163, 217, 223, 228, 307–308
Spelling device, 16, 282, 295–297
Spinal cord lesion, spinal cord injury, 29, 39, 125, 145, 150, 160–163, 167, 171, 175, 216
Steady-state visual-evoked potential (SSVEP), 10–11, 14, 17, 21, 33, 115, 123, 138–139, 141–147, 150–151, 260, 282, 306, 313–314, 317, 327, 366–367, 369–371
Stepwise linear discriminant analysis (SWLDA), 103–104, 106–108
Stroke rehabilitation, 157, 167, 374
Supplementary motor area (SMA), 38, 54, 84–85, 148–149
Synaptic potential, 31, 38
Synchronization, 11, 35, 37–38, 52, 81, 85–86, 116, 141, 148–150, 157, 180, 216–218, 231, 245, 270, 275, 285, 294, 306
Synchronous BCI, cue-paced BCI, 14, 80, 85, 91, 146, 177, 323–324

T

Targeted muscle reinnervation (TMR), 20–21
Temporal resolution, 6–7, 204
Thalamo-cortical system, 51
Time domain, 32, 37–38, 108, 229, 307, 339
Time-frequency maps, 55, 177
Time-locked, 31–32, 48, 103, 115
Transcranial magnetic stimulation (TMS), 50, 52, 55

V

Virtual reality (VR), 16–17, 72–73, 89–91, 298–302, 365, 374
Visual attention, 10–11, 370
Visual cortex, 11–12, 138–139, 230, 314, 317
Visual evoked potential (VEP), 32–33, 106, 137, 142, 230

THE FRONTIERS COLLECTION

Information and Its Role in Nature
By J.G. Roederer

Relativity and the Nature of Spacetime
By V. Petkov

Quo Vadis Quantum Mechanics?
Edited by A.C. Elitzur, S. Dolev, N. Kolenda

Life – As a Matter of Fat
The Emerging Science of Lipidomics
By O.G. Mouritsen

Quantum–Classical Analogies
By D. Dragoman and M. Dragoman

Knowledge and the World
Edited by M. Carrier, J. Roggenhofer, G. Küppers, P. Blanchard

Quantum–Classical Correspondence
By A.O. Bolivar

Mind, Matter and Quantum Mechanics
By H. Stapp

Quantum Mechanics and Gravity
By M. Sachs

Extreme Events in Nature and Society
Edited by S. Albeverio, V. Jentsch, H. Kantz

The Thermodynamic Machinery of Life
By M. Kurzynski

The Emerging Physics of Consciousness
Edited by J.A. Tuszynski

Weak Links
Stabilizers of Complex Systems from Proteins to Social Networks
By P. Csermely

Mind, Matter and the Implicate Order
By P.T.I. Pylkkänen

Quantum Mechanics at the Crossroads
New Perspectives from History, Philosophy and Physics
Edited by J. Evans, A.S. Thorndike

Particle Metaphysics
A Critical Account of Subatomic Reality
By B. Falkenburg

The Physical Basis of the Direction of Time
By H.D. Zeh

Asymmetry: The Foundation of Information
By S.J. Muller

Mindful Universe
Quantum Mechanics and the Participating Observer
By H. Stapp

Decoherence and the Quantum-To-Classical Transition
By M. Schlosshauer

Quantum Superposition
Counterintuitive Consequences of Coherence, Entanglement, and Interference
By Mark P. Silverman

The Nonlinear Universe
Chaos, Emergence, Life
By A. Scott

Symmetry Rules
How Science and Nature Are Founded on Symmetry
By J. Rosen

The Biological Evolution of Religious Mind and Behavior
Edited by E. Voland and W. Schiefenhövel

Entanglement, Information, and the Interpretation of Quantum Mechanics
By G. Jaeger

MIX
Papier aus verantwortungsvollen Quellen
Paper from responsible sources
FSC® C105338

Printed by Books on Demand, Germany